D1085631

TAKING THE NATURALISTIC TURN
or
How Real Philosophy of Science Is Done

SCIENCE AND ITS CONCEPTUAL FOUNDATIONS

David L. Hull, editor

TAKING THE
NATURALISTIC TURN

O R

How Real Philosophy of Science Is Done

CONVERSATIONS WITH

William Bechtel, Robert N. Brandon, Richard M. Burian,
Donald T. Campbell, Patricia S. Churchland, Jon Elster, Ronald N. Giere,
David L. Hull, Philip Kitcher, Karin D. Knorr Cetina,
Bruno Latour, Richard Levins, Richard C. Lewontin, Elisabeth Lloyd,
Helen E. Longino, Thomas Nickles, Henry C. Plotkin,
Robert J. Richards, Alexander Rosenberg, Michael Ruse,
Dudley Shapere, Elliott Sober, Ryan D. Tweney, and
William C. Wimsatt

ORGANIZED AND MODERATED BY

Werner Callebaut

The University of Chicago Press

Chicago and London

WERNER CALLEBAUT is associate professor of
philosophy in the Department of Social Sciences, Humanities, and
Communication of Limburgs Universitair Centrum (Belgium)
and assistant professor in the Department of Philosophy
of the Rijksuniversiteit Limburg (The Netherlands).

THE UNIVERSITY OF CHICAGO PRESS, Chicago 60637
THE UNIVERSITY OF CHICAGO PRESS, LTD., London
© 1993 by The University of Chicago
All rights reserved. Published 1993
Printed in the United States of America
02 01 00 99 98 97 96 95 94 93   5 4 3 2 1

ISBN (cloth): 0-226-09186-4
ISBN (paper): 0-226-09187-2

Library of Congress Cataloging-in-Publication Data

Taking the naturalistic turn, or, How real philosophy of science is
    done : conversations with William Bechtel . . . [et al.] / orga-
    nized and moderated by Werner Callebaut.
        p.   cm.—(Sciences and its conceptual foundations
    series)
    ISBN 0-226-09186-4 (cloth).—ISBN 0-226-09187-2 (paper)
    1. Science—Philosophy.   2. Science—Social aspects.
3. Scientists—United States—Interviews.   4. Philosophers—
    United States—Interviews.   5. Sociologists—United States—
    Interviews.
    I. Bechtel, William.   II. Callebaut, Werner.   III. Series.
Q175.3.T35   1993
501—dc20   92-24193
                                                         CIP

⊗The paper used in this publication meets the minimum requirements of the American National Standard for Information Sciences—Permanence of Paper for Printed Library Materials,
ANSI Z39.48-1984.

To my parents

CONTENTS

3. TOWARD A NEW THEORY OF SCIENCE: NEW DIMENSIONS, FEATURES, AND APPROACHES / 72

PART TWO · DOING IT

ANALYTICAL TABLE OF CONTENTS

1. TURNING NATURALISTIC: AN INTRODUCTION

Naturalism as a philosophical movement claims that whatever exists or happens in the world is susceptible to explanation by natural scientific methods; it denies that there is or could be anything which lies in principle beyond the scope of scientific explanation. Although naturalism is firmly rooted in the philosophical tradition (materialism, empiricism), a thoroughly naturalized philosophy of science is only being developed now.

P A R T   O N E · T A L K I N G   A B O U T   I T

2. BEYOND POSITIVISM AND HISTORICISM

Philosophy of science as currently practiced is a reaction against a reaction. Until the early 1960s logical positivism reigned supreme; its primary concern was the "rational reconstruction" of (textbook presentations of) scientific theories. A rather heterogeneous group of scholars led by Thomas Kuhn subsequently imposed itself, urging for a new rapprochement between the history and philosophy of science. Most of the protagonists in this book react to both of these movements while continuing to be inspired by them. They regret the positivists' and Popperians' alienation from both the history of science and science as currently practiced but want to keep up their high standards of clarity and rigor. They are all in favor of a historically and empirically (sociologically, psychologically) informed image of science, but at least some of them reject the relativistic implications of historicism and want to save the "rationality" and "objectivity" of science by redefining them appropriately.

3. TOWARD A NEW THEORY OF SCIENCE

The positivist view of science was drawn almost exclusively from a papier maché model of classical physics. The emerging field of science studies is much more diverse and heterogeneous in comparison. A variety of philosophies of special sciences such as evolutionary biology and cognitive psychology are now booming, thus acknowledging the factual disunity of the sciences. This "explosion" of general philosophy of science and the need for a real involvement in scientific subject matters is illustrated by the intellectual biographies of three of the participants in this book who, in addition to a shared interest in biology and its philosophy, are or have been involved in physics, engineering, mathematics, and economics. Despite the increasing compartmentalization of philosophy of science, some general features emerge in relation to its naturalization: the rehabilitation of psychologism, i.e., roughly, the view that factual thinking is relevant for any normative theory of thinking (logic in particular); and, related to this, the abandoning of the analytic/synthetic dichotomy with respect to statements. The chapter ends on a presentation of philosophy's major contender for a new view of science, the sociology of science.

4. PHILOSOPHY OF SCIENCE REVISITED

A naturalistic approach sheds new light on three key issues in the philosophy of science: explanation, reduction, and realism. Three accounts of explanation emerge as candidates to replace Carl Hempel's deductive-nomological account: causal-mechanistic explanation from the bottom up, top-down explanation as unification, and the view that any explanation is the answer to a why-question. With respect to reduction, the "horizontal" reduction of an older theory by a successor theory is to be distinguished from the "vertical" reduction of a theory about phenomena at a certain level by a lower-level theory, which may involve such complications as "downward causation." Realism re-entered the philosophy of science in conjunction with the historical wave. More recently the realism-constructivism controversy has superseded the older debate about the rationality or relativism of science; it occasions a majority of philosophers to defend some variety of realism against the anti-realism of contemporary empiricists in philosophy and increasingly also against the actual or presumed anti-realistic implications or presuppositions of the new sociological approaches to science studies.

5. NEW ROLES AND TOOLS FOR PHILOSOPHERS

The separation between science and philosophy is of fairly recent origin, and many a naturalist would like to see it disappear again. It was reflected in the positivists' distinction between "context of discovery" and "context of justification," which the historicists challenged and the naturalists have given up in favor of the view that "discovery is everywhere." The latter view may—but must not—be taken to alleviate the philosophers' justificatory burden. Most naturalists agree when being compared to Locke's "underlaborers." A more flattering way of saying almost the same thing is that they are therapists with respect to scientific method and research strategy. When it comes to taking a stand in public, as in the creation/evolution debate in the United States in the early 1980s, quite a few naturalistic philosophers are more hesitant. New roles take new tools as well. The "semantic" or "model-theoretic" view of theories takes structure rather than deductive quality as the key to scientific explanation. Various of the participants in this book consider it a promising instrument to scrutinize the fine structure of scientific theories. Whether a thoroughly naturalized philosophy of science leaves room for methodological, normative considerations is a matter of debate.

PART TWO · DOING IT

6. PHILOSOPHY OF BIOLOGY

The new philosophy of biology may be regarded as an exemplar of the goods a philosophy of science that is in tune with the science may deliver. After some general and historical considerations, Darwinian evolutionary theory—a conceptual minefield *par excellence*—is interpreted as a theory of forces not altogether unlike Newtonian mechanics. Evolution is defined as heritable variation in fitness, which allows its advocates to rebut the familiar but misguided charge that evolutionary theory is but one grand tautology. On this basis and taking into account some basic social-constructionist insights, the intricate dialectic of organism and environment is laid bare. Some of the ensuing methodological traps, e.g., concerning the nature/culture distinction, are also exposed. Considerable attention is being devoted to the controversy about the units of selection and evolution, in which the causal analysis competes with the analysis of variance (ANOVA) approach at various levels of mathematical sophistication. Another intriguing question is whether the metaphysical consequences of the Darwinian "revolution" have really been taken into account in contemporary biology and social science.

7. EVOLUTIONARY EPISTEMOLOGY

From Darwin's day, scientists and philosophers have speculated about the evolutionary origins of mind and behavior, which are intimately related. "Evolutionary epistemology" (Donald T. Campbell) is the attempt to explain animal and human cognition, including science, in terms of the selectionist model. It harbors two distinct programs: the investigation of the evolutionary basis of perceptual and cognitive mechanisms in living creatures by straight biological means ("bio-epistemology"), and the analysis of science and scientific change as an aspect of culture and cultural change by means of evolutionary analogies or in terms of a generalized version of natural selection. Two particular versions of the second program—which is much more controversial than the first one—receive considerable atttention: Hull's evolutionary account of the social and conceptual sides of scientific change in terms of credit maximization, and Richards's natural selection model of historiography. These evolutionary accounts are contrasted with Popper's attempt to save the objectivity of science by his "Three Worlds" theory, which has some undesirable features.

8. COGNITIVE APPROACHES TO SCIENCE AND PHILOSOPHY

While sociology of science is most indebted to the historicist line of work emanating from Thomas Kuhn, psychologists and at least some naturalistic philosophers expect that cognitive psychology and ultimately even neurophysiology will prove of at least equal importance. The "procedural turn" in science studies urges naturalists to devote attention to nonlinguistic (e.g., visual and practical) features of scientific work that were neglected for too long. The experimental-psychological study of the thinking of scientists, which focuses on the internal dynamics in the minds of the actors with a concern for ecological validity, exemplifies this. Connectionist models are opening new vistas in many areas, including the modeling of the social animals that scientists are and the representation of brain input and activity in terms of vectors. Contemporary views on the mind/brain issue and in particular surviving forms of dualism, varieties of the identity theory, functionalism, and eliminative materialism (which wants to revise if not eliminate "folk psychology") are reviewed and the question whether machines can think is briefly discussed.

9. DEVELOPMENT, LEARNING, AND CULTURE

Development (in the sense of ontogeny) and cognition are among the major unsolved problems in biology. Molecular biology and computer modeling begin to shed new light on the question of what would happen to a minimally complex organism in the absence of selection, given the constraints on its development. The "developmental lock" model highlights the "generative entrenchment" of acquired features and leads to a redefinition of the "innate/acquired" dichotomy. Many creatures do not just develop, they learn as well. Both development and learning are gender-sensitive issues. A multiple-level model of evolution positing an identity of process across the nested levels of genetic evolution, variable epigenesis, individual learning, and social learning is presented and contrasted with the emerging evolutionary psychology approach which remains tributary to sociobiology. The notion of fitness is reconsidered with a view to extending its use to the modeling of cultural evolution. The similarities and differences between biological and cultural evolution are reviewed.

10. PHILOSOPHY MOVES ALONG

In the last chapter, the emphasis is on future developments. Evolutionary ethics, the application of evolutionary theory to the normative realm of morality, is becoming fashionable again after almost a century of marginalization; two varieties of it are discussed. Another concern is metaphilosophical: the foreseeable consequences for the practice of philosophy and for philosophical education of the current explosion of philosophy of science. A question that has been lingering without getting appropriate attention heretofore is to what extent the rapprochement between naturalism and historicism that is currently suggested by many is a natural or an artificial one; some provisional answers are suggested. Yet another topic is gender studies. Though generally speaking they have become a big thing, philosophy of science has barely been touched by the feminist perspective yet, but that situation may change soon. In the final section, some philosophers overcome their reluctance to make predictions and tell us where they think their field is going.

PARTICIPANTS

For a biographical note on each participant, see the section indicated.

William Bechtel, Georgia State University / 8.3

Robert N. Brandon, Duke University / 9.5

Richard M. Burian, Virginia Polytechnic Institute and State University / 10.6.1

Donald T. Campbell, Lehigh University / 7.1.1

Patricia S. Churchland, University of California, San Diego / 8.4.1

Jon Elster, Institute for Social Research, Oslo / 5.2.1

Ronald N. Giere, University of Minnesota / 4.3

David L. Hull, Northwestern University / 6.6.1

Philip S. Kitcher, University of California, San Diego / 3.4.2

Karin D. Knorr Cetina, Universität Bielefeld / 4.3.6

Bruno Latour, École Supérieure des Mines de Paris and University of California, San Diego / 3.6

Richard Levins, Harvard University / 9.2.4

Richard C. Lewontin, Harvard University / 9.2.4

Elisabeth Lloyd, University of California, Berkeley / 5.5.1

Helen E. Longino, Rice University / 10.5.3

Thomas J. Nickles, University of Nevada at Reno / 2.3

Henry C. Plotkin, University College London / 9.1.2

Robert J. Richards, University of Chicago / 10.1.1

Alexander Rosenberg, University of California, Riverside / 3.3

Michael Ruse, University of Guelph / 9.3.2

Dudley Shapere, Wake Forest University / 2.7

Elliott Sober, University of Wisconsin, Madison / 6.5

Ryan D. Tweney, Bowling Green State University / 8.1.2

William C. Wimsatt, University of Chicago / 3.2.1

ACKNOWLEDGMENTS

Parts of the exchanges with twelve out of the twenty-four participants in this book are edited transcripts of BRT 3 *Symposium* radio broadcasts (Belgian Public Radio, Brussels). They are reproduced here courtesy of the Flemish section of Belgian Radio and Television. When I introduced my proposal to do a series of programs on evolutionary and naturalized philosophies of science to *Symposium* producer Frans Boenders, Frans (an anarchist in Feyerabendian and in other ways) responded promptly by allotting me plenty of studio time without further ado. Thanks a lot, Frans, for taking the risk! I should also thank the manager of the University of Chicago's radio station WHPK, Kent Yeglin, and the Modern Language Center at Harvard University for generously providing studio time and personnel.

The realization of this project would have been impossible without the direct or indirect financial support of several institutions: many of the exchanges recorded in this book took place in the margin of conferences or while I visited campuses to present papers (additional information is mentioned in the biographical notes on individual participants). The Belgian National Fund for Scientific Research offered grants which allowed me to work in the United States in the spring of 1985, and to present and discuss work on evolutionary and naturalistic epistemology in Newcastle, Australia, in July 1987, in London, Ontario, in June 1989, and in Evanston, Illinois, in July 1991. The Department of Social Sciences and Humanities of Limburgs Universitair Centrum, Belgium, supported travel to England in August 1985. The Evolutionary Epistemology study group at the same institution supported traveling to various campuses in the United States early in 1986 and again in the spring of 1990; to the meeting of the History of Science Society in Cincinnati, December 1988; to Paris (second Burian interview) in the spring of 1990; and to the World Conference of Philosophy in Nairobi in July 1991, where I presented materials that are incorporated in the Introduction and in the Postscript. The Department of Philosophy at Rijksuniversiteit Limburg, The Netherlands, funded traveling to the Philosophy of Science Association meeting in Pittsburgh in October 1986, to Paris in October 1989 (Latour interview), and to Marburg in April 1990 (second Knorr Cetina interview). The Committee on the Conceptual Foundations of Science at the University of Chicago; the Departments of Philosophy and Social Relations, Lehigh University; the Department of Sociology, The University of North Carolina at Chapel Hill; and Wake Forest University each provided accommodation or travel grants. Four interviews were made possible by the participants' presence in Belgium at the invitation of the Belgian Society for Logic and Philosophy of Science. Special thanks are due to the Plant Physiology Lab at the University of California, San Diego, headed by Maarten Chrispeels, for kindly providing a philosophical intruder with extremely useful logistics (an office corner and, above all, a Macintosh II computer) in June 1990.

I am also glad to express my gratitude for the wonderful hospitality and encouragement of Steve and Bea Chorover (Boston), Herman and Cathérine Höfte-Bourgouin (San Diego), Ludo and Annick Peferoen (Palo Alto), Rachel and François Rosenfeld (Chapel Hill), Tom Ronse, Jacqueline Goossens, and Mark Garrett (New York), Mike Sexton (San Diego), and Peter and Anne Taylor (Berkeley)—to mention only "nonparticipants" in the book. Talking about people who aren't but *could* have been in the book, my conversations with Nancy Cartwright and David Depew in sunny California, with Elihu Gerson in rainy San Francisco, and with Stan Salthe in misty-mysterious Ghent left more lasting marks than any of them probably realize. It was comforting to discover that Massimo Stanzione, a Roman friend I hadn't talked to for too long, has been moving along quite similar intellectual lines (Stanzione 1990). Linnda Caporael *is* present in the book, but only minimally and at the questioning end of things; my profoundest gratitude goes to her and Glen Culbertson for their wit and friendship and for making me feel at home in the Big Apple. Thanks to Sue Martinelli, Jeff and Nancy Ramsey, Sahotra Sarkar, Jeff Schank and his wife Brenda Hinton, and all other friends once or now living in the Windy City. There I confronted America for the very first time; it remains (sorry Linnda!) my favorite. Many thanks also to my anonymous readers and to all the people at the University of Chicago Press who, in various ways, eased this book into its final shape. My indebtedness to my twenty-four interviewees is immense.

At the home front, my student Igor Douven read a complete, almost final version of the manuscript; he and Olaf Diettrich (a physicist and a European Community official in Brussels) pointed out various lines of argument in need of clarification and suggested a number of stylistic improvements. My unnaturalized colleague Tannelie Blom helped me resist some quick and dirty varieties of realism and naturalism. José de Cózar, a visiting graduate student, biologist-philosopher Bernard Feltz, and my colleagues Jan Martin, Ton Nijhuis, and Jo Wachelder were more or less guilty complices. At another level, Guido Van Steendam, who sees things biophilosophical on a grander scale than other Europeans, was too. Marleen Vara made various transcripts from often noisy tape and typed hundreds of pages of manuscript. Sylvia De Ridder spent many hours proofreading, compiling the bibliography and index, and, most importantly, took care of my morale when needed. Among my former teachers in philosophy at the University of Ghent, I owe much to Marc De Mey, who recognized the relevance of the cognitive approach to science studies even before the sociology of science came to the fore; to Etienne Vermeersch and Diderik Batens, who tirelessly extirpate wishy-washiness wherever they meet it; and to Jaap Kruithof, for his moral and political courage. I am most grateful to my admired professor Leo Apostel, who was a radical naturalist at an unsuspected time (Apostel 1953). A student of the positivist Carnap, the constructivist Piaget, and the "new rhetorician" Perelman, he is the living proof that postmodernism can be transcended; I hope he will recognize some of the influence of his teaching here. Finally, Robert Halleux (Liège University), whose classes in the history of science I attended long after I finished my formal education, conveyed to me the *gut feeling* of an old saying I had only been echoing like a parrot before: that philosophy without history is empty.

Jazz pianist McCoy Tyner's ways of coping with complexity and integrating heterogeneous cultures, although inimitable, have been an inspiring example throughout.

Thanks, you all!

1

TURNING NATURALISTIC
An Introduction

THIS BOOK aims to introduce you to a number of topics, developments, and controversies in recent science studies, i.e. (in alphabetical order), in the history, the philosophy, the psychology, and the sociology of science. The idea that has guided me in structuring and editing the raw materials for this book—transcripts of extended discussions with twenty-four scientists (biologists and psychologists) and historians, philosophers, and sociologists of science—is that what has been going on in science studies in the last fifteen years can be best understood and made most sense of by relating it to the *naturalistic turn* in the philosophy of science—actually the *re*turn of naturalism, after the interregnum during most of this century of linguistically oriented philosophy. The naturalistic perspective implies that *matters of fact* are as relevant to philosophical theory as they are relevant in science. Post-Kuhnian philosophy of science is so thoroughly impregnated by developments in the history and sociology of science that it may be considered part and parcel of the emerging interdiscipline of *science studies*.

WHAT IS NATURALISM?

The *Encyclopedia of Philosophy* defines naturalism as "a species of philosophical monism according to which whatever exists or happens is *natural* in the sense of being susceptible to explanation through methods which, although paradigmatically exemplified in the natural sciences, are continuous from domain to domain of objects and events" (Danto 1967:448). In the same vein, *ethical* naturalism holds that "there are no values in the world that are not reducible to or explainable away in terms of the naturalistic conceptual scheme of things" (Adams 1960:14); whereas in the philosophy of social science, naturalism is "the doctrine that there can be a natural scientific study of society."[1] Naturalists refuse to be impressed by the various arguments that have been offered in favor of a bifurcation of *Natur-* and *Geisteswissenschaften*, e.g., in terms of presumed differing cognitive interests such as "nomothetic explanation" ver-

---

1. Thomas (1979:1); also Giedymin (1973); Bhaskar (1979); Little (1986).

1

sus "idiographic description" or *verstehen,* respectively. Instead of trying to develop a grounding for science in metaphysics—a kind of "theology of science," some would say—naturalized epistemologists and philosophers of science picture themselves as demythologizers and removers of the transcendent (Kurtz 1990)[2] if not of the transcendental.[3] They "find themselves on equal footing with psychologists, sociologists, and others for whom the study of science is itself a scientific enterprise" (Giere 1988:12). Since the content and—at a slower pace—the methods of science are being altered continuously, what is considered "natural" today may no longer be so tomorrow, and vice versa (cf. the current debate in cognitive ethology over the ascription of intentions to animals).

Naturalism denies, then, that "there exists or could exist any entities or events which lie, in principle, beyond the scope of scientific explanation" (Danto 1967:448). Although historically, as Kurtz (1990:12) emphasizes, its two main sources in the philosophical tradition are *materialism* in metaphysics and *skepticism* and *empiricism* (including its modern variant, *experimentalism*) in epistemology, current naturalism is not usually regarded as exclusively committed to either reductive materialism[4] or a

2. Von Ditfurth (1987) invokes evolutionary considerations in an attempt to secure a transcendent realm "beyond the cosmos" against the attacks of eliminative materialists. But if evolutionary epistemology makes it highly plausible that reliable knowledge of invisible aspects of the physical world is within the reach of us humans (cf. 7.1.6)—that is, that human knowledge is not limited to the "mesocosm" (Vollmer 1975) of our everyday experience—it would seem that both science and any naturalized philosophy worth the name will have to remain silent forever on matters "transcosmic" (Oeser 1987), even if some readers of Hawking (1988) have thought differently.

3. The sociologist Niklas Luhmann (1990a:15–16), for one, equates a "naturalized" epistemology with a "de-transcendentalized" one. He also relates the polarity "transcendental/natural" to the *reflexivity* of a theory, an issue to which we will return several times in this book. Irrespective of any further assumptions a theory makes concerning consciousness, reason, subjectivity, etc., Luhmann suggests, it may be regarded as transcendental if it does not allow that the *results* of inquiry impinge on the *conditions* of knowledge—"wenn sie nicht zuläßt, daß die Bedingungen der Erkenntnis durch die Ergebnisse der Erkenntnis in Frage gestellt werden" (1990a:13; cf. Giere at 3.6.5). Although most naturalists shun transcendental arguments, naturalism as defined by Danto does not logically preclude their use. Thus Lorenz's (1982) biological rendering of Kant's doctrine of the a priori allows of a transcendentalist interpretation—*pace* Wuketits (1987:48); cf. also Perovich (1986) on the possibility of reconciling the project of discovering stable synthetic a priori propositions with a recognition of the tentative and changing character of science. A well-known defender of transcendental arguments within a naturalistic framework is Roy Bhaskar (1979, 1986, 1989), who wants to answer the question, "What properties do societies and people possess that might make them possible objects of knowledge for us?" A milder form of transcendentalism is being defended by the "scientific realist" Richard Boyd (e.g., 1981, 1984).

4. See the responses of Dewey, Hook, and Nagel (1945) and Edel (1946) to Murphy's (1945) critique of naturalism. These authors' wariness of matters metaphysical (cf. Oliver 1949)—epitomized in their pro-behavioristic stance both in- and outside psychology (cf. Kurtz 1990:24f.)—is in line with the spirit of the analytic philosophy dominating at the time. In contrast, most present-day naturalists think of ontology and epistemology not so much as "impedimenta" but as "ways of giving precision to the world-view growing out of science" (Roy Wood Sellars 1944:694). Thus, on Quine's ([1951] 1953:45) influential interpretation of naturalism, "ontological questions . . . are on a par with questions of natural science." For a recent forceful statement of the position that naturalism does not equal reductive materialism, see Richards (1987).

narrowly defined empiricism (cf. Strawson 1985).[5] The philosopher who is held mainly responsible for the current renaissance of naturalism,[6] Willard Van Orman Quine, is an "ontological relativist" (Quine 1969a) to whom professing materialism seems "grotesquely inappropriate" (Skolimowski 1986:481). And Quine's reliance on the maxim of "relative empiricism"—"Don't venture farther from sensory evidence than you need to" (Quine 1973:138)—should not obscure his conviction that the total field of science "is so underdetermined by its boundary conditions, experience, that there is much latitude of choice as to what statements to reevaluate in the light of any single contrary experience" (Quine [1951] 1953:42–43).[7] More generally, what unites the naturalistic philosophies of a John Dewey, a William James or, closer to us, a Wilfrid Sellars, despite their obvious differences, is the view that there is nothing "given" to us cognizers in nature. We, who are a part of nature, help *construct* her. The consequences of our knowing efforts therefore have all of the liabilities of any natural process: our knowledge is uncertain and fallible, but also corrigible.

If naturalism was historically, and remains today, first and foremost an American movement (with roots in Britain and Germany as well),[8] its chief philosophical contender, analytic philosophy, originated in Germany with Gottlob Frege's anti-psychologism and gained momentum on the Continent (Wittgenstein and the Vienna Circle) and in Britain (Russell). It is a major paradox that the philosophical movement which, together with Marxism, paid most lip service to "scientific philosophy"—logical empiricism—turned away from the actual study of science[9] and even committed itself thoroughly to a foundational project which turned it into yet another "prima philosophia" (Quine), i.e., a philosophy purporting to be logically and methodologically prior to science.[10] As such it has no counterpart in the more mature sciences.[11]

---

5. What is inconvenient about this classification is that it leaves out philosophers like Aristotle or Spinoza, who have also been labeled as precursors of naturalism.

6. Cf. Kurtz's (1990:11) premonitory remark, written in 1974: "Lest analytic critics exult in naturalism's demise, let it be reminded that many histories of analytic philosophy are now being written, which may be a symptom of its own impending burial, and which only points out the fate that all philosophical movements seem to share: they are at the mercy of the winds of fashion." The life cycle metaphor is actually misleading here; it suggests more discontinuity than there actually is. Koppelberg (1987) and others have argued rather convincingly that Vienna Circle member Otto Neurath is to be credited for many of the accomplishments usually associated with Quine (see also Uebel 1991).

7. This is the so-called "Duhem-Quine thesis"; it will be discussed in 2.6.3.

8. See, e.g., Richards (1987); Bhaskar (1979); Desmond (1989).

9. The logical empiricists were only interested in "rational reconstructions," i.e., rather peculiar idealizations of limited aspects of scientific *discourse*, not in science as an empirical phenomenon.

10. Giere (1988:23–24) points out that this did not happen surreptitiously: logical empiricism was *explicitly* "foundationist."

11. This is not to deny that foundational crises may be beneficial for a scientific discipline, as anyone vaguely familiar with the more recent history of mathematics knows (cf. Stegmüller 1976, 1979). (I add the qualification, "more mature" sciences, because it would seem obvious that permanent foundational debates, as in sociology, are a reliable indicator of the immaturity of a discipline.) My point is that the foundational efforts of the logical empiricists, mainly because of their self-limitation to linguistic analysis

Small wonder that by the 1970s, many people's patience with analytic philosophy had run out (P. S. Churchland 1986a:ix). That is how various special philosophies of science which now flourish (the philosophy of biology, the philosophy of economics, the philosophy of psychology, etc.) took off.[12] There are also intricate links between naturalism and evolutionary thinking, which are represented in many chapters of this book. Early naturalists like Dewey and James were much taken with evolutionary ideas, and today certain tenets of an evolutionary epistemology seem irrepressible to many epistemologists and to most philosophers of science.

A question that many people—humanists and others—press upon the naturalist is: "Is naturalism 'scientistic'—an ideology in disguise?" (e.g., Sorell 1991).[13] I would like to invite the reader who is tempted by a relativistic answer to this question to ponder the following—undoubtedly "scientistic"—observation by Hull (1988a:26):

> Science *is* one of the major ways that people in Western and Westernized societies today establish their beliefs, but it is neither the only way nor merely the way that they do so. What science has in its favor is that it beats all other ways hollow. There is no contest. Totalitarian nations of every political stripe can safely suppress artistic expression, but they find it very difficult to suppress freedom of inquiry in science, that is, if they want to remain in power. Although life in a society without artistic productions might not be worth living, it is possible. It is no longer even possible without science.

Of course, science is not everything and does not know everything. Of course, science doesn't solve all of our problems—it creates many new ones. The joy many scientists feel when demolishing some Great Idea might even have something perverse about it, as Robert Musil suggested in chapter 72 of *The Man Without Qualities* (1953–60).[14] Yet if "success in the knowledge game" (Hull) is a prerequisite for *Homo sapiens* to thrive, the point of wanting to avoid *this bit* of scientism escapes me. And once this much is granted, a moral and political implication follows, as I came to realize by talking to Donald Campbell, a major evolutionary epistemologist. When I interviewed Campbell for this book, he told me this anecdote:

("rational reconstruction"), differed in kind from those of the mathematicians who tried to surmount the *Grundlagenkrisis* of their discipline by means of various unification programs at the turn of the century, or the biologists around MacArthur in the 1960s who wanted to develop a biological analogue to the Bourbaki group in algebra (Hull 1988a:223).

12. Kitcher (1992a:59) writes that naturalism in psychological cloth "re-entered epistemology quietly" through Gettier's objections to the traditional analysis of knowledge as justified true belief (see 7.5.1). While it is true that solutions to Gettier's puzzling cases—which, from a scientist's perspective, are irrelevant (cf. Shapere 1984:244)—have prompted analyses of the causal processes that generate and sustain belief (Goldman 1986), the challenge of the historical and, later, sociological waves of science studies to the very survival of the philosophical enterprise (as traditionally understood) should not be underrated either.

13. Remember Husserl's poignant question in *Die Krisis der europäischen Wissenschaften und die transzendentale Phänomenologie* (1962): Is science universal—rather than just the way we in the West happen to establish certain beliefs?

14. This theme has recently been explored by historian and philosopher of science Isabelle Stengers (Stengers and Schlanger 1991:176–78).

I was sitting next to Popper at lunch (this was in Spain in 1984) and asked him, "Why are you so hard on the relativists? After all, you say, 'We don't know, we can only guess'" and you agree that the facts which 'falsify' theories are but conventions agreed to among the scientists working in the field. Why don't you accept my distinction between 'epistemological relativism' and 'ontological nihilism'?"[15] Popper said to me, with his warm, fatherly attitude, but nonetheless dead serious: "Donald, if you are any kind of a relativist, you are my enemy." That was my last face-to-face visit with him. The specter of relativism was so horrifying to him.

Bunge (1983:8) and Bradie (1989), among others, have pointed out that Popper may not be as naturalized an epistemologist as is often suggested (e.g., Kurtz 1990). Yet what unites Sir Karl and the indubitably thoroughly naturalized Campbell is the fear that "relativist sophistry, combined with career opportunism, might lead to fundamental changes in the motivation of scientists, and spell the end of that precious and precarious social system of science we both so admire" (Campbell 1988c:376). That is again a form of scientism, and one that many participants in this book, including myself, might be willing to agree on. By furthering our understanding of the social system of science, we may produce a sociology of science which is on the whole relativistic, but which can nonetheless contribute indirectly to science's preservation.[16] "L'arroseur arrosé," as the French say!

About the Format of This Book

Depending on whether one is modern, postmodern, or nonmodern (Latour 1990a, 1991, 1992a), one may or may not be nostalgic for the legendary "days when a single lover of wisdom could hope to understand everything then known" (Hull 1984:11)— but *gone* they are for sure. Our only alternative, Hull stresses, is "a sort of social 'omniscience.'" This seems equally true in science (Kornfeld and Hewitt 1981) and philosophy, if not in social communities at large.[17] The unusual format I have chosen to present the results of my exploration of current naturalism to the reader reflects the "distributed" state of our knowledge (Chandrasekaran 1981; Callebaut and Van Bendegem 1982). Because it is unusual, it requires some explanation.

The keyword is *serendipity*. I started interviewing philosophers and sociologists of science for BRT (the Flemish branch of Belgian Public Radio) broadcasts in Belgium in 1984 and continued to do so when I went to the United States during a sabbatical leave in 1985. The dollar being worth twice what it is today, the honoraria were a welcome addition to my travel grant. When, after my return home, I came to realize

15. "Ontological nihilism" is a term which Campbell believes he negotiated with Quine during his William James Lectures at Harvard University in 1977: "I would have used 'ontological relativism' otherwise, but Quine (1969a) had used the term 'ontological relativity' in another way."

16. Cf. Papineau's (1988:55) optimistic assessment of our predicament: "It is perfectly possible for human beings to reflect on whether the belief-forming practices they engage in are reliable for producing truths, and to reconstruct themselves and their communities in the light of such reflections."

17. See the speculations of Kochen (1979) on the emergence of a "community mind."

what nice and up-to-date material I had on all these kilometers (or miles, if you prefer) of tape—in part from listeners' enthusiastic reactions and requests of additional information—the idea to commit to paper what would otherwise be gone with the radio waves took shape. But what kind of book? My original plan to write a sort of handbook, in Dutch (my mother tongue), on developments in philosophy of science after Lakatos and Laudan—that is, where existing books left off—was abandoned quickly. Although *I* felt that such a book would be useful, it was uncertain there would be large enough an audience to make the effort worthwhile. And most importantly, any publisher, invoking this uncertainty, would insist on a short manuscript.

I also thought vaguely of the more obvious possibility of writing a straight "interview book," but rejected this idea too. It was and still is my impression—though I add immediately that not everybody agrees on this—that on the whole, work representative of the current generation of philosophers of science is less individualistic than that in other periods or other realms of philosophy, in the plain sense that people tend to react to one another's work rather than to create elaborate systems. In a book consisting of separate interviews much of this cooperative or "distributed" aspect would be invisible. To some extent this objection may have been a personal rationalization, masking my realization that notwithstanding the high quality and great usefulness of some previous books of interviews with scientists, in philosophy one does not get much credit for such a "journalistic" endeavor—if it doesn't positively harm one's career. When an important philosopher whose collaboration I solicited refused to be interviewed with the argument that what I was doing sounded like "a cheap and easy way to get a book written," I lost any remaining illusions on this score.

Then one day, while rereading some transcripts (Public Radio bureaucrats require transcripts of all text broadcasts, which then disappear in their files), it occurred to me that I had been quite insistent in my questioning at times, pushing people in the secrecy of their study rooms into saying things about their colleagues' work they wouldn't say in public. (Some of them—but only a very few—may have *pretended* to be hesitant, of course.) It was clear from the results in most cases that, after a while, fatigue or some other factor greatly diminished their original resistance! The *Aha-Erlebnis* was unmistakable: *I should mix the interviews,* making the whole manuscript as interactive as possible. People to whom I mentioned this wild idea were skeptical, to put it mildly. But I was going to show the world that it could be done. And as soon as I started torturing the cut and paste functions on my text processor, trying to fit together the pieces of an immense puzzle that existed only in the vaguest of outlines in my mind, I was hooked. It is for you to judge whether I succeeded in communicating some of the thrill I experienced when composing the puzzle into the end product. And if you think I did, credit the participants in the project—my twenty-four *sine qua nons.*

In the process of compiling the book I stumbled over obstacles and confronted some gaps that looked insurmountable. That is when I started to write to people individually with requests for additional information, which most of them provided kindly if not always promptly (who was this obscure Belgian philosopher anyway?). By January 1990 I had a rough and ready version of all chapters which included a number of additional questions, most of which had arisen from my "puzzle laying." This manuscript was circulated to all participants except for a couple who still were not in the game by then. I urged everybody not only to answer my additional individual questions to them, but also to react to other participants' statements and views—within pretty tight space constraints, as my interviewing experience had taught me that there are philosophers who take their time. I also gave people the opportunity to defend themselves when they had been attacked. The response was overwhelming, with some people turning in more than thirty pages of new text, updating, revising, and expanding their earlier statements, but always keeping in mind the conversational style. Coping with the new information, fine-structuring the text, informing people about changes somehow pertaining to their statements, and the more usual editorial business such as writing the chapter introductions and compiling the extensive reference list completed the work.

As far as my selection of interviewees is concerned I can be straightforward. At one level, the cast of characters included is arbitrary. I only talked with people *I* thought were interesting, whose work *I* was more or less acquainted with, and who were within *my* reach (geographically and so on). This already excludes three non-empty classes of relevant spokespersons for the naturalistic perspective. There *are* important people whose work I didn't know in time, and there are bound to be invisible workers (that is, invisible to me) who have equally important things to say. I *know* I missed some opportunities for lack of alertness, insistence, or sheer energy. To my surprise, only two people I solicited declined to be interviewed (one I already mentioned; the other because of illness). Two additional criteria that played a role in the selection of participants were *diversity* and *overall coherence*. As these two tend to be conflicting, they had to be balanced. On the one hand, for quite obvious reasons I did nothing to include people—even quite famous ones—whom I expected would be mostly repeating things already said. Some of my original positive choices were highly contingent, but then the ensuing negative choices imposed themselves logically. Neither implied a preference: I grasped the opportunities that offered themselves, but *it could as well have been the other way around*. On the other hand, keeping in mind Campbell's (1969) "fish-scale model of omniscience," I also did not try to include people I perceived as, on the whole, "out of touch," intellectually and/or socially, with "mainstream" naturalism, *if such an animal exists*. Which brings me to my concluding remark in this context. To those who think one needs to (and is able to) define the *essence* of a thing one is interested in—in this case, naturalism in current philosophy—and who are bound to feel uncomfortable with the contingency of my actual selection of interviewees ("Why should we pay attention to just this assemblage of personages and not some other?"), I say: I had to start somewhere. It is my bet that the *type-specimen* method from biological taxonomy, where a single specimen is se-lected to be the reference specimen (Hull 1988a), may work here as well.[18]

One final caveat: As the coherence of the mosaic of views presented here had to be provided in part by my "questions," occasional reshuffling and addition of sentences that were never used in the original interview (primarily such that convey bibliograph-ical information) was inevitable. But I am confident that it will be easy for the reader who is attentive to such things to identify such editorial interpolations. In the same vein, I assume full responsibility for all bibliographical additions in the participants' "answers" and for all information provided in footnotes.

18. If this is cryptic here, it should become clear later on (see 6.6.1).

PART ONE

Talking about It

2

BEYOND POSITIVISM AND
HISTORICISM

THIS CHAPTER is primarily concerned with locating and delineating the views of a number of philosophers of science who are now most active on the scene vis-à-vis the views of their intellectual elders, the logical positivists (or, for others, Popper) and the historicists. What, from our present vantage point, do we find valuable in the views of both groups? What can we learn from their "mistakes"? Where do we find them altogether misguided? *How different are we today?*

2.1 THREE PHILOSOPHICAL GENERATIONS

Philosophy of science as currently practiced is a reaction against a reaction. Until the early sixties the *logical positivism* (*logical empiricism, neopositivism*) of Rudolf Carnap, Hans Reichenbach, Carl Hempel, Herbert Feigl and their followers reigned supreme.[1] The positivists' primary objective was an examination of "logical patterns" exhibited in the organization of scientific knowledge and of the logical methods whose use they took to be "the most enduring feature of modern science," despite "frequent changes in special techniques and revolutions in substantive theory" (Nagel 1961:viii). Karl Popper is often included in this category, although neither he nor his admirers like to be called positivists.

The roaring 1960s were dominated by a reaction against the "peaceful unanimity" (Wimsatt 1979:352) of the positivist era—the so-called *received view* of science. What united the rather heterogeneous group of people involved in this critical movement was their discontent with the alienation of the positivists' "rational reconstructions" of scientific theories from their own, historically informed picture of scientific activity. The *historicists* urged for a new rapprochement between history and philosophy of science, suggesting that "history is philosophy teaching by examples" (Laudan 1979:40). This movement included Norwood Hanson, Paul Feyerabend, Thomas Kuhn (who would eventually eclipse all the others), and Stephen Toulmin.

---

1. The capsule positivist in Laudan's (1990a) imaginary conversation on key controversies in the philosophy of science answers to the name of "Rudy Reichfeigl."

The protagonists in this chapter react to both of these movements. They see themselves as rejecting at least some of the central tenets of neopositivism while also resisting certain conclusions about the inadequacies of positivism reached by the historicists. Historian-philosopher Robert J. Richards (University of Chicago; see biographical note at 10.1.1), for instance, sees "a convergence toward a view that is held
by a growing number of philosophers and historians of science: one that tries to unite
the best of both logical positivism and historicism." He hopes that his own model of
science, which takes scientific change to be in fundamental ways like the evolution of
biological species, "emphasizes the strong features of those perspectives" while cor-
recting their "errors."

Unlike biological inheritance, intellectual inheritance may involve more than two
"parents," and cultural generation length is variable (Boyd and Richerson 1985:7–8).
Many of the third-generation philosophers whose ideas are discussed in this book were
raised as positivists and would not want to deny that legacy. When I asked Robert N.
Brandon (Duke University; see biographical note at 9.5) which school of philosophy
of science he belongs to, he said he did not know: "My training was in analytic philosophy and in some way I am very sympathetic to positivists; I think there is a lot to be
said for them." Brandon is a good friend of Marjorie Grene, who has influenced current thinking in philosophy of biology to a considerable extent. "Marjorie thinks the
word 'positivism' is a dirty word. She just rails and rails against positivism." For
Brandon, however, positivism is "a movement in philosophy that had a number of
good points to it and failed on a large number of details."

William C. Wimsatt (University of Chicago; see biographical note at 3.2.1) studied
philosophy at the University of Pittsburgh, a stronghold of positivism if ever there was
one. He recalls being "a hard-core reductionist and positivist" even before he got into
philosophy: "My paradigm of science, when I started out in physics and then engineering physics, was classical mechanics: clearly defined, powerful results from few
assumptions, axiomatic and so on." Bill's father was a biologist. "I admired his work
a great deal but I thought biology was too messy. It was a lot of fun but I wasn't really
interested in that." Wimsatt's present preoccupation with *complexity* and *messiness*
took shape during his undergraduate years. In combination with his keen interest in
the latest scientific developments ("Why always be twenty years behind the cutting
edge?"), it prompted him to drastically revise his earlier positivism. His interest in the
history of science arose later and rather independently of his scientific concerns.

Thomas Kuhn is generally regarded as the pivotal figure in the movement that
eventuated in the demise of positivism and the present vigor of historical and sociological approaches to science. After Kuhn, philosophy of science would never be the
same. Yet Kuhn's influence on third-generation philosophers was not usually a direct
one. Ronald N. Giere (University of Minnesota; see biographical note at 4.3) notes
that the work of Kuhn and the other historicists was perceived as "the new development." Giere remembers that "it was a hard time to know what you were doing in

philosophy." He mainly credits Kuhn for having sensitized philosophers to the importance of history for the *testing* of philosophical theories, i.e. for having revived *naturalism* in postwar philosophy of science. According to Giere, "Kuhn is the most important figure in this whole history. I think that to say so is now correct. But I would not have said this twenty years ago or even ten years ago."

Likewise, Thomas Nickles (University of Nevada at Reno; see biographical note at 2.3) thinks that "Kuhn, more than any other single person, is responsible for philosophers now taking history seriously." That implies "getting off the a priori, analytic style of philosophy of science that one associates with the positivists and Popperians." Kuhn came at the best possible time: "I think the interest in history was coming anyway. Had Kuhn never lived or never written, I think it was coming. But he no doubt was a dramatic stimulus that hastened the first wave of empirical information—historical knowledge—to wash over recent philosophy."

The exception confirming the rule is Alexander Rosenberg (University of California, Riverside; see biographical note at 3.3). He was trained by someone who was in strong reaction to positivism, Peter Achinstein. Achinstein taught him "all of the mistakes that the positivists made, both the technical mistakes—why their program was impossible—and the basic philosophical mistakes." Rosenberg left graduate school an ardent anti-positivist. He subsequently met with philosophers and "people who thought they were philosophers: classicists, literary critics, theological thinkers." Rosenberg felt they were "so woolly-minded and so unrigorous, and had so little understanding of science and so low a respect for it," that his experience with them turned him into a "positivist manqué": he wishes "positivism were right; unfortunately, it's not right, and we have to make the best of it."

2.1.1. *Dialectics or a Family Affair?*

Callebaut: Third-generation philosophy of science is sometimes seen as a kind of dialectical synthesis of positivism and historicism, Philip.

Philip Kitcher (University of California, San Diego; see biographical note at 3.4.2): I agree with that. Obviously, if you consider this not from the perspective of dialectics but from that of a marriage, you've got some offspring who resemble one parent more than the other! But some seem to have found a genuinely new combination.

Callebaut: Fifteen years ago, Lindley Darden (1977:242) wrote:

In biological evolution organisms cannot choose which genes to retain, nor can they strive to introduce particular new genes. However, we, as philosophers, can choose which things we wish to inherit from our intellectual parents and which things we wish to eliminate as non-adaptive in our current environment. And, some of us who began our philosophical careers after 1968 look back at the positivists as grandparents, with a generation of Kuhnian and Feyerabendian excesses in between. With that distance from the logical positivists, we can even choose whether we wish to retain the same stock of problems and questions to disagree about or whether we wish to introduce new problems instead.

She implied that one of the things distinguishing the second from her own—your—generation is that the historicists shared with the positivists a common stock of philosophical problems, while providing, of course, radically different answers (cf. Toulmin 1969:51).[2]

Kitcher: I'm not sure that's true actually.

Callebaut: I anticipated you wouldn't agree with such an assessment, which can be read as diminishing the revolutionary character of Kuhn's departure from the received view.[3]

Kitcher: No, I don't agree with that. I think there is considerable continuity in questions from the first to the third generation and considerable discontinuity in questions from the first to the second. I think of Kuhn as setting a quite different agenda of questions for the philosophy of science than had been set, say, by Hempel and Carnap. All of a sudden there were these large questions about how one resolves the big disputes in science. Hempel once said to me, "We had never posed the problem of scientific rationality in general because we had always assumed that large change was just small change blown up"—a kind of microevolution/macroevolution thesis![4] If one understood the confirmation of small-scale hypotheses, which could be formulated easily in a first-order language, then one would automatically have understood the confirmation of major new scientific theories, and there would be nothing new there except scale.

Callebaut: What seems to single out the third generation, among other things, is a

---

2. Likewise, it has been suggested that Kuhn's alternative to the received image of science in important ways *negates* the received view (Hacking 1981b:1–4). According to Hacking (1983:6), one can easily generate a list of contrasts between positivism and Kuhn's brand of historicism "simply by running across the Popper/Carnap common ground and denying everything."

3. In the words of Suppe (1977:233), the received view tries to present "a general logical or structural analysis of all theories which is epistemically revealing of theories and their connections with the evidence of experience and observation." Hacking (1981b:1–2) refers to the following nine tenets or ingredients of the "image of science" (Kuhn) of the first generation (the parenthetic references are to the main discussions of these issues in this book): 1. realism (4.3); 2. demarcation of science from other kinds of belief (2.1.3, 2.2.2, 3.4.1, 3.6); 3. science is cumulative (7.3); 4. the observation/theory dichotomy (2.6.7, 3.5.3); 5. foundations (2.3.3, 5.3, 5.4.1); 6. theories have a deductive structure (5.3.2, 5.5); 7. "scientific concepts are rather *precise,* and the terms used in science have fixed meanings" (2.6.1); 8. the justification/discovery dichotomy (5.3); 9. unity of science (2.3.2, 4.2).

4. According to the *punctuated equilibria* model of Gould and Eldredge (Eldredge and Gould 1972; Gould and Eldredge 1977; Eldredge and Cracraft 1980; Stanley 1981; Eldredge 1985), macroevolution, i.e., evolution above the species level or of the higher taxa, "proceeds by the rare success of . . . hopeful monsters [cf. Goldschmidt 1940], not by continuous small changes within populations" (Gould 1977:30). Orthodox defenders of the modern synthesis hold that, to the contrary, macroevolution poses no special threat to the received theory of microevolutionary processes, which they view as compatible with both gradualism and punctualism (Stebbins and Ayala 1981). Cf. Urbanek (1988) for a discussion of a similar opposition among evolutionists in the former Soviet Union.

shift toward new problems and new areas; for instance, a move away from physics toward biology and psychology. What constituted the most radical break, according to you, Ron: the transition from the first to the second generation or that from the second to the third? I am trying to locate the views of the third generation: Are they somewhere in between the first and the second generation (that would amount to a kind of middle-of-the road position, half-way back to good old positivism)? Or are they *beyond* the second generation, maybe constituting an even more radical departure from positivism than historicism meant?

Giere: You have to see the temporal breaks. It was logical empiricism up until the 1960s. There was a whole bunch of critics of logical empiricism about that time. Hanson, Feyerabend, Toulmin; Shapere (somewhat younger); Scriven's work on explanation. *There was a group there, and Kuhn in a way was more radical than the others were.* There was a loosening up of logical empiricism in various ways, but I think fundamentally not *radically* different. Part of it was a shift from formal methods to informal methods. In a way, the later Wittgenstein . . .

Callebaut: . . . the Wittgenstein of the "language games" of the *Philosophical Investigations* (1953), who taught at Cambridge and was a major influence on the "ordinary-language" school in philosophy . . .

Giere: . . . against the younger Wittgenstein . . .

Callebaut: . . . the admirer of Russell's logical atomism, who wrote the *Tractatus logico-philosophicus* (Wittgenstein [1921] 1961; cf. Peterson 1990), a book that influenced the Vienna Circle to a considerable extent.

2.1.2. *Truths about Nature Don't Come in for Free*

Giere: The critics of positivism were influenced by the later Wittgenstein. You got a more informal, ordinary-language, approach. But I don't see this as so radically different. *At the time* it looked very different, of course. Scriven made his whole reputation by attacking Hempel and other empiricists. In that sense, it is not that radically different. Kuhn turned out to be more important, in part because he *was* radically different.

Callebaut: Patricia, you studied at Oxford. In your book you approvingly quote Feyerabend's dismissal of the ordinary-language approach of a "closed society" founded on a 1950s myth.[5] What was that "powerful myth," actually?

Patricia S. Churchland (University of California, San Diego; see biographical note at 8.4.1): He referred to the common idea that by reflecting on the nature of language—on how you speak and how I speak and how he speaks, but not necessarily on how folks in the market speak—one would be able to reveal certain deep truths,

---

5. "The more people differ in their fundamental ideas, the more difficult will it be to uncover . . . regularities. Hence analysis of [language] usage will work best in a closed society that is firmly held together by a powerful myth as was the philosophy in Oxford of about ten years ago" (Feyerabend 1963, quoted in P. S. Churchland 1986a:272).

which would turn out not only to be truths about our conceptual framework but truths about the universe. And after all, it is the nature of reality that we want to understand. It's all very well to be interested in conceptual frameworks if you want to know the way somebody thinks. But what we really want to know is whether our beliefs are true—whether our conceptual scheme is good enough. There was this wonderful kind of ambiguity where these didn't seem to be the right questions to ask. Rather, it appeared that *if you got the conceptual truths, then the truths about nature came in for free*. For example, if you do the conceptual analysis on "know," then you will have discovered something about the nature of knowledge as it actually is in the brain.

Callebaut: What were the main differences between ordinary-language philosophy and logical empiricism?

Churchland: Most ordinary-language philosophers at Oxford did not have much regard for science. By contrast, his understanding of science was the enormously important thing about Quine: he gave examples and arguments to show that science could change how we ordinarily think. He argued that there was nothing a priori or sacred about introspection or intuition. *Everything* was revisable, everything was up for change, including common sense. In that respect, the people who came out of the logical-empiricist tradition were much more willing to see the significance of science in changing our beliefs than people within the Oxford tradition. On the other hand, logical empiricists seemed more committed than I am to the sentential paradigm of representation and to the logic-machine model of reasoning.

Wimsatt: I quite agree with Pat's general characterization of the antiscientific flavor of the ordinary-language philosophy of the late 1950s and early 1960s. Thank God I had lots of science before I ran into it. But the Feyerabendian gambit of modern eliminativists like Pat and Paul is not the same as Quine's.[6] Quine surely thought that, in principle, anything was revisable, but he didn't mean that all things are equally up for grabs. About twenty years ago, at the height of Feyerabend's popularity, there was a big ruckus about "quantum logic" and giving up the law of excluded middle because elementary particles appeared to ignore it (sometimes they seemed to be neither here nor there), but those radical revolutionists who waved Quine-Duhem banners (including Feyerabend and Putnam) were premature.[7] What happened? Just because particles are fundamental doesn't mean we

6. In contemporary philosophy of mind, eliminative materialists ("eliminativists" for short) contend that there are no mental phenomena proper: Mental talk commits us to saying things that are literally false, and should therefore be banned and replaced by brain talk. (For an introductory discussion, see Bechtel 1988b:102–6; cf. 8.4.1.) In his earlier work, Feyerabend (e.g., 1962, cf. 1975, 1981) defended the view that in scientific revolutions, an old theory is often replaced by a new one ("replacement view"), as both theories are "incommensurable" (see 2.6).

7. According to the French conventionalist philosopher Pierre Duhem ([1906] 1953:185), "the demonstrable value of experimental method is far from being . . . rigorous or absolute": the conditions under which it functions are much more complicated than empiricists have tended to suppose, hence "the evalu-

have to bend logic to fit their ways. We regard quantum logic as a rather quaint special case, and go on keeping excluded middle for our ordinary logical inferences because it is more conceptually fundamental than that particular behavior of elementary particles. Even if we were to adopt quantum logic as fundamental for elementary particles, common sense or Bohr's principle of correspondence would require us to explain why macroscopic objects and concepts seem to follow classical logic, so we still wouldn't be rid of it. Eliminativists shouldn't count coup on well-entrenched higher-level concepts, regularities, laws, and theories so quickly. Bridge builders, airplane designers, and even satellite jockeys can safely ignore the Lorenz-Fitzgerald transformations . . .

Callebaut: . . . as we will here!

Wimsatt: Newtonian mechanics is effectively true for almost all purposes, and relativity theory has not made it obsolete but just fenced it in, confined it to a limited domain.

2.1.3. *How Different Was Kuhn?*

Callebaut: The historicists denied that the "logic of justification" which was continually referred to by the positivists (cf. 5.3) was of any relevance for real science. Hanson (1958, 1969) turned his attention to a presumed "logic of scientific discovery" instead. Lakatos (1976) did something similar when he talked about "rational mathematical heuristics." But Kuhn's approach was altogether different in that he denied *both* that there exists a logic of justification and a logic of discovery. Didn't his success also have to do with his reacting more radically and more negatively to positivism than others, Ron?

Giere: Yes, Kuhn really denied these basic things. Whereas . . . take Toulmin, for example: he didn't deny that science was rational. He just wanted a different account of rationality. He had this thing, "You cannot reduce rationality to logical algorithm!" (Toulmin 1969). So he gave a new theory of rationality, a new picture of science as a rational activity.[8]

Callebaut: In this sense, Toulmin's position is quite comparable with Shapere's (see 2.2.2).

Giere: On the other hand, one of the reasons why Kuhn became more important—in

---

ation of results is much more delicate and subject to caution." Cf. the Quine quote in note 36. On the differences between Duhem and Quine, see the thoughtful essay by Hattiangadi (1989:468f).

8. Lakatos, who traced back Toulmin's rejection of "logicality" to Wittgenstein's influence on his brand of pragmatism, objected strongly to this move: "Toulmin has for many years been trying to bring about a counter-revolution in logic. He is recommending us to give up one of the most marvelous progressive research programmes in the history of human knowledge—mathematical logic—which provides the most efficient weapons of objective criticism mankind has ever produced. And he recommends us to replace it by woolly, 'élitist', Wittgensteinian 'inference tickets'." ("Understanding Toulmin," in Lakatos 1978b:235–36n.). "Élitism" is Lakatos's label for the view held by Kuhn, Polanyi, and others (*pace* a skeptic like Wittgenstein), that science is or can be progressive, but that there can be no demarcation criterion, viz. "an explicit, universal criterion (or finite set of norms) for progress or degeneration" (111).

hindsight, Kuhn appears as the major figure; the other people paled—was that he also had a strong *positive* theory. He was not just criticizing logical empiricism. All his criticism was indirect. Instead he had an alternative view. The other people didn't really. Shapere did much later. Toulmin did ten years later.

Callebaut: Kuhn also makes for quite easy reading.

Giere: Right. I reread Toulmin (1972) a couple of years ago, and it is still hard to see the forest for the trees. This was an accident: Kuhn was forced to write that book short. If Kuhn had been left on his own, he would have written a four or five hundred page book that would not have been nearly as successful. But he was forced, because he wrote for the *Encyclopedia of Unified Science,* to do it just like Darwin was forced to write the *Origin* (1859): quickly and short (see, e.g., Ruse 1982, Hull 1988a)! That made a tremendous difference!

Callebaut: One reaction to Kuhn was that people did shift away from logic to history (cf. Giere 1973; Manier 1980; Burian 1987c).

Giere: In fact you got people much more concerned with the history of science. Of my generation, I take Laudan (e.g., 1977, 1984) to be a major person in this respect; he incorporates the historical part of Kuhn. He is, of course, against logical empiricism; but he is quite adamant in showing that science is a rational activity. That is very much what logical empiricism was trying to do: *demarcation* of science from nonscience. For them, that is a major problem, and that problem stays. This is also why—this is an historical accident—most philosophers of biology also do history of biology. The reason is that when philosophy of biology began to flourish, right after Kuhn, for many different reasons, those two things happened to go together.

Callebaut: Was it really just an accident, Bill? Wasn't the marriage of history and philosophy of biology also conditioned by the fundamental *historicity* of the new subject matter?

Wimsatt: I think both Ron and you are right. Kuhn made history more respectable, but history (in the person of Darwin, as well as through his theory) casts an enormous shadow in evolutionary biology, which is where most philosophers have started in the last twenty years. Biologists tended to claim Darwin as their patron saint and implied that their enemies were "non-Darwinians." This naturally leads back to the texts for defense. There was also a growing Darwin industry among historians of science, and philosophers quickly jumped on. Finally, in biology at least, it is (*contra* Kuhn) often the people who forget their history who make the mistakes, as I found out through Wade's work when I rediscovered blending inheritance in modern models of group selection. Similar stories could be told for modern genetic reductionism and for the way developmental biology was left out of the evolutionary synthesis (cf. 9.1).

Callebaut: Couldn't one also say, Ron, that people like Scriven continued to view physics as the paradigmatic science, *even* when they were busy showing that the

philosophy of physics cannot be extended to other disciplines in any simple way—
that, for instance, historical or biological explanation cannot be made to fit Hempel's deductive-nomological scheme (cf. 4.1), so that you had to "weaken" the Hempelian constraints? Or do you see people like Scriven as really urging us to depart from the physics paradigm?

Giere: There is a little bit of that. Scriven (1959) and Toulmin (1961, 1972) wrote about evolutionary theory in biology. But it was not big, and it was "Can these models of explanation be applied to biology?" This was not the major thing then. Scriven did not change the problem. He thought that *explanation* was a major problem, and that it was just that Hempel got the wrong view. So he did not change the problem. He had a new account of it, but he did not say that explanation is not a problem. Toulmin was writing about "the nature of theories" as a major problem; it is just that a theory is not to be thought of as a set of axioms. It is not a change of problems; it is a change of methods: informal language, ordinary language so to speak, versus formal language.

2.2 IN PRAISE OF THE RECEIVED VIEW (YES)

Callebaut: Tom, you are one among several participants in this book who believe that, all things considered, there is a great deal to be said in favor of the positivistic view.

Nickles: We owe philosophy of science as a profession to the positivists. Were it not for them, philosophy of science in anything like its present form would not exist as a scholarly activity, I think.

Callebaut: In some quarters, the word "positivism" has become synonymous with "whatever we don't like." Even people who are otherwise wary of "whiggish" interpretations of history (see 2.3.4) are liable to oversimplify the philosophical past.

David L. Hull (Northwestern University; see biographical note at 6.6.1): The received view was very good, and those of us who criticize it are never fair to it. We present a parody of it and then knock it down; but that's the way the game is played.

Wimsatt: The similarity we see in the positivists' views at this distance is basically an instance of the old saying, "All Orientals look alike." I have a friend who is Japanese, who said: "All Occidentals are inscrutable." There is a kind of cultural relativity here. I think a similar thing is responsible for a lot of the apparent univocality we see now in the positivist views.

Nickles: Philosophers notoriously disagree with one another. The very structure of their discipline is built upon critical attack and disagreement. This was as true of the positivists as it is today. They were among their own best critics. There is a tremendous pressure for novelty and "progress" in philosophy as in all academic disciplines. Combine this with the usual generational phenomenon of the offspring condemning their parents and we can also understand why criticism of the positivists has sometimes been excessive.

2.2.1. *Clarity and Rigor*

Callebaut: Wittgenstein and the logical empiricists completed the *linguistic turn* in philosophy which Gottlob Frege inaugurated in the nineteenth century (Rorty 1967; Hacking 1975). The Vienna Circle's manifesto, *The Scientific Conception of the World* ([1929] 1973), struck me by its almost obsessive preoccupation with *communication*, with the *language* of science and of "scientific philosophy." Linguistic rigor may be one aspect of analytic philosophy which continues to deserve emulation.[9]

Nickles: It is largely due to the positivists that we have high standards of rigor. The original positivists were well trained in the sciences, under such luminaries as Max Planck and Ernst Mach; and much of their early work—I am thinking of someone like Hans Reichenbach—was involved with analyzing in great detail the scientific work of the day. The theory of relativity in particular had a tremendous impact on the early thinkers and did much to fashion their philosophical views. So we do owe to the positivists a great deal indeed. Unfortunately, as positivism aged, it became more and more concerned with merely formal matters. The new naturalists and historicists want to regain the old interest in the content of scientific claims and practices.

Elliott Sober (University of Wisconsin, Madison; see biographical note at 6.5): Although I reject the doctrines that the positivists developed, I have the highest admiration for the standards of clarity and argument and intellectual honesty that they brought into philosophy. They were willing to state things clearly enough, so that their views could be discussed and criticized. And quite frequently they admitted that they were wrong and tried to revise their views. This is, I think, one of the most beneficial things that has happened in philosophy in this century: *the idea that by emulating the—shall we say scientific—approach to the activity of philosophy, we can hope to make progress.* Some of the methods the positivists used—the use of logic and probability as a way of analyzing and clarifying scientific activity—I think remain useful too.

William Bechtel (Georgia State University; see biographical note at 8.3): I think I share something with Rosenberg that I should tell you. My undergraduate training was in religion, but in fact I felt rescued when I discovered A. J. Ayer (1936); that is, I found most of contemporary theology, which I was studying, to be gobbledy-

---

9. In this context, it seems to me significant that in the recent debate concerning the status of "African philosophy" (e.g., R. A. Wright 1984, Oruka 1990a), advocates of the use of logic and rigorous critical analysis like Peter Bodunrin, Paulin Hountondji, and Odera Oruka have been labeled as "neo-logical positivists" by a traditionalist defender of sage philosophy (Campbell Momoh) (Oruka 1990b:7).

gook and found the positivists to be such a breath of fresh air of concrete, articulated arguments that in part from that I decided I wanted to do philosophy.

Ryan D. Tweney (Bowling Green State University; see biographical note at 8.1.2): Kuhn must have been hard to take after that—he's almost a prophet of the messy, by contrast. Coming out of experimental psychology, I at first had trouble seeing what all the fuss was about when I read Kuhn!

Nickles: I too had something of a Bill Bechtel conversion as a callow youth. By the time I arrived at graduate school, I announced to the director of my program— Thomas Kuhn—that my goal was "to mathematize everything mathematizable!" The director's teaching did not fully sink in until some years after graduate school, when I was a young colleague of Dudley Shapere. I then found myself making a historical turn and have become increasingly historicist since then—which is not to say that I am a flaming relativist. Does that mean that I have embraced all the woolly enemies of the old-line positivists? I certainly hope not!

Callebaut: Clark Glymour (1980:7) has labeled the historical approach "the new fuzziness." Tom and Karin, both of you want to object to that.

Nickles: Well, one can find a certain amount of rubbish in all approaches, of course. I don't think an avoidable fuzziness is endemic to historical models of science. On the other hand, historicists are inclined to reply that if the world is messier than the positivists would like, it's not our fault!

Karin D. Knorr Cetina (Universität Bielefeld; see biographical note at 4.3.6): I don't think it's only a problem of these historical and sociological studies being a new area of research. I have a conception of the *ontological flexibility* of the world, which becomes very obvious if you go into a laboratory and see the complexity of the things going on. But it's more than that; it's a theoretical point. *Social reality works in a funny way.* Let me give you an example: the category of "thinking." We have the conception that thinking is located in the mind. AI (artificial intelligence) tries to model what goes on in the mind. Both scientists and the general public generally believe that the main ingredient of scientific work is thinking. But an activity like thinking, in terms of the *effects* it has, can be located in a number of other places. For example, it can be an outgrowth of social interaction, or conversation, or, as Clifford Geertz (1973:ch. 3) has argued, it can be a public activity. So an activity like thinking is not just located in the mind; it can be relocated. *Where* it is relocated—where you actually find it mainly happening—is an empirical and contingent question. You cannot predict where it is going to happen. That is what I mean by ontological flexibility: things can be found in a number of places, and cultures are made up by reshuffling the location of things. That's just one example; there are other issues which are hidden there.

Nickles: I also don't think philosophy should be defined as clear thinking, period. A better definition is that philosophers tackle the interesting, deep, and difficult problems that have no routinized, disciplinary treatment. In particular, the historicists introduced some deep problems of a sort that don't have routine solutions. In Herbert Simon's terms (1973b), these are very ill-structured problems, so far. People are struggling to develop new vocabularies, new perspectives, new modes of analysis. So we cannot expect to have everything set out with crystal clarity from the beginning. To sacrifice all this for a premature, overly principled concern for clarity would be to give up philosophy. Sometimes we have to choose between absolute *"cognitive* significance" and *significance.*

2.2.2. Rationality Lost?

Callebaut: In several of your papers (collected in Shapere 1984), Dudley, you argue that throughout history there have been two major attempts to understand what science is: "global presuppositionism" and historicism. Global-presuppositionist approaches, in your view, aim to capture science in terms of some kind of *essence;* that is, they try to find some "unchangeable defining characteristics of science."

Dudley Shapere (Wake Forest University; see biographical note at 2.7): For example, they try to state what the unalterable scientific *method* is that might be applied in the study of nature to gain knowledge.

Callebaut: Metascientific principles such as methods are often supposed to govern the substantive level of science (e.g., Rescher 1977, 1991). You deny that such a distinction between separate levels can be meaningfully made; even statements of methods, goals, and normative guiding principles you take to be themselves (based on) factual statements. There are not really two different things interacting (Shapere 1984, 1987a).

Shapere: Other people have said that science has certain *metaphysical presuppositions* that it is the job of the philosopher to try to discover; and those presuppositions are not discovered by scientific method but are the conditions for the very existence of science.

Callebaut: You think the logical positivists held a similar approach to the philosophical understanding of science.

Shapere: Science, according to them, was to be understood in terms of the techniques of *logic.* Logic would lay down the general, unalterable defining characteristics of what it is *to explain,* to be a piece of *evidence,* of the degree to which evidence can *confirm* or *falsify* a theory, and questions like this.

Callebaut: I've always found positivism quite puzzling insofar as the status of philosophy vis-à-vis science is concerned. The Vienna Circle (1973:19) was "antimetaphysical." They professed that "there is no such thing as philosophy as a basic or universal science alongside or above the various fields of the one empirical science," and they wanted to leave no room for special philosophical assertions. Yet their tenet, "there is no way to genuine knowledge than the way of experience," was immediately followed by an important qualification: "Nevertheless, the work of 'philosophic' or 'foundational' investigations remains important in accord with the scientific world-conception." In retrospect, one can see logical empiricism as *a first philosophy in new clothes* (Gibson 1986:148; Hooker 1987:75–76). Descartes' *epistemological turn* had made the theory of knowledge prior to science. In the end, the *linguistic turn* did little to change this priority of philosophy over science. Dudley, you are convinced that positivism has failed to deliver the goods. What do you see as the major flaws of global presuppositionism in general?

Shapere: Nobody has been able to pin down a description of an unalterable scientific method that is independent of the content of scientific belief. Nobody has been able to state metaphysical presuppositions of science and to justify them in any way that would make them independent of science and preconditions for science. All such claims to being presuppositions, for instance that science must be *deterministic* if it is to be explanatory, have been overthrown in the course of science, to say nothing of other objections to them.

Callebaut: The historicists saw that you had to understand science itself, in its development. Now when it comes to the issue of the rationality of science, you feel strongly that Kuhn and especially Feyerabend have thrown away the baby with the bathwater (cf. 2.6). Do you want to side with the positivists in *that* respect?

Shapere: The global-presuppositionist side wanted to account for the rationality of science, for the idea that there really is a way of deciding between competing hypotheses.

Callebaut: That much you accept.

Shapere: But they had the wrong approach to trying to do it. You can't do this by laying down unalterable defining conditions of science. There is a kind of *absolutism* to the global-presuppositionist views that I reject.

Callebaut: Michael, you agree with Dudley that the relativism that comes with historicism and sociologism[10] is a plague to be eradicated at all cost.

Michael Ruse (University of Guelph; see biographical note at 9.3.2): As I read Kuhn's *The Structure of Scientific Revolutions* (1970, first published 1962), I felt—as a lot of people feel, but Kuhn has always denied—that it was a pretty *subjective* book,[11] portraying a relativistic view of science. But at the same time, the thing which overwhelmed me was the use of the historical examples. I've always felt very strongly that if you're going to do philosophy of science—already I was into reading the science—you've got to know a lot of the history too.

Callebaut: Your first go at the history of evolutionary biology was motivated by your concern to vindicate the rationality of biology, that is, to turn the tables on Kuhn.

Ruse: I wanted to look at the history of the Darwinian revolution and show that in fact *it was rational and cumulative;* that one *did* get from the pre-Darwinian "paradigm," if you like, to the Darwinian paradigm in a connected fashion; that there were continuities and that it wasn't a question of different, incommensurable paradigms—in other words, that Kuhn was simply wrong (cf. Miller 1991, concern-

10. E.g., Brown (1984); Margolis, Krausz, and Burian (1986); Nola (1988); Laudan (1990a).
11. Cf. Scheffler (1967); for a vindication of subjectivity, contrast with Mitroff (1974).

ing the history of physics). The way I felt I could get at Kuhn was by showing that he was *historically* wrong. That, I felt, was the soft underbelly of Kuhn. I don't mean that in a pejorative sense; but that was the way I wanted to do it.

Wimsatt: Kuhn liberated us to do not only history of science, but also science. History of science is after all just science looked at after it has happened. There are some things in science that need to be looked at while, and even *before,* it happens, in ways continuous with the interests of historians, psychologists, and sociologists, and most of all of the scientists! This is particularly true if you're interested—as the positivists weren't—in *scientific discovery* or *problem solving.* You have to be there because people don't keep records of most of their mistakes—at best, they will keep records of their bigger and more long-standing mistakes. I agree with Michael that incommensurability was grossly exaggerated (it still is by modern eliminativists) and that when it occurs, it usually does for different reasons than Kuhn talks about. But Kuhn freed us generally to look at other things than the logical structure of science, including its functional and sociological dimensions. I think his account of "normal science" (Kuhn 1970, 1977) is perhaps even more revolutionary than his account of "revolutionary science" (Kuhn 1987)— and probably closer to the truth . . .

Callebaut: . . . *pace* Popper (1970). I guess that you are suggesting, Bill, that Kuhn not only caused us to give up one sort of rationality—the a priori, *categorical* sort—but that he also opened our eyes to the more elusive but much more important thing that has sometimes been called *instrumental* rationality (Simon 1982; Giere 1988), which is embodied in the scientific enterprise as a collective undertaking. Even if Kuhn himself didn't put it this way—and it is clear enough why he wouldn't at the time—we can see and say now that the *Structure* was really about the instrumental rationality of scientific communities. Another thing, of course, is that in philosophy, "rationality" traditionally goes hand in hand with "progress." Karin.

Knorr Cetina: Some philosophers of science have always felt obliged to defend the idea of the "progress" of science. That's where they come up with these "justificationary" type of philosophies, which *presuppose* that progress exists—that things work in a certain way—and then try to explain how the rationality which they have already presupposed looks in practice (cf. McMullin 1988).

Callebaut: Harry Collins (1981b:217) has referred to this as "TRASP": "Knowledge cannot be explained by reference to what is true, rational, successful or progressive."[12]

Knorr Cetina: The situation has changed to some degree because some of the newer

---

12. And Paul Tibbets (1986:40) delights in emphasizing that sociologists who focus on the construction, deconstruction, and reconstruction of scientific knowledge claims (*SKCs*) "simply refuse to play the game of justification, evidence, and facticity regarding *SKCs* within the rules and parameters posited by contemporary philosophers of science."

philosophers of science use a lot of history of science, and some even go in the laboratory, as you know.

Callebaut: Alex, you are even more adamant than Michael that the relativists' present triumph is unwarranted. You continue to be a "demarcationist."

Rosenberg: The positivist claim that science is objective knowledge—*the only kind of objective knowledge*—and that it is objective knowledge because of its predictive content, its contact with reality, is right. Any philosophy of science must come to grips with that fundamental fact about science by contrast with other intellectual activities.[13] To the extent that I believe that, and repudiate conventional interpretations of Kuhn, and practically everything that Feyerabend has had to say (or all the morals that Feyerabend draws from his studies in the history of science), to that extent I am as close to a positivist as you are likely to find today.[14]

Callebaut: Ryan, contrary to the overwhelming majority of "rationalist" philosophers, you have investigated human rationality *empirically,* as a psychologist. How do *you* assess the issue of the rationality of science?

Tweney: I take issue both with excessively rational approaches (Carnap, Popper . . .) *and* with excessively irrational approaches—for example Latour, or the kind of thing that people make out of the work on heuristics by Tversky and Kahneman,[15] which reduces it all to biases. I am very critical of naturalized attempts to understand science which fail to acknowledge that there is a rationality in science. Science is not entirely irrational. Many irrational factors enter, to be sure. But ultimately there is a cultural mechanism that produces good models of the real world. And *we somehow must understand how science reaches such models, using partially irrational means.* In that sense, I'm sympathetic to Rosenberg's views.

2.2.3. Objectivity Relocated

Callebaut: Helen, like Alex, you don't want to give up objectivity, but you define it in social terms rather than in terms of reference. In the preface of your book, *Science as Social Knowledge* (Longino 1990:ix), you write that you "abandoned a negative goal—rejecting the idea of a value-free science—for a positive one—developing an analysis of scientific knowledge that reconciles the objectivity of science with its social and cultural construction."

Helen E. Longino (Rice University; see biographical note at 10.5.3): We need to be careful about what objectivity means. The idea that has been rejected is that it is a kind of *exact representation of natural processes.* But there is another kind of

---

13. Clendinnen (1989) uses the view that scientific theorizing serves empirical prediction to counter the antirealist arguments of van Fraassen and others (cf. 4.3).

14. Opposing, on the one hand, *rationalism,* and, on the other, *relativism* (Shapere, Ruse) or *realism* (cf. Rosenberg's "contact with reality") can be quite misleading in a thoroughly naturalized perspective (Giere 1988).

15. Tversky and Kahneman (1974); Kahneman, Slovic and Tversky (1982).

objectivity, a kind of opposition to some understanding of subjectivity, that is important to retain for science. We try to develop some *nonarbitrary* account of natural processes that doesn't simply impose our own wishes for how the world is on our descriptions of the world. That's what the social account of objectivity is intended to do. If we accept the underdetermination thesis and accept the role of background assumptions in the structure of inference (cf. 2.6), there is a problem. Why don't we have tons and tons of background assumptions and why don't we have a complete, messy diversity in the sciences? We don't have that diversity . . .

Callebaut: . . . Complete diversity would also go counter to the very point of doing science . . .

Longino: . . . So the problem is: how is it possible to *minimize the role of individual subjective preferences* in the sciences, to produce objectivity in that sense? That problem can only be solved by understanding science as a set of processes involving *social interactions*. The particular interaction that I point to is *critical* interaction: the activity of criticizing some research program (both the description and analysis of data and the assumptions that give the data evidential relevance), which can be best done from outside the perspective of the research program itself, whether it involves a participant in the research program trying to step outside or someone from a different research program analyzing the program.

Callebaut: Isn't this part of your view very much like Popper's highly idealized picture of science as an open, critical community of scholars (e.g., 1970),[16] threatened by . . .

Longino: . . . totalitarianism? Sure, in some ways Popper held up the same kind of ideal for communities of inquiry. I did acknowledge Popper at one point.

Callebaut: Of course, he is also a realist . . .

Longino: . . . and that's why I would distinguish my view from that of Popper, who seems to think there are real, autonomous, conceptual relations which we can know, even if we can't know that a theory is true;[17] whereas when you take my account, you see how in fact scientific inquiry can't be held to result in true theo-

16. Thus Popper (1957, vol. 2:216–17) wrote that "no amount of political partiality can influence political theories more strongly than the partiality shown by some natural scientists in favour of their intellectual offspring. If scientific objectivity were founded . . . upon the individual scientist's impartiality or objectivity, then we should have to say good-bye to it. . . . There is no doubt that we are all suffering from our own system of prejudices. . . . Ironically enough, objectivity is closely bound up with the social aspect of scientific method, with the fact that science and scientific objectivity do not (and cannot) result from the attempts of an individual scientist to be objective, but from the co-operation of many scientists. *Scientific objectivity can be described as the intersubjectivity of scientific method*" (italics mine). Sir Karl went on to insist that "this social aspect of science is almost entirely neglected by those who call themselves sociologists of knowledge."

17. Popperians do not deny that human thought is a social product—they even concede that this is "trivially true" (Grove 1980:174). However, in order to secure the objectivity of the *contents* of human thought, they posit an autonomous "World Three" of objective knowledge, in addition to the material world ("World One") and the subjective world of mental experience ("World Two"). We will see at several occasions that many people for various reasons refuse to pay the price (to use a favorite expression of Bruno Latour's) for this rather Hegelian move.

ries or to be about some domain of reality totally independent of human, and mutable, conceptual schemes. How we analyze truth in this view becomes highly problematic, and I don't want to get into it. But the sense of objectivity as somehow adequate to or representative of the object—I think we attributed that characteristic to science to the extent that it was objective in the second sense: that somehow the methods of science involved minimizing of individual subjective preferences. So these two senses, I think, have been quite linked up.

Callebaut: Your view requires that there be variation, in relevant ways, among people's ideas about nature. If everybody thought the same way, they could easily become victims of a collective illusion. More specifically, aren't you actually required to presuppose a competitive relation between people (cf. Hull in 7.2.1) here? It takes at least two parties to disagree!

Longino: I'm not presupposing competitiveness, but rather a *commitment to certain kinds of goals.* In order for the community to make any progress in reaching those goals, diversity is required because of the problems underdetermination poses. On my view, *subjective preferences that characterize the collectivity don't get eliminated.* But that seems to me perfectly descriptive of what happened in the history of science! I'm not at all perturbed by that. I do think one of the worries that we can pose for this view comes from thinking of some future utopia, when we might not have interests in opposition or be antagonistic to one another and hence might not have the diversity resulting in criticism.

Callebaut: Your view actually takes more than diversity; alternatives must get a chance to be *voiced;* you refer to Habermas's idea of "dominance-free communication" in this context.

Longino: Yes, it takes much more than diversity.

Callebaut: We have had centuries of deviating views . . .

Longino: . . . which never made contact. That's why there has to be commitment to goals that criticism or alternatives are made relevant to, in order that there be critical *discourse* and not just criticism; that there be *interchange* and not just the expression of divergent points of view. Some of those goals have to be the product of values that are shared across communities. There is a problem that I haven't quite worked through here: in doing science, we are driven by a desire to describe the world, however it exists—whether it is characterized by laws or order or disorder. I have a belief about our own cognitive capacities that has the consequence that if we do succeed in describing the world, we can't know that we've succeeded. All we know is that there is a kind of local match between our descriptions and some natural processes.

Callebaut: How do we know that?

Longino: Because it works. We make predictions and they turn out to be the case. And then we think we know something about the underlying processes, but that, as I am saying, we don't really know. We know just the immediate match. And that may be local and temporary—we don't really know that either.

Callebaut: This sounds perfectly Achinsteinian to me, very much like Rosenberg's earlier point about the demarcation of science from other intellectual activities.

Longino: But I'm also saying that in order for these social interactions to give scientific inquiry its characteristic objectivity, one of the values that has to characterize these communities and that forces a kind of interaction has to be a commitment to some concept of truth. We have to be committed to the *hope* that the descriptions are in some way really descriptive of the world.[18]

2.2.4. *The Price of Extending Science*

Callebaut: People involved in empirical science studies often use the label "positivists" in the sense of "the bad guys." You are a spokesperson for an important approach to science studies, Bruno, and yet you object to this. You feel the positivists had a point.

Bruno Latour (École Supérieure des Mines de Paris and University of California, San Diego; see biographical note at 3.6): I do not agree with this, because there is a deep positivism—in Comte's sense—in science studies: *science can't be extended further than its own means of proof and instruments.* So in a sense, the old definition, even of logical positivism, is not very different from what we in science studies do. If by "positivism" we mean that it is always a practical or a logical limit that is imposed on the extension of science, then the people who study the networks of science are positivists in Comte's sense. After all, Comte was a good sociologist of science, in the sense that he made the link between states of science and states of society.[19] So I don't think positivism should be made a bad name. What is absurd in the positivism of the period between the two world wars is this *logical* aspect. But you can very easily sociologize the notion of positivism, so that you get a nice network definition of the idea that *a fact doesn't get out of its network,* and that *you have to pay the price of the extension of science* (cf. Latour 1987, Callon 1989). What we don't like are the guys who extend the science beyond, without paying the price. In that sense, we are good disciples of Comte.[20]

---

18. That's why Longino admits that in her conception of science "there still is a problem about how to characterize truth in a way that enables us to understand it as an ideal that shapes certain features of scientific practice, even though it's an ideal that for all we know can't be realized." She sees that as still a philosophical problem she hasn't successfully worked out (yet). "It may be just a problem of trying to talk about the relation between ideals and practices; I'm not sure."

19. Auguste Comte (1798–1857) described the intellectual history of mankind in terms of three stages: the theological, the metaphysical, and the (our) positive stage. His theological stage corresponds with the rule of priests and warriors, his transitory metaphysical stage with the preponderance of lawyers and philosophers with little authority. In the positive stage, scientists and industrialists were to rule society (cf. Saint-Simon, whose secretary Comte was). Contemporary views on the link between science and society are expressed in Cozzens and Gieryn (1990).

20. Latour's compatriot, the biologist and philosopher Henri Atlan, adduces a number of reasons for keeping the realms of science and other "rationalities" separate in *À tort et à raison* (1986) and *Tout, non, peut-être* (1991).

2.3 ALIENATION OF PHILOSOPHY FROM SCIENCE: A 1990s RECAP

Callebaut: The positivists considered science in a purely formal way. They tried as much as possible to assimilate extensional logic; that is, to restrict the nonlogical language of reconstructed science to referential notions.[21]

Nickles: They tended to view methodology as a priori in the sense that philosophical statements were about logic and language and had no empirical content. Methodological claims were certainly not to be tested empirically, against scientific practice.

Callebaut: As a result, the substantive content of science was neglected by them . . .

Nickles: . . . and so was the empirical study of scientists in action, whether historical or contemporary. Accordingly, the audience for methodological studies became still less clear. If the intended audience was ultimately scientists themselves, then why did methodologists rarely take into account the working scientist's point of view when facing genuine problems?

Hull: The received view was very inference-oriented, logic-oriented, formalization-of-scientific-theory-oriented (cf. Suppes 1968), and really, *the medium became the message.* You introduced formalizations to help you understand a scientific theory, but you spent 90 percent of your time solving a problem that arose from an axiomatic system and only 10 percent dealing with the theory. What is called for in this case is a greater emphasis, not on science as we think it *should* be practiced, but on science as it *is* practiced.

Callebaut: As a psychologist of science, Ryan, you find it difficult to make sense of a position to the effect that it cannot be the aim of philosophy of science to make testable statements about science or human rational behavior (Cohen 1981; cf. Stich 1985, Smokler 1990). Yet this is an implication of the positivists' understanding of the role of philosophy as a *foundational* enterprise devoid of empirical content—a "first philosophy," as Quine likes to call it.

Tweney: Right. It has never seemed strange to me that there should be a naturalized epistemology. In fact, the opposite position has always seemed strange to me. I continually had difficulty understanding philosophical approaches to science when they become excessively formalistic and move from what is really going on in science. That work has rarely been very meaningful to me.

Callebaut: Would you agree with the following: First, to be scientific, you *must* invoke some kind of idealization. Galileo's law of inertia is a classic example; it assumes, counterfactually, the vacuum and the absence of friction or any other forces acting on a body. I emphasize this because philosophers of the nineteenth

21. For a thoughtful discussion of what refraining from the use of intensional (i.e., meaning) notions implies for the theory of science, see Hooker (1987:63f.). Note that approaches that want to limit the language of science or philosophy to extensional language are sometimes called "naturalistic" (e.g., R. J. Nelson 1984, 1987).

"Foundationalism fails; all justification is at bottom social."

Thomas Jacob Nickles
(b. Charleston, Illinois,
1943) is Foundation Pro-
fessor (?!) at the University of Nevada, Reno,
where he began to teach
in 1976. He was a philos-
ophy and mathematics
undergraduate at the Uni-
versity of Illinois at
Champaign-Urbana. He
continued his philosophi-
cal education at Prince-
ton, where his disserta-
tion on The Structure and
Interrelationships of Phys-

ical Theories was supervised by Carl G. Hempel. He
taught philosophy of physics at the University of Chicago
in 1975 and has been a Senior Visiting Fellow at the University of Pittsburgh Center for Philosophy of Science,
1982–83.

When Nickles set out to investigate the "logic(s)" of
scientific discovery, his avowed hope was to salvage at
least some of the "rationality of science" which the waning
logical-empiricist enterprise no longer had a grip on (1977,
1980a, b). In this sense he was siding more with Simon
than with Hanson at the time. But his long-term project led
him to widen his philosophical horizon to include even
Rorty today (although he doesn't like to hear that emphasized) and to explore ever deeper layers of "generative
methodology" in the history of science. Unlike many
fortune-seekers of previous generations in Nevada, he
seems to be finding real gold, and lots of it.

Nickles's thing about justification being local and social is something he shares with several other participants
in this book. Somewhat self-defensive, he explains: "What
the (relevant local, critical) community lets you get away
with is not unreasonable, as far as that community is concerned. But I don't want to leave you with a highly relativistic picture of isolated enclaves of people doing anything
they damn well please. The local groups operate in specific
historical contexts interlaced by rich layers of tradition,
and these groups are imbedded in larger endeavors that retain some leverage on them." There are, for instance, the
craft traditions that Polanyi and Kuhn spoke about. "Communities maintain themselves by special kinds of training
and licensing, among other things." Even at the most abstract level, Nickles insists, "once certain logical rules of
reasoning are discovered and become socially licensed as
reliable, obligatory, or 'real' to the communities—be they
Aristotelian rules, Frege-Russellian, or something else—
these rules take on a certain life of their own, not in Plato's
or Popper's heaven, but in the sense that allegiances to
them are transmitted from past to future communities. The
justificatory constraints can become quite general, although specific applications may still vary among local
communities." (Cf. Richards at 10.1.1.) More generally,
Nickles thinks that philosophers have been "too cosmopol-

itan, too universal" in trying to find arguments that will
work equally well at any time, place, and situation. He
feels sorry for them to some extent: "It is partly a question
of audience. In trying to address all people at all times,
they have ended up addressing no one in particular." One
has to keep "the causal teeth" in talk about rationality. "A
possible objection to our claim does not become a real objection, to be taken seriously, unless we fear that our critics
in the community have the resources to make the objection
stick. Otherwise, the objection is merely 'philosophical'."
Here Nickles finds some of Latour's work relevant, "although he might not like the way in which I use him!"

Nickles is aware that one may see a tension between
his views about social justification and his claims about
"discoverability" ("Discoverability looks like a throwback
to an old-fashioned, foundational view of justification,
whereas social justification appears to be a soft-headed,
mushy-mushy view of justification"). But he waves this
line of objection away. "Discoverability is not a linear, Euclidean justification scheme based on a priori truths." In
one sense, "it is not based on anything"; he endorses Wim-
sattian talk about robustness, multiple connections, and
mutual support. In another sense it is based on social permission. "The background knowledge on which discoverability arguments depend is not established by proofs from
foundations but is just that body of results the community
thinks reliable enough to use—or not seriously objection-
able." A discoverability argument can depend on premises
that are not really "established" at all but which no one in
the community wants to contest. Nickles's favorite example is someone who certainly still *thought* he was using
a Euclidean model of justification, Isaac Newton. "In the
Opticks (1952:20), Newton lays out several discoverability
arguments from 'axioms' he has stated plus particular empirical phenomena he has discovered. But what does Sir
Isaac mean by an axiom? Does he mean an a priori, self-
justifying truth? No. He explicitly says that he is taking as
an axiom what the community of people working in optics
agree are the correct principles of the subject." Nickles
concludes that "everything requires a community stamp of
approval or at least a tacit *nihil obstat*, even the laws of
logic. They don't just drop out of the sky anymore than the
word of god does."

Nickles is the author of fifty or so articles in major
philosophical journals and collections and has edited
Scientific Discovery, Logic and Rationality and a twin volume of *Case Studies* (1980a, b). He is preparing books on
generative and bootstrap methodology.

The original interview was recorded on September 1,
1985 (to Tom's amazement, on an old but robust Nagra,
made in the USSR and owned by Belgian Public radio) at
Harry Collins's conference on Experimentation in the Sciences in Bath, where we met for the first time. We talked
more extensively while sighting Lake Tahoe, the Bonanza
farm, and gold digger ghost towns when I visited Tom and
his family in Nevada in the spring of 1986. The original
materials were substantially revised and amplified.

and twentieth centuries tended to neglect the crucial role of idealization in science. Yet the falsity of a model or theory needn't be a handicap—it can in fact be a powerful tool in the construction of better ones (Wimsatt 1987; Wimsatt and Schank n.d.).[22]

Tweney: I agree.

Callebaut: But second, the "rational" reconstructions of positivist philosophers, especially the later ones, are going far beyond what one finds in science in terms of distance from reality.

Tweney: Yes, they frequently go beyond that to the point where they've produced something that doesn't look like the real world of science at all. I have the most respect for Lakatos, because in reconstructing the history of mathematics he made very clear that it was a *reconstruction* as opposed to an account of what actually happened. That's rare among philosophers; they frequently will take a very idealized account of what happened and pretend that it's real, and that somehow they can account for science on the basis of such historical fictions. I dislike that. The point of it escapes me; I don't understand it. Much of philosophical inquiry into the nature of science seems to be instead philosophical inquiry into the nature of *imagined* science, which has no contact with the real world of science and so strikes me as irrelevant to what I want to understand.

2.3.1. *"We Live Forward"* (John Dewey)

Callebaut: The positivists and Popperians were only interested in the *final results* of scientific research—theories—and the logical relations among those results, to the expense of the ongoing *process* of research . . .

Nickles: . . . the tinkering,[23] the activity of the laboratory and of the theoreticians' desk or coach, and how all this ultimately gets turned into scientific proposals and claims and still beyond that into finished publications. There, I think, they missed a great deal about what is most interesting about cognition. Epistemology should apply at least as much to the process of inquiry as to the final products and in fact can inform our view of those products in that once one studies the process in certain ways, one comes to a quite different view of the products.

Callebaut: "Research is work, not sterile formulae or disembodied ideas," as Star and Gerson (1986:163; see also Gerson 1983) put it. Faults in, say, a theory, are "diagnosed" and then—one hopes—"fixed" (Darden 1990). As a corollary to the

---

22. Erathostenes (± 273–± 192 B.C.) is often regarded as the pioneer of the conscious use of idealization. When measuring the diameter of the earth, he assumed that the sun's rays in two distant places are parallel (i.e., that the distance of the sun is infinite), knowing this to be false. Or, to take another example, in Archimedes' law ("a lever persists in an equilibrium when the moments of forces acting on it are equal") it is assumed that the nonexisting lever is weightless. See Krajewski (1977:17f.), Cartwright and Jones (1992), and, on the limits of useful idealization, Shrader-Frechette (1989).

23. Campbell (1974a) and Knorr (1979), among others, have stressed that like the "tinkering" of biological evolution (Jacob 1977), research is opportunistic.

previous, we could say that by its very nature, the reconstructionist tradition in analytic philosophy of science is a philosophy after the feast, a reflection that comes after things have been settled.

Nickles: The positivists found their problems and solutions "technically sweet." To be fair, nearly all philosophy of *any* kind, and not just logical positivism, has been retrospective, backward-looking. If methodology of science is to be of much interest to working scientists, it must be more forward-looking and more discipline- and problem-specific. So far nearly all philosophy of science has been looking at a sort of *track record* of a theory, at what is already largely in the published literature or at least in the circulated papers.

Wimsatt: I quite agree. When I was an undergraduate and started reading about evolutionary theory, there was almost no philosophical stuff around, save for two very good books, Morton Beckner's *The Biological Way of Thought* (1959), and Tom Goudge's *The Ascent of Life* (1961). Beckner had a lot of original stuff in there, much of it verging on theoretical biology as it was done in that period; but it was thoroughly atemporal. Goudge's book was hardly a positivist tract, but he announced somewhere near the beginning that while there was a lot going on in evolutionary theory currently, philosophers should wait until the cards were down, and so he was going to talk only about past theory. I thought, "This is bizarre; can't you study it while it's going on?"

Nickles: Right. Heuristic appraisal or fertility assessment—the thing that matters most to scientists—has scarcely been noticed by philosophers. The main concern of most working scientists, the question they want to ask of a theoretical viewpoint or a claim or a proposal, is not "What have you done for me?" or even "What have you done for me *lately?*" but "What can you do for me tomorrow?" or "What resources are available to me that I can muster in order to write a new research grant, in order to gain new findings?" or "What future problems am I going to be able to attack fruitfully, and make progress in my field, and hence advance my own reputation, career?" and so on. The past *is* important to scientists, but mainly as a basis for future action. To the extent, then, that they are concerned with tidying up past theories—proving rigorous theorems and that sort of thing—it is not that they have the tremendous interest in the "foundations" of their discipline that philosophers think they *should* have and the philosophers themselves traditionally have had. It is more a matter of *reculer pour mieux sauter:* you regroup and pull back in order to better spring forward.

2.3.2. Unity of Science, An Idol of the Philosophical Tribe

Callebaut: Yet another feature of positivism that was vigorously attacked was its insistence on the unity of science: that there should, ideally, be only one science about the one real world; that psychology should be reduced to biology, biology to chemistry, chemistry to physics, and physics probably to mathematics (see 4.2).

How does a third-generation philosopher assess the traditional reductionist agenda?

Nickles: I think the positivists carried this program too far in two respects. First, they tried to assimilate all the branches of science to one—physics. In fact it was not even real physics but an idealized logical model of what they thought physics *should be like*. But second, they assumed that a single field, say a branch of physics, would have the same logical character and the same methodological procedures during its entire history, regardless of the stage of maturation and the particular problems being addressed.

Callebaut: Kuhn's *Structure* was originally published in the *International Encyclopedia of Unified Science,* of all places. One of the ironies in this is that the very spirit of historicism, which pictures the scientific enterprise in terms of paradigms or other "closed circles" (Munz 1985), runs counter to the positivist idea of the unity of science.

Nickles: Most philosophers of science have fallen into this generality trap . . .

Callebaut: . . . and so has Kuhn himself—as Munz correctly observes—when he introduced his sweeping stage theory.[24]

Nickles: They think philosophy should issue in general models, general methodological claims, general methodological rules, general logical procedures, which hold for all sciences, for all times.

Callebaut: If there *were* any such, it would be terribly interesting, of course!

Nickles: Perhaps, if you go to a high level of abstraction, you can find a few broadly true statements about science. But there is only so much you can say at that high level of abstraction, and after a while it gets rather boring to keep repeating the same things over and over. It seems to me that much of the essential activity of scientific inquiry is lost if one seeks a sort of least common denominator of that sort, something that will fit all sciences at all times.

Callebaut: The quest for the absolute, for something that never changes, seems typical of Western philosophy, which has long insisted on being a *philosophia perennis.*

Nickles: If I might use Francis Bacon's image here—it is certainly an idol of the tribe of philosophers to stress unity and generality. Philosophy has traditionally been a subject concerned with high generality and optimality. Kant not only engaged in the exercise but pointed out its dangers: that it was a tendency that philosophers had to curb lest they get into very severe trouble. Nearly every philosopher since Plato has sought to optimize something, to search through all of "possibility space" for the greatest good, the best of all possible worlds, the One True Theory, or what have you (cf. Slote 1989). As if historical context imposed no constraints on one's

24. Cf. Miller (1991:105): "Historical studies have shown that Kuhn's structure of scientific revolutions is relevant at best *sub speciae aeternitatis* as a convenient schematic of gross historical features or in discussing matters sociological. Incommensurability describes science as a tower of Babel, which is far from the case because scientists work simultaneously with many different theories."

access to the space of possibilities, and as if being satisfied by nothing but the best was an economical way to solve problems and get on with your work.

Callebaut: In *The Manufacture of Knowledge* (Knorr-Cetina 1981a; cf. 1981b), you talked about a new unity of the sciences, Karin.

Knorr Cetina: My point was that since we found all these interpretations, negotiations, etc., to go on in the "hard" sciences, it became obvious that the "hard" sciences are more like the social sciences than we believed. At least since the sixties, there has been a movement which claimed that the social sciences are very different from the natural sciences: hermeneutic, interpretative, dealing with actors rather than objects, etc.—you know the argument. But when you became aware of the importance of interpretativity (and even hermeneutics!) and of underdetermination in the natural sciences, the natural sciences become more like the social sciences.

Callebaut: But that's only part of the story?

Knorr Cetina: What I also want to make you aware of now is that there is a new sort of disunity. I've just completed a comparative study of particle physics and molecular genetics, which also includes medical research to some degree. There you find that the technologies of experimentation, the procedures of problem solving, the inference strategies—to use a traditional vocabulary—are very different in these two areas. There is a disunity of the sciences in terms of their *epistemic cultures;* and that has to be worked out (Knorr Cetina 1993a). There is not this great divide between the social sciences on the one hand and the natural sciences on the other. You have to have a much more differentiated picture of the sciences. There is not "Science" as such; there are many different sciences in terms of their epistemic procedures and also of their social organization (but that we always knew to some degree). The epistemic procedures are very different in different areas. That I find very interesting.

2.3.3. *Doing It Professionally*

Callebaut: Tom, you also relate the positivists' preoccupation with unity and generality to the *professionalization* of philosophy of science. Could you spell out what you mean by that?

Nickles: Historians have analyzed the professionalization of many fields over the last thirty years. Some of them now regard professionalization models as rather *passé,* but I think they can still be instructively applied here.

Callebaut: To many people, professionalization sounds like a good thing . . .

Nickles: . . . and in many ways it is: it raises standards, it enables the community to agree on a certain, very large background of assumptions and techniques, in order to focus more sharply on particular problems of interest.

Callebaut: But a field undergoing professionalization also faces typical problems.

Nickles: Right. If you are going to be a legitimate professional discipline, you have to carve out your niche, your own space in the intellectual universe, you have to be to a certain extent *autonomous* from, irreducible to other disciplines. Here the philosopher-logicians faced a tricky problem. How can philosophy of science be parasitic on science as a subject matter, as it obviously is, yet at the same time be autonomous in the sense that working scientists as such do not do philosophy of science and philosophers do not do empirical science? If you're going to have a discipline with professional integrity, you also want a discipline to be *integrated,* to have a unitary subject matter, not to be just a catch-all grab bag of rags and tags from here and there; otherwise there will seem to be no particular reason at all for grouping these things together and calling them a professional discipline.

Callebaut: How do these general ideas apply in the case of the development of philosophy of science?

Nickles: By making philosophy of science or "methodology" into an a priori, normative discipline for the logical analysis of scientific language, the positivists were able to solve the autonomy and integrity problems very neatly, in one swoop. In being logical-analytical, the field declared its autonomy from the empirical sciences themselves and from history and other empirical studies of science.[25] And this orientation simultaneously solved the integrity problems by offering a general, unified, formal treatment of all the sciences, one that seemed important or necessary, even *foundational.* Philosophy of science was demarcated as a distinct field with a coherent, unitary domain of problems calling for a general "theory"; only here the general laws were laws of logic instead of empirical laws of nature.

Callebaut: Symbolic logic was indeed of paramount importance.

Nickles: Intellectual integrity and autonomy were enhanced by the facts that symbolic logic was a very fertile, new discipline, that logicians were obviously the people best equipped to rigorously analyze the language of science and that, as they defined their discipline, their professional competence did not depend on mastering vast amounts of empirical information either from the sciences or about scientific activity. In fact, the intrusion of empirical information, say from the psychology of perception, was deemed worse than irrelevant—it was circular, since it was the "foundational" task of the logic of science to appraise and validate scientific activity, not vice versa. Of course I don't mean to suggest that the positivist founders themselves saw their situation in terms of this sociological, professionalization model.

Callebaut: What about the relation between professionalization and unity of science?

Nickles: Well, this professionalization orientation presupposed the unity of science and placed a premium on establishing it.[26]

25. Thus Carnap (1949:408) took care to demarcate the empirical study of the "body of actions carried out by certain persons under certain conditions" from what he thought of as the proper concern of the philosopher, viz. the "logical analysis of the body of accepted scientific theories"—what Lakatos would later call the "body scientific" (Worrall 1976:1).

26. Aside from the "perhaps considerable" intellectual merits in favor of unity, Nickles thinks, there was this professional pressure, because unity was a key to the self-definition of the field: "If you cannot

2.3.4. *Whig Philosophy*

Callebaut: Historians often talk about, and usually condemn, "Whig history"—basically, the reading of the past in terms of contemporary categories.[27] You call positivism, Popperian philosophy, and much philosophy in general "Whig philosophy" or "Whig methodology." How, do your think, can the notion of Whig history be carried over to methodology?

Nickles: In two ways. First, professionalization also creates a pressure to be general, especially in philosophy, which is the field of general claims par excellence. But given the diversity of the sciences, just about the only way you can say things that apply to all sciences at all times is to leave out all the interesting stuff. General methodology becomes a "least common denominator." It reduces very rich and diverse researches to a few simple patterns—inductivism, hypothetico-deductivism, and the like—just as Whig history reduces historical development to a few simple patterns or causal influences (Nickles 1986). By the way, historicists as well as positivists have been guilty of this.

Callebaut: What is the second way?

Nickles: Methodologists can be very whiggish in their treatment of the history of science. This may sound trivial, but it goes deeper. It applies even to study of contemporary science. You don't have to jump from century to century to be whiggish. Most people are whiggish about the course of a single scientific investiga-

---

show or presume that a logical treatment of all these sciences can proceed fruitfully, in more or less the same way, then you are in trouble, since the coherence or domain integrity of your own discipline is compromised. So we should not be surprised that 'unity of science' became a key issue at philosophy of science symposia." The situation is different now: "By now the field is big and mature enough to tolerate a good deal of diversity. In fact, there is now a premium on subfields like philosophy of biology to be different from philosophy of physics. But in the early days the positivists and Popper had to give a pretty straight and narrow demarcation of science. Any field that looked weird or did not fit was attacked and read out of science by philosophers of this persuasion." Hence the strong debates about "idiographic" (descriptive) historical-type sciences—the *Geisteswissenschaften*—versus the "nomothetic" (explanatory) *Naturwissenschaften* from the turn of the century on.

27. In *The Whig Interpretation of History* (1931)—"a delightful little book," to Nickles' mind—the British historian Herbert Butterfield was originally talking about political history, beating on other historians for being insufficiently historically sensitive. In Nickles's committed summary, "The Whigs were supposedly on the good side, on the side of progress, fighting for the future, trying to bring about the world we have today, thank God; whereas the Tories were deliberately putting up obstacles to progress." Basically, Whig history is reading the past in terms of the present instead of seeing the past in its own terms; the words *Whiggism* and *presentism* are used as synonyms in science studies. "It is treating historical actors as if they faced our problems and had access to our categories. It is praising them as brilliant and morally good insofar as they anticipate our ways of thinking and acting, and ridiculing them as stupid or bad insofar as they differ from us. Whiggism is antihistoricism, and historicism is antiwhiggism. Whiggism takes as just *there* things that actually have historical origins and thus leads people to give circular explanations, to explain something in terms of that thing itself. Whigs often suppose that we have some sort of absolute historical privilege, that we stand outside of history and can judge it, unlike our predecessors, who were locked into the limited vision of their historical contexts." Nickles, who had to leave out the subtleties here, thinks that Butterfield (1931) and Kuhn (1977) are still the best places to begin.

tion, say from a scientist's hitting upon a new problem to her finished publication or the eventual mention of the finding in a textbook. The positivists and nearly everyone else studied the logic of the final papers and then drew naive conclusions about how the work must have been done—how history *must* have proceeded to reach this point. This is reading the product back into the process and has everything to do with placing proper emphasis on context of discovery and adopting a forward-looking viewpoint. I should add that scientific work itself can and even should be very whiggish (Nickles 1990b). Many historicists don't realize that Butterfield (1940) himself later rejected extreme anti-whiggism and recognized that whiggism is invaluable for adapting historical precedents to contemporary problems and programs.

Callebaut: I found neat Elliott Sober's (1985a:867) distinction between the *explanatory* interest of the historian—since the present didn't cause the past, interpreting the past in terms of the present is dangerous—and the different issue of what *meaning* we assign to the past—"seeing the significance of the past may well essentially involve seeing it in terms of the present."

Nickles: Also, whether whiggism is good or bad depends on whether you are studying history or making it (Nickles 1990b; cf. Hull, "In defense of presentism" [1979], in Hull 1989:205–20)!

Callebaut: Bill, you have written something similar in the context of an appeal to draw more heavily on psychological accounts of learning, pattern detection, and matching in accounts of theory change: You criticize Feyerabend and Kuhn for treating scientists' tendency to overestimate continuity of theories through time (e.g., by sticking to words for concepts whose meanings have changed) largely as "a misleading and deleterious occurrence" (Wimsatt 1979:365).

Wimsatt: Kuhn clearly saw a role for Whig history, and he was right (cf. Brush 1974). Why would scientists always write such bad history with a presentist bias, unless it served a function? There would be no reason in principle why one couldn't write a "futurist" history—how Democritus might see modern quantum theory, or Empedocles or Aristotle might see evolution. It would all be speculative fiction, of course, and would seem pretty strange. The fact that Whig history doesn't seem strange to us indicates that it is serving a function—in fact at least three functions: (1) it gives us heroes or role models, and at the same time confers credit (cf. 7.2.1); (2) presentist versions of older theories, e.g. Galileo's or Newton's, are used to solve simpler problems first;[28] (3) which gives encouragement to proceed and motivate—in the sense of giving a direction to or

---

28. Mayr (1990:302) invokes the progressiveness of science ("in a succession of theories dealing with the same scientific problem each step benefits from the new insights acquired by the preceding step and builds on it") as contradistinguished from political change (which may be haphazard) to conclude that "Butterfield was ill advised in his literal transfer of the whig label from political history to history of science." Needless to say, relativists who deny or prefer to remain agnostic about scientific progress will find Mayr's argument question-begging.

qualitative "ballpark" constraints on—the limiting forms of the answer in more complex theory.

Callebaut: All of this nicely illustrates the various ways in which the scientist can use history (cf. Mayr 1990). But why, precisely, should the history be *whiggish?*

Wimsatt: Primarily because the function is to teach results and methods economically while pointing the way towards current theory. There's no point in using older methods or results if they have been replaced or refined—and this actually explains why successional reductions are intransitive (1976a). But the older problems and (presentist) successes can provide easy routes into the subject matter of the modern theory. This also explains the remarkable fact that modern science instruction (scientific "ontogeny") often roughly follows a very whiggish temporally ordered reconstruction of the history of the discipline (scientific "phylogeny"). There are literally thousands of detailed exceptions, of course, but there is definitely a correlation, and that has to do with (and indeed helps to constitute) the cumulative character of most scientific disciplines.

2.4 WHO KILLED LOGICAL POSITIVISM?

Callebaut: In his intellectual autobiography, Sir Karl boasts that he "must admit responsibility" for the dissolution of logical positivism (Popper 1976:88). He adds that he did not do it on purpose, and that his "sole intention was to point out what seemed to me a number of fundamental mistakes." Not everybody agrees that Sir Karl should be convicted of murder here.

Richard M. Burian (Virginia Polytechnic Institute and State University, Blacksburg; see biographical note at 10.6.1): In no way was it Popper.

Brandon: It was at least as much suicide as murder. Almost all the serious problems of positivism were raised from within. I think it was suicide to the extent that the patient was so weak; then the murder was easy!

Nickles: To their credit, the positivists themselves have been their own best (or worst!) critics, as I said before. They themselves developed their views considerably over time, and it was a highly critical community.[29]

Giere: My view about Popper saying he killed logical empiricism is: that's just crazy! *Popper is a minor variant on logical empiricism.*[30] You can maybe promote that to a middle-level variant . . .

---

29. Popper is actually aware of this line of argument. Referring to an article by John Passmore which Popper (1976:88) agrees "correctly ascribes the dissolution of logical positivism to insuperable internal difficulties," he persists, in his usual way, that most of these difficulties had been pointed out *by him,* especially in the *Logik der Forschung:* "Some members of the Circle were impressed by the need to make changes. Thus the seeds were sown. They led, in the course of many years, to the disintegration of the Circle's tenets."

30. Contrast this with the more-or-less-ex-Popperian Feyerabend (1981:330) about Popperians: "Vor noch nicht zu langer Zeit schien es (in Deutschland scheint es immer noch), als gebe es in der neueren Philosophie nur drei (oder vier oder fünf) Philosophen, nämlich: Popper$_1$, Popper$_2$, Popper$_3$, (Popper$_4$ und Popper$_5$)."

Callebaut: . . . A Kantian variant maybe . . .

Giere: . . . Well, I think Popper is very positivist. So I don't see him as that great critic of logical empiricism anyway. There is a famous (probably unpublished) quote of Carnap in which he said that he had discovered that distance is not a symmetrical relation, because the distance between him and Popper was very small, but apparently the distance between Popper and him was very big. I think Carnap was right: the difference is very small, but Popper exaggerated it.

Callebaut: What, then, *did* kill positivism? (but see Sarkar 1992b).

Giere: Both internal and external things. Even look at Hempel. Logical empiricism was going into a rococo phase. If you look at Hempel's own writings on explanation, they get more and more elaborate. There is his lovely paper on "Empiricist criteria of cognitive significance" (in Hempel 1965), where in the end he embraces a Quinean holism. So, internally, logical empiricism was getting squishier and squishier.

Nickles: By the 1950s, too many of the problems philosophers were discussing were problems that arose mostly out of positivistic logical and linguistic techniques—or alternatively, ordinary language techniques.

Hull: The main objection to the received view is that the amount of complexity you had to introduce to make little improvements in it was getting too extreme. It was epicycle upon epicycle. It had become so sophisticated and so complicated that the explanation was getting more complicated than what was being explained.

Burian: It is a case rather more like the death of the Ptolemaic theory: it killed itself by its own epicycles. That, I think, is a better answer than ascribing it to some dragon slayer.

Callebaut: Was the received view perceived at the time as moving even further away from real science, as David is implying?

Giere: My own experience at the time was colored by my work on probability and induction. I very early concluded that philosophers were wasting their time working on inductive logics based on the first-order predicate calculus. I found the mathematically grounded theories of statisticians much more relevant to questions in the sciences themselves. But mainly I just felt that logical empiricism was spinning its wheels, and I think a lot of people were coming to the same conclusion.

Callebaut: The English translation of Sir Karl's *The Logic of Scientific Discovery* (German original published, as Sir Karl never tires to repeat, in October 1934, dated 1935) was published as late as 1959, that is only three years before Kuhn's *Structure*. Were people in the U.S. aware of Sir Karl's existence before that?

Giere: Hardly. I'd say hardly.

Callebaut: Did they rather regard him as a political philosopher? His *The Open Society and its Enemies* (1945) and *The Poverty of Historicism* (1957) had been around for some time.

Giere: That's right. And almost nobody read the *Logik der Forschung* in German. I, in fact, had a xerox copy of the German edition, which I got from Max Black. I read it in German, more or less. It was not widely read. I don't remember Popper being on one's reading list. Let us put it this way: After the *Logic* came out in 1959, Black, who had his antennae out for the field, thought that Popper was a good dissertation topic because there were going to be people interested in it. He wouldn't have thought that earlier.

Donald T. Campbell (Lehigh University; see biographical note at 7.1.1): In my first two papers on evolutionary epistemology (1959, 1960) I had not been aware of Popper's existence. After I had written them, I read a review of *The Logic of Scientific Discovery* (Popper 1959) by Stephen Toulmin in *Scientific American*—a very biased review in that he spent the whole of it tooting Norwood Hanson's horn. Nonetheless it got me to be aware of Popper's volume; and when I read it, I was very enthusiastic about it. I was sitting in on philosophy colloquia and the like, so I knew that Popper was in the States, at the Minnesota Center and at Indiana. Arnold Levinson (a young philosopher who had co-taught my Knowledge Processes course at Northwestern) and I were more nearly his local hosts when Popper came to Northwestern than was the philosophy department as a whole. He was a difficult man about tobacco and everything, so we did most of our talking with him in his hotel room. But there was a banquet at which it was agreed there was no smoking, and at that banquet Paul Schilpp, who was a member of the Philosophy Department, announced that his Board had authorized him to make his next volume in the Library of Living Philosophers on Karl Popper. That croneyism with Schilpp was the only way I got into the book.[31]

Callebaut: I suppose it would be exaggerated to say that as far as philosophy of science is concerned, it was Kuhn who really made Sir Karl important.

Giere: I see them as somewhat independent—maybe reinforcing each other in the sense of making each other important.

Callebaut: Kuhn mentions Popper in the *Structure*.

Giere: It's true that Kuhn mentions him at that time, but I don't think Kuhn was terribly influenced by Popper especially. I actually started out writing a dissertation on Popper, in 1964. I read everything that Popper had published up till then— which was a lot easier then than now!—and thought about it. But then I just gave it up. I'd say I liked his *conclusions* in many ways better than I liked the other empiricists'—but I couldn't stand the arguments! And the idea of the dissertation was to analyze the arguments. I just thought, "I don't want to do this." So I just chucked it, and I decided to write on a more general topic, on the relation between prediction and confirmation. So it was more a standard type of topic, which wasn't hooked to any particular persons.

Callebaut: It's such a tremendous contrast: Listening to all you guys, it is as if Sir Karl has hardly existed. But mention his name in countries like Germany or the

---

31. Campbell is referring here to what was to become his classic paper, "Evolutionary epistemology" (1974a).

United Kingdom, and many people, most notably the scientists, will tell you that
they consider him to be the most important philosopher of science in our century.[32]

Burian: But Popper was taken as *the* authority on methodology by a large number of
philosophically naive biologists until quite recently; his work affected the actual
practice of many biologists. Mention Popper among philosophically naive biologists up until five years ago . . . Equally, Popper was very important in the actual
practice of many biologists. Look at the debates in ecology over strong hypotheses
and things of that sort. There is no question that he was taken very seriously and
played an important role. But again, he is not the cause of the death of positivism.

Ruse: I've always been repelled by Popperianism, because you've got to be a disciple
to be a Popperian. That's not my cup of tea. It's a pity really, because Popper
himself has important things to say.

2.5 A DECISIVE TRANSFORMATION OF OUR IMAGE OF SCIENCE

Callebaut: The new approach to the philosophy of science that emerged in the late
1950s and early 1960s was not just *historical;* it was strongly *historicist.*

Shapere: It was very strongly historicist. For example, people held that the true
understanding of science is to show how the methods—plural—have evolved over
the history of science; or that the rules by which science operates develop historically. The history of science was at that time a relatively new discipline in the
sense that it had just been professionalized, mostly since the end of the Second
World War.

Callebaut: The *Structure* begins with the words: "History, if viewed as a repository
for more than anecdote or chronology, could produce a decisive transformation in
the image of science by which we are now possessed" (Kuhn 1970:1). The "image
of science" Kuhn and other historically informed authors opposed was not so much
the "received view" of the positivists as the familiar, biased "textbook" image of
science. (The positivist and the textbook image of science are, of course, not
unrelated.)

---

32. Thus for Sir Peter Medawar, "Popper is incomparably the greatest philosopher of science that has
ever been" (Magee 1973:9). And according to Sir Hermann Bondi, "there is no more to science than its
method, and there is no more to its method than Popper has said" (ibid.). Michael Mulkay and Nigel
Gilbert, who have investigated the influence of Popper's methodology on scientific practice empirically,
conclude that "one reason why these and similar public testimonials to Popper are frequently produced by
eminent men and why such expressions of approval give a misleading interpretation of this philosopher's
actual influence on scientific practice" is the "generality" of his methodology, its "lack of interpretative
particularization" and "independence of institutionalized social relationships." These factors "allow individual scientists considerable freedom to conceive of their own actions as Popperian in character and to
attribute their intellectual success to the effectiveness of the Popperian approach." (Mulkay and Gilbert
1981:407). The mundane view that scientists believe whatever philosophers tell them to believe about
methodology is contested, as far as Popper's influence is concerned, by Berkson and Wettersten (1984),
who insist on reading Popper-the-methodologist as a psychologist (Popper was a student of Karl Bühler)
rather than as a logician, despite his antipsychologism (cf. Wilkins 1989).

Nickles: The historicists' own approach consisted in taking some empirical information about scientific activity seriously and getting off the a priori, analytic style of philosophy of science that one associates with the positivists and Popperians.

Callebaut: Ron, you stress (1985a) that Kuhn, in arguing for a role for history, may be seen in retrospect as one of the first to have held a naturalized view of philosophy of science. Naturalism means, in Quine's (1986:430) words, "banishing the dream of a first philosophy and pursuing philosophy rather as part of one's system of the world, continuous with the rest of science." You more or less identify a naturalized philosophical theory with a *testable* theory.

Giere: I'd say that naturalism in the philosophy of science is a research program, an approach—which is why I subtitled my (1988) book *A Cognitive Approach*. But it is *scientific* in the way that other scientific research programs are. Hopefully the approach can generate more specific hypotheses that are genuinely testable, but they probably will not be easily testable. I don't think philosophical theories should be scientific in the sense that philosophers provide theoretical "foundations" for specific theories. *I don't think scientific theories have foundations.*

Callebaut: Let's do some Whig history while Tom is not watching us. How do you assess the Kuhnian "revolution" from your *present* perspective, Ron and Bill?

Giere: For me now the most important thing is Kuhn's naturalism, the idea that the central task of the philosophy of science is to develop a theoretical understanding of how science actually works—not to show that science is "justified" or "rational" or "progressive." Second, the more specific idea of focusing on the cognitive practices of individual scientists, which Kuhn did with his emphasis on "exemplars" and individual judgment . . .

Wimsatt: . . . and in my case, I add: socialization, and learning how to solve puzzles, and (though I don't agree with his use of Gestalt switches), looking at nonverbal dimensions of scientific activities.

Callebaut: Same question to you, Don.

Campbell: We all recognize in Kuhn, Hanson, Feyerabend, Toulmin, and—I add—Michael Polanyi (e.g. 1962; cf. Kane 1984), the effective critiques (concentrated in the years 1958–62) of positivism that toppled the establishment. Yet were we to ask for the *first* item of post-positivist epistemology, we might find general agreement that it was Quine's "Two dogmas of empiricism" (1951). Kuhn cites it in his preface (1970:viii) and mentions his own status as a Junior Fellow at Harvard. He fails to mention that Quine was a Senior Fellow. Both types of Fellows enjoyed a private club together, and no doubt frequently talked. Regarding Quine's paper, Kuhn mentions only the section "The puzzles of the analytic-synthetic tradition," not the final fifteen paragraphs I find so dramatically modern: epistemologically relativist, and historicist.[33] Kuhn very likely was more influenced by both

---

33. At this point Campbell insisted that we "reprint these paragraphs here to remind us of that great emancipation proclamation." I refer the reader to chapter 1, where I have briefly summarized their content.

Popper and Quine than he acknowledges. These last fifteen paragraphs of the "Two dogmas" are tremendously emancipating; they are relativistic in the extreme. I can't help but think that even though Kuhn doesn't refer to the second dogma or the final paragraphs, they opened his mind to the possibility of a foundationless theory of science (cf. Gochet 1986, Gibson 1988).

Callebaut: Karin, with the benefit of hindsight, how important would *you* call Kuhn?

Knorr Cetina: The intellectual exchanges on positivism we had in Vienna included a critique of Parsonianism and functionalism in general. Because of the critique of functionalism—which was very widespread at the time—I had turned away from traditional sociology of science even before I went to Berkeley. It didn't seem to have the kind of thing to offer which I was looking for—not an alternative quantitative study of scientific institutions, but something really new. And there, Kuhn has been an encouragement. I didn't go to the laboratory to test some hypothesis by Kuhn or to add something to his account. But he provided this basic incentive to look at the content of science and take it seriously. That had not been part of the traditional sociology of science. It didn't offer the kind of alternative one was looking for. Kuhn has encouraged us, he has somehow *empowered* us, made it possible for us to go into the laboratory and look at the content of science. I don't believe one would even have considered that as reasonable before Kuhn. It took Kuhn to make that obvious to us. It was always obvious to historians, but it hadn't been obvious to sociologists. On the other hand, in terms of taking Kuhn's *substantive* results and extending them or working with them, the English tradition—Harry Collins, for example—is much more indebted to Kuhn than the Europeans are or were (cf. Barnes 1982). I had my own agenda when I went to the lab which had nothing to do with testing Kuhn's hypotheses or extending his findings.

Callebaut: I find it curious, to say the least, that many of the people I've talked with for this book maintain that Kuhn didn't have much of an influence on them.

Burian: My own personal history is quite different. I did get the *Structure* the year it was published as a gift from my father. At that time I was switching out of physics and mathematics into either history of science or philosophy of science. While in one sense the book did not influence my positive views strongly, in another it was a source of very strong reaction, doubly: a reaction against some of the logical-positivist views in their abstraction from actual practice, but also a reaction to the *physics-centered* character of the way in which Kuhn thinks.

Callebaut: What do you mean?

Burian: Well, for one thing, as Kuhn and the logical positivists parsed the issues about the reference of theoretical terms, the logical structure of a theory supplied necessary conditions governing reference: anything that failed to meet the constraints of an axiomatic system could not be referred to by means of that system. It struck me that there are highly theoretical disciplines—genetics and evolutionary theory were candidates—in which this did not seem to be true, partly because there were not axiom systems of this character and partly because there was successful reference to theoretical entities. In that case, given the historicism which Kuhn was pushing—which I was sympathetic to—and given the failure of his metascientific views to be adequate to those kinds of disciplines, it seemed to me necessary to do comparative work. I had no biology in college, but my father was a research ophthalmologist, and by osmosis I knew something really different was there. And so, although it took me about eight years to start it seriously, I had, by the time I was finishing graduate school, the notion that I would certainly spend some serious time looking at one or two biological disciplines as a way of setting a comparative background to physics for conceptual change questions. In this sense, despite the fact that I did not accept many of his specific views, Kuhn was a great stimulus to my work.

Wimsatt: Not only that, Dick; you could read it for enjoyment. After Carnap, or Hempel, or even Nagel (who had a few more details), it was such a breath of fresh air—*the scientists were doing things.* It wasn't quite biographical, but there were *people* there, not explanation schemata, and not systems S, in conditions C, going into a G-state. (I say that having committed many of the latter abominations in my early career!) I think it's no accident that first-person locutions began to re-enter philosophy shortly thereafter.

Churchland: Somehow Kuhn is not accorded quite the cachet of certain other philosophers or historians of science, and Feyerabend should be much more highly regarded for his early work. The early work is especially insightful, and he was way ahead of so many people. The same is true of Kuhn. But you're not supposed to say either was a great figure. I think Kuhn in particular *was* a great figure. He was tremendously important in dislodging the myths of logical empiricism and showing us that there was so much to be learned from the history of science. Some criticism of Kuhn is useful, but a lot is nit-picking; it's just not very interesting to me.

Callebaut: Can't we say, with hindsight, that when people like Scriven (1959) were criticizing logical empiricism through a defense of evolutionary biology, they were unwittingly making a kind of *naturalistic move,* as we would now call it? I mean, in the spirit of the time it is not obvious, as far as I can see, that such a move would have impressed any hard-headed positivist. For the positivist could always say: "On the one hand, we have our logical reconstructions, on the other, there are scientific theories. You can't bridge the gap between those two so easily as you, our critics, are suggesting."

Ruse: I think that's a good way of putting it, although I've never really thought of it that way. They were, I suppose you might say, taking the naturalistic approach against a sort of rather theoretical philosophy. Although I've always thought of it as not exclusively a theoretical philosophy, in that it was itself naturalistic—grounded in and growing out of physics. Nagel, to take one example, knew a lot of physics. He'd done some biology but was not that interested in it. But, overall, I think your point is well taken. My own inclinations were to be naturalistic in the sense that I wanted to work with the science itself. This turned out to be my Ph.D.

thesis, which was the basis of the first book I published, *The Philosophy of Biology* (1973).

Callebaut: What made Kuhn, a physicist turned historian of physics, embark on his project—I guess we should call it a *philosophical* project—of fleshing out a new image of science?

Kitcher: The thing that got Kuhn started—and he said this on a number of occasions; I think the more he says it, the more he sees the significance of it—was his experience of being unable to capture the past. He started from the idea that justification is always justification in a particular historical context. This is something that he was one of the first people to introduce to the philosophy of science. It's a profound insight; it is now one of the received insights of philosophy of science. He has this idea of ongoing bodies of belief and justification being relative to this body of beliefs that occur at the immediately preceding stage. (The topic of justification will be discussed in some detail in 5.3.)

2.6 UNDERDETERMINED INCOMMENSURABILITIES

Callebaut: The feature of Kuhn's theory that the philosophical profession always objected to most vigorously was the "incommensurability" he ascribed to different viewpoints in science (Kuhn 1970, 1983; cf. Feyerabend 1962, 1975).

Sober: Kuhn took the view that succeeding scientific theories are incommensurable with each other, that there is no possibility of rationally comparing them. Put in a very extreme way, the idea is that each scientific theory lives within its own world and really has no contact with others. I think of this extreme view—which I think Kuhn has since repudiated but nevertheless is thought of as characteristically Kuhnian even now—really as a form of *idealism* or tending toward a *coherence theory of truth;* and I do not think there is a philosophical defense for that.

Bechtel: I did not have high respect for Kuhn at the time I was studying philosophy. When I first read Kuhn at Chicago, people were far more interested in the responses to Kuhn: Shapere (1969), Scheffler (1967), and so on. The focus was on the incommensurability problem and how to surmount or to avoid it as a serious kind of problem. In fact there was one paper that I wrote as a seminar paper for Manley Thompson. It was relating Kuhn and Quine and using Quine's notion of observation sentences in an attempt to overcome some of the incommensurability problems in Kuhn. Manley despised it so much he refused to give it a grade. I wasn't going to do anything further with it. But after leaving Chicago I talked to Eric Stiffler, who was proposing to write very much the same kind of paper. We collaborated and wrote a paper that subsequently was given at a PSA meeting (Bechtel and Stiffler 1978). But even there the view was, "Kuhn presented a problem; we've got to answer it," as opposed to taking Kuhn's own contribution to understanding science seriously. And I certainly did not appreciate Kuhn's treatment of normal science or the things I have found valuable in Kuhn in the last five years. I had the same philosopher's skepticisms of Kuhn (although I was probably less informed than other philosophers at the time) but did not see Kuhn as a major contributor to a progressive philosophy of science.

Callebaut: Incommensurability seems to lead straight to epistemological relativism. Dudley, you were very much impressed by the problems posed by historicism at the beginning of your own career in philosophy.

Shapere: It was said by some that each period or even each area of science had its own paradigms, its guiding theories in terms of which evidence, method, explanation, etc. were all formulated. Those things were shaped and even determined by *background presuppositions*. At various stages there would be revolutions which overthrew those presuppositions and replaced them (cf. Hesse 1980). But there seemed to be no way to make any judgment as to the relative merits of the competing presuppositions. If you say that the methods of reasoning in science evolve over the development of science, aren't the methods that we use at one stage subject to being overthrown, and therefore untrustworthy at *any* given stage in the development of science?

Callebaut: Richard, you did a dissertation on Feyerabend under Sellars.

Burian: To this extent the influence of Kuhn had to work very deeply in what I was doing . . .

Callebaut: . . . so I take it that incommensurability must have been unbearable to you as well.

Burian: I am some form of scientific realist, I hope not too weak a realist (1984, 1987b, n.d.). I do not think that the most recent, most "advanced" study contains the truth and nothing but the truth. But somewhere behind the theoretical forefront, we consolidate our knowledge of theoretical entities and processes—and learn to work with them—with genuine (but not infallible) epistemic security. One of Kuhn's ideas that I think must be corrected in the long run is that if you look at the logical structure of a theory, you have something like the "main source" of the paradigm. Thus, the Newtonian paradigm is going to involve $F = ma$ in a very strong sense and so on; the relativistic paradigm has corresponding equations. I think there is a sense in which that is right and a sense in which it is wrong. The sense in which it is wrong involves the supposition that *that* particular logical structure becomes definitional for the kinds of entities that science claims there is. *There are many avenues toward reference.* The sense in which it is right has to do with the deepening and intensifying of an experimental practice that lets us have access to causally relevant features of the world, features that behave in ways that match such equations to a good (or excellent) degree of approximation.

Callebaut: In your paper, "Ontological progress in science" (Burian n.d.), you express the view that consolidated knowledge, some distance behind the theoretical forefront, provides a kind of normative criteria.

Burian: Some of these may be fairly routine in character, others quite esoteric. Thus, there have been something close to a hundred reasonably distinct "theories of the gene." These theories attempt to characterize entities the reference to which is secured by practices in the laboratory and the field, practices developed in interaction with theory but not irrevocably wedded to any particular theory of the electron or the gene. The criteria that ensure consistency of reference can break down (and they often do so at key turning points in theory construction), but by keeping them some distance back from the theoretical forefront, such breakdown can be made less likely. The relevant criteria are not going to be the same for quantum entities as they are for macromolecules or full scale metazoan organisms or solid state bodies of the sort that we use in computers. Two or three levels back, we can see progress most in the *scope* of the entities we can view—from galactic to microscopic—and in the kind of *precision* that we have about their behaviors. That endures through the kinds of conceptual change to which Kuhn helped draw our attention. The frequent preservation of reference through conceptual change shows, I think, that the notorious "incommensurability" that arises in theoretical change does not cut as deeply as Kuhn thinks it does.

Callebaut: Kuhn's own model of scientific revolution*s* will be evaluated later (2.6.4 and 2.6.5). Traditionally, *the* Scientific Revolution of the seventeenth century or thereabout was taken to have met certain sufficient or even necessary conditions—the "mechanization of the world picture" (Dijksterhuis 1961), the transition from a "closed world" to an "infinite universe" (Koyré 1958).[34] More generally, scientific disciplines were traditionally considered as reaching a certain moment in their history when a shift in "philosophic background" (Koyré 1954) put them on the road to modern science—Gillispie's (1959) "edge of objectivity." Thus it was said, for instance, that "Lavoisier is the father of modern chemistry." (Contrast Daston [1992] on the history of the concept of objectivity.) In this respect, Dudley and Don, your respective views depart from both the traditional and Kuhn's revolutionary view.

Shapere: There is much more continuity than a lot of people have thought (Shapere 1989b, 1991c, n.d.b.).

Callebaut: Kuhn (e.g., 1977), following (the flipside of) Koyré (1958, 1968, 1973) and others here, would accept that much as far as the continuity of pre- and post–Scientific Revolution science is concerned. But that is looking at the very long run, whereas you are now talking about developments on a shorter time span.

Shapere: My view is that Kuhn's idea of incommensurability is incoherent: it would simply make no sense to argue that theory *A* (or its conception of mass, or its explanation of heredity) is incommensurable with theory *B* (or its conception of mass or explanation of heredity), unless *A* and *B* were comparable as both being theories (conceptions of mass, explanations of heredity). That does not mean that

---

34. On the traditional view, (modern) science emerged with the Scientific Revolution only. One could wonder, in that frame of thought, if science as it came into existence at that time does still exist today. Ziman (1985) argues it doesn't; cf. also Funtowicz and Ravetz (1990).

everything is built on what went before in a rational way. One has to understand the irrational processes in the development of science too. But it does mean that there have been changes and innovations that have led to the establishment of a body of beliefs—although not by a linear progression, at least overall a body of beliefs on which increasingly we have been able to build. And I would even say that there were certain approaches, methods, and standards, in addition to substantive beliefs about nature, that developed over the sixteenth to the eighteenth century that were importantly different from what had been done generally before; for instance the introduction of a *piecemeal approach to inquiry* as something done by a continuous tradition of investigation.

Campbell: The "doubt/trust ratio," as I call it, is normally in the order of 1 to 99 percent.[35] This is in line with the Quinean view that we disturb the totality of our beliefs as little as possible.[36] In a Kuhnian revolution, the doubt/trust ratio may drop to 10 to 90 percent. But in the history of physics it never gets below that. After the 1919 eclipse observations, British astronomers and physicists moved from 5 percent adoption of Einsteinian general relativity to 99 percent adoption. That *palace revolution* was made by depending on Newtonian physics (Moyer 1979)! So I also do reject the incommensurability and the notion that progress only exists in the periods of complacency.

Wimsatt: Yup—and this is why I said earlier that Quine didn't believe that all changes were equally likely, or even (forgive the thought) equally possible. Some things are just hooked up to too many other things (*deeply generatively entrenched,* in my lingo) to expect them to change. Of course we may be wrong about this. Sometimes there are ways of making changes in deeply entrenched things which don't make waves in a "bad" way in the other things that depend upon them.

2.6.1. *Meaning and Reference*

Callebaut: The issue of meaning and reference is closely related to incommensurability. Feyerabend and Kuhn have been criticized for relying on quite naive assumptions about the meaning of terms and of the propositions in which they occur.

Sober: Part of the heritage we have from the historicists' reaction against positivism is in fact to try to go back and do some of the work in the theory of *meaning,*

---

35. See in particular "Descriptive epistemology: Psychological, sociological, and evolutionary" (from the William James Lectures of 1977 at Harvard University), in Campbell (1988a:477f.).

36. "The totality of our so-called knowledge or beliefs, from the most casual matters of geography and history to the profoundest law of atomic physics or even of pure mathematics and logic, is a man-made fabric which impinges on experience only along the edges. . . . A conflict with experience at the periphery occasions readjustments in the interior of the field. . . . But the total field is so underdetermined by its boundary conditions, experience, that there is much latitude of choice as to what statements to re-evaluate in the light of any single contrary experience. . . . A recalcitrant experience can . . . be accommodated by any of various alternative re-evaluations in various alternative quarters of the total system, . . . but . . . our natural tendency is to disturb the total system as little as possible" (Quine [1951] 1953:42–44).

in *scientific change* and *confirmation theory,* but without the philosophical commitments of positivism, to see what the upshot of the inadequacy of positivism really is.[37]

Callebaut: The idea of classical empiricism was that you must first be clear about your meanings and then go out into the world and test the propositions that you construct in using those meanings. You oppose this "hierarchical" approach to what is going on in science, Dudley.

Shapere: This is much more of a bootstrap process: we develop ideas that are open and vague and ambiguous, and through testing those ideas we are often forced to alter them, and so on.

Callebaut: You were one of the very first philosophers to strongly oppose the classical theory of meaning, which has come to be called the *description* theory of meaning because it suggests that the meaning of a term is given by some description of the term's referent (see, e.g., Hooker 1987:350).

Shapere: The idea of meaning as something to analyze independently of scientific investigation, and the distinction between meaning and reference as something fundamental—what I call the Frege-Russell tradition in philosophy—have had a long, and very disastrous, influence on the interpretation of science.

Callebaut: These are tricky issues. May I ask you to try to explain, as simply as possible, where you are coming from and how you differ from much philosophy on this matter?

Shapere: One of the major approaches in the 1950s became the idea of the dependence of meaning on the language or theory—Carnap's idea of *meaning-in-L* or even *meaning-in-T* (e.g., Carnap 1963). This was an idea that is ironically similar to the Kuhn-Feyerabend type idea of *meaning as being determined within the context of a certain world view.* Those views, it seemed to me, led to problems of incommensurability.

Callebaut: If meaning is always relative to a theoretical context, then in different theoretical contexts you have different meanings, which are difficult or even impossible to compare.

Shapere: Carnap's later views of meaning were very similar to those of Kuhn, and the problems were very similar. In the early 1960s, my intuition (and this idea has become incorporated into my larger view) was that a term like "electron" can be taken as a *transtheoretical* term, that it was *independent* of the particular theories of the electron.

---

37. For instance, the positivists had a clear view of what it means for two *theory presentations*—say, one formulated in terms of temporally extended physical objects, another in terms of time-slices—to be *equivalent* (i.e., to amount to the same theory). This is one aspect of the important issue of the *individuation* of theories, which will preoccupy us later (sections 7.2.3 and 7.3.5). Depending on whether a postpositivist—say, a realist—adopts the view that the meaning of theoretical terms is fixed solely by the role they play in the theory or by something over and above the theory (see below), he or she will have to adopt a simpler or more complex notion of theory equivalence (Sklar 1982).

Callebaut: There are two ways in which, according to you, philosophers have gone in relation to that idea.

Shapere: Let me read you a passage on this from Hilary Putnam's paper, "Explanation and reference," not because he refers to me but because I want to bring out the very different way in which Putnam went from the way that I have gone. Here is what he says: "with a few possible exceptions . . . realists have held that there are successive scientific theories about the *same* things: about heat, about electricity, about electrons, and so forth; and this involves treating such terms as 'electricity' as *transtheoretical* terms, as Dudley Shapere has called them (cf. Shapere 1969),"—I think it was earlier than that—"i.e. as terms that have the same reference in different theories" (Putnam 1973:197). He goes on to discuss the concept of particle: "The main technical contribution of this paper will be a sketch of a theory of meaning which supports Shapere's insights." What in that paper Putnam went on to do was to develop the idea that later became the *causal theory of reference;* that is, the idea that what uses of the term "electron" over a stretch of history have in common is not a common meaning (thus rejecting the Carnapian-Kuhnian approach) but rather a common reference (cf. Miller 1991).

Callebaut: Still you don't find Putnam's solution to the problem acceptable.

Shapere: No. Putnam's approach, which has become popular among philosophers, was simply another direction within the Frege-Russell tradition. In that tradition the important concepts were ones from the philosophy of language, but this time taking reference instead of meaning as crucial. One of the many things wrong with this "reference" approach is that with terms like "electron," determining that there is a common reference cannot be done by pointing, but rather depends on the *properties* that are ascribed to the electron. That seems to me to be a fatal objection to the causal theory of reference as a means to tackle problems of incommensurability.

Callebaut: What singles out your own view of a transtheoretical term like "electron"?

Shapere: In common with the causal theory of reference, my view holds that the properties that are ascribed to electrons are *nonessential:* any of them can be changed. But where I differ from the causal theory and from the Frege-Russell approach generally is in holding that the continuity of research in science is neither guaranteed by common meaning or common reference, but rather by *reasons.* What is common that enables us to say that the theories (or the "concept") are "about" the same thing—that they are *successor theories* of what the electron is— is that there is a chain-of-reasoning connection between the *particular* theories or concepts, in which the properties ascribed to electrons are changed for reasons. The "concept" of "electron" which spans the entire tradition is thus better seen as a *concept-schema,* in which the term "electron" refers to a series of reason-related theories (concepts), the detailed properties ascribed to the electron being filled in by the specific theories within that series. That is a concept of "a concept" that circumvents the problems to which the ideas of meaning and reference have given birth. It also brings to the fore the necessity of looking at real science and its history. Meaning and reference are to be understood in terms of the content of science and its rational changes, not the other way around.

Wimsatt: The "meaning change" route to incommensurability favored by Feyerabend and Kuhn caused a lot of trouble. Shapere did a marvelous demolition job on this in the late sixties, although when I reread it last year, I thought he was not interpreting his opponents in the most charitable light. Philosophers were still too close to the dominance of the linguistic tradition, and tended to regard a difference in meaning as blocking any possibility of communication. Empirical and meaning changes did not mix. That's nonsense. The real problem was the *holistic theory of meaning*—which covered everything which wasn't clearly empirical (if anything was!). A far-reaching difference in empirical results, experimental protocols, research designs, methodology, theory, or meaning *can* always cause trouble, but *usually* it doesn't, and even if it causes trouble initially, people *usually* find ways to get around it. One problem with a lot of philosophy of that period is that philosophers seemed to regard words like "generally," "often," and "usually" as unintelligible because there wasn't a quantifier for them, so "usually" usually got read in terms of the existential or universal quantifiers, which immediately trashed some otherwise sensible remarks.

Callebaut: Please give an example.

Wimsatt: "Most terms usually depend for their meaning and reference on a few other terms" is a sensible statement about meaning. It is sensibly in-between the correspondence and coherence theories, and people worrying about meaning holism— whether they are for or against—should think about it. Philosophers seem not to take this intermediate view seriously, choosing instead to substitute ($\forall$x) for the outermost "most" and to read "a few" systematically as either "($\exists$!y)," yielding the correspondence theory, or "($\exists$y)," yielding an apocalyptic coherence theory. No wonder we can't make any progress! The critical issue for any of these kinds of changes is, "Can the *source of a disagreement*—about meanings or about anything else—be localized?" If so, we can do piecemeal engineering to fix it and go on. If we can't localize the source we can't go on, and we must junk the whole thing. That's when you get what looks like radical incommensurability—which doesn't come automatically with meaning change, and is orders of magnitude less common than people used to think.

2.6.2. *Lexical Structures*

Callebaut: Philip, how does Kuhn himself come to grips with the phenomenon of being unable to formulate, in the contemporary idiom, the precise content of the past? I mean both the Kuhn of the *Structure* and the later Kuhn (e.g., 1977, 1983, 1987, 1991, and n.d.) who was more influenced by Quine's ideas on the difficulties involved in translation.

Kitcher: Kuhn proposes that there are these psychologically real elements—he now calls them *lexical structures*—which are present in the scientists' past as well as in the present, and which undergo shifts in the course of scientific development. And in response to the kinds of objections that Hartry Field (1973), Hilary Putnam (1975b), myself (1978) and others have raised against this notion of radical conceptual change, Kuhn is now prepared to say that in extensional terms—in terms of reference—one can specify the referents of the old terms in the new language. But what one cannot do is capture the lexical structure. And because one cannot capture the lexical structure, one cannot translate. Because one cannot translate, one cannot appreciate the way that the world is structured from the point of view of the old discipline. And because one can't appreciate the way the world is structured—because these things are so radically incommensurable—and because Kuhn does not believe one can give sense to the idea of a structure to the world apart from someone structuring it, he claims that the *world* changes. That, I think, was the argument of the *Structure* . . .

Callebaut: . . . reconstructed with the benefit of hindsight.

Kitcher: And in more recent work that he has done—in the lectures that he gave at Johns Hopkins University (Kuhn n.d.) and in the material he gave as part of our institute (see 3.1.1), I think he is really getting the argument down and is getting it straight. And one can see how much of that line of reasoning is common nowadays; common, I should say, because of his earlier efforts. There is a sense in which, if we take history seriously today, it is in large measure because of him.

2.6.3. *How Society Enters Theory*

Callebaut: Many philosophers of science who have been influenced by Kuhn or Quine see the underdetermination of theory by data as a central problem. What they mean by that is, roughly, that no matter how much data one has, in principle, any number of theories might be compatible with those data, so that one never seems to be able to have adequate justification for a theory.

Nickles: The way philosophers have presented this problem it almost seemed as if science should be constantly paralyzed. Philosophers tend to be highly skeptical that a scientist would ever have sufficient reason to get on with his work, because the grounding of any particular theory would always be grossly underdetermined by the facts and by any other constraints on theory.

Callebaut: Don't you think, then, Tom, that underdetermination is as serious a problem as is often claimed?

Nickles: Yes and no. On the affirmative side, it is even worse than usually claimed because underdetermination is a *double* problem. Most philosophers look only at the retrospective problem of justifying those claims already on the table, but the early pragmatists and a few contemporary sociologists such as Andy Pickering (1984a, b, 1990) see the openness or underdetermination of the future as equally important. The question is, how do scientists go about cutting down the wide space of possibilities to one or two specific projects? Ironically, Kuhn, who supposedly glories in the underdetermination problem, dropped some strong hints about how to handle both retrospective and prospective underdetermination—or underdetermination in context of justification and in context of discovery, if you will.

Callebaut: How so? Can you be more specific?

Nickles: Start with justification. Kuhn can be read as retreating from a realist, "Truth Now" account to a sort of pragmatism in which the *solved problem* rather than the true theory becomes the unit of achievement in science. Scientists are often more confident that they have an adequate problem solution than that they have an absolutely true theory. This line of thinking was further developed by Larry Laudan in *Progress and Its Problems* (1977). But second, Kuhn's whole account of normal science, and especially his discussion of exemplars—exemplary problem solutions as models for further research—suggests a promising approach to the underdetermination of the future, the "what do I do next?" question. Pickering and others are trying to work this out in more detail. Now I personally think Kuhn's normal science is too monolithic, too rigid and inflexible, and thereby makes both underdetermination problems a little too easy to handle; but his stress on the local contexts of research and the constraints they impose on thought and action are very important.

Callebaut: There is a difference, Helen, between the view of underdetermination that comes out of Kuhn and the kind of underdetermination that you argue for in your book (Longino 1990), which is much more like Duhem's thesis.

Longino: In Kuhn's work, we call a theory underdetermined by the data because in his view of paradigms and theory-ladenness there is no theory-independent way of describing the data. So theory choice can't be data-driven; it has to be driven by other factors. Now one could still tell, I think, an internalist story with that view, although you can see how room could be made for the beginnings of an externalist view once you abandon the idea that theory change or theory choice is exclusively data-driven.[38] But it's quite easy for me to see why Kuhn would still remain an internalist, if you're arguing that the ideas develop purely in response to and in the context of scientific practice, guided by the kinds of values Kuhn talks about—accuracy, breadth of scope, and so forth (cf. Latour 1983). It seems just a matter of deciding what kind of values are going to be guiding theory choice. The kind of underdetermination thesis that emerges from Duhem—or that I saw emerging from a consideration of the nature of inference and logical relations between theory and evidence—I think again can permit both an internalist and externalist approach, but I think an externalist approach becomes much more compelling in the version of underdetermination that claims that assumptions of some sort are

---

38. Cf. Slezak (1989a:587) who, on the other side of the politico-intellectual fence, points out that underdetermination does not establish the externalist thesis automatically: "The alleged indeterminacy of content and meaning appears to open the way to a social determination of the contents. However, even if this Rorschach view is conceded, social causation is still not thereby established. However they might be construed, the inkblots are not *caused* by the subjects' perceptions."

required to mediate the relation between data and hypothesis. Assumptions are required to give a set of data evidential relevance to some theory. Once we make that claim, it becomes clear that those background assumptions can be the vehicles on which cultural ideology or social values "ride into" the rest of science. At least that's my view.

Callebaut: It worries you that the now widespread awareness of the relations between science and society (Cozzens and Gieryn 1990) "has not yet had much impact in the philosophy of science." You also write that your aim is "to show how social values play a role in scientific research by analyzing aspects of scientific reasoning" (Longino 1990:3).

Longino: I should say it didn't occur to me for a while that that was actually the case, until I became concerned with these debates in the sciences that had clear connections to the cultural, social, and political context in which the debates were taking place, that the values of the context were expressing themselves in and shaping the inferences that were being made.

Callebaut: Could you say something more about the relation between background beliefs and underdetermination in your view?

Longino: The background assumptions can be understood in a completely internalist way. Think, for example, of Glymour's (1980) bootstrap theory: he thinks of what he calls the "auxiliary assumptions" as just parts of the theory that facilitate inference, that enable the theorist to assign evidential relevance to a set of numerical, quantitative data. But when one takes that philosophical view and applies it to the analysis of particular cases, and asks oneself, "OK, what are the evidential relations here? What is the data? What are the hypotheses? What is required in order to take the data as relevant to the hypothesis?" one comes to see that background assumptions that are in some way encoding cultural values are at work in a particular research program. I don't think it's a view that one would come to automatically, simply by considering the role of background assumptions. (That may be simply because of my educational history. I wasn't educated to think about the sciences as shaped by social context; I was raised as an internalist.) But once we admit that background assumptions are necessary, there is no way to exclude value-laden assumptions from playing a role.

Callebaut: Wouldn't it make sense to unpack various things that have been lumped together under the label "background assumptions?" For instance, it might well be that reductionistic thinking as we find it in the sciences ("It has to be point-at-able!"—cf. 4.1.1) has a very deep rooting in our culture; or think of Holton's (1973) recurrent "themata." At the other extreme, there may be background ideas that are specifically tied up with a particular piece of research. And there may be many things in between.

Longino: "Background assumptions" is a very general term for me. I haven't really made any particular discriminations about the level of embeddedness in the culture, although I have tried to distinguish the different roles value-laden assumptions can play in scientific research. You're right . . .

Callebaut: There is a political side to this as well: the chance to uncover assumptions that are deeply rooted in the culture and shared by all the members of that culture are likely to be minimal . . . You would have to extend your plea for "alternative science" to nonwestern cultures.

Longino: Well, I'm perfectly happy to do that! In fact cultural values can be more or less local, more or less deeply embedded. Western cultures consist of a variety of interacting and intersecting subcultures characterized by particular values, and we can more easily see the differences between these *sub*cultures and articulate the values that characterize a particular subculture. As you say, it is much harder to recognize the values that we might all hold in common. It's as important to be able to recognize and understand the ways in which they may be shaping our understanding of the natural world as it is to understand the local ones. It's going to take rather more extreme affirmative action, I think, to make those visible to us.

Callebaut: The current educational practice in which the best universities in the U.S. attract an increasing number of first-rate students from Asian countries (and elsewhere) would rather seem to be going in the other way.

Longino: That's something people in communication talk about: that the so-called "free flow of information" really is a flow of information from the center out to the periphery, which changes the values of the periphery to produce a kind of global uniformity. Speaking as a political person—an oppositional political person—I find that very dangerous. As you say, the same thing happens in the universities and science education: people from third world countries are coming to European, American, Japanese universities and acquiring a certain scientific culture in which their own values are subordinated, erased, made invisible. So those cultures are being disabled as a source of criticism for our world view.

2.6.4. *The Essential Tension between Tradition and Innovation*

Callebaut: Fifteen years after the *Structure* appeared, Kuhn published *The Essential Tension* (1977), by which he referred to the tension between tradition—dogmatism, if you like—and innovation in science.

Nickles: Kuhn's early idea about the "essential tension" was very fruitful.[39] After all, problems themselves are virtually defined by a complex structure of constraints on what counts as an adequate solution; so that it is largely the context, the background, the tradition, which makes problem definition possible in the first place. By the way, this gives us a sense in which we can perfectly well say that not just *solving* problems, but *finding* good problems in itself is an achievement in science.

---

39. Nickles is referring here in particular to Kuhn's paper, "The essential tension: Tradition and innovation in scientific research" (reprinted in Kuhn 1977:225–39), which was originally published in 1959.

Because to find a problem and to have the community recognize it as a good problem implies that there is a body of background claims, practices, and constraints that are considered settled by the community. Problems are *achievements* in their own right, something we have learned.

Callebaut: In the *Structure,* the distinction between innovation and tradition is to a large extent equated with that between "revolutionary" and "normal" science, respectively (cf. 7.1.4).

Nickles: Yes, by the time he wrote the *Structure,* Kuhn had somewhat spoiled the point. Instead of having a continual tension between tradition and innovation, a simultaneous pull in different directions, Kuhn now divided the process into two successive temporal stages, so that one has "normal science," in which tradition carries all the weight, and then one has "revolutionary" or "extraordinary" science, viz. scientific periods in which there is a dramatic break.

2.6.5. *Stage Theories*

Callebaut: Philosophers of science weren't much aware of Popper's *The Poverty of Historicism* (1957). "Historicism," in Popper's sense of an all-embracing ideological doctrine rooted in so-called "historical laws," purported universal statements that are actually nothing more than existential and hence "unfalsifiable" statements in disguise, has been barely linked up with Kuhnian historicism as an account of the scientific enterprise in terms of "closed circles" (Munz 1985), that is, "self-perpetuating mutual admiration societies whose social systems prevent reality testing, stifle innovation as heresy, and suppress disconfirming evidence" (Campbell 1979b:182).

Giere: That's right. Also, when I became more and more aware of the thinking of Kuhn as a generic type, of his stage theory, I often wondered why this was not more criticized when it first came out, especially in the American context. I mean, Marx had a stage theory, and in the American context surely nobody thought that was any good. There was Walt Rostow's *The Stages of Economic Growth* (1960) and nobody thought that was any good either. Nobody thought that Piaget (1977) was any good; and there were Kohlberg's "stages of moral development" (1973) which nobody thought was any good. Yet Kuhn comes by with a stage theory of science and nobody jumps on it just for being a stage theory. I find amazing that it was not criticized for this reason.

Hull: In the early chapters of the *Structure,* Kuhn sets out a stage theory of scientific change and then at the end compares it to biological evolution. I never took his stage theory seriously. The history of human culture is littered with stage theories. Marion Blute (1979) lists a couple dozen. I was puzzled that Kuhn would add another to this long list, but I never complained in print because at the time I was not working on the more global issues in philosophy of science.[40]

---

40. Sociologist Döbert's (1981:75f.) purported "good reasons for social scientists to proceed in a different manner than biologists and to be preoccupied with the construction of stage models of social evolu-

Callebaut: Did nobody criticize it?

Giere: There were people who made minor criticisms, like I. B. Cohen, who had no impact on philosophers. He did not like philosophers and they did not like him. He said, "Look at history, there are cases of revolution with no crisis. You clearly got a revolution, but it is missing one stage." You could find those kinds of criticisms in the literature, but they were not wholesale. That is not to say that this is the *kind* of theory that is just no good.

Hull: Since then, of course, numerous authors have shown that episode after episode in the history of science does not fit Kuhn's scenario. More importantly, from the perspective of biological evolution, one should not expect scientific change to go through any set sequence of events. Scientific theories no more undergo cyclical development than do species. Nor is biological evolution progressive (1988c). Granted, at first glance, one would think that biological evolution *must* have a direction, towards increased articulation, increased specialization, increased *something*, but biologists have not been very successful at discerning this direction. As far as I can see, one of the major differences between biological and scientific change is that science does clearly progress, while any progress in biological evolution is anything but clear.

Callebaut: Don, you are more sympathetic to Piaget's stage theory.

Campbell: I am in favor of an ordinal or stage Piagetianism that does not claim that a child has a fixed level for all possible tasks.

Callebaut: In Campbell (1987a), you rather strongly associate his dialectic of "assimilation" and "accommodation" with the Kuhnian succession of "normal" and "revolutionary" science (cf. 7.1.4).

Campbell: I am very sympathetic to a Kuhnian interpretation of the history of science in which there are periods of complacency with basic theory and periods in which the theories' inadequacies are being worried over (cf. Piaget and Garcia 1983). Given our predicament, we have to pick up the strong, main effects first. If we can't do that, then nobody is going to get off the ground at all. If it's all higher-order interactions, we are never going to know anything. If all of the stars were planets, we would never have solved the problem of planetary motion. It's because most of the stars were not planets that we eventually learnt to distinguish between

---

tion before beginning to specify evolutionary mechanisms in detail" do not withstand scrutiny. His functional argument for the persistence ("a very longstanding tradition") of stage models—that they allow "people at least to think that certain events have come to an end, that certain events will definitively take place and perhaps that their own suffering is a necessary precondition for the better time to come" (76)—confuses object level (a feature of society, i.e., the object domain of sociology) and metalevel (a feature of sociological theory). And the asymmetry he suggests between biological evolution as basically irreversible and societal evolution as prone to "regressions" (i.e., the reproduction of ancient practices), apart from being problematic (cf. our discussion of Wimsatt's "developmental lock" model in chapters 7 and 9), is besides the point: neither sustained "directed change (toward adaptation, differentiation, more complexity and so on)" (73) nor, for that matter, the total independence of the mechanisms of mutation and selection (cf. p. 77) are preconditions for useful evolutionary modeling, as Van Parijs (1981) has brilliantly shown.

two types of stars, between each of them individually, etc. Euclidean geometry had to come before non-Euclidean geometry. Newton clearly had to come before Einstein (I'm not sure whether quantum mechanics had an ordinal relation to those two). My feeling is that we have to start in an oversimplified fashion (if *that* is not working out somewhat well, we are never going to get anywhere). And when that is stabilized, we can develop more complexities. Assimilation followed by accommodation—that seems to me to be a true picture of how human individual cognition works, and, as far as science is concerned, a much better way of putting it than talking about "adhocery" and the "selection of research programs" in the Lakatosian tradition.

2.6.6. *The Nonhistorical Basis of Historicism*

Callebaut: Until now we have discussed various philosophical aspects of Kuhn's new image of science. But the *historian* Kuhn has been criticized as well. Historians notoriously think of their colleague Kuhn as a . . . philosopher. And in a paper written a decade ago, we read that his historical claims, which (on your naturalistic interpretation) ought to provide Kuhn's image of science with an empirical foundation, "are little more than imaginative illustrations of his position rather than items for historical support for it" (Kourany 1979:55).

Giere: *It is a crazy thing, but the historical school in the philosophy of science has produced very little good history.* The historians of science have gone on producing history more or less unaffected by the philosophers. Larry Laudan once gave a paper, sort of complaining, "Why don't historians take seriously the theories of science that philosophers are developing?" My response was, "Well, if we develop some decent theories they might, but we don't yet have any that are worth their using!" History of science is actually flourishing and very interesting, and the philosophy of science, I think, is now maybe less flourishing. And *history of science is now flourishing, not because it has paid attention to philosophy; it is flourishing because it caught up with the mood in history, generally, which is the move to social history and away from intellectual history.* Its real inspiration has come from history and methodological things going on in history. Philosophy of science has not provided them with much to hold on to.

Callebaut: Kuhn barely fits in the movement toward social history. In fact, considering his predilection for the role of theoreticians to the detriment of the role of experimentalists (Franklin 1986; cf. Ackermann 1985, Galison 1987), and more specifically his predilection for the role of the great revolutionaries in the history of science, one could even claim that, in a way, he continues to appeal to heroes in history!

Latour: No, I wouldn't say that of Kuhn, because a *collective* hero is not . . .

Callebaut: . . . Let me put it another way: In the development of historiography in general, there has been a shift from the history of ideas toward social history after the war. Less so in the historiography of science, at least of previous generations. Kuhn is still very much in line with the older history of ideas . . .

Latour: . . . sure . . .

Callebaut: . . . and much less so with social history. In fact, I am probably still being too cautious.

Latour: He is the pope of a church he does not belong to. Kuhn has been made the hero of a discipline which he doesn't like at all. When he received the Bernal Prize of 4S,[41] he was at once thanking the Society for the award—it would be difficult to do otherwise—and saying that it is all a big misunderstanding and that he is not for social history at all.

Giere: *Kuhn has not much influenced the history of science; Popper has not much either.* Kuhn himself is an intriguing person, because in some sense he is the most influential person in the whole field of science studies and at the center; but at the moment he is homeless. He has not really been accepted by philosophers, or by historians either.

2.6.7. *To Give the Shadow Back to History: How Different Are the French?*

Callebaut: I'm trying to track differences between the Anglo-American and the French situation. In the American case, it seems to me, you had history of science on the one hand, philosophy of science on the other, both professionalized quite recently,[42] but at any rate older, as a profession, than the sociology of science. Most of the philosophers knew little of the history, at least not until quite recently. And the historians didn't like, and still don't like very much, the philosophers' approach.[43] The Americans think you should go at history "without any preconceptions." Thus, according to Kuhn (1980), what applies to scientists—say, that their observations are always theory-laden—does not or should not apply to the historian studying the scientists. However, in the French situation, it seems to me that the history is very philosophical, and that the philosophy—a kind of critical rationalism, not in the Popperian sense of course, but in the sense of *critique,* as you have talked about it recently (Latour 1988b)—is very historical. Take the work of Bachelard, for instance. If you look at the people Kuhn refers to as having influenced him, they are all French, and they are all very philosophical *and* historical.

41. The Society for the Social Studies of Science.

42. With respect to historiography of science in the United States, Kuhn (1979:121) has written that "until about twenty-five years ago, only half a dozen people were employed in the United States and Canada as historians of science. Three or four times that number published occasionally in the field or attended meetings of the History of Science Society. But their primary association was with other academic fields, mostly the sciences, or else they had been drawn to history of science by a vocational or avocational concern with book collecting."

43. Wartofsky (1979:119f.) has an historically well-informed discussion of the "full absurdity" of the contemporary debate on the relation of the history of science to the philosophy of science.

Latour: There is a misunderstanding between the French and the American tradition here. First, in France there is only one historian of science coming from history (Bowker and Latour 1987). All the others are coming either from philosophy or from the sciences. And of course, you are right: compared to what in the U.S. or England is called "philosophy of science," they are very highly historicized. I mean, there are figures and names and dates and events in their talk. That is the source of a lot of misunderstanding. Because in *that* sense, they always take history seriously. Now, it depends on *which* history you take seriously. It's not the history of historians; it's a philosophical history of *the advent of rationality*. And that makes a big difference. In that sense, Canguilhem, for example, is much more congenial to Kuhn or to a historian of science in your sense, except that he is not at all in favor of a naturalistic turn. The reason why you have to tackle history—Canguilhem (1988) says it explicitly—is *to give the shadow back to history without which the advent of reason would be without depth;* it is just to add a supplementary glory to the history of reason. So, *yes*, we in France do take history seriously. But, *no*, we don't take history seriously, because we have deprived it of all historical characteristics: circumstances, agitations, actions, uncertainty, and so on. So in a sense, it is very difficult to map the debates of the Anglo-American tradition onto the French, because it's very different. But it has one thing which is the same, exactly: it is that epistemology in France—which is the highest way of doing philosophy, even for the scientists—is defined by the exclusion of history, and society, and culture.

2.6.8. Beating Kuhn on His Own Ground

Callebaut: Michael, you were trained as a philosopher but at some point in your career felt it was necessary to take history seriously—as seriously as the historians of science themselves.

Ruse: During the seventies, I got very excited with the idea of history of science. I'd always enjoyed doing history at school, although I'd only done it until sixteen; but I loved it! And this was an excuse to do it. I felt that if I was going to write a book to critique Kuhn, the only way that I could do it would be by taking historical examples and beating him on that ground. People like Shapere had written good analytic criticisms of Kuhn. That really wasn't my *forte*, then or now. My strength would be to show that Kuhn was wrong about the history. And since Kuhn hadn't really dealt with biology, that was the area that I was going to do. I realized that if I was going to do it properly, I had to do the history of science as well as the historians. I went on sabbatical in 1972–73 to Cambridge, England, to join Robert Young's unit. I'd read Young's papers (1971, 1985) and thought that it was the best stuff on Darwinism going. In a way it was a bit disappointing, because my inspiration there, Bob Young, had finished with Darwin and was off into radical politics. I'm a John Stuart Mill type, liberal socialist at some level, but I've never been a Marxist; so I felt somewhat alienated there. But there I was, able to work on Darwin and the Darwinian material.

Callebaut: There is an ambiguity here. Kuhn uses history to put forward a certain view of the philosophy of science—think of his plea for "a role for history." I take it that by attacking his history, you intended to get at his philosophical standpoint.

Ruse: Oh yes, definitely!

Callebaut: Can you spell out what it was you were opposing?

Ruse: Kuhn offered a relativistic analysis of the history of science. If you swing from one paradigm to another, you are going beyond both reason and the facts! I never agreed with critics like Popper that Kuhn was simply making science totally irrational. I still feel that what Kuhn was arguing was that there is an *arationality,* and, as I said before, an important element of *subjectivity,* as you go from one paradigm to another. Coupled with this—certainly in the *Structure* —he wants to deny scientific progress at *some* level, even though at another level he agrees with it. At an important level, if you change paradigms, it is wrong to say that one paradigm is better than another. You can never go back in paradigms. But if a paradigm is better than another, you must have an independent Archimedean point where you can turn to it and say, "Well, this is how I judge that this is better than the other," and you can't do this. So, all in all, I found Kuhn to be confusingly relativistic, and I wanted to counter his philosophy.

Callebaut: You mentioned before that you felt that you could get at Kuhn by showing that he was historically wrong—that here was his "soft underbelly." I gather that is easier said than done.

Ruse: I remember sending my first paper to John Greene, the historian of biology. He wrote back, "You can't do history of biology this way, you've got to do it seriously!" That was true, and I still feel that way.

Callebaut: So you spent time in the Darwin archives in Cambridge.

Ruse: I got very interested in people like Whewell and Herschel; I felt and still feel that the regular historians underrated them because they didn't have the philosophical background and interests that I had. In a funny sort of way, my book *The Darwinian Revolution* (1979a) didn't come out quite as I'd anticipated. Books have lives of their own. By the time I'd finished it, the final chapter was forty pages on Kuhn, Lakatos, and these things. Just before I sent it to the publisher I realized it was too long, and by the time somebody got to the philosophy they would probably have had *enough.* So I just cut the last chapter off! Although my book was started as a critique of Kuhn, the critique of Kuhn was left on the floor. And I've never ever had anybody say to me, "That was a nice book but, you know, it ought to have had a philosophical chapter at the end!" If somebody said to me, "But Kuhn's right on the history," I still think I would want to say, "My book shows that he's wrong!" But I never mention it anywhere.

Callebaut: Maybe your decision to leave out the final chapter was the right thing to do after all. Look at the reception of Bob Richards's book, *Darwin and the Emergence of Theories of Mind and Behavior* (1987): most of the critics have more reservations about the two philosophical appendices than about the actual historical content of the book itself!

Ruse: I know!

Callebaut: David, I have taken for granted until now that we understand what it means for a philosopher of science "to be historical." But, as you have pointed out, that is not obvious at all.

Hull: The problem is: what do you add to the word "change" when you put "historical" in front of it? What would "nonhistorical change" be like? People like Kuhn are obviously trying to say something. I think you can have change which is just like ping-pong balls in a box, the sort of change that you have with enclosed molecules. Yes, there is change, but the sequence is irrelevant, the order is irrelevant; it is just change. But in historical change, there is some *grouping* of change; it is a sequence of events which are *locally constrained*. At one moment, the situation constrains the possible outcomes in the immediate future but not the distant future. *Given enough time and enough intervening variables, you can get from anywhere to anywhere; but not instantaneously.* So I think what we mean by historical change—in the Western world, that is—is *something like an evolving species.* The order in which the contingencies come into play does matter. Both in biology and in conceptual development in science, we have the same sort of historical phenomena. As a result they have to be analyzed in the same general terms.

Callebaut: Your own evolutionary account of science will be discussed in chapter 7. At this point, I'd like to ask you to react to the continued resistance of many people—I am mainly thinking of humanistic circles and to a lesser extent of social scientists—to any approach to (human) history along biological lines.

Hull: There are some obvious explanations. Humanists feel threatened when topics that have traditionally been part of their province are addressed by scientists. Humanists perceive scientists as being crass. Scientists don't seem to understand what humanists are getting at. They transform subtle, significant issues into those that are scientifically tractable, ignoring all the deeper problems in the process. The scenario is replayed within science as well, when biologists attempt to "biologize" some area of the social sciences or when physicists attempt to take over biology. The people whose subject is being "reduced away" don't like it.

Callebaut: How about you yourself?

Hull: In my work I am not trying to reduce anything to anything. I am attempting to produce explanations that are sufficiently general so that they apply equally to areas that have been previously treated as distinct, the way that Toulmin (1972) suggests. In this connection, I treat biological species and conceptual systems as lineages. Lineages are integrated by descent, a special form of causal connection. What gives rise to what is primary, while similarities are secondary, and "secondary" does not mean of no consequence whatsoever! Objections raised to treating either species or conceptual systems as historical entities stem from commonsense intuitions, but the intuitions in the two cases tend to be different. For example, anyone who knows anything about biological evolution knows that the development of particular species is very largely a matter of contingency. Certainly general laws govern biological evolution, but given what we know of these laws, evolutionary biologists have been forced to acknowledge that they can make only short-term predictions about particular species. There may be some general laws governing conceptual change in science, but we do not know what they are. However, if conceptual change is anything like biological change, contingency will play a significant role (see Rosenberg 1992b and Hull's 1992 response; Gould 1990; Lewontin 1990). I don't think that this contingency bothers historians of science, but it does bother some sociologists and a lot of philosophers.[44] If we view Darwin's theory or Newton's theory as historical entities, then the development of these theories includes large doses of contingency. Students of science find this feature disturbing, profoundly disturbing.

2.6.9. *Historicism without Kuhn?*

Callebaut: I would like to step back for a moment and try to get an overall picture again. Tom Nickles told me he was convinced that even without Kuhn, the interest in history would have come. You, on the other hand, are telling me that people at first didn't know how to react to "the new development." Isn't there a conflict here, Ron?

Giere: In fact, my view of that period is that there were two or three major problems in philosophy of science: explanation, confirmation, and to some extent the nature of theories also. That's it! Kuhn was just coming in at that point. It was not clear what to do with him.

Callebaut: So for instance, Michael Polanyi (1962), whom Kuhn refers to when discussing the "tacit knowledge" of scientists, and who had been active for a very long time—he was involved in a famous fight with the Marxist sociologist of science John Bernal before the war . . .

Giere: . . . Polanyi was just regarded as somebody outside the profession. Or take Percy Bridgman.[45] You see, these were not in the circle of professionals; they were not publishing in philosophy.

Callebaut: Bridgman did influence the logical empiricists' testability criterion.

Giere: Yes, but you discussed Carnap, you wouldn't have discussed Bridgman. You

44. Blom, Callebaut, and Nijhuis (1989) identify various methodological problems related to the "irruption of the contingent" in historiography and the social sciences.

45. The physicist Percy William Bridgman (1882–1962) was professor of mathematics and natural philosophy at Harvard. His name is linked with operation(al)ism, the view that scientific discourse should be cleansed of operationally undefinable concepts, which influenced the logical-empiricist picture of science—and psychology—to a considerable extent but has since fallen into disrepute (cf. ch. 5, n. 3). Bridgman was awarded the Nobel Prize for physics in 1946 for his work on the properties of matter under extremely high pressures.

would not write a dissertation on Bridgman, you would on Carnap, or Popper. These three big problems—confirmation, nature of theory, and explanation—that was it. That's where the action was!

Callebaut: A last attempt to push this query about the "inevitability" of Kuhn: Does it make sense to put all these people in the same bag, to call them "the historicists"? What unites them in addition to their being dissenters?

Giere: There was a grouping of people who were appealing to history. I don't know exactly why that happened. I've often thought that it must have been an accident: why didn't they appeal just to contemporary science? Hanson did to some extent with his book, *The Concept of the Positron* (1963). I just don't know about that. One could have appealed to contemporary science and have done the same thing. But that was the influence: Toulmin was interested in history, Hanson was interested in history, Kuhn of course was interested in history. So what happened is that the major focus of criticism of logical empiricism shifted to the historical ground. *That was probably of itself a historical accident.* Logical empiricism was getting more rococo. And there were internal critics. You could view Quine as an internal dissenter in some ways. I think people were looking for something new.

Callebaut: Quine is rather different from the people who are appealing to history. When did he enter the picture?

Giere: In retrospect, Quine's criticisms began in the early 1950s. I remember reading Quine (1951) as an undergraduate. It did not seem to me as powerful a critique as it in fact was. Then there was his "Epistemology naturalized" (1969b). At that time the program of naturalized epistemology seemed to me so obviously circular that I could not take it seriously. Partly that is because I was then strongly influenced by the Midwest group in the philosophy of science, people like Herbert Feigl (Minnesota), Adolf Grünbaum (Pittsburgh), and especially Wesley Salmon (then my colleague at Indiana). It was only in the late 1970s, when I independently became convinced that the traditional epistemological project in the philosophy of science was fundamentally misconceived, that I began to appreciate what Quine had been saying. But others were more strongly influenced by Quine much earlier.

Callebaut: Wasn't there also a generational aspect involved, as Tom Nickles suggested earlier?

Giere: I think no intellectual movement can go more than about three generations. You have the founding generation. You have the immediate followers, who are enthusiastic. But by the time you are the follower of the follower, it gets less exciting; and at that point people start looking for something new. Who wants to be a follower of a follower? So I think a lot of people were looking for something new to do.

Callebaut: What about the French situation, Bruno?

Latour: You've got to realize that Kuhn, in the French context, is not very extraordinary; because all the collective aspects are in Bachelard; the notion of a paradigm is also in Bachelard, who speaks of the "trade union of the collective workers of the proof" . . .

Callebaut: . . . and in Foucault . . .

Latour: . . . yes, in the whole French tradition. What Kuhn lacks, though, viewed through the spectacles of French epistemology—which is not my point of view— is that he puts aside the very philosophical questions concerning the advent of reason. That's why the French philosophers think he is just another sociologist whom they don't like. *If by "naturalistic turn" is meant the avoidance of all the big philosophical questions of the past, then to the French epistemologists the whole thing means nothing.* Why do they like Duhem or people like Meyerson or Poincaré and in fact the whole French tradition? It is because they think philosophy of science is a good place to ask the real questions of philosophy; the good, big questions about certainty, truth, and what have you. If by "naturalistic turn" you mean that you are going to be without preconceptions, that makes the French laugh—because they know that theory is *always* there. The big misunderstanding, I think, is that *you never have to convince the French that facts don't speak for themselves.* Most of the enterprise of the Anglo-American tradition has been a fight against empiricism. But to us—and here I am putting myself in the same tradition—this fight is always already won! It's *enfoncer des portes ouvertes,* as we say. Therefore, it is always impossible to map exactly the English-speaking tradition onto the French tradition.

Callebaut: Let us jump over the Atlantic again. The University of Chicago has always had a historical tradition, in the broad sense of "historical." In his autobiography, Rudolf Carnap notes that in the Department of Philosophy at Chicago, where he taught from 1936 until 1952, great emphasis was laid on historical carefulness as well as on "a neutral attitude." The staff insisted that a philosophical doctrine be understood *immanently,* inasmuch as a criticism from outside would not do justice to its peculiarity. Carnap felt that such an attitude was "useful and proper for the purpose of historical studies, but not sufficient for training in philosophy itself." [46]

---

46. "The task of the history of philosophy," Carnap (1963:41) went on, "is not essentially different from that of the history of science. The historian of science gives not only a description of the scientific theories, but also a critical judgment of them from the point of view of our present scientific knowledge." Carnap thought the same should be required in the history of philosophy, since "in philosophy, no less than in science, there is the possibility of cumulative insight and therefore of progress in knowledge." And he laconically concluded: "This view, of course, would be rejected by historicism in its pure form." As an illustration of "historical neutralism," Carnap mentions a Ph.D. thesis on the ontological proof for the existence of God: "From the thesis and the oral examination it became clear that the candidate knew that later philosophers, e.g., Kant and Russell, had rejected the proof; but for him this fact seemed merely one more example of the old rule that any assertion of a philosopher is rejected by some other philosopher" (ibid.). Carnap goes on noting that on some occasions, he "was depressed to see that certain philosophical views which seemed to me long superseded by the development of critical thought and in some cases devoid of any cognitive content, were either still maintained or at least treated as deserving serious consideration" (42). For a contemporary discussion of the conflict between "relevant" and "anachronistic" historiography, see Pitt (1986).

Is historicism, in the sense of taking the history of a science seriously, if not in Carnap's sense, something you simply take for granted here, Bill Wimsatt?

Wimsatt: It's something I take for granted now, but I didn't when I came to the University of Chicago. When I came here I was interested in current science. I didn't see any particular distinction between the history of science and current science other than time. But, reacting specifically to Carnap's remark—and the remark, undoubtedly by Richard McKeon, which stimulated it—I don't think that any perspectives or "attitudes" are totally neutral (though some are unbiased for certain purposes, and others may be more or less biased for those purposes, and the orderings may shift for other purposes). Such neutrality as there is results from shared purposes (some of which may be so general as to be universal to members of a discipline), and from interactions and intercommunication, through which we discover our respective biases, the existence of other perspectives, and how to choose the right ones for the right problem. Also, the contrast posed between historical "immanent" understanding of a doctrine and some unspecified, presumably presentist-internalist "objective" vision is too simple. There are multiple levels of organization in nature, and we need to understand natural objects at each of them. Similarly, any historical event can be viewed from a variety of perspectives, time scales, and degrees of grainy detail, each with different degrees and dimensions of detachment. We need at least to take a sample of these perspectives (etc.) to "span" the problem at hand. After all, we go to history with different kinds of questions, for which different techniques, indicators, and scales of resolution are appropriate. Picking one such and saying that's how we should approach problems would be like legislating that astronomy should be limited to a certain spectral line (say the lowest energy carbon transition in the optical domain). Wouldn't that be silly?

Callebaut: It sure would. How pervasive was the historical dimension when you were a student at Chicago, Bill Bechtel? I am thinking of Carnap's (1963:43) remark, in his autobiography, that in the Gothic style buildings of the University of Chicago he sometimes had the "weird feeling" of "sitting among a group of medieval learned men with long beards and solemn robes." (Cf. also the "genuflections at the altar" of U. of C. sophomore Herbert Simon [1991a:47].)

Bechtel: That was still present, but in a much healthier way than in Carnap's time. The medieval cathedral was gone. There are schools like that in the States now, where the goal is purely to work within some classical tradition. No, the attitude toward history of philosophy at Chicago when I was there was that it was an incredibly important resource for dealing with contemporary philosophical problems. That was manifest in Alan Donagan's use of Descartes, Manley Thompson's use of Kant and of Peirce, Gewirth's use of Aristotle.

Wimsatt: After I got here, in large part in conjunction with the Biology Common Core course that I teach, I wanted to do more stuff with a historical component. Dudley Shapere, who I suppose has been more than anyone else responsible for hiring me, explicitly wanted me to do history of science. At that time that really did not move me much. I think it was rather the biology course. I worked more of the history of science into the course after he left. Often you can't understand the orientation of a theory unless you understand it historically.

Callebaut: You have come to feel that philosophy of science is inseparable from studies in the history of science and from studies in current science, and that those kinds of studies of science are complementary.

Wimsatt: Hindsight, which you have in the history of science, gives you a very useful perspective on where things are going: you can see things go to completion, and so on. But current science allows you to focus in much more detail on the structure of the discovery process. And since in fact discovery processes, problem-solving processes, and heuristics are among the core foci of my interests (e.g., 1986c, 1987)—the other being in general, "How can we have a simple theory of a complex system and get away with it?"—I'm very interested in this. Most of my work in the history of science has been either to trace the influence of ideas in guiding the formulation of later disputes in the direction of their later resolution, but also to look in some detail at the problem-solving heuristics that were used in those periods—that is, to get evidence for kinds of timeless theories of scientific change, or optimal scientific change, or whatever.

2.7 POSITIVISM AND HISTORICISM BACK TO BACK

Callebaut: Dudley, you have tried to develop a view of science and scientific change that is neither absolutist nor relativist. How would you summarize it?

Shapere: We learn in science through the application of what we assume, at a given time, to be proper methods in search for our goals. Sometimes these come into conflict with what we find in experience. (Yes, despite the post-Kuhnian relativist turn, there *is* something in experience which is independent of our presuppositions: no matter how much selection and interpretation we impose, a result of experience or experiment can still disagree with what we expected in the light of those presuppositions.) Depending on the character of the conflict, that can lead us to change our methods and our goals as well as our substantive beliefs about nature. This self-corrective feature of science is something that older philosophers used to emphasize, but which has been very largely forgotten.

Callebaut: But you mark off your own notion of self-correction from that of previous generations of philosophers.

Shapere: The view used to be that only the substantive beliefs could be self-corrective, but they did not have any feedback to the "independent" methods or goals. The self-correctiveness that I have in mind includes methods, goals, the conception of problems, rules, and standards. Even the conception of what facts are, and of what counts as an observation, can alter in the light of the beliefs we come to in the study of nature.

Callebaut: You have used the notion of "bootstrapping" to refer to the continual feedback between the substantive picture of nature developed in science and our conception of the goals, methods, problems, and standards of knowledge. The transcendental question of traditional epistemology was: "Don't we have to have guiding principles to begin with in our inquiries?" Would you agree that all approaches to our experience require *some* sort of presuppositions?

Shapere: That I take to be a central doctrine of mine.

Callebaut: You find the analogy with the Kantian categories quite misleading . . .

Shapere: . . . as though somehow these ideas can in the first place be distinguished sharply from other ideas which are not guiding principles and which are somehow to at least a certain degree immunized from inquiry. I don't doubt that they have in fact often been immunized, because the Parmenidean idea of permanence, for instance, became an *ideal of explanation.* And philosophers and other people thought that that was because one had an idea of what inquiry is, or what an explanation is, and that this was something that was prior to entering into the inquiry process and whose principles could be established prior to and independent of the inquiry process. We have learned that this is not the case, that even such ideas are subject to investigation and rejection. So the idea that they are immune, even though it might be historically correct for early stages, is something we have learned not to respect.

Callebaut: This brings out something very central to your approach: your insistence on the contingencies and on the openness of scientific evolution.

Shapere: I claim it is the function of philosophy to show that the possibilities are *open* rather than, as in traditional philosophy, to define the—preset—*scope and limits* or the fundamental necessary assumptions of science. In the light of what we have learned about human evolution, in the light of the pervasive depth of the alterations in our thought brought about by scientific inquiry, and in the light of the wholesale failure of philosophers to establish alleged necessary preconditions, we must show how there are no such limits, either way. We have to find out what the character of science is: how we are to go about finding out about nature, how we are to go about understanding it, and what we have when we have understanding.

Callebaut: The other part is then, of course, what actually happened, contingent as it has been shown to be in the "in principle" part.[47] Another pivotal notion in your analysis is the notion of "internalization." What do you mean by that?

Shapere: There comes to be made gradually a distinction between those beliefs which can serve as background information on which to build new ideas, methods, and so on, and those which can't. That is, *science becomes internalized in its reasoning processes.* A distinction is made between that which is relevant and that which is irrelevant. *What is relevant become scientific reasons; what is irrelevant*

---

47. An application of this view to the problem of intertheory reduction in science will be discussed in section 4.2.

"Philosophy is the last refuge of superstition."

Susan Mullally Clark

Dudley Shapere (b. Harlingen, Texas, 1928) is Z. Smith Reynolds Professor of the Philosophy and History of Science at Wake Forest University. He received all his degrees from Harvard University. Before going to Wake Forest he taught at Ohio State University, the University of Chicago (where he founded the Committee on the Conceptual Foundations of Science), the University of Illinois, and the University of Maryland. Shapere has also held visiting appointments at Rockefeller and Harvard. From 1966 to 1975 he served as Special Consultant (Program Director) for the National Science Foundation's History and Philosophy of Science Program. For four years, 1974–77, he visited a large number of colleges and universities as a Sigma Xi National Bicentennial Lecturer. During the academic year 1978–79 he was a member of the Institute for Advanced Studies at Princeton, where he visited again in 1981 and 1989.

In the editorial preface of his book, *Reason and the Search for Knowledge* (1984), Robert Cohen and Marx Wartofsky wrote that an impressive characteristic of Shapere's studies of science has been his "dogged reasonableness." It is crucial to understand his notion of a *reason* to get a grip on his enterprise. For Shapere, a "reason" is first and foremost "a consideration that we have found to be successful, coherent, and relevant to the particular subject-matter we are dealing with in a particular investigation. We had to learn that *that* is what to count as a reason, and how, specifically, to understand the 'concepts' of 'success,' 'coherence,' and 'relevance'." For instance, if one wants to answer the question, "Why do we believe that the energy of a star is produced in nuclear reactions in the core of that star?", the reasons for that are going to be physical and astronomical reasons that we have found to be relevant. It is as simple as that, Shapere stresses. What we have done over the course of the history of inquiry—which is much longer than the history of science—has been "to build up bodies of relevant considerations, relevant relations, in which some of the beliefs are 'well-established,' in a sense that we have also had to develop over the course of inquiry."

Two facets of our best current scientific beliefs are especially relevant as the framework within which to conduct philosophical inquiry. First, our knowledge of human evolution, which proceeded in a context of trying to cope with an immediate environment. All human ideas must have originated and developed from such a context. In terms of such a frame of thought, any idea that the human mind has access to truths that are independent of investigation or somehow transcend it are just "hangovers of superstition." But, secondly, the scientific ideas which we have attained by now depart very far from anything, in myth or early philosophy, that emerged from those primitive, everyday beginnings. Our problem, then, is "how we ever *got* to our rather weird present views, and the source and extent, if any, of justification of those beliefs."

As the scientific enterprise evolves, it becomes more and more segregated from the environment of irrelevant beliefs—*internalization* is of paramount importance. "You aren't adjusting to the total body of social beliefs; as time goes on, you're distinguishing more and more clearly what counts as a reason from the rest." And *we learn how to learn:* "We learn also *how* to think about nature, how to talk about it, what kinds of questions to ask about it." The kinds of questions that are asked in science thus transcend our personal or social or cultural interests, the more so as science develops historically and the process of internalization becomes more sophisticated. In one way, Shapere insists, "this evolution is more Lamarckian than Darwinian."

Shapere takes himself "to be doing *real* naturalistic epistemology—and as *doing* it, not just saying we *ought* to be doing it." In his view, though, the study of reason*ing* is not the central problem in understanding science; the central problem is to understand reason*s*. If we ask why we do believe that the world is so-and-so, logic won't enlighten us. Also, at present, he thinks, psychology and the study of AI have little to contribute to our understanding of the processes by which we have arrived at our justified beliefs. "Perhaps some day they may be able to illuminate the processes, even perhaps raising specific doubts about the entire process, or showing why it works." For now, he takes the relevant knowledge in such fields to be inadequate to serve as a basis—much less as the sole basis—for a naturalistic approach to understanding the knowledge-seeking enterprise.

Shapere's numerous publications on the philosophy and history of science, and particularly physics, include *Philosophical Problems of Natural Science* (1965), *Galileo: A Philosophical Study* (1974), and *Reason and the Search for Knowledge* (1984). He is preparing another book, *The Mechanical Philosophy of Nature* (n.d.a). Shapere is a member of the editorial boards of *Philosophy of Science* and the *Journal for the History and Philosophy of Science*. He is a Fellow of the AAAS.

A first interview with Dudley was recorded on August 31, 1985 at the conference on Experimentation in the Sciences in Bath, England. We pursued our conversation on several days in October 1986 at his home in Winston-Salem and corresponded extensively afterwards. The awe-inspiring "view from somewhere" of the Milky Way in Dudley's office on the Southern-autumn-colored Wake Forest campus (the Reynolds family's old polo playground) reminded me that there are vistas beyond the biological—as far as we know.

become external factors. The distinction between what is internal and what is external is not made in terms of an unalterable line of demarcation discovered a priori, but is itself a product of this historical development.

Callebaut: According to De Mey (1982), the way the internal/external distinction is made is itself a function of the developments in a discipline; and this circumstance has far-reaching implications for the analysis of science. In particular, he thinks it mortgages all attempts at externalist explanations of science. These implications, as well as some other aspects of your work, Ron, will be discussed in subsequent chapters. To round off this historical review of philosophy of science: What, in a nutshell, is happening in post-Kuhnian philosophy of science?

Giere: I think that is wide open now. I think that the historical school in the philosophy of science, which I said Laudan and other people are an example of, that also is coming to an end. People do not know what to do anymore. Lakatos died in 1974, which hurt a lot. The Lakatosian view was just fixed where he left it off; nobody has been able to take his place. So there is a question of what comes after. That is why it's open.

2.8 Three Waves of Empirical Information: A Sneak Preview

Callebaut: To conclude this chapter I propose to reinsert the preoccupations of philosophers of science into the broader developments in the burgeoning field of empirical science studies. Were philosophers of science in the 1960s aware that Robert Merton's sociological views on the "ethos of science" (dating from the late 1930s) mirrored their own rationalist and objectivist methodology?[48] And were they aware of the complementarity of their own internalism and the Mertonians' externalist emphasis on "science as an institution" (Merton 1973:268)?[49]

Burian: I think I'm not a proper witness to evaluate things in this area.

Callebaut: That seems to me significant in itself!

Burian: I myself became aware of Merton—both of his puritanism thesis and his reward system stuff—in the late sixties. I did not study it seriously.

Giere: If I just take myself as a measuring instrument for what was going on in philosophy of science, I'd say that nobody in philosophy of science was paying

---

48. Merton took the social structure of the scientific community at large to be characterized by "universalism," "commun(al)ism," "disinterestedness," and "organized skepticism" ("CUDOS"—Ziman 1985), four norms (also called "institutional imperatives" or "mores") internalized by the scientist-members and representing the community's "moral consensus." Deriving from "the goal" of the scientific enterprise ("certified knowledge") and its "methods" (cognitive and technical norms such as logical consistency, predictive power, etc.), CUDOS was regarded by the Mertonians as entirely functional to the advancement of science.

49. For a discussion of these two aspects of the relationship between traditional sociology of science and "received-view" philosophy of science, see, e.g., De Mey (1982) and Laudan (1984:ch. 1).

any attention to sociology of science. People were only interested in the two or three problems I mentioned earlier.

Callebaut: Tom, you have used the wave metaphor in talking about the naturalistic turn. When did the sociology of science—or the sociology of scientific knowledge (SSK), as many prefer to say today—become important to philosophers?

Nickles: After the historical wave washed over us in the 1960s, a second wave began gathering momentum in the mid–1970s, originating at the science studies units in Edinburgh, later Bath, Paris, the Low Countries, and elsewhere. As your question suggests, Werner, this was basically a sociological wave and it heightened concern about underdetermination by focusing on scientific controversies. Some of this work was historical, e.g., Steve Shapin and some of the other Edinburghers (Barry Barnes, David Bloor). Harry Collins and Trevor Pinch at Bath decided to study contemporary controversies in physics—claims about the detection of gravity waves, solar neutrinos, and the like. Since much of the new-wave literature strongly attacked philosophers, they could not ignore it for long. So their early engagements with sociologists were negative in tone—dismissive or defensive— and there is still a lot of that today. More positively, the new case studies presented large amounts of empirical information about science that naturalistic philosophers felt they needed to consider. Lately the rhetoric on both sides has cooled and some degree of constructive cooperation seems possible. Meanwhile, a third wave of empirical information is coming from psychology, e.g., the studies of reasoning, judgment, perception, and the like. The next step is for philosophers themselves to become participant observers of scientific life. Ronald Giere has done a bit of this, not to mention the several philosophers who double as working scientists or vice versa.

3

TOWARD A NEW THEORY OF SCIENCE
New Dimensions, Features, and Approaches

AS LONG AS positivism reigned supreme, philosophy of science was really a philosophy of "rationally reconstructed" *physics* disguised as general philosophy of science. Current approaches to science display much more diversity and heterogeneity in comparison. The emerging field of "science studies" is *multidimensional* in at least two senses: a variety of "philosophies of . . ." unambiguously acknowledge the factual disunity of the sciences; and alternative accounts of the study of science are being articulated that explicitly aim to compete with, if not to replace, philosophy of science (whether traditional or naturalized).

Special philosophies of various special sciences now flourish.[1] The philosophy of (evolutionary) biology and the philosophy of (cognitive) psychology in particular are booming.[2] In this chapter, the reader is given a flavor of the expansion (some prefer to call it an *explosion*) of general philosophy of science through an *aperçu* of the (auto)biographies of three philosophical "voyagers," as I like to call them because of the ease with which they move back and forth between the various scientific disciplines they investigate. Wimsatt moved from engineering to (the philosophy of) biology. Rosenberg has embarked on a twin career in the philosophy of biology and the philosophy of economics. Kitcher, besides also being a philosopher of biology, publishes extensively in the philosophy of mathematics and on matters calling for general public awareness.

Despite the increasing compartmentalization of philosophy of science, some general *features* seem to emerge in relation to its naturalization. Here we will pay attention to the rehabilitation, after a century of dismissal, of psychologism, and on new ways of dealing with the distinction between analytic and synthetic statements and their consequences for the status of logic and mathematics.

Finally, philosophy of science is being convulsed by competing *approaches*. After being shaken by a wave of historical studies in the 1970s, sociology of science in

---

1. The notion of "special science" is discussed by Fodor (1974) in his plea for "the disunity of science as a working hypothesis." Cf. also Stegmüller's (1979:41–49) distinction between "general" and "special" philosophy of science.

2. They will be at the center of most of the discussions in Part Two.

72

particular has come to the fore in the last decade as philosophy's major contender for naturalistic explanations of current scientific discourse and practice, and to some extent even of what happened historically (Shapin 1982; Latour 1988a, 1989a; Pickering 1992). The prominent use of methods devised in anthropology—such as ethnomethodology—is one of the outstanding characteristics of this new approach, as is its embrace of radical contextualism and other sorts of relativism. Some of the contrasts between the sociologist's and the philosopher's way of accounting for science will be highlighted in chapter 4 in an exchange on the *realism* issue between Knorr Cetina and Giere. Here, some of the most salient claims of STS (science and technology studies) are scrutinized in exchanges with the *social constructivist* Knorr Cetina and the *actor-network theorist* Latour.[3]

Taken together, these new approaches should enable the student of science "to climb inside the heads of the members of the group which practices some scientific specialty during some particular period; to make sense of the way those people practiced their discipline; to isolate the forces which caused changes in their goals, concepts, and techniques; and finally, to discover when and how these changes were assimilated" (Kuhn 1979:123).

3.1 PHILOSOPHY OF WHAT?

Callebaut: Tom Nickles warned against the "generality trap" philosophers seem to be prone to more than ordinary folks. But what does the "generality" of philosophy of science really amount to?[4] Elliott, how do you explain that physics is no longer regarded today as the science to study *par excellence* by many philosophers with an inkling for science?

Sober: The positivists took as one of their main paradigms of a scientific theory Einstein's theory of relativity, and their philosophical problems and the views that they developed about them were often keyed to that single theory. To a lesser degree they considered quantum mechanics. But issues in biology did not interest them very much; specific issues that are internal to psychological theory mattered to them almost not at all.

Rosenberg: Well, if you look at the *Encyclopedia of Unified Science* (Neurath, Carnap, and Morris 1955–1970), even back in the thirties and forties there were articles on economics and biology and other disciplines . . .

Callebaut: . . . But in retrospect we can say it was basically a philosophy of physics

---

3. Yet another enterprise which is rapidly gaining momentum is the *cognitive approach* to science. It is complementary to the sociology of science in that it seeks to elaborate an adequate model of the individual (and social?) psychology of scientists. The cognitive approach will be discussed in chapter 8.

4. Twenty years ago, Harvard zoologist Ernst Mayr—who doubles as a historian and triples as a philosopher of his field (Mayr 1982, 1988)—lodged this complaint: "I have some five or six volumes on my book shelves which include the misleading words 'philosophy of science' in their title. In actual fact each of these volumes is a philosophy of physics, many physicist-philosophers naively assuming that what applies to physics will apply to any branch of science" (Mayr 1969:197).

(of a rather peculiar kind). The "application" of logical-empiricist views to, say, biology—I'm thinking of the work of someone like Woodger (1937, 1952)—now make us smile—or cry. The thing may have been *intended* as general theory, but the methodology that was put forward . . .

Rosenberg: . . . was drawn exclusively from physics . . .

Callebaut: . . . and other fields had to fit that model. You agree with that. So a fundamental problem of older work in philosophy of biology was that to the extent that it was done by people working in the positivist tradition, like Woodger, they had a very difficult time.

Rosenberg: Yes, absolutely.

Sober: What has happened since the demise of positivism is that philosophers have gotten interested in the *details* of particular scientific theories. In the 1930s, philosophers of physics were interested in relativity theory and quantum theory and that has continued to the present. Only more recently have philosophers of biology really gotten into the details of evolutionary theory and other theories in biology; and similarly with the philosophy of psychology, I think. The demise of positivism allowed this proliferation to occur because it was no longer necessarily a given that all scientific theories were the same; there could be problems internal to a scientific theory that might be of philosophical interest.

Wimsatt: There's a paradox here. About sixteen years ago, I wanted to use the paper, "Reflection on fundamentality and complexity" by the physicist Max Dresden (1974). It is a marvelous blast against the traditional Nagel-Hempel analysis of reduction, pointing out the slop, approximations, hand waving, and the like between quantum mechanics and any other parts of physics that are supposed to be derivable from it. What puzzled me was that he should bother to talk about the analyses of philosophers—most physicists don't. When I called to ask his permission to use it in class, I asked about this. He said that he used to eat lunch at Iowa with a psychologist "who was always telling me that psychology was becoming more like physics. So I asked him to tell me what psychology was like. I didn't recognize anything. So I asked him to tell me what he thought physics was like. I still didn't recognize anything." He described it to me, and it sounded like Carnap of the *Aufbau* (1929). "So I wrote this paper to tell him what physics was really like." Dresden, by then executive director of the Brookhaven National Accelerator Lab, later told me that the psychologist was Kenneth Spence.[5]

Callebaut: So your point is that the positivists didn't have their physics right either . . .

Wimsatt: . . . or were right at most about the smallest corner of physics, namely disputes about the foundations and interpretations of the foundational theories of relativity and quantum mechanics, with maybe a little statistical mechanics mixed

---

5. Spence was probably the most dogmatic behaviorist of his day, and one of the four subjects of the marvelous Krantz and Campbell social-psychological study, "Leadership styles in science" (Krantz and Wiggins 1973; Campbell 1979b).

in. No particle physics, no solid state stuff, no astrophysics, no thermodynamics (or nonequilibrium thermo, which was really heating up in the mid-sixties when I was a graduate student), no magnetohydrodynamics, no quantum chemistry, no supersonic aerodynamics (or aerodynamics of any sort), no cosmology, and surely not any meteorology. . . . I could go on indefinitely. So what physics were they representing? Only foundational physics, and not terribly representatively at that. We've been a little broader in philosophy of biology.

3.1.1. *Finding One's Niche on the Generality Continuum*

Callebaut: Philip, at the University of Minnesota's Center for Philosophy of Science you were involved in a project that aimed to find a "new consensus" in the philosophy of science (1986–87). One of the topics this institute investigated was quite traditionally labeled "The Structure and Function of Scientific Theories." What came out with respect to the generality issue?

Kitcher: There are people—I think Adolf Grünbaum would fall in this category—who think it is possible to talk about theories in any field in roughly the same way; so *it didn't matter that traditional philosophy of science concentrated on physics.* I extrapolate from something Adolf said to me: that he found that he was able to use the ideas he extracted from physics perfectly well when he turned his attention to psychoanalysis (Grünbaum 1984, 1986). On the other hand, there are people—Alan Nelson, a philosopher of economics, is one prime example, Paul Churchland and Ned Block are others—who think the example one takes is quite important, and that maybe there are different needs for representing the states of different disciplines. I believe there is a certain general way of looking at the state of a science at a time, but I think the way in which one extracts from that state what might be called *the theory of the science at that time* is variable. And I also think you can represent different parts of the science of a time differently for solving different epistemological problems.

Callebaut: One trend in post-positivist philosophy of science has been to argue against any would-be "universal" picture of science and to argue instead that physics may be and in fact is different from chemistry, chemistry from biology, biology from the social sciences, and the like. Bill, you are well known for your quite sophisticated views on the reduction of one piece of science to another (e.g., Wimsatt 1979, 1986b). Do you also see generality as essentially a "trap," as Tom Nickles calls it?

Wimsatt: Yes, to some extent, but I don't agree with it nearly as strongly as I suspect many of the new philosophers of science would. I do think that there are general themes; they may not be *absolutely* universal. One of these themes is *the structure of problem solving.*

Callebaut: Please elaborate.

Wimsatt: For example, reductionistic approaches in any discipline share a family of methodological approaches—often adapted differently to different local contexts, and with different weightings on the importance of the different kinds of problem-solving heuristics. But I don't know a single research strategy called "reductionistic" in any area that doesn't use the "near-decomposability" heuristic—analyze a complex problem (system, structure, task, statement, . . .) by breaking it down into parts, getting whatever counts as solutions for the parts, and then attempt to get a solution for the whole in terms of an articulation of the various partial solutions. (This is one of Herb Simon's major insights.) It doesn't *always* work—which is why it is a heuristic. Sometimes we haven't chosen good boundaries for the system or its parts, and sometimes maybe there aren't any good boundaries, in which case this strategy will fail.

Callebaut: Couldn't you say that because it fails sometimes, it isn't (absolutely) general?

Wimsatt: This is true, but there is more generality than you think, because you may be able to give quite general conditions for when you would expect this strategy to break down. Indeed, you may even be able to use the pattern of breakdowns to identify the strategy used (1980a, b)! So I'm not quite ready to give up on generality. It's just that we've probably been trying to generalize on the wrong things, and I expect that the generalities we find will be much more contextual and limited in scope—but generalities nonetheless because we will be able to characterize those scopes fairly accurately and in an explanatory fashion.

Callebaut: An issue intimately related to generality concerns the dialectic between *empirical* and *theoretical* investigations of science and the way this is reflected in the division of labor between philosophical and other approaches to science. As an anthropologist of science, Bruno, you strongly object to the common self-image of philosophers—the naturalized specimens included—as "the synthesizers of the case studies."

Latour: The thing I don't like is this idea that philosophers of science have that they are going to *unify* the field studies—as if the duty of philosophers was to put some order and large-scale principles on little guys doing the empirical stuff. I mean, *we are not little guys doing empirical stuff*. We are ourselves tackling all the big questions of philosophy. But many among us, it is true, are very ignorant and distrustful of philosophy. They usually have never read anyone older than Wittgenstein. For them, philosophers like Leibniz, Plato, or Nietzsche are dead.

3.1.2. *History and Philosophy of Science: What Difference Do They Make?*

Callebaut: Do you think, Bob, that incorporating the historical and philosophical dimensions actually makes a difference as far as *practicing a science* is concerned? I am thinking in particular of disciplines like evolutionary biology and cognitive psychology, where there seems to be—at least at some level—a real symbiosis of the science and its philosophy and history.

Richards: Evolutionary biology is one of those peculiar scientific disciplines which is intrinsically historical. Unlike other areas in science, it has a number of conceptual problems, and the theories that it advances have strong conceptual components; so that they are simply not attempts to empirically describe the world in a rather bold fashion. The practitioners of contemporary evolutionary theory have—perhaps because of the nature of their own science—become terribly interested in both conceptual philosophical problems and historical problems. One only has to mention people like Steve Gould, Richard Lewontin, and Ernst Mayr to recognize that philosophical and historical considerations inform the development of their own views.

Callebaut: Aren't these people quite exceptional? They are highly visible and, in addition, fully engaged in ideological debates. Whereas for most practicing biologists—isn't it business as usual for them? Do *they* bother about philosophy or history at all? It would seem to me that at that level, the situation in biology doesn't differ all that much from, say, that in physics.

Richards: It is certainly true that the preoccupations of Gould, Lewontin, and Mayr are not exactly the same as those of someone who is in, say, a large university, in a biology department, teaching or doing research in evolutionary biology. For example, Lewontin's (1974c) own brand of materialism has led him to discover in the logic of nineteenth-century evolutionary theory a comparable materialism. My book (1987) was an effort to correct this perception, to show that most of the Darwinians of the last century were spiritualists. And both Gould and Mayr argue that evolution is nonteleological and nonprogressive, and they interpret Darwin, the patron saint, as holding the same views (contrast Richards 1988). I believe that theirs is dubious history, but excellent historical rhetoric for their own scientific theories.

Callebaut: It's the Whig history debate once again.

Richards: Now when you have the leaders of the field making such historical arguments, it must have an impact lower down, and it must begin to alter the views of the practitioners of less stellar magnitude. Graduate students in evolutionary biology in most universities are concerned with the kinds of questions that Gould, Lewontin, and Mayr have introduced. In my own classes in the history and philosophy of biology there are a fair number of graduate students in evolutionary biology who take courses in other areas of the university—Bill Wimsatt's courses in philosophy of biology for example—so that that particular discipline has been *infected* by philosophy and history of science!

Callebaut: There is another aspect in which the history and philosophy of science plays a role in contemporary evolutionary biology: it would seem that Darwin has never really died. How would you explain this?

Richards: In defending, or rebutting, a position in contemporary scientific theory of evolution, Darwin's name is quite frequently invoked indeed. It would appear to be a winning argument if you can point out, as for example Edward Wilson (1975) does, that Darwin himself was moving toward, and in fact had formulated, a quasi-sociobiological view (see 9.2.3). This gives a kind of authority to what Wilson is doing; at least he so regards it.

Callebaut: You actually do something similar yourself in your plea for an evolutionary ethics (see 10.1).

Richards: On the other side, one way of rebutting the kinds of moves someone like Wilson makes in biology—applying evolutionary theory to society—is to suggest, as some have (e.g., Lewontin 1976), that what Wilson is doing is not *echt* Darwinism, is in fact Spencerianism.

Callebaut: So there is both an appeal to the "patron saint" of evolutionary biology and to his work to give some kind of validity to an enterprise.

Richards: This goes on all the time in science in one fashion or another. It is a little more obvious, I think, in evolutionary biology. And by the opponents there are attempts to undercut that authority by suggesting that someone advancing a so-called "Darwinian" point of view would, if he were thinking correctly and historically about the problem, discover that it is not Darwin at all whose case he is moving, but someone like, perhaps, Ernst Haeckel. In this way, historical considerations play a large role in the kinds of disputes and their arbitration that go on in evolutionary biology. This is one of the ways in which historians can play a— rather modest—role in contemporary discussions, if only as referees to point out when one side is playing loose with the evidence.

3.2 First Voyager: From Engineering to Biology

3.2.1. *Discovering Complexity and Messiness*

Callebaut: We are now ready to look at the rather untypical intellectual career paths of some "third generation" philosophers of science. Bill, you started out studying physics and engineering physics at Cornell, and what you perceived as its clash with the reality of science did not leave you indifferent.

Wimsatt: When I was an undergraduate in 1962, I audited a course on magneto-aerodynamics in the Graduate School of Aerospace Engineering at Cornell. I took it because I very much admired the professor, Ed (Edwin L.) Resler. Actually it met at the same time as a classical mechanics course I should have gone to.

Callebaut: Good boy!

Wimsatt: In any case, he came in the first day and covered the board with equations. He said: "These are the equations that apply to this system" and wrote down 22 equations: Newton's laws, the electromagnetic field equations, the hydrodynamic flow equations, the aerodynamic equations that allow for partial compressible flow, chemical kinetics and diffusion equations for ions. A plasma is an ionized gas, and so you have diffusion equations. A lot of these equations were partial differential equations. He said: "There are 22 equations here and 22 unknowns. What do we do in this case? In principle it's all solvable. But what we do actually is we let 18 of the parameters go either to infinity or to zero, and then scratch our heads and try to figure out some interesting values for the 4 remaining parameters, and put it in the computer and let it run over the weekend . . ."

Callebaut: . . . This was thirty years ago . . .

Wimsatt: ". . . and then we get this huge pile of printout, scratch our heads over the results, and try to figure out either what went wrong or what other sets of parameters to look at." I was both delighted and scandalized; scandalized because here was research at the frontiers of physics, and it did not look at all like my conception of physics, or it turned out not at all like the conception of physics promulgated by the positivists of the time.

Callebaut: Then you worked for a year in engineering.

Wimsatt: I worked for NCR in their adding machine division. One of the most interesting problems I had was a cam design problem. They had an order from a Swiss bank for 3,000 check printing machines. These are machines in which somebody puts in a check that needs a magnetic code imprinted on the bottom. In order to do this, the machine had to advance about 3 inches of carbon ribbon tape each time. It was a cam-activated thing and the tape was breaking. So I said: "Aha! This is a simple problem in classical mechanics. What I've got to do is to design a cam such that the maximum value of acceleration is at the minimum." Since $F = ma$ that ought to minimize the force on it, and make it least likely to break.

Callebaut: So you set out to design this cam?

Wimsatt: It was actually a marvelous problem of application, because the machine was old (designed in 1919), originally manual, now electrically driven, and there was a complex set of linkages that went from the drive shaft—up through about 5 or 6 linkages—to the cam. In principle, it was a trigonometric problem: you just figured the angles things swept through, and then you could compute the displacement at the cam as a function of degree of angular rotation of the shaft. What we did was called *layout work,* doing engineering drawings of the parts which were 8 times the normal size, cut the parts out as paper dollies, and pin them to the appropriate places in a large drawing of the machine. We then rotated them through the angles, as they would in the actual machine, and moved the other things according to the constraints determined by how they were pinned. This way you got the displacement at the end of the last link as a function of angular rotation of the drive shaft, which you can use to construct the appropriate cam curve which will minimize the acceleration of the ribbon. That would be fine if the shaft rotated at constant velocity. But because of different loads on it at different points in the cycle it didn't. So we took high-speed motion pictures that gave the actual velocity profile of the shaft and then imposed all of that on our paper dolly transformations.

Callebaut: A nice example of model building in engineering before the computer!

Wimsatt: A marvelous example of analogue computation, which I'd like to describe in more detail sometime. When I had this thing just about finished, I thought, "Now this is funny. I've had three years of engineering physics and physics, and I

William Church Wimsatt (b. Ithaca, New York, 1941) is professor of philosophy and evolutionary biology and a former chairman of the Committee on the Conceptual Foundations of Science at the University of Chicago. He studied physics, engineering physics, biology, and philosophy at Cornell University and at the University of Pittsburgh.

If you don't interrupt him and bring up another subject about which he also has a thousand things to say, Wimsatt can talk ad infinitum about the people who have influenced him intellectually. "In no particular order, they are, first of all, Frank Rosenblatt for the examples he gave me of modeling a complex phenomenon using relatively simple (sometimes not-so-simple) models. Then, chronologically, Herb Simon would be next. Campbell is another person who has influenced me strongly. I met him after I came here through a common admiration for George Williams's (1966) work. Dick Lewontin and Dick Levins have influenced me immensely." Of those five people, Simon, Lewontin, and Levins have had the strongest influence. "I came to Chicago—originally I had a job offer here—in the summer of 1969, but I took in preference a postdoc with Dick Lewontin—not only over different job offers, but also over a two-year postdoc at Rockefeller. Those were days in which jobs were more common and postdocs were rare. And Lewontin influenced me a great deal. The year and a half I spent in his lab was certainly intellectually the most exciting time in my life." Wimsatt acquired all sorts of things from Lewontin that took a long time to percolate out. "My first paper on the units of selection controversy owes a great deal to Dick. But I didn't write it until nearly ten years after I'd stopped being a postdoc!"

Richard Levins, Wimsatt thinks, may have influenced him even more deeply in that his own conception of scientific procedures—as being primarily "heuristics all the way down" rather than being modeled on any formal or axiomatic structures—is probably due to Levins more than to anyone else. Levins may have been instrumental in leading him "to appreciate Simon more deeply" (although Simon on the one hand, and Levins and Lewontin on the other, are "on opposite sides of the political fence"). Wimsatt had read some Simon, particularly the latter's paper, "The architecture of complexity" (reprinted in Simon 1981a), the first year he was at Chicago. Then Levins gave a biology seminar on complexity. "I was enormously bemused by the title. I mean, it wasn't 'Mathematical Mod-els in Population Genetics,' it wasn't 'Mathematical Ecology,' it wasn't 'Speciation'—it was 'Complexity'! How can a course like that be given in a biology department? And he wasn't just reading biologists: he was reading stuff all over the place! We read a lot of W. Ross Ashby, for instance. Levins is that kind of guy. He has a very broad intellectual scope, and he is an incredible idea generator." Wimsatt was "fortunate enough" to co-teach a couple of courses with Levins.

With the exception of the monograph, *Why Use Models in Biology?* which is forthcoming (Wimsatt and Schank, n.d.), Wimsatt has not published any books yet. But several of his papers have been truly seminal; in particular, his paper on teleological explanation and functionality, which elaborates a chapter of his dissertation (Wimsatt 1972), some of his papers on reductionism (1976a,b, 1980a, 1981a), the paper on robustness and reliability (1981b), and, last but not least, his "developmental lock model" of generative entrenchment (n.d.a; also Glassmann and Wimsatt 1984, and Schank and Wimsatt 1986), which he is currently expanding into a book. Wimsatt is one of those academics who are writing and rewriting continually but who seem to care rather little about where their output eventually ends up—in his case, typically in one of a number of huge piles which permanently threaten to come down on a visitor's head and which together with more or less empty Pepperidge Farm cookie boxes and dozens of coffee cups, fill up the floor space of what must formerly have been an office room. (I once saw a picture of Jean Piaget's office, which was quite a mess; but compared to Piaget's, Wimsatt's workroom is—well, awesome.) I learned this when I first visited the man—he was my main destination on my first trip to the United States—under the roof of Gates Blake Hall on the University of Chicago campus in the winter of 1985. The door was open but I didn't see anybody. After I had timidly announced my presence ("the philosopher from Belgium who would like to discuss your work on reductionism"), I heard someone muttering (it turned out that Wimsatt was squatting on his heels under a table): "Let's see if I can find some stuff here that you can read tonight and that we can discuss tomorrow"! Only after he had got hold of "the stuff" did he look up to say hello. During the three months I stayed at Chicago, he would give me "some other stuff" of his—invariably unpublished—on an almost daily basis, and this is no overstatement. Most amazingly, somehow he always managed to find what he was looking for. That winter, the notion of "(chaotic) complexity" acquired a new meaning for me.

Our original conversation took place in Bill's office on March 21, 1986. It was much amplified in 1990 and 1991 as he scanned several versions of the book manuscript on his computer, adding comments here and there and replying to additional questions I had asked him or had typed in.

know that they have advanced courses in cam design which are normally taught in the fifth year. Why so late in the curriculum? The problem can't be this simple, can it?" So I went out over the weekend and took out a book in the engineering library. On the first page it said: "The first principle of cam design is: Minimize the jerk derivative." So I said: "What's the jerk derivative?" It turned out to be the derivative of acceleration, the third derivative of displacement. I thought, "Is Newtonian mechanics false? How can this be? This isn't relativistic velocity, nothing like that. It's not a problem in quantum mechanics either, so what's wrong with $F = ma$?" Then I realized that what was false was the implicit Newtonian theory of materials that dealt with rigid bodies or only perfectly elastically deformable bodies, and so on. And it turns out that for real materials, what you want to do is minimize the jerk derivative. It is jerks, that is, changes in the rate of acceleration, that make things break. *This is a principle that turns out to be true across very wide areas.* It is even true for the design of human safety equipment.

Callebaut: Can you give an example?

Wimsatt: You can take a quite heavy force on you as long as you can set yourself for it. But if it *suddenly* changes, your nervous system can't respond fast enough and you jerk; and that causes problems. So in designing car restraint systems they worry about the jerk derivative.

Callebaut: So you had gotten that far through your engineering and you had never heard about the jerk derivative. Why not?

Wimsatt: Well, I'd heard about all sorts of things, including some quite applied things; we had done drafting work, machining, casting, and so on. My feeling as to why this is not taught is that it is was *too applied.* In particular, it is too applied—I subsequently learned—in the sense that it is not derivable from any of the fundamental theories. (You can see why I resonated to Max Dresden's paper!) So they sort of ignore it and teach it, if at all, in a very advanced-level course, even though it is a far more general rule and law in the behavior of materials than any of the ones we were talking about. The solid state physicist who filled in the missing pieces for me said that they don't have a complete explanation in any single case, and if they did, it wouldn't do them much good, because it would be quite different for every different kind of material. So what is this—a "special science" (right in the middle of physics!) complete with functional characterizations, and Levinsian sufficient parameters or supervenience (cf. 4.2.1)? I love it! So it wasn't taught, not really because it was too applied, but because they were too embarrassed that they couldn't derive it, or explain why it worked!

3.2.2. Cybernetics and Computer Simulations

Callebaut: Those two things—the experience with Resler and designing this cam—kind of got to you.

Wimsatt: After a year out working in engineering I came back to school with a new major in philosophy (I had taken two logic courses and a philosophy course earlier). I actually got interested in biology in a course in ancient philosophy taught by Richard Sorabji; I got interested in Aristotle's problem of teleology. My initial response to this was, "This is something a little cybernetics can handle very easily," and I wrote papers on that. But I was sort of hooked; I soon realized it wasn't quite that simple. Even in that first year I decided that teleology in biology really required close attention to evolutionary theory; and I would read some, and by then I was hooked!

Callebaut: You then moved on to the University of Pittsburgh.

Wimsatt: At that time it was an exciting new place, full of heresy. Actually the heresy it was full of would now look like relatively modest deviations from established positivism, but it was a lot more radical than Princeton, which looked at me like I was crazy when I said that I wanted to get a masters in biology while I was there in philosophy. The people at Pitt thought that that was a great idea. People were running up and down the halls and bursting into one another's offices, saying, "Hey! What do you think about this?"—a very exciting place to be. It has since, I think, become rather more conservative; it hasn't moved as far as some other places that have kept moving. The price of success is that you get established and less open to radical change. In any case, somewhere in that period I changed my attitude from worshipping simplicity and formal elegance and having a nice base of set-up things from which you can derive everything to reveling in *complexity* and not only not being bothered by it but positively *loving* it! I came to think that if something was simple enough to be axiomatizable, it probably wasn't too interesting.

Callebaut: You are not implying that this was in the air in Pitt at that time?

Wimsatt: No, no! It was a personal experience, I think deriving partly from these two things and partly from something else: In my senior year at Cornell, I took Frank Rosenblatt's course in "Brain Models and Mechanisms," which was about perception, neurophysiology, early mathematical brain models, and so on. I just marveled at the complexity of this machinery. (Rosenblatt, by the way, was out of favor by 1970, due to an elegant but roundly unfair book on his "perceptrons" by Minsky and Papert [1969].) They have since recanted, and he has become the patron saint of modern parallel distributed processing models; but too late—he died in a sailing accident in the early 1970s.) I'd always worshipped complex machines, especially self-organizing ones, and brains are great for that.

Callebaut: How about others, like airplanes, for instance?

Wimsatt: That's why I wanted to be an engineer: to build complicated machines, particularly aircraft. But on the other hand, before I had had this sort of metaphysical bias toward simplicity and elegance and so on, and had stayed away from biology basically because it was awfully messy. Somewhere in that period my enthusiasms changed. And I think I brought that to Pitt. I don't think I found it there, and I think when I left Pitt it probably left with me! I would say that I was actually very well treated at Pitt, but I was probably listening to a different drummer and didn't internalize a lot of what was going on there. I studied various aberrant logical systems: tense logics, deontic logics, and all of that. I even showed that all of the then known systems of deontic logic would also work as logics of function. I also did a similar number on all of the then known "criteria of intentionality"—that Chisholm (1957) stuff was getting well into epicyclic decadence even then. But my interests didn't really jell there. I didn't even bother to publish it.

Callebaut: Meeting Lewontin has also been very important to you, as it has been for many other people.

Wimsatt: The last year I was at Cornell, Dick Lewontin came and gave a talk, and he was describing some of the simulations he'd done with a computer. This was in 1964, and by that time he and White had done the *Moraba scurra* case (Lewontin and White 1960) that I've used in my research in the units of selection controversy (1980a, 1981a). He was doing a simulation of the evolution of the t-allele in mice, which was, as he first argued in print (Lewontin 1965), a case of group selection. I was impressed by all that. I was even more impressed by the use of the computer to do these simulations. That probably had an early formative influence. I was also enormously impressed with Lewontin, who was a scientist who seemed to be a natural philosopher in the best sense: he raised conceptual problems, he loved discussing things in those terms, and he loved to argue.

Callebaut: How did you get involved with philosophy of biology as practiced by professional philosophers?

Wimsatt: When I went to Pitt, because I was the only person interested in the philosophy of biology, Adolf Grünbaum, after my first year there, asked if I would act as sort of an advisor for the lecture series and help getting people. The first year we got Tom Goudge, who gave a very nice discussion of evolutionary theory, but he was sort of conservative. He reflected one of the biases in positivism that I didn't like. He says in his book (Goudge 1961), "Look, there is a lot going on in evolutionary theory, but I am going to focus on those things that are pretty well settled." It was context of justification versus context of discovery all over again, and I thought that that really wasn't a good idea. It was in fact a lot more fun, at least, to look at what was really going on. And why always be twenty years behind the cutting edge? So the next year, Grünbaum asked me what philosopher we should get to talk about biology, and I said, "Why don't we get a scientist—you get scientists to talk about physics." (It might not have occurred to him that an original biologist could have been interesting to a philosopher.) As to who should do this job, Lewontin was the natural. When he came I acted as his guide for the two days, and talked with him nonstop—about eighteen hours, I think—and at that point I decided I wanted to do a post-doc with him. I didn't tell him until later, when I had talked or conned him into being an outside reader on my dissertation. I was the first of Lewontin's philosophy post-docs, starting in the summer of 1969—the first of many.[6]

---

6. We will discuss several of these people's work in chapters 6 and 9.

3.3 Second Voyager: From Economics to Biology, Back and Forth

Callebaut: Alex, apart from philosophy of biology, you have another major interest, the philosophy of economics. Those two fields would seem to be quite distant from one another. Actually, I personally don't think they *are* so distant, but at first sight they look like completely different things. What got you going, first in the philosophy of economics?

Rosenberg: In fact I was finishing my first book (1976), making it ready for publication, the year that I taught my first philosophy of biology course. As an undergraduate I was a physics major. I made a conceptual mistake which led me into philosophy. I acquired the false view that only teleological theories were explanatory.

Callebaut: Bob Richards (1973) thought something similar as a youth who was still under the spell of Aristotle.

Rosenberg: By the time I figured out what the conceptual mistake was, it was too late to go back into physics. During my last year as a philosophy student I was required by my university regulations to take an economics course, which I did, despite the fact that I had no interest in economics. I walked into the class and discovered that I *loved* it. It was the most wonderful subject; it was intellectually really very interesting. So when I went to graduate school in philosophy of science, I started to audit courses in economics. When I was ready to write a dissertation, I was thinking of writing on Kant. It was pointed out to me by the chairman of my department and by other people that no one had ever written on the philosophy of economics . . .

Callebaut: . . . in the U.S. in that period . . .

Rosenberg: . . . and that if I wrote a dissertation on this subject it would mark me out as unique. And since I was somewhat interested in the subject and had found another dissertation on Kant that pretty much scooped me (that said what I wanted to say), I decided I would write on the philosophy of economics. It was the smartest thing I ever did, because it was very easy to write a dissertation. In fact I finished my doctorate in three years. I wrote a long dissertation on the philosophy of economics. I immediately was able to convert four of the chapters into publications. I got a good job; within two and a half years I had tenure because I had produced so many articles already. And within four years I had sufficiently lost my disgust with the thesis to be able to dust it off and send it to a publisher, and it was published.

Callebaut: After that book was published . . .

Rosenberg: . . . I shifted to other subjects. I became very interested in *causation* . . .

Callebaut: . . . causation as a general philosophical notion . . .

"A lot of people have had a lot of wild ideas."

Alexander Rosenberg (b. Salzburg, 1946) is professor of philosophy at the University of California, Riverside. He studied philosophy at City College, New York, and Johns Hopkins University, where he received his Ph.D. in 1971. Before joining the Riverside faculty in 1986, he held academic positions at Dalhousie University in Canada and Syracuse University, New York. He  was a fellow of the Minnesota Center for Philosophy of Science. In 1974 Rosenberg visited the Institute for American Universities in Aix-en-Provence, France, where he spent time with the epistemologist Gilles-Gaston Granger, now an Honorary Professor at the Collège de France, whose austere rationalism and structuralism influenced him. He received many grants and fellowships, including a Guggenheim, and is a member of the editoral boards of several important philosophical journals.

Rosenberg is a prolific writer whose publications cover matters epistemological—e.g., with Tom L. Beauchamp, *Hume and the Problem of Causation* (1981); the philosophy of biology—*The Structure of Biological Science* (1985); the philosophy of economics—*Microeconomic Laws* (1976), *Economics: Mathematical Politics or Science of Diminishing Returns?* (1992a); and of social science in general—*The Philosophy of Social Science* (1988a). He wants to interfere with and reform the social sciences, "instead of just standing back and respecting their history and trying to learn what lessons they taught us." He was taught "how the bad old positivists made prescriptions that were misplaced, injudicious, and uninformed." But after some years of "toeing the antipositivist, Kuhnian-Feyerabendian line," he realized that "philosophy has a prescriptive role to play with respect to the backward sciences. Somebody's got to tell them what the matter is and what they're doing wrong!"

In a book that made quite a splash, *Sociobiology and the Preemption of Social Science* (1980), Rosenberg inquests why there has been relative lack of progress—in the sense of practical, pragmatical success—in social science. The explanation, he claims, is in "the intentional character of the vocabulary of the social sciences, which can't be expressed in nomological generalizations" (we have independent reason from the philosophy of psychology to believe this: Dennett 1969; Rosenberg 1988a). Rosenberg does not expect the softies—deconstructionists, rhetoricians, and what have you—to listen to his message. "But I have the right to form an opinion, and having formed it, to express it." This is not intellectual arrogance, for on the intellectual market everybody is free: "It is rather a reflec-

tion of having thought about this problem against certain constraints. Other people have come up with different answers, because they've thought about other problems against other constraints. *You pays your money and takes your choice.*" Without intentionality one can't answer the questions the social sciences set out to answer—questions about human action and social institutions. "If we are never going to answer those questions, let us stop wasting money trying to answer them!" Rosenberg agrees that his is a pessimistic view: To many of the really interesting questions in the social sciences, we're never going to get answers, "no matter how hard we try!" The factual matters "are so complicated that we can't provide good general answers—answers that have sufficient predictive content, that would enable us to *ameliorate, or at least prevent the deterioration, of the human condition.*"

In his programmatic book, *La Raison* (1955), Granger proposed to "decontaminate" science to get rid of ideological and irrational motives. In a similar vein, Rosenberg, who has occasionally defended sociobiology against some of its fiercest critics (1987, 1988e), castigates the biologists Levins, Lewontin, and Gould because they have an "ideological axe to grind." He thinks Lewontin and Gould are "a special problem in American philosophy, because they have assimilated philosophical and scientific questions to psychological, ideological, and sociological ones—especially Gould." Gould seems to Rosenberg "incapable of understanding the difference between the causes of a particular belief and the justification of that belief." Thus, in *The Mismeasure of Man* (1981), Gould purported to show that IQ testing "was simply false nonsense because its causal origins were in discreditable racial beliefs and attitudes of eugenicists in the twenties." The "serious irrelevancies about the social origins of various ideas" Gould has introduced in the philosophy of biology distract philosophers from the consideration of the truth or falsity of propositions of their pedigree, Rosenberg maintains. For example, "he thinks it is a matter of some relevance to our assessment of a Darwinian gradualism that Darwin read Malthus and was brought up in a period of increasing ascendency of the British middle classes in the period of the Industrial Revolution." Rosenberg thinks "all that is quite interesting as history of science, but it is irrelevant to the truth or falsity of Darwin's views." Postpositivist "stories" about abduction and the historical origins of scientific views will not deflect Rosenberg from his rightful course: "Again, I am a positivist manqué. I still think there is a distinction between the context of justification and the context of discovery." (Cf. Lewontin's reply in his biographical note at 9.2.4.)

The original interview with Alex was conducted at the Pittsburgh Hilton on 25 October 1986 on the occasion of the joint meetings of the History of Science Society, the Philosophy of Science Association, the Society for the History of Technology, and the Society for the Social Study of Science.

Rosenberg: . . . and I wrote a book on Hume and causation (Beauchamp and Rosenberg 1981). I started work on the philosophy of biology, and I stopped the philosophy of economics. I continued to be interested in the philosophy of social science, but although I published these papers and published the book, the subject was dead; there was nobody else doing any philosophy of economics. So I dropped it. I already had a job; I had published enough in this area to get tenure. I wanted to work in other areas of interest to me: metaphysics, biology . . .

Callebaut: Just an interpolating question: How would you explain such lack of interest? Isn't it the case that economists, on the whole, tend to generate their own philosophy "from within?"

Rosenberg: Well, in Europe and in England there has long been a tradition of methodology in economics, and Sir Karl Popper has been involved in it. But in America there was very little interest in economic methodology at that time, in the late sixties and early seventies. Even among economists there was very little written about methodology. There was Friedman's *Essays in Positive Economics* (1953) (cf. Rosenberg 1972), and people thought that was enough; and that was the end of it!

Callebaut: There were some methodological discussions in the *American Economic Review,* involving people like Machlup (1964), Samuelson (1963, 1964, 1965), and Simon (1963). Some of these people were discussing—or even *defending*— the crudest empiricism. They didn't seem to be aware of what had been going on in philosophy of science at large in the last couple of decades. When I wrote about their debate in my own dissertation, I was criticized by one of my readers for devoting too much space to it . . .

Rosenberg: . . . It was very primitive indeed. The only decent philosophical contribution was a paper Ernest Nagel gave to the American Economic Association (1963), and almost nothing else. There wasn't a single book on the subject published in this country from 1938 until my book was published. In 1938, Hutchison's book was published in this country, and that's all. Everyone was satisfied with Friedman. Indeed my first paper was a critique of Friedman's views. There was literally nothing for me to cite, in writing my dissertation, later than . . . Oh, there was Papandreou's very silly book. Andreas Papandreou, before he became Prime Minister—he is the son of a former Prime Minister[7]—was a Professor of Economics at Stanford, at Penn, and at Rochester, I think. And under the influence of—I don't know who, maybe it was Friedman—he wrote a book called *Economics as a Science,* published in 1958, about hundred pages stuffed with set theory. There has been no serious mention of that book anywhere in intellectual circles since he wrote it, except for ten pages in my book, in which I show what nonsense it was.

7. Andreas Papandreou was the Prime Minister of Greece from 1981 to 1989.

Callebaut: It was basically on the concept of a model, and on the difference between economics as theories and as models.

Rosenberg: That's right. Well, I can hardly now reconstruct what it was. It was a very quick and partial interpretationist approach to economics: theoretical predicates are defined implicitly by their relations to observational statements. The only other person who ever cited it is Blaug in his *The Methodology of Economics* (1980). He cites it in one sentence and the next sentence says that Rosenberg has shown this to be nonsense.

Callebaut: I had to read Papandreou's book when I was studying philosophy.

Rosenberg: Really?

Callebaut: Yes, as about the only thing on economics, available at the time, that was written from a formalist or logical perspective.

Rosenberg: It was worthless. So while I was interested in economics, interested in the subject, there was no feedback in the seventies. So I stopped. But then something happened: Dan Hausman wrote his *Capital, Prices and Profit* (1981), and Hollis and Nell published their *Rational Economic Man* (1975). Hollis is a well-known English philosopher of science and Nell a well-known Marxian or classical economist. Together they wrote a book in which they argued that neoclassical economics rested on positivism; positivism was a species of empiricism; rationalism was correct; therefore neoclassical economics was false, and Marxian, classical, Ricardian economics could be deduced as synthetic a priori truths. That's hardly a view even worth taking seriously (cf. Rosenberg 1978a)!

Callebaut: Well, playing the devil's advocate I could say that Ludwig von Mises's (1949) praxeology rests on the same neo-Kantian basis . . .

Rosenberg: . . . Oh yes: von Mises thought that the truth of Austrian economics was synthetic a priori. *A lot of people have had a lot of wild ideas, and that is one of them!* On the other hand you can understand why he would hold that view. Because the propositions of, say, rational choice theory were clearly not analytic. Yet no one ever hunted up evidence to test them, and everyone thought they were evident in introspection. That's what Lionel Robbins said already in 1935, and I think to some extent that view is correct.

Callebaut: Anyway, then it snowballed.

Rosenberg: Suddenly I discovered I was the grand old man—at the age of thirty-five—of a whole subject! The other thing that happened, of course, was that normative ethics and political philosophy suddenly came back with a bang with the publication of Rawls's *A Theory of Justice* (1971). Rawls's book had a fair amount of economics in it, and Nozick's (1974) had a certain amount of economics in it. Suddenly the demand for people in philosophy who had some training in microeconomics increased tremendously. The education of philosophers in economics increased tremendously. At that point it started to become interesting to do work in this field again.

Callebaut: What got you into your other major interest, the philosophy of biology?

Rosenberg: Partly it was accidental. My wife is a biologist, and I long wanted—and failed—to convince her of the importance and interest of the philosophy of science to biology. In 1975 I went to the University of Minnesota, to the Center for the Philosophy of Science, to teach. Just at that time David Hull and Michael Ruse had published their books on philosophy of biology (Ruse 1973; Hull 1974). I thought, "Ah! This is a perfect opportunity to teach this subject," which I had never taught before; I hadn't even studied biology since high school and looked down my nose at biology, because I was interested in physical science. But at least I could bring my philosophy of science into contact with my wife's specialty, and show her the importance of the philosophy of science. Well, I failed miserably to convince my wife, but in the process I discovered this new and very interesting subdiscipline, the philosophy of biology, and I saw obvious mistakes in Hull's book and in Ruse's book that I felt I had to correct . . .

Callebaut: . . . We'll return to those later (in 6.1) . . .

Rosenberg: . . . So I started writing papers. And very soon thereafter, I found myself part of the relatively small community, at that time, of people writing in the philosophy of biology.

Callebaut: May I interject an anecdote here? I once attended a paper Herbert Simon gave at the Piaget Archives in Geneva. This was to an audience that consisted mainly of psychologists. The Department of Economics had also invited him to lecture, as he was there. Now this was a Jekyll-and-Hyde kind of experience, at least to me. I found it rather schizophrenic: here were really two guys talking to two different audiences, using completely different vocabularies, rhetorics, and styles of reasoning! I wondered how this man managed to do this. So I'm asking you, who are involved in these two rather different enterprises: do you also have to change clothes—or at least hats—to walk from one room to the other?

Rosenberg: No, no! I find, for one thing, that my work in philosophy of biology has had an influence on my work in philosophy of economics, and that, of course, of all the disciplines, the mathematics of economics—the quantitative aspect of economics—has had the most profound effect on population biology and on ecology. And it has helped me in no end in understanding biology that I did as much economics as I did as a student and subsequently.

Callebaut: In many of your papers, you have argued that the explanatory style of evolutionary theory and the explanatory style of neoclassical economics are almost identical.

Rosenberg: Now I don't think that this fact does any credit to economics. In fact it is a large part of the explanation for some of the vacuity of economics and the difficulties associated with testing it. Economists, for example McCloskey, think that the fact that economics is like evolutionary biology insulates it from criticism. Little do they know that it simply reflects the fact that economic theory is subject to even more serious criticism than evolutionary theory! The criticism made of evolutionary theory is often mistaken because it is based on too narrow a view of what evolutionary theory is; whereas *there is nothing more to economics than that part of evolutionary theory which is subjected to criticism.*

Callebaut: So generally you don't find that you have to be schizophrenic about this?

Rosenberg: I think that the groups, the circles within which I work in economics and philosophy, never overlapped at all when I first worked in this area. Now there is a slight overlap, partly, I think, because people in each of these areas have begun to read work in the other area. Also biologists have seen the importance of General Equilibrium Theory (e.g., Weintraub 1985) for ecology, and economists have seen the attractiveness of evolutionary styles of thinking in economics. So that has brought some of these two circles closely together. But generally I found no difficulty adopting quite a consistent voice and a consistent philosophy about both of these subjects.

3.3.1. *The Biological Undercurrent in Economics*

Callebaut: There has always been a biological, or rather, *organicist* undercurrent in economics; first in classical economics, and then in the work of people such as Alfred Marshall, who had these biological metaphors but did not really get them working (Schlanger 1971; Mirowski 1988a, b). It has been suggested that was because Marshall lacked the mathematical machinery to do so at the time (Boulding 1981:86–87). And again, Herbert Simon, in his Nobel Prize Lecture (1979b), also said something to the effect that economists should look at biology.

Rosenberg: There were two things he said about that, I think. Most of the people who have tried to apply biology have either done it the way that Nelson and Winter (1982) have done it (cf. R. R. Nelson 1987, Dosi et al. 1988), which has produced nothing very interesting—to economists at any rate. Or else they have applied biology as a rationalization for the methodology of their subject in the way that Alchian (1950) and Friedman (1953) did it (cf. Rosenberg 1992c). Or they have tried to do some kind of biologization of economic choice, and that has certainly not been a success either.

Callebaut: Who are these people?[8]

Rosenberg: Offhand I can't even think of a name. It was quite long ago. But I think that only to this extent is there a real import of biology, and that is that *any kind of innovation, whether it is an economic innovation or any kind of innovation in decision or in human action, involves variation and selection.*[9] I think that is what

8. As Herrnstein (1988:36) points out, economists frequently invoke perceptual or cognitive limitations to explain failures to maximize; they usually do not take limitations in performance to have any deeper meaning. Yet "an organism's perceptual and cognitive capacities are likely, on evolutionary grounds, to be adapted to the processes that guide its behavior, just as its digestive system is adapted to what it eats or its locomotor system to where it lives." Organisms "seem to be perceptually and cognitively adapted for matching rather than maximizing."

9. This is the central theme of evolutionary epistemology and, more specifically, of Campbell's general selection theory (see chapter 7).

Simon is talking about when he says that economists should pay attention to biology: they should understand that most innovations are the result of blind variation—not foresighted variation—from which we then make selections (cf. Simon 1983). And to that extent, biology or evolutionary thinking is going to be unavoidable in any account of human behavior. Because if human behavior is innovation, it is always doing something new, doing something for the first time. But at least in the short run, I don't see any special appeal of biology for the social sciences.

Callebaut: Georgescu-Roegen—Paul Samuelson once called him "an economist's economist" . . .

Rosenberg: . . . but his interest has always been thermodynamics, physical processes and economics, and that's much different, and also harder to understand (Georgescu-Roegen 1971; cf. Mirowski 1989).

Callebaut: Agreed, but today there are a number of people claiming . . .

Rosenberg: . . . that biology is thermodynamics (Brooks and Wiley 1988; Weber, Depew, and Smith 1988); but we haven't enough time to go into that (cf. Rosenberg 1988d)!

3.3.2. *No Consolation for the Lakatosian Economist*

Callebaut: Going back to the philosophy of economics, it is striking that the kind of methodology that is usually referred to by many of the people in this field (e.g., Latsis 1976; Hausman 1984) is Lakatosian. Lakatos's (1970, 1978a) "Methodology of Scientific Research Programmes" (MSRP) emphasizes the importance of *novel predictions,* which fits in nicely with the modal economist's disregard or straightforward disdain for model realism. I am referring, of course, to the so-called "F-twist" (Friedman 1953), which can be profitably contrasted with Simon's (1963) criticism from a radical empiricist perspective.

Rosenberg: The attractiveness of the MSRP is that it is a very convenient and pat solution to all the problems in the philosophy of economics. You make them go away by waving the ritual of "protective belt," "positive heuristic" and "hard core." You pack all the problematic claims into the hard core, where they can't be examined or assessed. Then you make broad claims about the "progressiveness" of the whole research program which the philosopher can't challenge, because he is a philosopher and not an economist and he can't get up in public and say, "Well, you economists are wrong about the progressiveness of your own discipline," because then they will get up and say, "What do you mean we're wrong, look at the journal articles!" and stuff like that. So it is a particularly convenient and attractive view.

Callebaut: What can one do about it?

Rosenberg: The first thing that we philosophers have to do is show that economists who employ it can get no real consolation from it, because *to show that something is a research program of the Lakatosian sort is just to show nothing about it,* since literary criticism satisfies the strictures of the MSRP. And then we turn our attention back to the good old questions: Is there really empirical progress? Is there really improvement in predictive strength?

Callebaut: What is your next big project in the philosophy of economics? What will your new book be on?

Rosenberg: What I want to do is come to a settled view about certain problems in the philosophy of economics on which I have been changing my mind a lot over the last ten or fifteen years, and also attempt to assimilate and systematize some of the literature that has been produced—the really good stuff by Hausman, Weintraub, McCloskey, and other people writing in the same area like de Marchi, and Doug Hands. But I'm not sure that I will come out with some nice and neat conclusion in the way that I have come out with some nice and neat conclusions about biology or the social sciences in general, or in my first economics book. So I can't very clearly say at this point how this is going to turn out. (See Rosenberg 1992a.)

Callebaut: Could you say something about your recent work on the status of economic theorizing?

Rosenberg: I wrote a paper, "If economics isn't science, what is it?" (1983a), which has created a firestorm or at least provided a convenient object of attack for economists. In that paper I argue that economics is a species of applied mathematics, that its status is rather to be assimilated to that of Euclidean geometry instead of empirical science—Euclidean geometry in the period when we didn't recognize that it was subject to an interpretation as a physical applied geometry and as a pure geometry, and it sort of straddled in the middle: synthetic a priori truths about space.

Callebaut: A view that could have pleased economists of a previous century.

Rosenberg: That's right, yes. Whereas if we view economics in the way we now view Euclidean geometry, we have to treat it as a body of empirical propositions that is plainly false and of very little predictive value, or a body of applied mathematics to which economists, especially those doing General Equilibrium Analysis, devote huge amounts of genius, which, however, doesn't get paid off, at least in any kind of immediate predictive applications, but which interestingly enough does get paid off in biology, in mathematical ecology, where the conditions for equilibrium, stability, and uniqueness of an equilibrium population, for example, are identical to the equilibrium conditions that the economists produced for market clearing.

Callebaut: That paper has brought you a fair amount of attack . . .

Rosenberg: . . . including a whole book written by Roy Weintraub (1985)—quite an excellent book, which, aside from attacking me, also advances our understanding of the history of General Equilibrium Theory very greatly. And no matter whether I ultimately decide that the view I argued in that paper is right or not, I think I did a good thing in producing that paper, because it really brought a lot of issues forward that hitherto had not been holding the attention of economists.

Callebaut: How does a positivist manqué assess the fashionable "rhetorical" approach to economics (McCloskey 1983, 1986; Klamer 1984; see also the contributions and discussions involving McCloskey, Mäki, Mirowski, Rappaport, and others in vol. 4 [1988] and subsequent volumes of the journal *Economics and Philosophy;* cf. also 5.6.2)?

Rosenberg: Those people, interestingly enough, are defending neoclassical and mathematical economics by appeal to the tools of the tender-minded philosophers: the hermeneuticists, the deconstructionists, the postmodernists. It is the mathematical economists that are kind of being the antipositivists, the antiformalists . . .

Callebaut: . . . on the metalevel.

Rosenberg: I don't understand why.

Callebaut: Do you think they have much of a future?

Rosenberg: I hope not! Maybe I should hope they do, because the longer they continue to be taken seriously, the more scope there is for me! (Cf. Rosenberg 1988b, c).

3.4 THIRD VOYAGER: FROM MATHEMATICS TO BIOLOGY

3.4.1. *No Consolation for the Lakatosian Mathematician, Either*

Callebaut: As Alex just argued, from economics to mathematics is but a small step. Philip, you diverged quite early over fundamental epistemological problems with Lakatos, a philosopher to be reckoned with when you were a student.

Kitcher: In Lakatos's long paper on MSRP (1970), for instance, you see somebody who starts out with a grip on the Popperian view that the fundamental epistemological problem is the *demarcation* problem . . .

Callebaut: . . . What distinguishes scientific knowledge from nonscientific knowledge (such as common sense or metaphysics)?

Kitcher: He is extremely ingenious and extremely clever and extremely sensitive. Having put himself into this straightjacket, he then goes through the most amazing contortions to do justice to insights and ideas that normally used to get neglected from a Popperian point of view.[10] So I think there is this fundamental mistake in

---

10. Campbell provides some biographical background that may help explain this. He gave a lecture at the London School of Economics on June 10, 1969. He recalls: "It was probably the last time that Popper was in that hall, because it was the last seminar in the series that he chaired before he retired. Watkins inherited Popper's professorship, I think, and Popper had enough power to create a second professorship for Lakatos. So they were both full professors. They had been squirming under Popper's tyranny. In spite of what he says about criticism being the secret of good science, anybody who criticized him had merely failed to understand him or was being willful. So it was a real tyranny working with him. When Lakatos wanted to tell me what he thought, he would speak in a whisper and look around to see if he was not being overheard. The Lakatos and Musgrave (1970) book in which Kuhn has a chapter at the beginning, another at the end, and wins the debate, temporarily, while Lakatos in his chapter has all this Popper₁, Popper₂, Popper₃—just suppressing it. All of that criticism on the part of both Watkins and Lakatos led to a real quarrel. There were such angry relations that Popper didn't come into the university any more. The uni-

Popperian epistemology as I see it (which is, of course, the inverse of what Popper would see as his great insight). It can be put in one of two ways: either that *the fundamental problem is the demarcation problem,* or that *there is no such thing as a positive justification.* Either way of putting it seems to me a basic error. But both Popper and Lakatos, working within this framework, have developed ideas and insights that everybody has to come to terms with. It is just this root central difficulty!

Callebaut: Do you see any deep relations between the set of questions one faces today in the philosophy of mathematics on the one hand, and in the philosophy of biology on the other? Or are there just two parts of you with little or no contact between them?

Kitcher: No, I don't think they are unrelated, because it was through my study of the history of mathematics that I first was led to formulate this framework within which I have been trying to come to terms with scientific change. It seemed to me, as I was working more and more on the problems which strike me as most acute in the philosophy of mathematics—"How does mathematical knowledge grow?" "What is progress in mathematics?" "What is rationality in mathematics?"—that I needed a general framework to represent the state of a mathematical field at a time. And as I began thinking about this, it came to me that *the existing ways of describing mathematics at the time were radically impoverished.* I decided than one needed a much more articulated notion of what the state of a science like mathematics is at a time.

Callebaut: And you have applied the same kinds of ideas within other areas of science later?

Kitcher: I've taken a look in particular at biology from this perspective. So I actually see all the work I've been doing as having a certain unity, and I hope that this becomes apparent in the book in which I try to give a general account of progress and rationality, the search for explanation, and so on (Kitcher and Salmon 1989).

3.4.2. *Explanation as Unification*

Callebaut: Did your work in the philosophy of mathematics also influence your work in the philosophy of biology with respect to the issue of unification? You have written quite a few things on unification (1976, 1981, 1985b, 1989).

Kitcher: Oh yes, this is a very important theme in my thinking about explanation, about the structure of scientific theories, and about scientific rationality.

Callebaut: What I mean is this: In mathematics, all major unification programs seem to have failed—and that is probably a mild way of putting it (see, e.g., Putnam 1979; Granger 1980; Benacerraf and Putnam 1983). Wouldn't mathematics pro-

versity continued to provide him with a research assistant who would go out and bring books two or three times a week. Popper and his wife Henny were out there in the little town of Penn, writing like mad with a research assistant. But I don't think that he went back to that seminar room after I gave my talk there."

vide us, then, in a sense, with an important counterexample to the logical-empiricist view on unification as the reduction of one theory to another (cf. 4.2)?

Kitcher: There may be a link there. I have to think about that. I think what you say is interesting and I haven't actually thought, previously, from this perspective. One of the things that I thought I had to do in coming to terms with the problems that interested me in the philosophy of mathematics was to reformulate the way in which I had been looking at traditional mathematical bits of theory. I had to introduce this multidimensional notion of a "mathematical practice."[11] In a similar fashion, I felt that it was necessary to reformulate the issues about reductionism in biology. Even though I've made a cornerstone of my views about explanation and scientific inference the idea that we try to find a *unified picture* of the way the world is working in a particular domain, and even though this is something that enters in both my account of mathematics and my account of science, in both cases what I have tried to do is break with the traditional ways of explicating this. So I would regard myself as not in any sense hanging on to the old unity of science view, or a kind of unity of mathematics view that I think is often found in impossible foundational programs. I am trying, I guess, to repose the issues in a way that will enable me to see what was motivating people who had been attracted by those views, and to try to capture insights that they had as well as insights of their critics.

Callebaut: You have told me that Quine was one of the major influences on your thinking.

Kitcher: Yes, but I should explain that. I can date my conversion from history of science to philosophy of science to the day, sometime in 1968–69, when I read the final section of Quine's "Two dogmas of empiricism" (1951). I thought that was one of the most liberating things that I ever read. I still think it is an absolutely brilliant piece of philosophical writing. It seems to me profound and challenging; and it led me to think about problems in the philosophy of science in entirely new ways.

Callebaut: You got your Ph.D. in 1974. The book based on it, *The Nature of Mathematical Knowledge,* appeared much later (1983). Why was that?

Kitcher: The book on philosophy of mathematics took me about ten years to write. It went through innumerable false starts, and I didn't know where I was going. I was immensely helped during the time I spent at Michigan, especially by talking with Alvin Goldman and Jaegwon Kim, who managed to get me straight on some of the things that I needed for the beginning. But it still took a very long time to write, and it was built up very slowly. My book on creationism, *Abusing Science* (1982), was written *after* my book on the philosophy of mathematics. The work on creationism that I did was done much more quickly!

---

11. "I suggest . . . that we view a mathematical practice as consisting of five components: a language, a set of accepted statements, a set of accepted reasonings, a set of questions selected as important, and a set of metamathematical views (including standards for proof and definition and claims about the scope and structure of mathematics)" (Kitcher 1983:163).

"We can have both theoretical ecology and natural histories, lovingly done."

Philip Stuart Kitcher (b. London, 1947), who is professor of philosophy at the University of California, San Diego, began life as a mathematician. But, feeling that he would never be a particularly creative mathematician, he soon entered the History and Philosophy of Science Programme at Cambridge, where he spent most his time studying the history of the philosophy of science with

Lisa Lloyd

Gerd Buchdahl. He found himself increasingly interested in philosophy of science, reading Popper, Quine, and Putnam. Kitcher obtained his B.A. from Christ's College in 1969, with first class honors in mathematics and in history and philosophy of science.

He then went to the Department of Philosophy at Princeton "with almost no real philosophical training." Princeton's highbrow atmosphere came as a shock, but he "somehow managed to survive it." By then his interests were mainly in the philosophy of mathematics and the philosophy of science. Two topics in particular, scientific change and scientific explanation, have continued to interest him ever since (see in particular Kitcher and Salmon 1989). It struck him that with the exception of the work of Lakatos and especially the latter's *Proofs and Refutations* (1976),[1] philosophy of mathematics had made very little or no contact with the actual practice of mathematicians, and Kitcher felt it was time that one try to remedy this. In his thesis, Mathematics and Certainty, which he defended in 1974, he looked at the philosophy of mathematics more from the perspective of a philosopher of science than from that of a logician. Lakatos had started to do this, but in a way that Kitcher—who by now had turned his back on Popper—found no longer congenial. His book, *The Nature of Mathematical Knowledge* (1983), further elaborates on this line of work (see also Aspray and Kitcher's *Essays on the History and Philosophy of Modern Mathematics,* 1988). (Kitcher is the co-winner, with the philosopher of physics Michael Friedman, of the 1986 Imre Lakatos Award.)

After leaving Princeton, Kitcher taught at Vassar College and then at the University of Vermont. He continued to work on history and philosophy of mathematics for a number of years, getting very interested in general topics in the philosophy of science. One of the most significant events that triggered his interest in the philosophy of biology was a semester at the University of Michigan in 1979, where he spent a great deal of time talking with Allen Gib-

bard, who had been a colleague of Bill Wimsatt in Chicago. From Vermont Kitcher moved to the University of Minnesota in 1983, where he was the Director of the Minnesota Center for the Philosophy of Science from 1984 to 1986. Intellectually, the "really big break" came when he spent a year (1981–82) at the Museum of Comparative Zoology at Harvard. "Steve Gould and Dick Lewontin just taught me an immense amount. And that was a very large influence on my subsequent work."

In addition to his book on the philosophy of mathematics and eighty or so articles and reviews, Kitcher, who enjoys a good scientific fight, has written two polemical books. *Abusing Science: The Case Against Creationism* (1982) expressed his concern with the rising influence of Jerry Falwell's Moral Majority on education and the general intellectual climate in the U.S. at the time. Kitcher conceived his "manual of intellectual self-defense" as a chase: "The Creationist is allowed to choose one battleground after another. Given each choice of battleground, I insist that the battle be fought on that ground. In every case, 'scientific' Creationism is defeated." In *Vaulting Ambition* (1985a), his prime target was the kind of sociobiology which seeks as its end product claims about human nature (e.g., Alexander 1979, 1987; Barash 1977, 1979; Wilson 1978). He remains puzzled by what he perceives as Wilson's cavalier treatment of this most serious of topics: "In his ant work Wilson is so careful, so clearly deeply devoted to understanding these organisms in a way that many really great biologists become devoted to their organisms. You can see this also in some of the discussions of other organisms in *Sociobiology.* But he becomes more careless when he comes close to our species." Kitcher got quite some flak for his tag "pop sociobiology" and for the "negative" or even "hostile" tone of his book. Sociobiology advocate Michael Ruse, for one, does not mince his words: "*Vaulting Ambition* it is a very clever book—in many respects, far more brilliant than mine (Ruse 1979b)—but Philip is fighting dragons which are already dead or would die of their own accord. The spirit of what I'm trying to do is positive, the spirit of what he is trying to do is negative." Kitcher defends his stance by pointing out that much of the problematical writings he was attacking were *designed* to be popular. As far as his "negative thinking" is concerned, he admits that critics of human sociobiology need a new image: "Intense as the joys of strangling infant sciences may be, they do eventually fade, leaving a desire to nurture some more healthy enterprise" (Kitcher 1987c:63). Some of his recent work (1988b, 1990a) exemplies the positive work he now has in mind.

A first interview was taped on March 28, 1986 in Philip's office at the Minnesota Center; it was our first serious conversation. (The sight he later offered me of the cheerful and scarcely dressed crowd skating on the Minneapolis lakes on that first lukewarm afternoon of the year is something to remember.) An update was recorded on June 1, 1990 in his UCSD office.

---

1. Originally published as a series of papers under the same name in the *British Journal for the Philosophy of Science* in 1963–64.

3.4.3. *From Existence to Constructability*

Callebaut: In your mathematics book (Kitcher 1983), you have torn to pieces apriorist epistemology.

Kitcher: I claim that all the classical programs that dominated the history of the philosophy of mathematics have been infected with a (usually unarticulated) apriorist epistemology. The first task in the book is to make explicit the demands of aprioristic epistemology, and to show that they are in fact unsatisfiable.

Callebaut: That's the critical section.

Kitcher: After that I am in a position to formulate in outline a new approach. According to my approach, mathematical knowledge does not get fixed by some kind of process which gives you a priori knowledge, foundational principles from which you then deduce everything else. Rather *it grows up generation by generation*. It gets passed down from the ancestral community to the offspring community, and gets extended. The task of the philosophy of mathematics is to understand this process: how it got started and how it goes on. The second section of the book takes up an issue which seems crucial to resolve if one is going to have an account of how this thing could have gotten going . . .

Callebaut: . . . What is it one is talking about in mathematics?

Kitcher: There is this famous quip of Russell's that "mathematics is this subject in which we never know what we're talking about or what we're saying is true." The mathematicians' response to that is: "Mathematicians don't care either!" But it seemed to me that I had to have some account of what mathematics is all about. I attempt to develop a position which I regard as *pragmatic constructivism,* where mathematics is conceived of, *not* as being about any special kind of objects, but *about kinds of ways in which people can interact with and operate on their world,* either physically or by structuring it mentally.

Callebaut: This sounds to me very much like Chihara's (1990) approach, in which all existence assertions are replaced by assertions of the constructability of certain entities. In the last part of your book you provide an account of the growth of mathematical knowledge.

Kitcher: I try to show how to characterize the state of mathematics at a given time and how to characterize the transitions through which mathematical knowledge grows. All this is illustrated through a study of the calculus from the seventeenth century to the work of Cantor and Dedekind in the late nineteenth century. For me, this is the most important part of the book and it points toward the project in which I am now engaged, namely that of giving a naturalistic account of the growth of scientific knowledge in general.

Callebaut: I would like to relate back Kitcher's remarks on constructivism to the general issue of *scientific change*. I was once struck by a remark of Ron Laymon's. He had been reading on the history of intuitionism, and he thought he saw a number of deep parallels between intuitionism and a more gradualistic view of scientific change than the one that Kuhnians have been propounding.[12] Dudley, would you agree that an emphasis on *constructibility* as a guiding principle makes a difference in this discussion?

Shapere: Yes, I can see how one could think that. Constructivism could seem to be parallel to the development of science in the sense that we develop, as we go along, newer and newer assumptions, presuppositions, and those are always subject to the development of further and further presuppositions. But I see that as not a very deep view. *Mathematics is fundamentally nonconstructive,* I think. The theory of transfinite numbers has become so integrated into physics, into our view of the universe, that nobody will drive us from that paradise, as Hilbert said of it, in reference to mathematics.

Callebaut: But the reason for going beyond accepted theory in mathematics . . .

Shapere: . . . is not constructive. The reason for going beyond current theories in pure mathematics is in trivial cases the denial of an axiom and its replacement by another one; in deeper cases, in finding proofs concerning problems which may be, and usually *are,* nonconstructive; for instance, the proof that four-dimensional spaces have alternative structures (Donaldson 1987). In these proofs one changes the structure of mathematics by finding deep relationships between different areas, such that this recasting of relations effectively creates a new branch of mathematics; new deep insights into structures that change the ideas of what those structures are, effectively. And then the question is whether those structures will have physical application. But that is another question. Again, on my view, we just don't know, because we don't know the way our physical theories are going to go.

Callebaut: A reaction, Philip?

Kitcher: My own version of constructivism is thoroughly pragmatic. Of course I don't believe that there are a priori limitations on the idealizations of our constructive powers that we should attribute in our mathematical stories. Indeed, I want to retain classical and contemporary mathematics, regarding them as idealized stories about the possibilities of reorganizing the world that have grown far beyond the simple types of construction. All this is spelled out in my book and in subsequent articles (1988a, d).

3.5 MAJOR FEATURES OF A NATURALIZED PHILOSOPHY OF SCIENCE

3.5.1. *The Return of Psychologism*

Callebaut: I now would like to attempt to identify some general features that are common, maybe not to all but to *most* of the new philosophical approaches to the sciences. A first feature that is (re)gaining prominence is *psychologism.* Psycho-

---

12. To avoid misunderstanding here I should add that Kuhn himself (e.g., 1977) has exempted the history of mathematics from the subjects to which he thinks his theory can be applied; cf. Hirsch (1985).

logism, as a view about the nature of logic and reasoning, was very influential until the late nineteenth century. It is minimally a naturalism claiming that the processes by which we *ought* to arrive at our beliefs are (or ought to be) informed by the processes by which we *do* arrive at our beliefs (cf. Kornblith 1985b).

Sober: Boole, for example—the British logician—described logic as giving the laws of thought. And Kant, of course, was a psychologistic philosopher of the first rank also.

Callebaut: Then it was eclipsed.

Sober: Frege reacted against "psychologism" in what many philosophers viewed as a totally devastating and conclusive way, arguing that questions of psychology have nothing whatever to do with whether logical propositions or principles are correct; and that the two subjects have nothing to do with each other—that logic, on the one hand, is objective, whereas psychology concerns what is essentially subjective. The positivists inherited this antipsychologistic viewpoint, and made a distinction—which is still quite influential—between the context of discovery and the context of justification. The context of discovery would involve questions about the psychological influences that lead a scientist or a creative thinker to come up with an idea (*"how you get there"*). The context of justification would involve questions of logic and methodology having to do with the justifiability and defensibility of that idea (*"once you are there, how do you evaluate the product of this free creation of the imagination?"*). The positivists took the view that there is nothing of any philosophical interest about the context of discovery, that this is a matter for psychology. On the other hand, as philosophers and logicians they could ask questions about the nature—the *logic*—of justification.

Callebaut: One effect of this long dominant position (through Husserl, antipsychologism influenced the phenomenological movement as well) has been the idea that philosophical inquiry can learn nothing whatsoever from psychological inquiry into the nature of rationality, and vice versa.[13] But now, mainly under the influence of Quine in epistemology, psychologism is making a *retour de force*. You resuscitated psychologism in your paper by the same name (1978).

Sober: The idea that I tried to develop in that paper is that Frege and the positivists provided no reason whatever for thinking that there can be no transfer of this sort. I argued that in case after case, one can understand cognitive psychologists like Piaget and especially psychologists who are interested in computer models of thinking, as trying to make precisely this sort of transfer—taking models of rationality, and trying to think of them not just as describing what we *ought* to do, but as describing what human beings *in fact* do. One of the things that is happening

---

13. Philosophers might state normative principles about how we ought to reason if we wish to avoid fallacies. But as a part of the heritage from Frege and the positivists, a very general reaction was that this has nothing to do with psychology; that psychologists cannot in any way exploit these normative theories in philosophy as part of their description of how the human mind works.

in the psychological literature now is a very fruitful interaction between these two points of view (e.g., Kornblith 1985a; Thagard 1988; Goldman 1992).

3.5.2. *How to Get Beyond the Purely Descriptive*

Callebaut: From psychologism to the sociologism that many students of science today embrace overtly (see below) is no big leap (for us, who are looking back, that is). In your view, David, are there means for the naturalistic philosopher of science to go beyond the purely descriptive?

Hull: The impression that naturalistic philosophy of science is purely descriptive stems in part from an ambiguity in the term "descriptive." In the broadest sense, "descriptive" is contrasted with "normative"—what *is* the case with what *should* be the case. Perhaps scientists working within the same research program argue over priority, but according to Lakatos (1971), they *should* not. The "should" in the preceding claim does not concern morality but rationality (cf. Lycan 1985). But within *this* sense of "descriptive," a distinction must be made between those claims that are "merely" descriptive and those that are *nomic* or lawlike. The contrast is between what *is* the case and what *must* be the case. Some people are highly skeptical of the notion of natural laws . . .

Callebaut: . . . Ron Giere, for instance (cf. 5.2.2) . . .

Hull: . . . but I don't see how we can understand science without distinguishing between statements such as "The top of this table is elliptical" and "Planets revolve around stars in ellipses." The former is merely contingent, while the latter refers to a basic feature of the universe. If the study of science is to take a naturalistic turn, then we have to at least entertain the idea of looking for characteristics of science that are nomically necessary, akin to laws of nature.[14] At the very least, we have to commit ourselves to some sort of instrumental recommendations like "If you want to succeed in doing *X*, then you best do *Y*." Such claims are not "merely" descriptive.

Callebaut: Helen, once we've given up analyticity,[15] it is difficult to see where the nomic character of laws could or should come from. Have we got anything left besides the contingent—what can but need not be?

Longino: Herakleitos—I'm certainly more partial to that view, I think, than to the idea that there are laws of nature. Although I suspect that there really is some third

---

14. The realist epistemologist Milton Fisk (1973:72) calls "ontological essentialism" the view that "natures are the grounds of the necessity of necessities" and that "every individual is subject to necessities and hence has a nature." On Fisk's construal, ontological essentialism is neither infallibilist nor apriorist ("whether an entity has a character contingently or by nature, justifying the claim that the entity has the character can be accomplished by an empirical procedure"). Hence, Fisk maintains, it is immune to Popper's (1976) objections to essentialism ("methodological essentialism," in Fisk's terminology).

15. I.e., Quine's (1951) "first dogma of empiricism." In this century, the class of "analytical truths" has been defined as the class of those truths which are necessary for reasons of logic in a wide sense (Swinburne 1984). Cf. 3.5.3.

perspective; it depends on what we mean by "everything is contingency." If we mean that there is no order in nature, I think that is wrong. *Certainly there is order in nature.* But I don't think we know what it is. We require a certain amount of order in order to be able to function; we *postulate* a certain order. And to the extent that there is a match between our postulations and what actually occurs, we have grounds for supposing that there is *some* degree of order in the natural world. But that match does not really entitle us to say that we've discovered "laws of nature" or that there *are* laws of nature. What the order consists of is, I think, beyond us. That strikes me as one version of a Kantian distinction between "things in themselves" about which we don't know and Nature which we construct at the intersection of our own cognitive grids or needs and experience. But that's a construction that involves both Nature and ourselves.

Callebaut: Dudley, how, without an explicit account of what kinds of the reasons (to use one of your central concepts) science is willing to use or accept in a certain domain and at a certain time, can one go beyond the purely descriptive? How can one avoid having to say, "This is how science has been using reasons; this is what we've got to accept as rational"? How does one *get to* normative standards in your view?

Shapere: A central aspect of my view is that normative considerations about how we should go about inquiry—*epistemically normative* principles—arise in the course of the development of science.

Callebaut: But these guiding principles seem to me to go beyond purely nomic generalizing. I think you want to go beyond that.

Shapere: We find reason to believe that a renormalizable gauge theory is the kind of theory to look for (cf. 5.6).[16] We find reasons for thinking that this kind of theory can be constructed in areas in which we don't have a detailed theory developed of

16. Quantum electrodynamics (QED) as developed from Dirac's quantum theory of the electromagnetic field (1927) accounted for the interactions of charged particles with radiation and with one another by describing electrons in terms of fields, in analogy to photons. According to QED, the vacuum is filled with electron-positron fields; and electron-positron pairs are real or virtual, as dictated by Heisenberg's uncertainty principle. The virtual pairs led to fundamental difficulties with QED, having mainly to do with the occurrence of infinite probabilities (e.g., the mass of a single electron was infinite according to QED). In the late 1940s Richard Feynman, Julian Schwinger, and Tomonaga Shin'ichiro independently proved that they could rid QED of its embarrassing infinities by a mathematical technique called "renormalization." Basically, renormalization acknowledges all possible infinities and then allows the positive infinities to cancel the negative ones; the mass and charge of the electron, which are infinite in theory, are then defined by their measured values. This solves the infinities problem, but only at the price of new inconsistencies. Feynman himself (1985:128–29) suspected that "renormalization is not mathematically legitimate." A *gauge theory* is a class of quantum field theory (involving quantum mechanics and special theory of relativity) that is commonly used to describe subatomic particles and their associated wave fields. The "gauge transformations" in a gauge theory leave the basic physics of the quantum field unchanged. This condition ("gauge invariance") gives the theory a symmetry, which governs its equations (cf. van Fraassen 1989, Part III: "Symmetry as Guide to Theory"). The structure of the group of gauge transformations entails restrictions on the way in which the field which the theory describes can interact with other fields and elementary particles. It is hoped that all physical forces can be encompassed in a single gauge field theory (cf. 4.2.4).

that sort. The immense success of this idea leads us to the normative principle that we ought to be looking for such theories in other areas. It is a hypothesis; but *normative principles I take to be hypotheses based on a generalization, but going beyond the mere descriptive* in saying that they have a normative force. They have a normative force in the sense that, because of their tremendous success so far, because they have led to unifications so far, we have reasons to think they may very well be applicable more generally. On the basis of considerations like these, which are factual, we go to the normative principle, "We ought to try to construct our theories in this form." "We ought to try to understand chemical substances in a compositional way" became a guiding principle (to use your term) in nineteenth-century physics, for example. And this has not only the force of what kind of thing we ought to do, but also of what kind of thing we ought *not* to do!

3.5.3. Rethinking the Analytic/Synthetic Distinction

Callebaut: Elliott, one of the main reasons you abandoned positivism was Quine's critique of the analytic/synthetic distinction. What does that distinction boil down to?

Sober: The analytic/synthetic distinction says that there are two kinds of beliefs or propositions. Analytic statements are true simply in virtue of their meaning. For example, "All bachelors are unmarried" might be thought to be an analytic statement. Philosophers have sometimes thought that "2 + 3 = 5" and in fact all of mathematics is analytic . . .

Callebaut: . . . the logical positivists, for instance.

Sober: Synthetic statements, by contrast, are ones that are not true simply in virtue of their meaning; typically, they are statements for which we have to go out into the world and seek empirical evidence. So the statement that "You are reading a book now" would be a synthetic statement; you could not know that this was true without getting sense experience, you could not learn that it is true just by thinking of the meanings of the terms.

Callebaut: What do you see as Quine's contribution to the analytic/synthetic issue?

Sober: Quine's major contribution, I think, has been a *critique* of this distinction, asking important questions about what really distinguishes so-called analytic and synthetic statements. And the view that he developed in reaction to the positivists was that in fact every statement, every belief we have, is in principle empirically revisable; that experience impinges on everything we believe, and even our beliefs about logic and mathematics can be revisable in the face of empirical evidence.[17]

---

17. Sober refers to Quine's main example, Euclidean geometry: "Until the end of the nineteenth century, it was thought to be an a priori, nonempirical body of propositions. With the advent of relativity theory, it became clear that Euclidean geometry makes empirical predictions about the nature of space, and in fact one can do experiments to find out about the nature of the geometry of space. So we now have the view that Euclidean geometry is not analytic. It is not even *true:* it is a synthetic claim about the structure of physical space which science has disconfirmed."

Callebaut: So far so good. But now you want to disagree with Quine's claim that *everything* we believe is empirically revisable (Sober 1981b); a claim which may also be related to his—official—endorsement of behaviorism.[18]

Sober: I agree in many ways with Quine's critique of the analytic/synthetic distinction; but I think the conclusion he reached in terms of his positive theory of this problem—the view that everything is empirically revisable—is defective, and that conclusion does not in fact follow from the perfectly correct criticisms he made of the positivist point of view. In particular I think one can argue that total revisability is not a biologically plausible view to take about the organ that we call the human mind. Every organ has constraints which structure the possible changes it can undergo. The human mind, I take it, has to be the same way. So I think there have to be constraints on how human thought proceeds, on the possible forms a human thought can have, which the human mind does not have the power to modify.

Callebaut: Could you be somewhat more specific?

Sober: What these constraints actually are is not clear now; this is a matter for extensive psychological and biological investigation. But the existence of such constraints, I think, is a very plausible working hypothesis. It is in that sense that I think of Quine's conclusion that everything is empirically revisable as being overstated. I doubt that the human mind is like that (cf. Wimsatt at 2.1.2).

3.5.4. *Immunization and Efficiency*

Callebaut: Dudley, like Quine you stress the openness to revision of all claims. How do you make sense, in your approach, of the idea of *immunization* in science?

Shapere: We have developed a body of beliefs which can be used as background for further inquiry, and we have developed criteria for a belief's serving in that body: our background beliefs must be ones which satisfy those criteria to a very high degree; they must be highly successful, free of specific doubts, and so forth. That body of background beliefs is taken for granted in inquiry; it is *immune* to criticism until we find specific reason to question it. In sophisticated modern experiments in every field, a great deal of background is required in the formulation of the problem and the experiment, and in the execution of the experiment and the interpretation of the results (1982).

Callebaut: On the other hand, there are ideas that one adopts as guiding principles

---

18. In Quine's (1969:82) much cited dictum, epistemology naturalized is "simply to fall into place as a chapter of psychology and hence of natural science." However, as Campbell enjoys pointing out, Quine's doctrine of the *indeterminacy of translation* can be given a plausible cognitivist rendering instead. The issue is too technical to be dealt with here, but I cannot suppress disclosing just this one passage. Campbell (for aficionados only): "I like to quote Quine from his essay 'Ontological Relativity' (1969b), where he goes over the Gavagai thing in briefer form, so you and I don't have to read all of *Word and Object* (Quine 1960). This quote here I probably used to confront him with: 'Such is the quandary over Gavagai (etc.)' (pp. 31–33). I call these 'hypotheses of translation' *cognitive* or *mentalistic*. And there is a specific disclaimer: simple conditioning or ostension takes guessing and response to ostension (cf. 7.1.5). You may quote me as remembering him, when he met with our class [Campbell is referring here to the William James Lectures he gave at Harvard], conceding that he was a cognitive psychologist."

in a more hypothetical sense: one does not know whether this is true or not and let's investigate it.

Shapere: Of course: often the solidly-grounded background beliefs must be supplemented by less well-founded ideas—"hypotheses" . . .

Callebaut: . . . in the sense scientists and lay people (as opposed to philosophers) use that word.

Shapere: Still the construction and choice of such hypotheses are also guided by our background information.

Callebaut: You emphasize the revisability of all claims. On the other hand, it seems to me that there is a legitimate preoccupation with speed and other forms of efficiency in scientific inquiry nowadays. For instance, one could ask, "What would an optimal strategy of immunization (of concepts and the like) in science look like?" That, I think, would be a perfectly legitimate question, although few people are willing to claim they have an answer to it (as yet). At least one naturalized epistemologist whose work more traditionally-minded epistemologists continue to take seriously, Alvin Goldman (1986), takes efficiency considerations such as "speed" of problem solving very seriously.

Shapere: I think it is very misleading to talk about "efficiency" as the grounds for distinguishing between what is going to be immune and what is not. If you are going to be efficient, you certainly would reject quantum field theory, for example, as a guiding principle, unless you define efficiency in very, very inefficient ways. Because the calculations that have to be carried out are so difficult and sometimes impossible that one often can't get down to details of problems. It is not the most efficient type of mathematics. So I am not sure what clarificatory force the idea of efficiency would have.[19]

Callebaut: I would like to relate the critique of the analytic/synthetic distinction to your previous criticism of the "levels view" of science (cf. 2.2.2).

Shapere: I take the concept of analyticity to have to do with bringing out what we're trying to say or think with regard to a certain concept. Philosophers at least since Hume have thought that that can—and should—be brought out before we undertake any inquiry concerning the referent of that concept. This alleged priority of setting your meanings is an example of a "levels" approach. Now, I do think it's possible to lay out precise definitions and their implications ("analytic systems") in the abstract—mathematics is an example of doing this. But when we're dealing

19. For Shapere, "the optimal strategies of immunization consist primarily in satisfying the criteria of success with regard to the domain of responsibility of the theory, and coherence with other theories, plus all of the other multitude of constraints that have been developed over the history of science." This is a way of marking out what we take to be immune; or conversely, and more importantly, he thinks, "what we take as adequate bases on which to build further inquiry. That is the positive side of what is only expressed negatively by a term like 'immunity.'" Focusing attention on immunity, he admonishes us, "might call our attention away from the revisability of *all* our beliefs and claimed knowledge; that (as far as we have reason to believe as of now) doubt might, in principle, always arise with regard to any of the beliefs (even that parenthetical one, if, as some results in contemporary physics suggest, constraints become so tight as to exclude all but one theory!)."

with the application of concepts in actual empirical inquiry, we rarely "know what we mean," if I can put it that way. We don't know everything we want to say about electrons, and what we do want to say may involve problems or the open possibility of alteration. Even if we could lay out the alternative modifications (the alternatives of what we might "mean" by "electron"), there would still be an openness. And *that openness in what we're saying about electrons, or about the concept "electron," is what makes the concept of analyticity inapplicable in the analysis of actual, functioning scientific concepts; and to say this is to recognize the dependence of functioning scientific concepts on empirical findings, both past and future.* There may have been some naturalness to the idea that there was a distinction between "meaning" and "the empirical" until around the later part of the eighteenth century, but after that the terrible flaws in the distinction became more blatant, until in the twentieth century it beats you over the head that one cannot anticipate the way nature is, or even the way *inquiry* should proceed (1987a).

3.5.5. *The Revisability of Logic and Mathematics*

Callebaut: Until now we have been talking about natural science only. What about the revisability of mathematics and logic?

Shapere: Mathematics has its own independent development, which takes place by such processes as the denial of the axioms of an area, but much more by generalization of methods in one area of mathematics to show the connections with others. Through this, new areas of mathematics develop and new mathematical techniques, which may or may not have application in science. The development of areas like group theory and topology, in all their generality, and of matrix algebra (e.g., Manheim 1964; Kline 1972; Wussing 1984; Monastyrsky 1987) has taken place to a large extent independently of science, though not always; and even though at some stages they have seemed irrelevant to science and in fact to be such that they could not have any application in science, they nevertheless so often have had such application. Of course I think the alternatives to standard ZF set theory[20] are too primitive yet for us to know whether, say, alternatives to the axiom of choice, for example, might have application in science. Nevertheless, when you look at all the ways in which very esoteric mathematics like fiber bundles or algebraic geometry have had extremely deep applications in science, one just does not know. And again, it is a contingent matter.

Callebaut: So you look on mathematics as very analogous to conceptual structures in general?

Shapere: We can set up any conceptual structures we want. Minor modifications are always possible. Major modifications are always possible in a conceptual structure. Analysis of conceptual structures is very important, but one has no way of knowing which conceptual structure is going to apply. And if one thinks that one *does* apply, the possibility of minor modifications can leave a potential ambiguity

20. Zermelo-Fraenkel set theory (see, e.g., Fraenkel, Bar-Hillel, and Levy 1973).

in that. That can mean it was not exactly that way that it should be applied, but, as we find out later, it should have been understood in another way. So that there are all of these different conceptual structures.

Callebaut: You don't accept, then, the complaint of some applied scientists that the development of mathematics is too "independent of sense observations or any other reality than symbol manipulation itself" (Cavallo 1979:12)?[21]

Shapere: I don't accept that! There is this recent proof that I spoke about earlier that there are radically different structures that a four-dimensional space can have, as opposed to all other types of spaces; whereas for all others, you can show that there is a certain fundamental unity in description of those. That would sound very esoteric, if we did not have physical theories right now that *talk* about the possibility that space-time at the beginning consisted of twenty-six or ten dimensions and that our space-time now does seem to be four-dimensional—and which this multitude of structures will fit. So often, very esoteric ideas in mathematics come into play. It is also not true that mathematics develops entirely independently of science, because mathematics has gotten a tremendous shove in the direction of analyzing certain problems *by* the structure of physical theories at various times in its history. So that does not seem to me to be right.

Callebaut: What about the revisability of logic?[22]

Shapere: Logic deals with how to lay out what is involved in our concepts, assuming them to be well defined. But our concepts in actual use in dealing with the world of experience and with science aren't well defined in the relevant sense: there is an openness to them that is a function of our ignorance. *To the extent that* we know what we're talking about, logic can apply; but in general, there are good reasons why and—in some cases—in what respects we can't "lay out" what we mean.

Nickles: Even if there does exist an objectively correct logical system or, say, a methodology of science *out there,* we still have to discover it as human beings and establish its reliability for our purposes in our own terms. It must be discovered and licensed by us chickens, using only the cognitive resources available to us. We are not gods, and we have no precognition or "prescience" (Campbell). So we just can't take a logical system or a particular methodology as given, be it positivistic, Popperian, Kuhnian, or whatever.

3.6 A Trojan Horse: Steps toward an Anthropology of Science

Callebaut: Anthropology of science may be seen as a Trojan horse. I refer to what you wrote about Robert Guillemin taking into his lab at the Salk Institute someone he took to be "an 'epistemologist' (Bruno—Dr. Jekyll) who subsequently turned

---

21. According to Cavallo, this is the most important factor that has contributed to the motivation of systems research.

22. Cf. all the recent fuss about relevant logic (Read 1988) and the presumed irrelevance of classical logic (Keene 1990).

into a sociologist of science (Latour—Mr. Hyde)"[23] (Latour and Woolgar 1986:274). What I mean is that the networks (to borrow your term) of historians and philosophers of science were already present at the time anthropology—as one of the new approaches to the sociology of science or knowledge—emerged. And that, of course, made a difference. Could you say something about what that difference was? And next I would like to ask you how you would link this to your being French.

Latour: You have to realize that Guillemin was also French. For him an epistemologist was someone who, in the tradition of Bachelard or Canguilhem, distinguishes—*demarcates*—science from nonscience altogether. On such a view epistemology is OK, because it helps the scientists to break with the past, to break with history, to weaken the link between "scientific facts" and society, and to strengthen the demarcation between science and nonscience. So "epistemologist" was OK and I had no trouble. In France, epistemology is very *agreeable* to scientists, and many scientists *dream* to be considered as epistemologists. Now the "sociologist" or "anthropologist" was an entirely different matter, because all the dirty things which French philosophy of science has excluded through the work of Bachelard, Canguilhem, and others—society, history, and so forth—come back in through that door. So although the French tradition differs from that of most of the other people you have interviewed, the demarcation between science and nonscience is much stronger in France than in the positivist American tradition. I'm not a specialist of the history of philosophy, but it seems clear to me that the French tradition from Bachelard to Foucault strongly opposes any link with history, even more so than the positivist tradition.

Callebaut: How about anthropology of science being a Trojan horse?

Latour: In France it's perfectly clear: The Trojan horse is to say, "We are not going to do epistemology and use examples from history to give shadows and depth of vision to the demarcation between science and nonscience. We are going to be *symmetrical* . . .

Callebaut: . . . Symmetry is an important topic we will have to return to . . .

Latour: . . . That's anathema! But not in the sense of saying, "We are not going to ask the big questions." *We want to ask the big questions;* and if by "naturalistic turn" is meant a sort of return of naive empiricism, then of course I don't want to take the naturalistic turn.

3.6.1. *California Blue*

Callebaut: Bruno went to La Jolla for his first case study; you went to Berkeley, Karin. Why was that?

Knorr Cetina: It's hard to answer. It was a series of accidents. I was stimulated to

---

23. In a note accompanying this sentence Latour and Woolgar (1986:285) add: "Which is the bad guy? Readers are invited to transpose identities here. The important point is that a metamorphosis occurred."

"Relativism is just one way of arguing against universality."

Bruno Latour (b. Beaune, Côte d'Or, 1947) has been associate professor of sociology at the Centre de Sociologie de l'Innovation, École Supérieure des Mines de Paris since 1982 and at the Department of Science Studies at the University of California, San Diego, since 1988. It will come as a surprise to some readers to learn that he was an excellent student of . . . philosophy,

having been ranked first, nationally, in the "agrégation de philosophie" in 1972. "Even worse" (these are his words!)—he holds a "doctorat de troisième cycle" in theology (Exégèse et ontologie, Université de Tours, 1975). He obtained the "habilitation à la direction des recherches" at the École des Hautes Études en Sciences Sociales in 1987.

Latour and Woolgar (1986:273f.) describe how Latour got involved with science studies by the roundabout of his fieldwork in the sociology of development in rural Africa, from 1973 to 1975: "Many features were attributed a little too quickly to the African 'mind'." As the study proceeded, "the established preference for far-fetched cognitive explanations over simpler social ones became more evident." A "terrible doubt" arose: "What would happen to the Great Divide between scientific and prescientific reasoning if the same field methods used to study Ivory Coast farmers were applied to first-rate scientists?" Robert Guillemin's previous invitation for Latour to carry out an "epistemological" study of his lab at the Salk Institute now was *gefundenes Fressen,* as it would allow Latour to find out the answer for himself.

Latour met Steve Woolgar at a sociology of science meeting in Berkeley in 1975, i.e., in pre-4S days (Latour was one of the founding members of 4S in 1977). He discovered that this was a field he "fortunately" didn't know about ("My ignorance saved me!"). "Steve helped me articulate the field study approach for an audience which had already debated a lot on the social construction of science." Latour himself came from an entirely different tradition "We had, and still have, no example in France of a field study. I had no role model at all. So I was greatly helped by the excellent knowledge of the field that Steve had."

Is Latour a naturalist after all, then? He hesitates: "I *took* the naturalistic turn, in a sense, when I abandoned philosophy in a recantation, doing fieldwork, first in Africa and then at Salk. Afterwards I realized the complete absurdity of the social sciences as far as their philosophical background is concerned, and went back to philosophy."

The way he would put it now is: "I did philosophy, abandoned it, found philosophy back, never believing a word of philosophy of science." What has he got against the latter? Latour insists that he has always had "too much sense of what the practice of science was to get any sort of interest in the philosophy of science." And also because the French philosophy of science he was acquainted with (Bachelard, Canguilhem . . .) "is more concrete and more loaded with practice" than its American counterpart. But *la vraie philosophie* remains important: "You need philosophy to defend the localities, the historicity of things and the networks, *against* the unification which is imposed on them by the social sciences." That's the way Latour views things now, granting that he is thus "speaking against part of his own work," especially *Irréduction* (1984), a philosophical book within his book on Pasteur, in which he was "trying to find a common language to travel through networks." Readers of *Science in Action* (1987) or other outlets of Latour's prolificness (see the bibliography) can judge for themselves if they find his position coherent.

When Latour returned from the U.S., he found a small trend of science studies in France to which Michel Callon was contributing. Thanks largely to Callon, the École des Mines group has gained some influence on science policy, which Latour finds "is now becoming very interesting in France." The economics of innovation is also on the move. The Boul Mich group invests a lot in international relations. "We have no choice; here, we do not have one single colleague in our own field. We are quite unknown in the intelligentsia—which as you know has a much higher status in France than in other countries—and have little influence in the journals." In general sociology, people "ignore entirely" what the sociologists of science and technology are doing. "They hate it when they hear about it. And they worship science. Their whole education is Bachelard. If you read Bourdieu, *Le métier de sociologue* (1968), it is 99 percent Bachelard. Not a word of field studies, not a word of history." Yet Latour remains stoic. Theirs is a shared predicament—"In France, everybody hates each other!"—and the group's constant contact with industry is at least getting them many grants.

Some critics have complained that much of the recent literature on science studies is "plagued by unjustifiable jargon" (Levy 1982:151) or abounds with "rhetorical flourish" (Kitcher). Latour admits that "the French are always a bit idiosyncratic" and identifies a scapegoat: "Bachelard was the first to talk gibberish, and now everyone—including, I'm afraid to say, Serres—writes terribly." But he defends himself, suggesting that he was saved by the English, by Steve: "I am trying to get back to the pre-Bachelardian way of talking and writing: clear and with examples."

Our main interview took place in Michel Callon's office at the École des Mines on October 9, 1989. An update was recorded on May 31, 1990 in Bruno's office at UCSD.

go to Berkeley by Aaron Cicourel, who I had met in Vienna as one of our visiting professors at the Institute for Advanced Studies during my time there. And of course I *wanted* to go to California. I was attracted by California as an environment which projects many symbolic futures, in a way. It symbolized the future of life styles, of architecture, of all kinds of things you do not find in Europe . . .

Callebaut: . . . The French philosopher Edgar Morin, who spent the fall of 1969 at the Salk Institute, has written a quite personal account of his Californian experience, in which he describes California as a "land in trance" and as "the bridge" (*tête chercheuse*) of "spaceship Earth" (Morin 1970).

Knorr Cetina: The good part about Europe is that it preserves some of its glorious past. But it's not that good at projecting alternative futures. The excitement was to go to a country where that happened. Also, Berkeley had the reputation of being an excellent intellectual environment. You can do many things there which would be much more difficult to do in Europe because of the greater restrictions and conventionalism.

Callebaut: Are you implying that it wouldn't have been possible in the 1970s to have someone grant you access to his laboratory in, say, Vienna?

Knorr Cetina: It might have been possible, sure. It's possible now. I have done two large laboratory studies recently in Germany and Switzerland/France (the CERN in Geneva).

Callebaut: Yes, but now that kind of work is better known.

Knorr Cetina: Well, the scientists still don't know these studies. They haven't changed. They don't really care. They don't read these kinds of materials.

Callebaut: They already know what's going on in the lab; they wouldn't be too surprised by what you people are reporting (cf. Latour and Woolgar 1986:274). How do you actually *get* the scientists interested, Bruno?

Latour: We train them.

Knorr Cetina: Also, they don't believe in the social sciences, so you have to convince them every time. It's not that it is easier now than it was. But the whole atmosphere in Europe . . . There were in California at the time and there continue to be different paradigms, a different discourse community. It's not by chance that ethnomethodology (Garfinkel 1967; cf. Barnes 1985b) and some of the microsociological approaches grew out of America and not of Europe. And many of them have been developed at some point in California. Erving Goffman, for example, had been in Berkeley before he went to Pennsylvania. Ethnomethodology, cognitive sociology (Cicourel 1964, 1975; Knorr-Cetina and Cicourel 1982)—all that comes out of California. You needed not only access to a laboratory, you needed a different theoretical framework and frame of mind to even contemplate doing this work and to be able to do it in a certain way. Microanalysis is very important, and microsocial theory. And that wasn't, and still isn't, very popular in Middle Europe and usually isn't done or understood very well there. The kind of empiricism which it includes, for example, is dismissed in these European countries . . .

Callebaut: . . . as we witnessed at this conference[24] . . .

Knorr Cetina: . . . So you needed a different setting, in a way, to be able to do that. I had the opportunity to go to a laboratory. The interesting thing was that I wanted to do it because I wanted to do something intellectually more promising than the work that was done in conventional studies of science, but I didn't really believe in it at first. I was so much raised on traditional philosophy and philosophical thoughts about science that I believed at the time I wouldn't *see* anything. After all, scientists *think*—right?—so there wouldn't be much to be observed. But I got hooked the first day I went to the lab. There was so much going on and there was such a tremendous surprise effect. It was so fascinating that I decided immediately to continue to do just that. Today I'm still fascinated by what goes on in these laboratories.

Callebaut: Paul Feyerabend—the Popperian-turned-guru (cf. Munévar 1991)—was already at Berkeley at the time you got there; you must have met him.

Knorr Cetina: Sure. I went to his lectures. He was supportive. But Feyerabend at that time was a curious fellow. I had already met him in Vienna before. It was very difficult to have an intellectual discussion with him. He didn't seem interested. He put his own philosophy into practice and he seemed interested only in pleasure. He was no longer interested in intellectual discussions with newcomers or with anybody, it seemed (Lakatos had already died). Still Feyerabend was a very interesting experience for me because he was very committed to his historical materials and he gave fascinating lectures on it. But you could not get into a discussion with him; he refused to do that. He was concerned about earthquakes in California, he was concerned about his health; he wanted pleasure and money for his presence at conferences, but he didn't want to engage in conversation.

3.6.2. *Free-ranging Scientists*

Callebaut: "Naturalistic turn" is a very ambiguous expression, and that's one of the reasons . . .

Latour: . . . you didn't give a definition . . .

Callebaut: . . . and why I chose it for the title of this book. I didn't, for instance, call it the "empirical turn." I think that in fact a lot of work in science studies today *is* based on hopelessly naive empiricist views about methodology, and I don't want to be associated with any of that.

Latour: The best term—if it exists; I am not sure—is: We study *free-ranging* scientists, not caged ones.

Callebaut: Wild type.

Latour: Wild type! In that very narrow sense, the expression "naturalistic turn" can be defended. What we do is not armchair epistemology or anthropology. *You go*

---

24. 1990 Marburg meeting of the Theory Section of the German Sociological Association on "Kommunikation und Konsens."

there. But the term which would be acceptable to me is, of course, the exact *opposite* of naturalization: what we do is a *denaturalization* of the whole activity, right? *Enfin,* for me, it's the anthropological turn, not the naturalistic turn. "Naturalistic turn" is used by people like Kitcher, who use it to make *models* that are the exact equivalent of the old models, except that it's now collective models instead of models of the individual scientist.

Callebaut: Philip calls his view "conservative naturalism." My own predilection for the term is—among other things (cf. ch. 1)—a way of stressing the historicization of the subject matter of philosophy of science: no logical analysis of entities which are eternal—or rather outside space and outside time—but a historical analysis of "situated" entities. It's a kind of evolutionary approach in many ways. Not Popper's evolutionary approach, evidently, because insofar as I can make sense of it, Popper's evolutionary epistemology (1972) is about things in heaven (cf. 7.4). In terms of my personal history, that's probably more important than Quine's replacement of logic by psychology.

Latour: The difficulty is that you can't use the word "history" to describe the new combination of philosophical questions and empirical case studies which is appearing now. Because what is interesting about the current redistribution of philosophy and empirical work—at least in France this is very clear—is that you realize that *what you mean by "history" is itself predetermined by the very definition of "science."* For example, in France it is very clear that the historians of, say, the Annales school never do history of science. They consider their subject matter as what passes, what evolves, what is transformed, what is arbitrary, more or less, or contingent, but always arbitrary in some sense—in opposition to *the things that don't pass.* And the things that don't pass, of course, are defined by the scientists. So the very difficulty is that history *as it is shaped* has been made for us by the divide between historians and philosophers of science in the first place.

3.6.3. A New Church Already?

Callebaut: A puzzling thing about the kind of case studies one sees appearing is that the moral that people draw from them (the debunking routine: "Look, it's all negotiation and all social construction!"), even if they claim not to do so, always seems to be the *same* one. Why don't people ever come up with radically different conclusions from their cases?

Latour: What is "the same one"?

Callebaut: Well, some approaches which are vindicated by a growing number of people—yours, for instance. A lot of people say "This is the way to do it"; so they go out to the field (actually a library, most often) and study their cases. *They consider themselves to be the little guys* doing the empirical work. And then they link it—more or less successfully—to the Framework, to the Theory. I can also turn it around and ask you, if you didn't like my previous way of putting it, "What can one do about this to prevent it from happening?"

Latour: That's a good question. *(Laughter.)*

Callebaut: I mean, I see science studies already turning into a new church!

Latour: No, it's a symmetry point. You're right that the little guys are accepting the position in which they are put by the divide between "empirical" and "theory." That's true; and lots of people in science studies have their share of the blame (cf. Callon and Latour 1992). They share the guilt, because they accept to be only empirical. I mean the English tradition especially: "We just do empirical work; we don't get into the big philosophical questions." They are of a crass ignorance as far as philosophy goes . . .

Callebaut: . . . or sociology for that matter.

Latour: They are proud of being ignorant of philosophy . . .

Callebaut: . . . Oh yes!

Latour: The only sort of thinker they might have read is Wittgenstein—Winch (1958) in the case of the English. They know nothing of the history of philosophy, not a word. And they despise it deeply. So they *accept* the empiricist position. And, of course, the trap, as you say, is that every empiricism is hooked up on a theory, which is usually a very badly thought-out framework, like Winch's for instance.

3.6.4. *Behind the Looking Glass Lies a Lab*

Callebaut: What do you say to the criticism that many social students of science are crass empiricists as far as their own methods are concerned, Karin?

Knorr Cetina: I'm familiar with the criticism. It usually doesn't pop up in the U.S. but it does in Europe. It has always struck me as rather incomprehensible, because certainly what Bruno or I have produced is not without theory—that should be obvious! What we use is microanalysis, and that strikes some people as being extremely descriptive. But what does microanalysis actually do? Let me use an analogy by Bachelard, who said once that "the miniature is a shelter for largeness." The way I see microanalysis is like a small door you have to go through; behind it, you discover a world that, like Alice's world in *Behind the Looking Glass* (Gardner 1970), is structured but is quite unfamiliar to us who come from the other side. Now there is nothing particularly small or exclusively descriptive about this world. It can be massively large. It's just that you have to pass through the microanalysis in order to discover it. So I see microanalysis as a kind of passing-through step you have to take in order to get to this other side of things.

Callebaut: Still, I know people in science studies who will themselves literally say things like "We don't care about theory! *We*—as opposed to you philosophers—stick to the facts!"

Knorr Cetina: That has always also been the ethnomethodologist's point of view: Theory is an interpretative device that some sociologists use and that society also needs to some degree. But since they want to refrain from (too much) interpretation, they refrain from doing theory and they attempt to articulate the empirical world without interpreting it. On the other hand, it is quite clear that in many of the microparadigms—even in Garfinkel (1967; cf. Knorr-Cetina 1983) or Goffman (1974)—it is not that there is no theory, no conception of things, no theoretical concepts in them—not at all. It's just that these are frequently *dissolved* in "empirical descriptions." There are different kinds of theories around. One prominent approach is to interpret one theorist in the light of another, as in literary criticism. There's also hypothesis construction and then coming up with a propositional account of what things are like. Certainly that is not the kind of theory one is after.

Callebaut: A concrete example may be illuminating here.

Knorr Cetina: A notion like "laboratory," which we use, is a very theoretical notion. The laboratory is not just a "new field of study" that has been discovered. The laboratory—in my view anyway—is a sort of *reconfiguration of the relationship between actors and objects;* a reconfiguration as compared to what you have in both the natural and the social order. In the process, both the actors and the objects change. From the reconfiguration and from the difference you implement through the reconfiguration, the difference between the laboratory and the social and natural order, you derive epistemic effects. So it is a very theoretical notion, and not at all just an empty space in which some things happen which you can look at. So I don't think there is no theory in these studies. It is just a different kind of theory. You also need a different kind of theory to deal with the ontological flexibility that you believe exists. When I say there is "ontological flexibility," that's a certain model of reality. When you are a constructivist you have to recognize that your own productions are also (re)constructions of things. Every construction paints a picture, a sort of model of the world. It may be hidden in a lot of empirical detail, but it's certainly there. So I don't understand the complaint, really.

Latour: The paradox is that both the philosophers of science and the little guys doing empirical studies share the same distrust for philosophy, and, I should add, social theory.

Callebaut: Why would that be?

Latour: That's because the turn is not finished. It's a very, very long move. The first problem was to get rid of philosophy of science. That is basically being resolved now. But philosophy is there, and all the important questions in philosophy and social theory are back in. The link between empirical field studies and theory isn't made yet. When I see sociologists of technology take seriously as a theory things like the SCOT[25] stuff or the EPOR[26] stuff . . . I mean, this is ridiculous! It's nothing; it's just a slogan. It's just a way of . . . in what sort of box do you put your reprints when you get them? You can have a box "EPOR" and another "SCOT." That's the only use they can have. But this has nothing to do with philosophy. It won't solve any of the philosophical questions.

25. Social Construction of Technology (e.g., Bijker, Hughes and Pinch 1987).
26. Empirical Program of Relativism (e.g., Collins 1981a, b; 1983, 1985, 1991).

Callebaut: So you are opposed to (most) extant philosophy of science, but not to philosophy per se.

Latour: The problem is that traditional philosophers, when I say such things, triumph. People like Kitcher will triumph and say "Of course, you need a theory!" . . .

Callebaut: . . . Karin agrees.

Latour: No, we need philosophy, which might be very, very different from what they mean by "model building" or a "theory of rational action," even if it is collective rational action (cf. also Longino 1989). The link between field studies and the big questions must be entirely remade *because* of the notion of locality, *because* of the notion of networks, precisely, *because* of the notion of history. *We have no idea, yet, what the link between philosophy and empirical studies is,* precisely because we have lived for so long with this idea that empirical studies are just there like small localities that the theories and the models have to reorganize.

Callebaut: What if we change the relation between the theory and the empirical evidence?

Latour: I think that is now being questioned. So I agree with you. There are two impossible situations: one being that the traditional sort of philosophy of science would still exist—I think that's disappearing now—the other that there would be empirical field studies satisfied with their ignorance of theory and not asking the big questions about philosophy—I think that is also impossible; it's too unstable a situation.

Callebaut: In terms of mobilizing resources (cf. Latour 1987, Star n.d.), there seems to be something funny happening here as well: When you do a case study without bothering about any philosophy, *you are calling on the science.* And of course, in the hierarchy of things, the science is much higher than philosophy. Maybe not in France . . .

Latour: . . . in France, epistemology is the Queen . . .

Callebaut: . . . but in most other places in the world. In a way, it's a good old positivist trick you are using there: *you* are able to mobilize the science. Jerome Ravetz has this thought-provoking paper (1980) in which he develops the idea that from the Vienna Circle to Feyerabend, so-called "philosophers of science" have not really been that much interested in the science; they didn't do the philosophy *of* the science. Rather, they had a message for society: that science is good (Neurath, Popper . . .), or that science is bad (Feyerabend) for you and for society at large. So you make up the image of science that best suits your needs. For instance, in the case of Popper, you invent falsificationism. "Falsification" is not something you derive from the common practice of science; rather, you make it up because it will allow you to proclaim that "communities must be open and critical" and all that. (Ravetz then turns into a positivist himself as he goes on to claim that with the advent of sociology of science, we are now finally addressing questions to science "as it really is.")

Latour: This is Ravetz, you say? The feeling that I often have is that the people who get the most pleasure and passion out of the practice of science are people like us, who are not teaching lessons . . . When I talk with people like Kitcher, they always see through the details of the sciences. The details are *there,* and Kitcher is very knowledgeable and serious about them. But they are *irritating* details. What counts is really the *lesson* you can get out of them, which is, I suppose, often a *moral* lesson. And people who respect the details are *bizarre,* not well organized. I mean, that's the good thing in anthropology, *ça va,* we have a tradition. But before anthropology was used as our main resource, before the anthropological turn—not the naturalistic turn—the whole question of locality, of circumstances, had to be solved, *philosophically* solved (cf. Star and Gerson 1986). And that's a difficult part for American philosophers, because they are living with a very impoverished philosophical tradition that is deeply scientistic.

3.6.5. Reflecting on Reflexivists Reflecting on Reflexivity—Full Stop

Callebaut: Impoverished or not, philosophical or not, it would seem that *any* naturalistic account of science must sooner or later face the reflexivity issue. If our explanation of what goes on in science is itself scientific, then that explanation must extend to itself, as the Edinburgh Strong Programme (SP) sociologists realized in their more scientistic moods.[27] Self-referentiality is a rather general problem (Luhmann 1990a, b).[28] Here I want to raise the question of reflexivity with respect to science studies only. Can, say, philosophy of science meet this SP requirement—and *should* it, Ron? If I may make a comparison here: It is quite clear that Kuhn-the-historian, whom you credit for being the pioneer of the naturalistic turn in philosophy of science (Giere 1985a), wants to exempt the history of science itself from the sort of analysis he applies to episodes in the history of natural science (Kuhn 1980).

Giere: I don't think Kuhn has a whole lot of patience with the reflexive argument. If you come back to him and say: "Look, would you apply your theory to yourself?" that is something I think he doesn't much worry about.

Callebaut: Can Kuhn be made reflexive by naturalizing him, in the sense of giving history the same status you give to the sciences?

Giere: There are two questions: "Can any naturalistic view do it?" and "Can Kuhn do it?" Kuhn doesn't worry much about this problem. But there is a simple way

---

27. On the general point, see, e.g., Luhmann (1990a); cf. Atlan (1986) on the limits of scientific rationality. The main tenets of the Strong Programme are listed in note 32.

28. Thus linguists have become conscious that research into language has to make use of language. The case of neuroscience is even more intriguing: "Brain research has shown that the brain is not able to maintain any contact with the outer world on the level of its own operations, but—from the perspective of information—operates closed in upon itself. This is obviously also true for the brains of those engaged in brain research. How does one come, then, from one brain to another?" (Luhmann 1990b:64).

to make Kuhn reflexive. You just say, "Yes, the study of the nature of science, if it is a science, would go through the same stages." That is what you'd have to say. And then, of course, he could say that the study of science and philosophy of science is now in a crisis, or a revolutionary stage . . .

Callebaut: . . . But there is also the Kuhn who says that his stage model only applies to natural science, not to mathematics, and definitely not to the social sciences (e.g., Kuhn 1977). There are no paradigms—in the sense of "exemplars"—in the social sciences, according to Kuhn. Maybe his model doesn't apply, then, to history either.

Giere: I don't know about that. My view on the general question is that, in principle, any scientific account has got to be reflexive; otherwise *you will get out of the system, and you can't do that.* So that if one has a naturalistic account of science, it has got to be subject to its own models.

Callebaut: Toward the end of their book *Opening Pandora's Box,* Gilbert and Mulkay (1984:188) write: "At first, it may have appeared that, like Pandora, we were heading for chaos. But, as in Pandora's box, Hope still remained; in our case, hope of creating order out of diversity. Although we emphasized that the multiplicity of voices with which scientists and other social actors speak makes traditional sociological objectives unattainable, we held fast to the assumption that interpretative regularities could be discerned behind the babble of tongues, if a suitable analytical approach could be devised." The approach they have in mind is *discourse analysis* (Mulkay 1979, 1985; Mulkay and Gilbert 1982a, b; Knorr-Cetina and Mulkay 1983). The end of the linguistic turn (Rorty 1967) is not in sight yet, it would seem, Karin.

Knorr Cetina: Mulkay originally thought that he was going to solve the reflexivity problem by studying discourse instead of scientific practice or action. The basic argument is that since much of our data is really verbal material, viz. scientists' accounts of what they are doing, we avoid the reflexivity issue by studying their information as *accounts*—not as an indication of what they actually do in practice. Now I don't think he solves the reflexivity problem that way, and he has come to recognize that in the meantime (Mulkay 1988, 1991). But it's a valid move to shorten the chain of steps you have between your assertions and the referent of your assertions when you take stories about reality, which is your data material, as stories *about* reality and not *as* reality. In that sense, discourse analysis is very useful.

Callebaut: Discourse analysis has gone further and has become a reflexivity program again.

Knorr Cetina: The issue has bothered Mulkay (1985) and other people like Steve Woolgar (1988a, b) and Malcolm Ashmore (1989). When you consider it as a reflexivity issue, then it becomes another question, and different kinds of things have been done with it. On the one hand it has become fashionable for people to outdo each other in terms of who is more reflexive than the other. You have to show that you are reflexive about the reflexivity issue, that you know that your own account is also just an account and not reality as it really is, and you have to invent devices to show that, like including dialogue in your presentation instead of just telling a straight story and thereby making clear that you are aware of the interpretative flexibility of things—issues like that. That part of it is fairly ridiculous, because *you can go on outdoing each other in terms of how reflexive you are.* It's a game, but it loses any interest after the first few steps because it's so ridiculous.

Callebaut: Reflexivity has been considered a sort of "aporia" from which reflexivists "cannot escape except by indefinite navel-gazing, dangerous solipsism, insanity and probably death" (Latour 1988c:155) . . .

Latour: Literature study in the sense of Mulkay's discourse analysis is absurd.

Callebaut: Why?

Latour: Because it's most superficial; it does not give the nonhumans' action, it only defines the rhetoric of humans *about* nonhumans. And in Gilbert and Mulkay's studies, there is never any voice of a nonhuman or a study of them.[29]

Knorr Cetina: You cannot arrive at an end anyway. The other thing is that some people in anthropology have turned the reflexivity issue into an industry of studying texts.

Callebaut: You refer to anthropologists who no longer go into the field but prefer to study other anthropologists' monographs instead.

Knorr Cetina: That is valid to some degree, because you presumably find something out about the textual strategies anthropologists use in presenting their material. That's an interesting subject matter, and you can go on doing that for some time. What the people who have done that so far don't realize, I think, is that it's not only a textual question. You would also have to look at what the anthropologist is doing in the field and incorporate that. Even in terms of the textual analysis to do you would have to become much more precise than they have been: You would have to draw on linguistic categories and all that. So that is an interesting thing, but in a way it takes you out of the reflexivity issue; because you use the reflexivity issue to open up a new field of study, namely the texts produced by social scientists, and then you study these texts in a fairly realistic way. I mean, what else can you do? I don't blame them for doing that; but it takes you out of the issue. So I think this attempt to analyze texts is a valid reaction to the reflexivity issue. But there are other reactions you can have.

29. In Latour's picture of things, the sociologist of science or technology must deal with "actants," which can be humans (the "actors" of conventional sociological analysis), or nonhumans (e.g., the bacteria unleashed by Pasteur), or even inanimate objects (cf. Callon 1986, 1992; Latour 1987, 1988a; Callon and Latour 1992). Relax for the time being, reader of these lines: you are not the first intelligent person to find this idea weird. But, it should all become crystal clear in section 3.6.9.

Callebaut: For instance?

Knorr Cetina: There is a difference, for example, between constructivism as it grows out of the Berger and Luckmann (1966) phenomenological tradition in the U.S. and our kind of constructivism, because of the reflexivity issue. In social constructivism, social reality is used as a resource. Sociologists use the resource for studying natural-scientific facts and natural science. But it's never used as a topic. The construction of sociology—that is, of sociological categories, sociological texts, etc.—and of social reality in general is not studied and documented in the same way as natural-scientific facts are studied and documented. Now what we have to do, of course, is to be reflexive enough to apply the constructivist program and method also to our own resources.

Callebaut: What that means is not obvious.

Knorr Cetina: It can mean many different things. For example, you can study how social actors in the laboratory and social patterns which you observe in the laboratory are also constructed. They are not merely there and to be used as a resource to explain how the natural objects are constituted; we must also explain, by drawing upon the natural objects and natural science issues, how social actors change their shape and become something different, and how they are constructed and deconstructed. Another version of it would be to ask the question whether, if you apply the constructivist program to sociology, you would have to draw the consequence that not everything is social, for example. I mean, reality could organize itself in ways which are nonsocial. If you have a more flexible constructivist viewpoint of your own sociological resources, you would have to consider again the possibility that not everything is social, and that reality might fashion itself in ways that are nonsocial in the sense of traditional sociological concepts.

Callebaut: Latour is also aware of this. But he now insists—probably for strategic reasons—on also calling "social" what was previously considered nonsocial (Latour 1988e).

Knorr Cetina: It's a tricky issue, but I don't think one can do that, really. The possibility of something being nonsocial ought to be admitted, I think, if we want to apply the constructivist viewpoint to our own resources and not just to our natural-scientific topic of analysis.

Callebaut: Isn't what you say also a way of avoiding tautology?

Knorr Cetina: Yeah, it's obvious that if you call everything social, then you find that everything is social. It's a tautology, as you say; it doesn't make much sense. You have to be more precise about what you actually want to call "social," at least, if you really want to apply the constructivist approach to your social resources too rather than reserve them for the natural.

Callebaut: Wouldn't most people agree with that now?

Knorr Cetina: That one should do that? *(Hesitates.)* Yes, I guess many people will agree, but still not many people are doing that. There's a lot of what I call "social constructivism" now in sociology of science—the Berger and Luckmann type of stuff—that looks at how things are negotiated in a natural science but doesn't apply the same point of view to its own theoretical devices or to society itself.

3.6.6. *Society Is No Better Explanation Than Nature*

Callebaut: Can you say a few things about some of the crucial concepts in *Science in Action,* Bruno? Otherwise I would have to make a *résumé* of all this.

Latour: You want me to summarize all of it? *(Laughter.)*

Callebaut: Yes! Of course not, but maybe you could provide a road map that would give people an idea of where to start reading Latour.

Latour: The key argument is that you have a reversal of foreground and background. If you have the universal as a rule and the local or particular as the exception, then you go to science to see that it's not completely universal, that there are local and historical events—like, I could say, *everyone* did before we started, but that would be arrogant. *We* reverse background/foreground, and we start positively—in that sense we are "positivists"—from the localities. Karin's constructivism is very much in line with that. So in that sense, *if the universal is always the local achievement of a network being built, background and foreground shift. The key argument is, of course, the network.* Because we are not relativists in the classical sense. All the people of the generation that preceded us were absolutely obsessed by trying to avoid relativism, for good reasons—largely because of the Nazis and the Stalinist hordes. But we are forty years later now, so we can change the arguments a bit by now. *Relativism is a big problem, again, only if you start from the universal as a rule.* But when you shift to our position, you get of course a completely different matter: you now have *localities being extended.* The whole question then becomes: *What price must be paid for the extension of the network?* To watch a given science is to show the price paid for the extension of the networks; how a locality can dominate another one. In a sense, what I did in science and in technology was a bit of the same sort of interest: what is the price to be paid for the extension?

Callebaut: Our readers may not be aware of the extension of your networks to non-human entities such as objects.

Latour: What is at stake—I didn't see it ten years ago, but now I see it clearly—is that we went fast from modernism to postmodernism. Most of the people you are talking with about naturalism are postmodernist, in the sense that they are *disappointed rationalists:* they know that the claims of rationalism can't be upheld the same way, but they want to save whatever they can. Now the important thing to realize is this: *We are not expelling rationality. We are not postmodern.* The key to what is happening now—it is a turn within the turn, a very important turn—is to realize that *we have never been modern* (Latour 1990a, 1992a). The whole modernity argument is about accepting universality as a *given* rather than to see universality as something which is *built.* Everything, then, that is said by way of criticism of our position—that it is relativist, circumstantial, idiosyncratic, and so on—becomes *positive*. There is not a hint of a critique of science in what we do; there is no denunciation of science; we have nothing against the sciences. The whole Critique movement in philosophy is actually modern. The question is *how locality is extended*. The turn in that is to realize that the whole debate about reason was based on another debate, about society. All the traditional philosophers of science were *protecting reason*—exactly like what you said about Ravetz's argument—not because they were especially interested in science but because they had an idea of what *society* was. This is what is now being reshuffled: society *itself* is being redefined.

Callebaut: So you attack on two fronts, so to speak.

Latour: Yes. Many people never understand us, because we are on two fronts: we are not extending society inside the sciences . . .

Callebaut: . . . as, for instance, Edinburgh[30] and Bath[31] try to do (see also Restivo 1983 with respect to mathematics) . . .

Latour: . . . but we are redefining science *and* society. Shapin and Schaffer's (1985) argument (cf. Latour 1990a, b, 1991, 1992a, c; Callon 1992) is so important because they show the archeology of that.[32] We now realize—retrospectively— that we never were modern and that we have always mixed the humans and the nonhumans together (Latour 1988f). I think that's philosophically the most inter- esting argument; and it has got nothing to do with disputes about case studies; it has been shown by good historical and philosophical works, like Shapin and Schaffer's, or Serres' (1987)—all these people digging past the distinction be- tween society and science. In that venture, the most interesting feature is the prin- ciple of symmetry that we devised here in Paris, which is not Bloor's principle of symmetry[33] but another principle: *that society is not a better explanation than*

30. E.g., Barnes (1974, 1977, 1981, 1985a); Bloor (1976, 1981, 1983); Edge and Mulkay (1976); Barnes and Edge (1982); Pickering (1984a, b); Shapin (1982).

31. E.g., Collins and Pinch (1982); Collins (1985); Pinch (1986); Collins and Yearley (1992).

32. In their book on Hobbes, Boyle, and "the experimental life," Shapin and Schaffer (1985) examine "the origins of a relationship between our knowledge and our polity that has, in its fundamentals, lasted for three centuries" (p. 343). From their historical case study they infer that "the problem of generating and protecting knowledge is a problem in politics, and conversely, that the problem of political order always involves solutions to the problem of knowledge" (p. 21), and predict that "the form of life in which we make our scientific knowledge will stand or fall with the way we order our affairs in the state" (p. 344).

33. Bloor (1976:5f.) wanted the "strong" programme of the sociology of science to adhere to four tenets: (1) *causality:* it should be "concerned with the conditions which bring about belief or states of knowledge"; (2) *impartiality* with respect to "truth and falsity, rationality or irrationality, success or fail- ure"; (3) *symmetry* in its style of explanation, in the sense that the same types of cause would explain both true and false beliefs; and (4) *reflexivity:* sociology's patterns of explanation should be applicable to itself (otherwise, "sociology would be a standing refutation of its own theories"). The first tenet involves a commitment to determinism and to the view that sociological explanation is a form of causal explanation— two controversial views (Trigg 1978; Latour 1988c; Slezak 1989a, b). The second and third tenets clash with more traditional conceptions (e.g., Laudan's) of the division of labor among philosophy and sociology of science. (Cf. Slezak 1989a:584: "The entire discussion would be unnecessary, were it not for the philo- sophical pretensions of proponents of SSK.") Thus, according to the "rationalist" philosopher-historian Larry Laudan (1977:202), who is echoing Merton (1959) here, "the sociologist of knowledge may step in

nature. That is philosophically the most interesting part, because it changes the relation between the empirical case studies (which are not empirical any more) and theory (which is not generalization any more). This whole idea of theory as generalization, I think, is wrong (cf. 7.2.6).

3.6.7. *Learning from Philosophers' Theories of Practice*

Callebaut: Like Bruno, you have been called an "eliminativist" *in re* philosophy of science, Karin (Guichard and Looijen n.d.; cf. Henderson 1990). Is it true that you would like to get rid of my ilk?

Knorr Cetina: It depends on what philosophy of science does. There is the kind of philosophy of science that is normative. Frankly, I don't quite understand that kind of thing. If you want to prescribe what scientists *should* do, you would have to know very well what is a successful strategy and what is not. But that is not something you can take out of your plausibility reasoning; you need an empirical basis for making such prescriptions. Now this normative kind of philosophy is no longer characteristic of the core area of philosophy of science. But in fields like philosophy of economics, there is a strong belief that you can work out what scientists should do and actually *tell* them what they should do. That kind of philosophy I don't understand at all. I think it has no right and no justification. Then there's the other kind of philosophy of science which produces "rational accounts" of, say, "theory choice" and things like that. Again I don't think scientists are irrational, or that science in general is irrational at all. There are many kinds of rationalities. Rationality is itself a complex issue, and with a simple-minded model of theory choice you cannot capture what is going on there. So that kind of philosophy of science is also more or less useless. On the other hand, I continue to have some admiration for analytic philosophy, because of its very precise and articulate way of expounding things. I don't believe that philosophers are *fools* or anything like that. There are many philosophers you can read with pleasure. Wittgenstein was a philosopher, after all, and many people in science studies cite him. There is something to be said in favor of Peirce's philosophy of pragmatism, to take another example. Even Heideggerian pragmatism has some interesting aspects. So I wouldn't throw out the baby with the bathwater.

Callebaut: What can a contemporary student of science learn from Heidegger?

Knorr Cetina: I really studied Heidegger when I was in Berkeley, where I partici-

---

to explain beliefs if and only if these cannot be explained in terms of their rational merits." In Laudan's view, normal scientific change, i.e., change which is not *disturbed* by extraneous forces, is rational, and its study the province of philosophy: "a body moving at constant velocity and a man behaving rationally are both 'expected states', which require no further causal analysis. It is only when bodies change velocity or when men act irrationally that we require an explanation of these deviations from the expected pattern" (1977:189). (The parallel between equilibrium situations in physics and the social sciences is also the subject of Samuelson 1975.) The boundary dispute between intellectual historians à la Laudan and radical sociologists is neatly analyzed by Lugg (1983), who argues that "the rational does not exclude the social."

pated in lectures by Hubert Dreyfus (cf. Dreyfus 1990), who has also written on Foucault (Dreyfus and Rabinow 1982) and on AI (Dreyfus 1979; Dreyfus and Dreyfus 1985). These were excellent lectures, in which Dreyfus managed to combine the Heideggerian point of view with Kuhn and Wittgenstein, with a sort of praxeological, "theory of practice" viewpoint. Of course, Heidegger's is a philosophical theory of practice (see, e.g., Zimmerman 1990). But you don't have to stay on the philosophical level. What I continue to be interested in—and that's why I used to like Bourdieu a lot—is a praxeological sociology. I don't mean the Marxist type of theory, which is also in a sense normative and critical, but one that considers the practice of what goes on in society to be the main level of interest. It's an antistructuralist point of view. For example, you don't always try to find the mechanisms *behind* things without considering what is on the surface. If you consider the practice of social agents—I wouldn't have a very restricted definition of social agent—to be the most interesting phenomenon, Heidegger (e.g., 1962) has a theory of practice, although it's not the kind of theory we would produce as sociologists. You can find interesting overlaps between Heidegger, Wittgenstein, and Kuhn in that respect.

3.6.8. *Picking the Losers*

Callebaut: Playing the devil's advocate, couldn't one say that either on the Edinburgh symmetry principle or on yours, there is still a basic problem. When I read your stuff on Pasteur and Pouchet (Latour 1983, 1988a, 1989), the best I can make out of the losers are *anti-winners:* the losers continue to be defined in terms of the winners, because you pick out the topic of a controversy study in terms of the outcome.

Latour: I think a good definition of "to win" is to be able to define the topic. I mean, one of the reasons why Pasteur won was because he defined the topics.

Callebaut: Campbell (e.g., in "Perspective on a scholarly career," in his 1988a) has this point that much if not most of what scientists do *fails.* The history is biased; "Man sieht nur die im Lichte" (Bertold Brecht). Is it really possible to do away with the whiggish bias, or does one always, *necessarily,* view history in terms of the winners?

Latour: I see what you mean. It is a critique of the sort of thing we did.

Callebaut: It's not necessarily a critique. But is it possible to reconstruct the networks of those that got nowhere?

Latour: Yes. It's very interesting actually. We have one guy here, Pierre Lagrange, who in his Ph.D. thesis is doing precisely that. There is a whole section in *Science in Action* exactly about that (ch. 5), but it's a difficult issue because it's precisely the one for which we have no vocabulary—people who are considered "losers" or "parascientific" or "prescientific." So it's very difficult to get back to them. But we have, finally, one guy here doing a study on flying saucers—how do people build the *fact* of flying saucers, the discovery of flying saucers?—in the same terms. So it's feasible. I agree that we didn't do it. It is fair to say that we didn't do it. But it's feasible in principle. It's the same for historians. Historians know how to do the history of little guys. We will learn from the history of failure and of failing scientists—little scientists, little engineers. I agree that we didn't do it. We accepted the winner's point of view, which is the same as the philosophers saying, "Yes, but try Newton! Try Hobbes!"

3.6.9. *The Return of the Object*

Callebaut: Still in the same vein: When I read you I sometimes get the impression that you take it as quite unproblematical that controversies get settled and that this implies that the losers adopt the winners' view. Now Hull—and that may be an artifact of his thinking in terms of populations—claims that if you just ask a hundred scientists what they agree and disagree about, they offer you altogether different lists of answers. *Is* there consensus formation to the extent that some people have claimed? I have the same problem with Kuhn, who often writes as if it is obvious that in the end people will agree on one solution. Is it necessary, in your terms, for a network to be elaborated to have this kind of convergence?

Latour: That may be my French bias.

Callebaut: Please explain.

Latour: In the case of Pasteur, that's typically a French bias, because of our system of having officially settled the controversy before a commission which has some semi-legal force in the twentieth century. In Pasteur's case the controversy *was* settled, because the Academy of Science met twice and settled the controversy. So it may be an artifact—it's very difficult to be half in and half out of a controversy in France. In general, I agree with you . . .

Callebaut: Maybe it's just schizophrenic; maybe people just have to pay *lip service* to the official solution but can continue disagreeing "privately."

Latour: I think that's a good point. The mechanism of closure never interested me that much in the past. Now I'm very interested in it, because I'm studying technical standards. In the case of standards it's very interesting, because you have hundreds of people who say that it would be better to have other standards. The losers *are* excluded physically, because they can't hook up the instruments or their machines.

Callebaut: Maybe it would be useful to have a kind of follow-up study to certain controversy studies: What happens when a new controversy emerges and people are going back to what happened before?

Latour: In my paper on Pasteur and Pouchet (1989) I mentioned that because of the argument that "in the end," Pasteur lost. In that paper I'm fighting this argument, because it's abstract. People say, "Well, in the end there is prebiotics which vindicate Pouchet more than Pasteur." But that's abstract, because it's a hundred and fifty years later. So it's retrospective. But I agree with you that the study of the precise ways in which the winning side is enforced is not very much studied, and that is because you would have to look at a lot of textbooks, at education, at norms and standards—which are very important—to settle and diffuse the final outcome of the controversy. But that's not our strong point!

Callebaut: Let us create some order here. We have distinguished three generations of philosophers of science before. Let us now — *"aller guten Dinge sind drei"* — introduce three generations of sociologists of science as well. First, there was Mertonian sociology of science. Jonathan Potter (1988:37) may call this "the dark ages of normative sociology of science," but in the U.S. that tradition is still very much alive and apparently doing well. Then came the proponents of the so-called "Strong Programme" of the Edinburgh school . . .

Campbell: I'm talking about the "Principle of Symmetry Edinburgh sociologists," because symmetry is a very important principle and the label "Strong Programme" does not say anything about it . . .

Callebaut: . . . and its Bath variant, Harry Collins's EPOR.[34] And finally Bruno Latour's actor-network approach arrived on the scene. In important ways, the École des Mines approach, which takes part of its inspiration from semiotics, departs radically not just from the first but also from the second generation in the sociology of science. As Collins and Yearley (1992:310) put it, "Whereas the Anglo-American tradition of epistemological thought is concerned with how we represent reality (what the relationship is between the world and our representational devices), the Continental tradition more naturally asks how anything can represent anything else."

Latour: The dispute in our field between symmetric sociology or anthropology of science and the classic approach of sociology of science—"classic" meaning the extension of society to nature . . .

Callebaut: . . . in Harry Collins's relativism, "the natural world has a small or non-existent role in the construction of scientific knowledge" (Collins 1981a:3; cf. Barnes 1981; contrast Galison 1987:10, 278) . . .

Latour: . . . is very interesting philosophically, because it is the position of the object which is in question (Callon 1986; Latour 1988b, 1990b; Serres 1992; cf. Dagognet 1989). The sociologists of science—99 percent of them—have accepted that if you grant too much to the realist (that is, to the object position), you are led away from the social. So they see no way of advocating the presence of the social but to deny the presence of the object. It is a complete mirror image. Now what we did—and I think that's philosophically interesting again—is to say there is

34. For Campbell, the principle of symmetry of his friends of the Strong Programme is an important one: "If people, including scientists, in the past have adopted beliefs we now believe are true, that was due to persuasion processes, propaganda, argument, and not proof. We need such social communication explanations equally for the adoption of theories we now think of as true as for theories we now think of as wild, such as Kelvin's on the age of the Earth." But Campbell distances himself from Edinburgh's and Bath's insistence on looking solely for *external* origins of scientific beliefs ("How did the social structure cause . . ."): "What we need is a sociology of science which will be predominantly internalist and deal with the social relations among the scientists and their mutual persuasion and propaganda."

nothing wrong with the object except for the *position* where the philosophers want it to be. Especially Kant. All these people are modern or postmodern; they all live under the shadow of Kant, which means that for them *the object is there, on the other side of the Transcendental Ego*. What I am trying to do—and here I am the disciple of Michel Serres—is to put the objects in the intermediary position *before* it's made clear that they are a thing or a subject (see fig. 3.1).

Callebaut: So your "object" has characteristics which are quite different from those of the object of the past?

Latour: It's an action, an actor, it has a will, it has force, it does all sorts of things. And it has *some* characters of the object; it's nonhuman. But it is also completely different from the Ego, or the collective subject, or the society of the past. To explain science without taking into account the nonhumans is a complete absurdity. It's an absurdity which I understand, because it's an absurdity caused by the absurdity of the other ones—the sociologists. And the debate between the sociology of science, on one hand, and the traditional position in philosophy is understandable. I mean, they play games like this tug of war between object and subject: If you pull the adversary away from the subject/society, then that means that you are closer to the object; and if the other camp pulls towards society, then that means that you are further from the object. *This is funny, but uninteresting.*

Callebaut: What do *you* find interesting, then?

Latour: What is interesting is to try to rehabilitate the nonhuman and to move along the other line. It's also interesting, of course, because it enlightens technology, which is a big issue. So that's our debate with Edinburgh. Our position is not largely accepted, so it's a majority versus a minority so far. But it will win in the end, because it's such a compelling argument: *You can't study scientists without studying the nonhumans.* Now nothing says the nonhumans are in the position of the objects, which is a position *given* to them by Kant. And again, Shapin and Schaffer's book is the occasion of making the decisive move here.

Callebaut: So your "object" is entirely different.

Latour: It is a quasi-collective; it makes the collectives. We find this argument in Serres' work: *the transcendence of Society and the transcendence of Nature are actually one and the same.* And we were completely wrong in thinking that we had to choose between the transcendence of Society or the transcendence of Nature—that is the tug of war—for there is only one. That entirely changes the argument. And since I have written this review of Shapin and Schaffer, I see all this much more clearly.

Callebaut: How about the symmetry point?

Latour: What I just said is linked, of course, to symmetry, because it is tied to a definition of actors who are not social actors, who are of course not occupying the position of the object. And we find symmetry a useful organ to try to dig this collective thing. *What* is the collective thing? What is a *quasi-object,* in Serres'

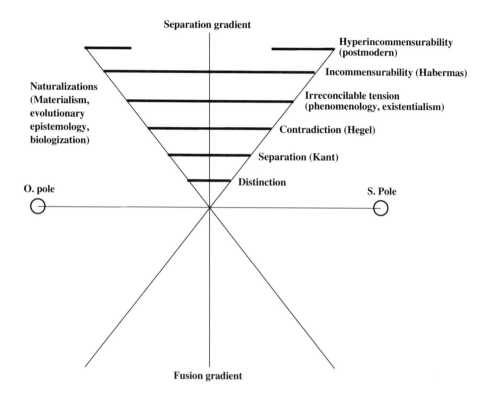

Figure 3.1. Latour's Overview of Modern History of Philosophy. The more quasi objects there are, the wider the gap between object and collective in the major schools of thought.

sense? All this has nothing to do with semiotics, with literature study in the sense of discourse analysis. *Symmetry is fabulous,* because it gives a whole vocabulary to study the nonhumans. The electron is never a thing discovered by Millikan nor a representation imposed by physicists or U.S. industry onto a thing-in-itself we will never be able to reach. Rather, the electron is an actor being defined in Millikan's labs and defining a large part of American society and industry as well.

Callebaut: We will revisit the electrons later (in 4.2.2).

4

PHILOSOPHY OF SCIENCE REVISITED

IT USED TO BE commonplace for philosophers of variegated plumage to announce, with a smile that was meant to be significant, that "real philosophical problems are never really solved—they have been with us since the dawn of Reason, and they are to stay with us forever." Even scientifically minded positivists earlier in this century could accommodate this verdict, claiming mischievously that as soon as a philosophical problem becomes solvable it metamorphoses into a "scientific" one. It took guts to advertise one's belief in the progressiveness of philosophy, even in philosophy of science.[1] Naturalists, of course, know better now.

At any rate, some key issues in the philosophy of science, though not necessarily "perennial" (science is, after all, of fairly recent origin), predated the wars between the positivists and the historicists, were then undeniably marked by these wars, but somehow managed to survive them more or less intact. Three such topics—explanation, reduction, and realism—will be followed up in this chapter because of their continued relevance. It should come as no surprise that they are intrinsically interrelated.

Scientific explanation was *not* an issue for nineteenth-century positivists and empiricists,[2] but it became *salonfähig* in logical-empiricist circles under the influence of "undercover" realist tendencies (van Fraassen 1980:153–57; McMullin 1984). Hempel's classic *Studies in the Logic of Scientific Explanation* (1965) consecrated this tendency. Today almost everybody agrees that Hempel's "deductive-nomological" (D-N) account of explanation—in which what is to be explained (the *explanandum*) is logically derived from one or several laws (universal statements) in conjunction with initial and boundary conditions—is a dead horse. Post-Hempelian approaches to explanation are multifarious, and a replacement view of comparable scope is not in sight. A first group of explanation theorists have moved away from Hempel's linguistic orientation to embrace a more or less radical realism. They talk of explaining an event by

1. Thus Wolfgang Stegmüller titled one volume of his multi-volume opera *Probleme und Resultate der Wissenschaftstheorie und Analytischen Philosophie* (1969–73).
2. For them, the only aims of science were to describe and predict—"to save the phenomena," in Duhem's phrase.

pointing to its insertion into the relevant part of the *causal network* of the world, or, more specifically, by describing the operation of causal *mechanisms* as a viable alternative to the covering-law approach in many cases. In Hempel's heyday, philosophers of science devoted considerable effort to fitting all sorts of "aberrant" forms of explanation into the D-N corset: statistical, inductive, functional, intentional . . . explanation—in fact the kind of things people in the natural and social sciences were most concerned with! Today some mechanism enthusiasts are tempted by a similar eliminativism (e.g., Elster 1983a, 1989b; Veyne 1984).[3] Others—most notably van Fraassen—present in sophisticated cloth a position that ultimately retreats to the comfort of good old conventionalism. Still others, like Kitcher, attempt to revive Hempel's "unofficial" view of explanation as *unification.*[4]

Reduction is, in a way, "the reflection of methodological and metaphysical commitments to the simplicity of nature." It is "the attempt to make precise the notion that under the buzzing, blooming confusion of nature, there is a small number of mechanisms or processes, and a small number of different types of things, which can systematize and explain the world" (Rosenberg 1989:252–53). Approaches to explanation and reduction—the best explored avenue to the *unity of science* in this century—are intimately related (e.g., Sarkar 1992c). On the standard model of intertheory reduction, due to Ernest Nagel and others, a theory is deduced from another theory by means of appropriate "bridge laws"—the reducing theory *explains* the reduced theory. This model was viewed as applying both to (1) "horizontal" reduction, in which a later ("successor") theory which is more general is used to deduce its historical antecedent as a special case, and (2) "vertical" reduction, where a lower-level theory about parts explains the behavior of entities ("wholes") at a "higher" level as described in a second theory (Nickles 1973; Wimsatt 1976a). Oppenheim and Putnam's somewhat neglected (1958) essay combined Nagel's model of reductive explanation with a mereology (i.e., a theory of the part-whole relationship), and—more interesting from the perspective of the current interest in levels of organization, "entitativity" (Campbell 1973), and robustness (Wimsatt)—with speculations about the origin and evolution of the various levels of nature. Nagel-type reduction withered in tandem with the D-N view of explanation, to make room, here too, for a plurality of new approaches, several of which will be discussed in this chapter.

Scientific realism—the claim that science makes possible knowledge of the world beyond its accessible, empirical manifestations (Leplin 1984)—was not much of an issue for the logical positivists, who regarded ontological issues as "metaphysical" and hence "meaningless." Yet, as McMullin (1984:9) has observed, it was only a step from the realization (by Hempel and others) that "theoretical terms have a distinctive and indispensable part to play in science" to "an acknowledgement that these terms

---

3. I assume throughout this chapter that finding explanations in either natural or social science is a desideratum. As so often, Latour (1988c:157) dissents; he urges us to "abandon the idea that offering an explanation is good for your health and inherently better than 'just story-telling.'"

4. Cf. Reichenbach (1951:6) on *generalization* as "the very nature of explanation."

carry with them an ontology, though admittedly an incomplete and tentative one." The realism issue has superseded the controversy surrounding "rationality" that was instigated by the historicists' denial of the fundamental tenets of positivism.[5] Many philosophers have reacted vehemently against the "ontological nihilism" which they think is implied by social constructivism in science studies—even if social constructivists deny this (McMullin 1988). This is why I have organized much of the material in the third and final section of this chapter around the polarity "realism/ constructivism."

One might think that with the success of the new empirical approaches to the study of science, the "linguistic turn"—which many regard as the hallmark of philosophy in our century—would be over. Some keen observers of the field actually see this happening. Van Fraassen (1980:56; my emphasis), for one, who is definitely not a realist, provokingly states that the "main lesson of twentieth-century philosophy of science may well be this: *no concept which is essentially language-dependent has any philosophical importance at all.*" That the realist Salmon (1984:91) concurs with van Fraassen would seem only logical. But we get an interesting paradoxical twist when someone like Thomas Kuhn—the patron saint of the empirical case study adepts— maintains, *contra* such philosophers, that problems like theory change, theory choice, and progress "all . . . have an *ineradicable linguistic* component" (1979:126). I am trying to provide you with a roadmap to this chapter, but I cannot help it if the terrain is so complicated . . .

. . . And such a wasteland. Twenty years ago, David Hull (1969:421) observed: "Not everybody wants to put Christ back into Christmas, nor science back into philosophy of science. Some philosophers view philosophy of science as nothing more than epistemology of the most abstract sort." Concerning explanation, Hull ironically concluded that "whether or not all (or any) scientists ever explain anything by deducing it from a law of nature, scientific explanation is deduction from laws of nature." Let us find out if things look better today.

4.1 EXPLANATION: THE HEMPEL HERITAGE

Callebaut: Philip, the general moral that could be drawn from your "New Consensus" institute (cf. 3.1.1) was—not surprisingly—that it is impossible to formulate a picture of science on which everybody would agree. But you found more agreement on some issues than on others. How did explanation come out?

Kitcher: There is a large measure of agreement that the Hempelian approach, which was enormously fruitful in raising all kinds of issues about explanation, is at this stage dead. There is a majority view which proposes a generic approach for deal-

5. See the collections by Hollis and Lukes (1982), Brown (1984), Margolis, Krausz, and Burian (1986), Pitt and Pera (1987), and Radnitzky and Bartley (1987); also Popper (1975), Barnes (1976), Nickles (1980a), Barnes and Bloor (1982), Newton-Smith (1981), and Shapere (1985, 1991c).

ing with the problems (Salmon 1984; cf. 1989). This is the view that somehow scientific explanation has to be understood in terms of the concept of *causation*. There are two minority traditions. One, represented by Bas van Fraassen (1980, 1985) and Peter Achinstein (1983, 1985), looks at explanation purely in pragmatic terms. Van Fraassen, for example, has a wonderfully lucid account of the pragmatics of explanation, showing how explanations are sought and given in different contexts, but he claims that this is all there is to the theory of explanation. Wes Salmon and I have criticized this on the grounds that it leads to a trivial theory of explanation (Kitcher and Salmon 1987).[6] The other minority position is the search for *global explanatory virtues,* represented most prominently by Michael Friedman (1974) and by me (1976, 1981, 1985b, 1989).

Callebaut: Are these divergences of any relevance to the practicing scientist?

Kitcher: The majority focuses on causal notions and mechanisms in explicating explanation. I think we are able to paint a picture (Kitcher and Salmon 1987; Kitcher 1989; Salmon 1989) of where things stand in the philosophy of explanation which may be quite useful for practicing scientists. We can show that the issue that divides that tradition from the minority tradition represented by Michael and me is one of how you respond to worries about causation. For scientists who don't concern themselves with that issue, it may be possible to find a perspective from which those two traditions come together.

Callebaut: So the biggest split you perceive is the split between those two groups, taken now as one, and the philosophers who think of explanation as not being thoroughly pragmatic.

Kitcher: And that is perhaps the way things have always been in philosophy of science. There has always been a bunch of people standing around, saying, "Science just describes, it doesn't explain"; although, of course, the views of someone like van Fraassen or Achinstein are much more sophisticated than that simple response. But they now occupy that role.

Callebaut: Some reconstructions of scientific explanation make use of causal notions, others—deliberately or not—refrain from doing so.

Kitcher: One way of making use of such notions is just to take them for granted; to say, "Causal vocabulary is all right." For example, some people are prepared to take a counterfactual conditional as a primitive. If you do that, of course you can get on a lot faster! There was a wonderful moment at an APA [American Philosophical Association] session in December 1985, when van Fraassen and I were co-symposiasts discussing Salmon's *Scientific Explanation and the Causal Structure of the World* (1984), and Wes responded. In the discussion period, David Armstrong said (*in Australian*): "You don't have to worry about this stuff about causation. I think your attempt to give an epistemology of causal claims in the

---

6. Gärdenfors' (1990) analysis of explanations, in which the "surprise value" of a sentence is a key notion, interestingly combines the causal and pragmatic approaches.

book is quite misguided. You just *see* causes!" That's amazing: Wes tried to explain very patiently why he wouldn't take that line! But within the group of those who take what I would call the "causal approach" to explanation, there are two different strands: people who want to use a nomic necessity operator, a counterfactual conditional or something like that, and other people who say, "No, you have to respond to Hume's worries about causation, specifically to the concern that causal statements be justifiable." Salmon is one of the latter.

Callebaut: This looks really like a split that is not going to be overcome in the near future.

Kitcher: It's also, I think, a split that is not of enormous concern when one is considering a lot of topics in scientific methodology. Paul Humphreys made the nice and useful point that from the perspective of the working scientist or the working methodologist this is an issue that can be postponed. This might be a more sophisticated way of working out Armstrong's point. *The ghost of Hume doesn't always have to be present to spoil the feast,* right? That's why I said that I think the minority tradition which looks for global virtues in explanation can be lumped together with the mainstream causal tradition—for certain methodological purposes at any rate.

4.1.1. *Mechanisms Not Laws*

Callebaut: Bill, when you studied Theodor Schwann's contribution to cytology (Bechtel 1984a, 1989b), you were confronted with *mechanistic explanation,* and this seems to me also central in your own work. Mechanistic explanation is gaining prominence, both in general analyses of explanation (Salmon 1984) and in social science (Veyne 1984; Elster 1983a, 1989b). But before that, it was to be found in the work of philosophers of biology (Kauffman 1972; Wimsatt 1974, 1976a; Brandon 1985b), including your own work. Like Wimsatt, you came to mechanism from detailed analyses of reduction. How are these two lines of thinking—Salmon's general philosophical approach and yours—to be related?

Bechtel: They come together. I've been citing Salmon for a long time. But my feeling has always been that it wasn't detailed enough. In particular, I've been interested in how scientists *develop* mechanistic explanation. I wanted concrete historical or contemporary cases; whereas Salmon was very analytic and only *suggestive* of the kind of explanation. I don't know how early it became clear to me that the notion of explanation I found most satisfactory involved identifying how the parts of a machine go together to enable the machine to perform a comprehensive process (1982a). Thinking about Claude Bernard probably helped me arrive at that notion.

Callebaut: Why?

Bechtel: In Bernard I saw a model relating levels of inquiry that was *nonreductionistic.* He insisted that you have to look at the parts of a system in terms of how they're contributing toward the whole in which they are embedded—not unlike

Campbell (1974b) with his "downward causation." Bernard's notion of the *internal environment* as a self-regulating environment clearly was a major step toward the kind of multilevel view that I now articulate.[7]

Wimsatt: I got Salmon's *Statistical Explanation and Statistical Relevance* (1971) while I was reading proofs of my long (1972) teleology paper, and wrote a five page addendum to show the relevance of his analysis—which I preferred to Hempel's statistical explanation model—to my account, which had causal probability statements in it. Paradoxically, I ignored what most philosophers took as his main message—to have claimed to have analyzed causal relations. I didn't think it needed much of an analysis, though I found the mark methods and the various forms of spurious correlations useful to talk about. Causality didn't bother me: it is everywhere, at least at higher levels of organization. We couldn't do without it! Like Armstrong, I guess, I think I can "see" causal relations, though I am also particularly aware of the rich array of "visual" illusions involving causality, and think we need to be constantly debugging our perceptual judgments.

Callebaut: What did you see in Salmon's analysis, then?

Wimsatt: I saw his structure of statistically relevant partitions of the reference class, which I took as the expected result of a classification of cases by the different causal factors that articulate to make up a mechanism. Kauffman's (1972) paper was the first one I know of to explicitly model the steps in making a mechanistic ("articulation of parts") explanation, and it still is a classic, although unfortunately unknown outside of philosophy of biology. (Philosophers of psychology, in particular, should read it!) That led me to write my "Complexity and organization" paper (1974) to deal with the problem real-world scientists have (this side of the "apocalyptic complete-in-principle science" that philosophers love so much) in dealing with systems for which we have a number of severally incomplete and only partially articulated perspectives on what they are doing. In my "reductive explanation" paper (1976a), I gave a cost/benefit account of the order in which we include causal factors in an explanation. This account has a natural affinity with the way approximations are used in an explanation, and explains why we don't include every statistically relevant partition.

Callebaut: A kind of pragmatics?

Wimsatt: Exactly so, but one which unlike van Fraassen's leads to a kind of realism, which I call "local realism." This led naturally to defining an "effective screening-off" relation (modelled on Salmon's screening-off relation) to deal with a critical problem in interlevel reductive and mechanical explanation: the fact that we most of the time use upper-level theories and explanations even when we have lower-

---

7. A Bernard-like notion of inner environment is also discussed in Simon's *Sciences of the Artificial* (1981:86), where he defends the middle-of-the-road position that "in the face of complexity, an in-principle reductionist may be at the same time a pragmatic holist."

level mechanistic accounts which we think are "in principle" adequate to explain the upper-level domain.[8]

Callebaut: Dudley, we obviously need an alternative to the D-N account of explanation. Many people today endorse the view that to explain is somehow to invoke a causal mechanism of some sort. As someone who is well acquainted with quantum physics, how do you assess this view?

Shapere: Causal mechanisms—you can't take that idea too much for granted, because they are ultimately quantum-field-theoretical (cf. van Fraassen 1985, 1989). The D-N view is certainly too limited. A good theory of explanation ought to make clear how we can say that chemistry, for example, is understandable in quantum-theoretical terms, in spite of the fact that beyond the most elementary sorts of structures and processes nothing in chemistry can be calculated or deduced in a rigorous quantum-mechanical way. We can't deduce or predict the detailed behavior of complicated atoms in terms of quantum theory. In developing a more adequate theory of explanation, we have to remember, first, that we *can* calculate *some* very simple chemical structures and processes in quantum-mechanical terms, though with approximations; and, second, that we have every reason to believe that the same principles can apply to the complicated molecules and reactions. Then we say that we understand the more complicated ones until we find reason to doubt that.

Callebaut: So we have to consider, in our analysis of explanation, the role of *approximations*.

Shapere: Yes, the role of paradigmatic (simple) examples and of what I call "conceptual devices," such as approximations, in science. A view of explanation truer to scientific practice would answer the following questions, among others: What leads us—in the sense of reasons—to say that chemistry is understandable in quantum-mechanical terms *despite* the fact that D-N conditions aren't realized? What is the role of the elementary cases, in which quantum-mechanical calculations *can* be done, in leading us to say we understand? What is the role of the sorts of approximations that are made in such calculations? (Why don't they hurt the conclusion? What, more generally, is the relation between approximation, truth, and falsity?) Something like this is going on: the approximations employed in the

---

8. Wimsatt explains that Salmon's "screening-off" relation, if applied to interlevel explanation (which Salmon did not envisage), "would have all of the upper-level variables 'screened off' as causally insignificant by the lower-level variables (something which some philosophers of physics and others of an eliminativist or foundationalist persuasion might like, but which is at odds with good sense and scientific practice)." In contrast, Wimsatt's "effective screening-off," interpreted mechanistically, "explains why and when we are satisfied with upper-level causal accounts, and when we feel the need to go to a lower level, without saying that if we have a successful lower-level theory, that all of the 'real' causes are located there." Wimsatt still thinks "it is the only account in the literature which can successfully make room for upper-level phenomena in a reductionist world in a way that fits in any detail the actual practice of scientists." Bechtel and Richardson's book (1991), he hopes, "will properly succeed much of this account—by giving a much richer picture of this process."

simple cases are not such as to dilute the quantum-mechanical explanation of those cases in any damaging way; there are arguments—these and their character must be analyzed—to show that, in all relevant respects, the more complicated cases are not different from these, *and* we can understand why we *cannot* calculate those cases from basic quantum principles.

4.1.2. *Children of Chaos*

Wimsatt: I couldn't agree more with Dudley here. As I argued in my (1979) paper— and I've seen a lot more cases since—I don't know of a single interlevel or compositional account of a higher-level domain of phenomena by lower-level mechanisms which does not involve approximations. To put this another way, the classical reduction model is a good approximation in some cases, but taken literally, it is everywhere false! But the situation is much worse than that. My student, Jeff Ramsey, did his dissertation on the role of approximations in the debates over the choice between the collision theory and the transition state theory in the development of chemical kinetics between 1920 and 1940 (Ramsey 1990). The latter theory was more explicitly connected to quantum theory, but had to use a lot more hand waving to do so—something which both theories did a fair amount of.[9] More generally, philosophers of physics used to chuckle at the inelegance of the other sciences, and the brute empiricism of "natural history." Now look at the menagerie of brutes emerging there, and the even bigger one being studied by searchers for rates of convergence of families of simple functions in the complex plane! *Chaos, fractals, and the computer have made natural historians of us all:* of physicists, biologists, economists—even of the mathematician (Hofstadter 1981). And to turn full circle, since it started there, even meteorologists are rapidly finding that many of these beasts are alive and well in various parts of the natural domain.

Callebaut: When, a couple of years ago, I visited Hendrik Tennekes, the research director of the Royal Netherlands Meteorology Institute, he gave me a copy of a paper, "The recently recognized failure of predictability in Newtonian dynamics," in which a highly respectable Fellow of the Royal Society, Sir James Lighthill (1986:35), granted that even in "wide classes of very simple systems" which satisfy the equations of Newtonian dynamics, predictability is impossible "beyond a certain definitive time horizon." Prigogine, for one, delights in citing this *mea/nostra culpa* paper of a renowned physicist.

Wimsatt: In this work, also, approximations have acquired an ontological significance, even according to Quine's original criterion (see, e.g., Sterman 1988). *Chaos theorists quantify over them!* What are we coming to? As I predicted some

---

9. Ramsey has found "multiple layers of conceptual, experimental, physical, and analytical approximations, intermingled with physical intuitions, interpolations, and extrapolations, curve fitting parameters (the 'steric factor' of the collision theory) that made good physical sense, but had to be allowed to vary over 6 orders of magnitude to fit the data—and lots of other similarly weird beasts" (Wimsatt).

years ago, the preference for a desert ontology of last generation's Compleat Metaphysician is being replaced with a preference for the ontology of the tropical rain forest . . .

Callebaut: . . . Bruno Latour and his philosophical mentor Michel Serres will like this!

Wimsatt: We are just discovering too many new kinds of things. Let a thousand flowers bloom? For the new realist ontology, that's clearly an underestimate, and the eliminativists of the world are about to be swamped!

4.1.3. *Distinctions*

Callebaut: Let us now look at various kinds of explanation—causal, functional, and intentional explanation—in some detail. Jon, you maintain that the only meaningful way to distinguish between various clusters of sciences—roughly: the physical, the biological, and the social sciences—is in terms of the sorts of explanation that they seek, rather than in terms of differences in object or method, as is traditionally done. Before discussing in turn each of the three sorts of explanation that you distinguish and assign to physics, biology, and social science—causal, functional, and intentional explanation, respectively—it will be useful to bring out the *differences* between them when applied to one and the same phenomenon.

Jon Elster (Institute for Social Research, Oslo; see biographical note at 5.2.1): Let us look at three pieces of human behavior. First, take the case of muscular cramps, caused by, say, lack of vitamins. This is a piece of behavior which has no meaning whatsoever; it can only be explained causally, by going into the physiology of the muscles and so on. Consider next reflex behavior such as shutting one's eye against strong light. This is a kind of behavior which in one sense has a meaning. It is a behavior that has been produced by natural selection, and it is meaningful in the objective sense of maximizing fitness, of making the organism adapted to the environment. It is not, however, *meaningful in the subjective sense;* that is, it is not intentionally produced by the agent for a subjective purpose, whereas most everyday actions that we perform, such as opening doors or writing letters and similar actions, are performed on the background of subjective intentions that we want to accomplish. That is a specific form of explanation that is unique to the social sciences: explaining behavior in terms of the *intended consequences.*

Callebaut: With respect to causation, you embrace a rather traditional Humean view; you endorse Beauchamp and Rosenberg's (1981) analysis.

Elster: I really think the ultimate philosophical analysis of those concepts is beyond my competence. Probably it is impossible to say something really valid about causality without knowing a lot about quantum mechanics—about which I don't know anything. So what I have to say about causal explanation and causality will have to rest at a relatively superficial level. At that level, I think the theory of causality proposed by Hume still remains the best and most powerful approach to causality and hence to causal explanation: causal explanation consists in producing laws to the effect that phenomena follow each other in regular, invariant ways, and particular instances of such succession are then explained by appeal to the general law.

Callebaut: There are a number of specific features of causal explanation in the social sciences which you think distinguish such explanation from standard physical cases of causal explanation.

Elster: If you look at the work of social scientists—economists or sociologists studying consumer demand, social mobility, and similar phenomena—you find that much more frequently than in physics they invoke the past, sometimes even the distant past, to explain the present. I think the goal in science, quite generally, is to explain the present by means of the very near past; that is, anything that happened in the remote past should just be captured or summarized by the traces left by those remotely passed events in the near past or in the present. But I think that in the social sciences, very often you must—not as a matter of theoretical necessity but as a pragmatic necessity—invoke the quite distant past in order to understand the present (Elster 1976, 1978, 1983a). This is obvious in cases where culture or tradition is invoked in order to explain present behavior; and it is also widely found in—as I mentioned—attempts to explain social mobility, where it turns out that it is very difficult to predict the future mobility of an individual just on the basis of his present occupation, class membership, or whatever. In order to predict future mobility you actually have to know a lot about his whole occupational history. That is something which, although it exists also in the physical sciences, is, I think, much less important than in the social sciences.

4.1.4. *What If?*

Callebaut: You also stress the importance of counterfactual reasoning in the social sciences.

Elster: Another aspect of causal explanation in the social sciences which, to my knowledge, has no analogy whatsoever in the natural sciences, is the importance of counterfactual statements about what could have happened but did not happen; for instance, to the effect that if the railroad had not been invented, the growth of the American economy would have been such that the American GNP in 1890 would have been a given percentage less than it actually was.

Callebaut: Why should this kind of thought experiment be necessary (cf. Murphy 1969)?

Elster: I think it is implied by the very structure of historical arguments that, even though historians do not often (or at least did not often) recognize explicitly what they were doing, implicitly they were conducting such counterfactual thought experiments. For instance, when a historian cites a number of causes of a given event, say, of the First World War, (s)he usually, explicitly or implicitly, singles out some of these causes as more important than others. Now the only way in which you can make sense of the notion that one cause is more important than another cause is by arguing that its absence would have made a greater difference than the absence of the less important causes. So we might say, with respect to the outbreak of the First World War: "In the absence of one cause, the war would not have occurred at all"; "In the absence of another cause, it would just have been delayed a couple of weeks." In that sense, the first is more important than the second.

Callebaut: Economic historians have gone to incredible lengths to reconstruct whole counterfactual economies.

Elster: Yes. For instance, Robert Fogel, who is here in Chicago, wrote about the impact of the railroad on American economic growth, and in the course of assessing the impact of the railroad, he had to reconstruct or invent a wholly hypothetical economy, where canals were built in places where they were physically feasible and economically profitable; agricultural activities were relocated in order to adjust to the new means of communication; Chicago, which was the center of much American trade at the time, would essentially consist of warehouses; an enormous number of warehouses would have to be built in order to store goods during the months of the year when the canals were frozen and therefore goods could not be shipped; and so on and so forth. I am pretty sure that in the natural sciences, you'll never find anything like this at all—thought experiments on this gigantic scale.

Callebaut: E. O. Wilson, for one, makes exactly the opposite claim, actually (see 10.6.8). What is going on here? Who sees a way out? Bill?

Wimsatt: I think that counterfactual reasoning is required throughout the sciences, but becomes more important either as the complexity of systems increases or as one is talking about historical sequences. Either introduces or increases the possibility of interesting alternative sequences or structures, and you want to know what would have happened if they had been realized. This is especially important in any area where selection or other optimization processes (if there are any!) are involved as a means of deciding the functionality of and explaining the existence of traits.[10] It is very important in evolutionary biology, if that's a natural science. I have a passage in my "False models" paper (1987) on why there aren't three (or more) sexes, and George Williams's book *Sex and Evolution* (1975) is one counterfactual thought experiment after another.

4.1.5. Paradoxical Functional Explanation

Callebaut: Causal explanation explains an event by citing an *antecedent cause,* whereas intentional explanation (to which we will return in the next section) explains actions by referring to their *intended consequences.* Many philosophers

---

10. Cf. Fetzer (1990). Both the "logic" of selection processes and rational choice theory, which describes maximization and optimization processes, are discussed below. On social selection, see, e.g., Van Parijs (1981), Nelson and Winter (1982), Faia (1986), and Elster (1989b). Basalla (1988) has a good discussion of artifactual selection as compared to natural selection.

have had difficulties with this feature of intentional explanation. Jon, you have called functional explanation "paradoxical" in this respect. Why?

Elster: In functional explanation, you explain something by pointing not to intended but to *actual* consequences; that is, you explain it by pointing to something that is brought about subsequently to the phenomenon itself. That seems puzzling, because we would like to think that the phenomenon must have an explanation at the very moment when it occurs. I think this general argument shows one thing: that functional explanation in terms of consequences can never apply to *one-shot* events. For instance, to explain, as Marx did, the English Revolution in the seventeenth century by virtue of the beneficial consequences this revolution had for the supply of workers for industry in the nineteenth century is implausible, because we must have an explanation for the revolution that can actually be used to explain it when it occurred (Elster 1985:431f.).

Callebaut: Nevertheless, you think functional explanation is not inherently implausible or useless.

Elster: No, because if instead of considering one-shot events—unique individual events—we consider a *pattern* of events—repeated similar events—it is possible to explain the pattern as a whole in terms of its consequences. That will essentially involve *demonstrating the presence of a feedback loop from the consequence back to the cause,* so that the phenomenon at one point of time has certain consequences which in turn have consequences of producing or reproducing the same phenomenon at a later point of time. That is the general structure of functional explanation; it explains phenomena in terms of feedback loops from their consequences back to the thing we want to explain.

Callebaut: As such, it is really only a more elaborate or special case of causal explanation in your view.[11]

Elster: As a justifying functional explanation, it cannot really claim to be something different from a causal explanation. To the extent that functional explanation claims to differ from causal explanation, it will have to be by virtue of a claim that we can use the consequences of the phenomenon to explain it, even when we do not know how the feedback loop operates.

Callebaut: For a meaningful functional explanation to obtain it will not do to point to *any* kind of consequence.

Elster: It does very specifically point to *beneficial* consequences, consequences of the thing you want to explain that are beneficial for someone or something. It could be that they are beneficial for the very object we are trying to explain, but it could also be that the consequences are beneficial for something else altogether. The general structure of functional explanation, in the sense in which it can claim to be something distinct from causal explanation, is that we explain something by

11. Cf. Nagel (1961:ch. 12, and 1977), or Wright (1976); contrast Cartwright (1986).

pointing to the benefits it produces for something or someone, and that someone might be the person or the group whose behavior we are trying to explain, or it might be a quite different group, or a quite different feature of society.

Callebaut: In biology, you maintain, the situation is totally different from that in the social sciences, where you find most functional explanation dubious[12] . . .

Elster: . . . because in biology we have a general theory—evolution through natural selection—that entitles us to assert that beneficial features of organisms—features that enhance the reproductive fitness of organisms—*are* there because they have these beneficial consequences. We don't have to demonstrate that in each particular case. We know that if we observe that a feature of an organism is part of a local maximum with respect to reproductive fitness, then that's why it is there; because we have this general theory of natural selection. But there is no comparable general theory in the social sciences that can allow us to explain phenomena by their beneficial consequences in the absence of a demonstrated feedback loop.

Callebaut: Many a philosopher of biology would disagree here, I'm afraid.

Wimsatt: Some, but not as much as you might expect. I would want to emphasize three things here. First, *a feedback process by itself will not do*. Thus self-extinguishing processes could involve negative feedback which leaves nothing to be explained—so maybe we explain the absence of a trait this way? Nor will positive feedback do it: we don't explain all positive feedback processes functionally, even if it does happen to benefit someone. (It would have to occur for that reason.) All chain reaction processes, indeed, all exothermic reactions which are initiated by heat and result in the release of more heat, are positive feedback processes, but they don't automatically justify functional claims or explanations. *Iterative selection processes are required*. They happen also to be feedback processes—so that's necessary but not sufficient for functional explanations (Wimsatt 1972).

Callebaut: So you agree with Larry Wright's (1976) selectionist reading of function statements . . .

Wimsatt: . . . and with his etiological claim for functional explanations, but not for

---

12. A favorite example of Elster's is taken from Pierre Bourdieu's book *Distinction* (1984), in which Bourdieu argues that certain habits of intellectuals can be explained functionally. Elster said, for instance: "Intellectuals like to play around with language. They violate grammar when they feel like it; they may even violate spelling. They experiment with language because words are their tools. One consequence of this, Bourdieu argues, is that the petty bourgeois, would-be intellectuals—for whom being an intellectual is just a question of going by the book—are effectively prevented from rising into the class of intellectuals. The aspiring petty bourgeois, would-be intellectual discovers to his or her dismay that it is a question of learning when to be able to *break* the rules." An objective consequence of the behavior of intellectuals is thus to keep contenders out of their class and thereby to preserve their monopoly power. Bourdieu goes on to argue *that the intellectuals' attitude towards language can actually be explained by the benefit of keeping rivals out of their sphere*. For Elster, this explanation is dubious, for Bourdieu does not in any way suggest what kind of feedback loop there will have to be for those consequences of the behavior of intellectuals to reinforce or maintain that very behavior. (The importance of *reinforcement* in social science explanation is highlighted in Van Parijs 1981, Vaughan and Herrnstein 1987, Herrnstein 1988, and Elster 1989b.)

function statements. Wright's etiological requirement—that to be functional, something has to have already been selected for in virtue of its (past) beneficial consequences—is responsible for at least 90 percent of the counterexamples to his analysis, which have made selectionist approaches appear to be easy targets— especially for philosophers of psychology who don't know any better. (Note that I did not say that all, or even most, philosophers of psychology don't know any better.) In any case, the counterexamples disappear immediately if the etiological requirement is dropped. I considered but avoided that route in 1972.

Callebaut: And what were your other points?

Wimsatt: Secondly, *things may have many functional consequences,* and it's important to choose the right one for functional attributions to be explanatory. That's why Gould and Vrba (1982) introduced the distinction between adaptations (things selected for in terms of their current functions) and exaptations (things selected for for some other reason, which happened to be preadapted to or become co-opted for their current function). Most new terminology doesn't catch on, but theirs has, and with very good reason. Thirdly, I think that the reason for functionalism's demise in the social sciences had to do with *Talcott Parsons's explicit avoidance of the evolutionary/selectionist turn*—paradoxical because Cannon and Henderson, from whom he got the basic ideas, both interpreted function ultimately evolutionarily.

Callebaut: What do you think went wrong there, specifically?

Wimsatt: Without this, Parsons lost an important source for the disambiguation and adjudication of function claims, and he tried to get more structure by looking for functional requirements, which hardened into his "four function" paradigm. Later on, too, his theory became more and more pan-functional. But that's the particular disease of functionalist theories—that are not tied down to selection mechanisms operating in specific contexts on specific systems—as they become overextrapolated and overextended after the first blush of success.

Callebaut: Moral?

Wimsatt: Pan-functionalism (which, in biology, is called "pan-selectionism") must be regularly purged from all functionalist theories. This is true even in biology, where we know a lot about the detailed constraints on what selection can do— limited by the availability of genetic variance for the trait, slowed down by positive linkage with bad genes or too loose linkage with other positively epistatic ("symbiotic") genes, genetic load limitations on rate of substitution of alleles, the rising sun of the neutral mutation theory (Kimura 1983; the pun is deliberate, as its founder, Motoo Kimura, and many of its brightest advocates have revitalized Japanese genetics), etc. Even in biology, models that are fruitful and look like they are working, particularly those that don't stay too close to the genetics—e.g., Harvard sociobiology of the late 1970s—rapidly acquire too much of a panfunctionalist character. That's why Gould and Lewontin (who both still have strong adaptationist flavors—what biologists don't?) wrote their "Spandrels of San

Marco" paper (1979). We all needed that—a bracing blast of ice water—but those who began immediately preaching "Adaptation is dead" misunderstood that paper.

Callebaut: What do you think are the consequences, Jon, of the asymmetry you perceive between biology and social science?

Elster: I don't think much of functional explanation as useful explanation in the social sciences. It can have its uses under certain, carefully limited conditions; but in general, what goes for functional explanation is just unsupported assertion or pointing to benefits while not specifying how these benefits are supposed to have any explanatory power. The specific social science explanation, the kind of explanation that makes the social sciences *different* from the other sciences, is intentional or purposive explanation.

Callebaut: Bill, once again you disagree.

Wimsatt: With concrete limitations on what selection can do, there is no reason why selectionist/functionalist theories shouldn't do well in the social sciences. I think in fact that they are growing again.[13]

4.1.6. *Good Intentions*

Callebaut: Jon, you endorse the view of Max Weber and other "methodological individualists" in (the philosophy of) social science that the basic building block for social-scientific explanation is individual human action (cf. O'Driscoll 1977; also Sober 1981c, 1985b).[14]

Elster: Out of these building blocks we can construct institutions, behavioral patterns, social change, economic cycles—all the aggregate phenomena that most of the social scientists spend most of their time studying. But the basis for these macroexplanations is always—at least in principle—microexplanation of intentional individual action. The thesis that macrophenomena, institutional phenomena, ought always to be explained, at least in principle, by individual human actions, intentions, and motivations is of course a very controversial one. Since Durkheim or, I guess, since Marx or probably already since Hegel, it has been contested by many, who argue that social wholes or collectivities have some kind of causal or explanatory logic of their own, which does not require reduction to individual behavior.

Callebaut: We don't seem to have a methodological collectivist in our milieu here. But in the absence of any other devil's advocates, Bill would like to go back to

13. In conversation, Wimsatt gives the following examples: "Human capital theories are all functionalist in character (complete with flagrant examples of pan-functionalism, you could fairly say—this case fits my above theory very well!). So are game-theory accounts (Luce and Raiffa 1957; Friedman 1986). So is all of Herb Simon's stuff—and it has been for a long time. (Read his *The Sciences of the Artificial* again carefully.) So, of course, is Campbell's work, Hull's, Richards'; I think Ron Giere is at least a closet case, and since Boyd and Richerson's magnum opus in 1985, we've even had a fair amount of population genetics-like theory for cultural evolution."

14. Elster (1989b:13) in fact calls methodological individualism (as opposed to methodological collectivism) "trivially true," granting immediately that "many think differently, however."

your earlier point about the distinction between functional and intentional explanation.

Wimsatt: Who said that intentional or purposive explanation is that different in form from functional explanation? I'd argue that one with Jon.[15] In addition, not all social scientists would agree with him in classifying intentional or purposive explanation as a radically different sort of thing than functional explanation. There's a lot I disagree with in that essay, but Merton's "Manifest and Latent Functions," the first real chapter in his *Social Theory and Social Structure* (1959), in effect classifies intentional explanations as kinds of functional explanations.

Callebaut: You insist that intentional explanation not be simply equated with explanation of human action by some kind of appeal to rationality, Jon.

Elster: There is a funny difference in language between philosophers and social scientists,[16] because when philosophers talk about intentions or intentional explanation, social scientists talk about rational choice explanations. Although these two do largely coincide, there are cases in which people act intentionally but not rationally. Perhaps social scientists have not been sufficiently aware of the importance of this class of motivations. The most prominent form of intentional but irrational behavior is *weakness of will* (Dunn 1987), when people act against their own better judgment, when people for instance accept a cigarette—which is certainly an intentional action—in spite of the fact that they have decided to quit smoking, and indeed think that all things considered, the best thing to do would be to refuse the cigarette. So there is a wedge that you can drive in, as it were, between intentionality and rationality; the two phenomena are not identical.

Callebaut: One can also drive in a wedge between rationality and optimality, which is not only used to explain utility maximization in economics but is also invoked in the context of least effort principles in physics, entropy in chemistry, and survival of the fittest in biology (Schoemaker 1991).

Elster: Economists very often assume that rational behavior, rational choice can be defined in terms of optimizing, in terms of choosing the best act within what they call the *feasible set,* that is, the set of acts that satisfy various physical and economic constraints. It is certainly often true that rational choice is maximizing choice (cf. Rachlin, Battalio, Kagel, and Green 1981). But I think that as a general proposition, the definition of rationality in terms of optimality or maximizing cannot be defended. In more general perspective, a more plausible theory is Simon's (1982) theory of *satisficing,* which essentially says that people in general do not search for the best solution to their problem; they search for a good or satisfactory solution, and when they have found one, they just stop searching.[17]

Callebaut: Please give an example.

15. See Wimsatt (1972); Wright (1976); Achinstein (1983); Bechtel (1986d, 1989a).
16. E.g., the philosophers Davidson (1970), Searle (1983), Dennett (1987), and R. J. Nelson (1987); the economist Becker (1976), and the game theorist Harsanyi (1976; cf. Elster 1986c).
17. See also Leibenstein (1976), Cherniak (1986), and Slote (1989).

Elster: When people go picking mushrooms in a forest, there is certainly in some objective sense one place which is the "best" for picking mushrooms. But if people don't know the forest, they cannot spend the whole day looking for the best place, because then the night might come before they have even started picking. So what people essentially do—and I think that is a rational thing to do—is to look around until they have found a place which they think is good enough and then stop looking and start picking. That is satisficing behavior, which does not necessarily lead to optimal results but to results which the agents think are *good enough*. Now there are many reasons why people might want to satisfice rather than maximize. But the central reason is that we cannot know, before we collect information, what the value will be of that information. We cannot know, if we allocate thirty more minutes to looking for a good place to pick mushrooms, whether that half hour will actually be worth it. If we knew the expected value of more information, then we could compare it with the costs of looking for information, and then go through the usual calculations and stop at the exact maximum. But since we cannot even know the value of more information, we just have to make a decision (Winter 1964).

Callebaut: This means that you do not accept the usual line of defense of advocates of the optimization approach to the effect that they can accommodate satisficing as a special case of optimization (cf. Simon 1979b, 1992a). What is the upshot of this for the ascription of rationality and intentionality to behavior?

Elster: Although many courses of action can be excluded as clearly irrational, it is in general impossible to pick out one behavior and say that it is the rational, the optimal behavior; there are many forms of behavior which are all equally good and satisfactory, and hence all rational. When we distinguish between intentionality and rationality and between rationality and optimality, we get a more true picture of human behavior than by insisting relentlessly on rationality as maximizing, as for instance the Chicago economists typically do.

4.1.7. *Chicago Addicts*

Callebaut: Could you give some examples?

Elster: One of the best-known current economists in Chicago is Gary Becker, who has produced a theory of addiction as rational behavior, where he explicitly tries to argue that we don't need to invoke weakness of will in order to explain addiction (Becker and Murphy 1988). To engage in addiction is, for certain people in certain circumstances, a perfectly rational, indeed maximizing form of behavior. It is not possible to go into the arguments here, but I think certainly for most addicts that argument is likely to fail. I am not saying that there could not be cases in which it works, but I think for addiction in general it does not work.[18]

---

18. Becker was awarded the Nobel Prize in 1992. For various other illustrations of the "imperialist" explorations by economists of phenomena such as crime, discrimination, sexual behavior, learning, and

Callebaut: This example is a case in point of your insistence on the difference between intentional and rational behavior.

Elster: Addictive behavior is certainly intentional—it is done for a purpose. But it is not rational, since it is ultimately self-destructive.

Callebaut: How about the distinction between rational and optimal behavior?

Elster: The Chicago economists would argue that at least with respect to firms we can expect to find optimal behavior, not just satisficing behavior (Friedman 1953; cf. Alchian 1950, 1977). The argument is that the market is based on competition, and firms that don't maximize will be wiped out by competition, so only the maximizing firms are left. This argument obviously rests on an analogy from biological theory . . .

Callebaut: . . . to which we will return in the next section.

Elster: It concedes that *subjectively,* firms may be satisficers rather than maximizers, but then goes on to argue that only the satisficers which happen to hit upon maximizing forms of behavior will survive in competition and remain in place. So as a matter of fact, they argue, only maximizing *behavior* will be observed, although it will not be derived from maximizing *intentions.* This argument has been very powerful in economic theory as a way of buttressing the maximizing assumption which seems quite implausible on the face of it, when you look at empirical studies of how firms actually make their decisions.

Callebaut: In *An Evolutionary Theory of Economic Behavior,* Richard Nelson and Sidney Winter (1982) have argued against this Chicago type of argument for maximizing (cf. also Winter 1964; R. R. Nelson 1987).

Elster: They say essentially that in economic competition—unlike in natural selection in biology—we cannot expect the processes to converge towards equilibrium. That is essentially because adapting in the market is like trying to hit a moving target; whereas firms that do not behave optimally tend to be eliminated. What is optimal changes so fast that firms in the process of being eliminated may be resurrected when market conditions change. This is one of the most central differences between natural selection processes in biology and in economics: that in biological evolution, the environment changes relatively slowly compared to the speed of adaptation, whereas in economic markets the environment changes so fast that most firms most of the time will be quite far from optimally adapted to it.

Callebaut: It would seem that Jon is really relying on a papier maché model of evolution here. Bill!?

Wimsatt: I think that actually says more about the state of evolutionary theory twenty-five years ago than it does about evolution. Few evolutionary biologists today would agree with Jon here (cf. Lewontin 1987).

---

political behavior and organization, see Becker (1976), McKenzie and Tullock (1978), Lévy-Garboua (1979), Elster (1986c), and Radnitzky and Bernholz (1987). See also Ghiselin (1974a) on the interrelations between economics and evolutionary biology.

Callebaut: What would be relevant developments to look into?

Wimsatt: In no particular order: (1) MacArthur and Wilson's *Theory of Island Bio-geography* (1967), which has implications for colonization of isolated patchy environments, that is, most environments for most species; (2) Dick Levins's *Evolution in Changing Environments* (1968), with its recognition of spatially, temporally, and spatio-temporally heterogeneous environments and the beginnings of theory for how to deal with it (see also Levins and MacArthur 1966); (3) Van Valen's (1973) "Red Queen" hypothesis—that species are caught in an ever-escalating arms race with their predators, prey, and competitors, and have to run as fast as possible just to stay even; (4) the species selectionism and punctuated equilibrium of Eldredge, Gould, and Stanley (see references in ch. 2, n. 4)—I'm not differentiating here among significantly different theories; (5) the development of theory for linkage disequilibrium for multi-locus systems, the awareness that different-level units of selection go to equilibrium at different rates, some of them quite slowly;[19] (6) the implications for unpredictability (for species) of chaotic interactions in ecological systems; (7) the enormous rate of species extinctions recently induced by man; (8) the biological implications of continental drift; (9) Alvarez's hypothesis that an asteroid collision caused the final dinosaur extinctions (Nitecki 1984); (10) Raup and Sepkoski's (1984) theory about periodic extinctions, and the "death star"/Nemesis hypothesis of Muller (ibid.) . . .

Callebaut: . . . talk of overkill! What is the upshot of all this, then?

Wimsatt: Even if any given evolutionary biologist doesn't believe in more than a third of all of this—different thirds for different players, of course!—the outcome is overdetermined: *This*, at least, is *not* a difference between economic and evolutionary processes. Indeed, the recognition that satisficing claims could be reformulated as claims that adaptation processes were not close to equilibrium convinced me (in the early seventies) that Simon's (1982) satisficing account should apply to all temporal optimization processes, and of course, to evolution. Not surprisingly, Simon (1977b, 1981, 1983) thinks so too—he's been there already for a long time.

4.1.8. *Variation versus Development*

Callebaut: As was already hinted at, the *ex post* maximization argument of the "Chicago boys" was inspired by a pattern of explanation which is typically used in evolutionary biology. Throughout history, economists and other social scientists have often flirted with biological analogies related to evolution and development. Your biological mentor, Dick Lewontin, has looked into the logic of these sorts of explanation to some extent, Elliott. With him, you insist on the importance of distinguishing between explanations concerned with evolution (phylogeny),

---

19. E.g., for adaptive radiations, tens of millions of years.

which invoke variation, and explanations concerned with a developing system (ontogeny).

Sober: One of Darwin's main contributions to our understanding of evolution was the elaboration of a new mode or pattern of explanation that had not been exploited very much. This is a pattern that Lewontin (1983a) has called "variational explanation." It contrasts with another, older pattern of explanation, which he has called "developmental explanation."

Callebaut: Before talking about the biological applications of these ideas, could you describe what the differences between variational and developmental explanation are in general?

Sober: Suppose we observe a bunch of children in a schoolroom, and we see that they all speak Dutch. We ask, "Why do all the children in the room speak Dutch?" What I want to describe is two ways of answering that why-question, one of which will be a variational explanation, the other will be a developmental explanation. This will give you a feeling for what the difference is. Suppose the *variational* explanation I offer you says the reason all the children in the room speak Dutch is that there was a selection process which only admitted into the room those children who could speak Dutch. By stating that that was the way individuals were introduced into the room, I have explained why the room contains only Dutch speakers. I will not have explained why John speaks Dutch, or why Susan speaks Dutch, or indeed why *any* of the individuals speak Dutch; but I will have explained why the population of children has a certain characteristic.

Callebaut: This would be one example of what Ernst Mayr (1959) has dubbed "population thinking." How would the developmental explanation of the same case look like?

Sober: From the developmental angle, you ask me, "Why do all the children speak Dutch?" And the answer I give you is: "Well, John was raised in a certain household and he absorbed a given language; Susan was raised in a certain household and she absorbed a certain language," and so on. That is, for each of the individuals I describe the developmental processes that produce the characteristic in question; and then I aggregate the individual explanations. The explanation for why all the individuals in the room speak Dutch is that each of them had a certain developmental process take place; a certain process of learning, a certain environment, and so on. Notice that the variational explanation offers no individual explanation at all and hence does not explain what happens at the population level by aggregating individual explanations.

Callebaut: So one way in which variational explanation and developmental explanation differ from each other is that the former is a population-level explanation, whereas the latter is an individual explanation which then explains (not unlike methodological individualism in social science—cf. 4.1.6) some upper-level feature by aggregating individual explanations. Is this the only important difference?

Sober: No. In the variational explanation it is also part of the story that I am telling that *the individuals do not change.* I described an admissions process: to get into the room, the individual has to speak Dutch. It is an assumption of that story that the individuals don't change their ability to speak a language as they enter the room. The individuals don't change; but the explanation nevertheless can explain why the population has a certain characteristic. Part of the essence of the developmental story, of course, is that each individual changes.

Callebaut: What is the carry-over of this contrast to Darwin's theory of evolution by natural selection?

Sober: The theory of natural selection is a variational model of explanation. We explain why a population changes by describing a selection process. The individuals in the population vary from each other; and selection eliminates some of them and keeps others.

Callebaut: How about change? Isn't change quintessential here?

Sober: It is perfectly possible in this story that the individuals do not really change. Each individual remains the same, but the population changes because some individuals are eliminated and some are kept. This is not the same as a developmental explanation, in which the population would change because all of the individuals get older or develop in certain ways.

Callebaut: What was Darwin's contribution here, specifically?

Sober: One of Darwin's insights was that one can explain evolution, or offer certain explanations of evolutionary phenomena, from a variational point of view. This was in contrast to one of the most influential approaches to the theory of evolution that Darwin was combatting, namely the theory offered by Lamarck, according to which evolution takes place by a developmental process in which individuals themselves change and develop in a certain pattern. According to Darwin, individuals do not have this program of change built into them. There is nothing inherent in an individual or in a population that forces it to change along one pathway as opposed to the other. It is all a question of what accidental characteristics of the environment there are and what selection processes that environment sets up. According to Darwin, *that* will determine whether a population goes in one direction or in another.

4.1.9. *No Genes Required: A Business Suit Will Do*

Callebaut: The distinction between variational and developmental explanation remains relevant when one moves from biology to the social sciences.

Sober: Some of the influences on Darwin were from outside of biology; Malthus was an influence, economists like Adam Smith were an influence as well. There has been a reciprocal interaction between biology and the social sciences, beginning with Darwin and extending down to our own time, where biological ideas have influenced the social sciences and also social-scientific ideas have influenced biology. This contrast between variational and developmental explanation has been carried over in some ways into the social sciences. There are theories in economics which try to explain changes in an economy on a variational paradigm. So one sees the development by economists of theories of the business firm, according to which a set of businesses are selected in terms of their ability to compete in the marketplace; the ones that prosper are fitter in the context of that process of competition (cf. 4.1.7). *It is important to realize that those uses within the social sciences of the variational-selectional idea do not require the idea that there is a genetic basis for the behavior at all.*

Callebaut: What is required, then?

Sober: All that is being used is the idea of variation, selection, and retention. You have a set of alternatives; these are selected among; and the best ones within that set—by whatever criterion of "best" you are using—are retained. There doesn't have to be any assumption about there being a genetic basis for the characteristics that are selected amongst. So in the business firm, you have different entrepreneurs differing from each other, being selected, and some prospering and some going bankrupt; but there simply is no assumption whatever that the good businessmen are good *because of some gene* they have; that is just not part of the model. *So the variational, the selectionist model of explanation does not have to have anything whatever to do with genetics.*

Callebaut: If only more social scientists would realize how true that is (much more on this in chapter 7)! On the other hand, there also exists an attempt to extend this pattern into the social sciences which *does* try to bring along with it not just the idea of selection and retention, but the idea of a "genetic basis" as well—sociobiology.

Sober: In sociobiology, the idea is not simply that variation, selection, and retention can be fruitfully applied to social phenomena, but the additional assumption is that the differences in behaviors that one is trying to explain themselves have a genetic basis; and that the selection process is in fact a process that modifies the genetic composition, and therefore the behavioral composition, of the population. This application of the variational idea is of course very controversial. It is important not to confuse it with other applications that don't in any way have anything to do with genetics or with sociobiological ideas at all.

4.2 REDUCTION: WHO HAS THE LAST WORD?

Callebaut: We will approach our next topic, reduction, via a brief historical detour. Physics in this century has witnessed the demise of determinism.[20] What do you think is the significance of the transition from a deterministic ideal of explanation to the probabilistic approach of quantum physics, Dudley?

Shapere: In a long essay on the seventeenth century (1991a), I try to show how much

---

20. Krüger, Daston, and Heidelberger (1987); Krüger, Gigerenzer, and Morgan (1987); Agazzi (1988).

of the debate between Cartesians, Leibnizians, and Newtonians on the proper interpretation of the four fundamental mechanical concepts of space, time, matter, and motion, and of what mechanical theory ought to be like, rests on ideas about *what an explanation ought to be like.* Determinism was a guiding principle, not because there was empirical evidence (the idea of empirical evidence had not become clear yet at that time), but rather because it was based on more general abstract or philosophical ideas, largely inherited from the Greeks (or before) about what an explanation ought to be. One could not have an explanation unless it explained every detail of experience and allowed the specific prediction or retrodiction of every detail of experience. That ideal of what an explanation ought to be and must be if the theory is to be explanatory is rejected in quantum mechanics, and that is of course one of the reasons why a lot of people resisted the acceptance of probabilistic explanation as being a self-contradictory expression!

Callebaut: What are the implications of the acceptance of indeterminism for mechanism and mechanistic explanation? There are people claiming that one has to redefine mechanism; that mechanistic explanations may continue to be what science should be looking for, but they can take forms unsuspected before.[21] Do you think that indeterminism and mechanism—the view that everything can be explained in terms of matter in motion—can be reconciled?

Shapere: The descendant of the mechanical philosophy of nature of the seventeenth through nineteenth centuries might be something like this: Even if physics is changed so that explanation is not in simplistic terms of four fundamental categories of mechanics (space, time, matter, and force), nevertheless the theories of physics, particularly of the very small high energy stage, are the ways to explain, and everything can be understood in those terms: chemistry, biology, I suppose also brain physiology and therefore psychology, maybe—I don't know what else—everything!

Callebaut: Now you are referring to the problem of reductionism, which occupied the positivists a great deal (Nagel 1961; cf. Schaffner 1967, Wimsatt 1979) and continues to interest philosophers today. Considering the diversity of scientific disciplines and specialisms one sees, it is rather obvious to ask whether and how various models, theories, or even whole branches of science are, or might be, interconnected. Reductionists think that physics is in some sense the most basic science, and that in some way all of the "special" sciences reduce to physics. Anti-reductionists picture the different sciences (etc.) as in some sense autonomous. Don?

Campbell: Popper (1974) has said that reductionism always fails but that the efforts are always productive. He gives rich historical details.

Callebaut: What do you think of the ambitious claims of reductionists, Dudley?

---

21. See, e.g., Brandon (1985b), who, in the vein of Grene's philosophy, tries to reconcile anti-reductionism and mechanism.

Shapere: Nagel-type views of reduction won't work, but that doesn't mean that biological entities, say, didn't arise from nonbiological ones in the history of the universe. It's possible to combine such origination with "autonomy" for biology by showing that the *need* and the *possibility* of using unique kinds of explanations in, say, biology can be understood in terms of, say, physics. But if that's true, it's true as a matter of contingent fact: what we have to do in order to decide what ideals of explanation we are to use is *examine nature to find out.* My view is just as consistent with our having—should that turn out to be the case—to have fundamentally and irreducibly different categories of explanation for different domains. It is perfectly consistent with the possibility that we might have gotten a different fundamental explanation for matter and for light, or for living and nonliving, or for living, nonliving, and rational. We've just got to wait and see what we learn!

Callebaut: This seems to bring out something very central to your approach: your insistence that one has to identify *general philosophical possibilities* or "in principle" possibilities—you claim it is the function of philosophy to show that they are open—rather than to define the *limits* (cf. Kant) or the fundamental "necessary assumptions" of science.

Shapere: Yes, the prevailing view since Kant has been that the function of philosophy is to lay down these positive or negative limits. My idea is that *one* task of philosophy is to show that there are no such limits, either way: that we have to find out what the character of our experience, or of nature, is, and how we are to inquire about it and understand it, and what the scope and limits of inquiry and understanding are. The other task of philosophy is to understand the nature and implications of knowledge, inquiry, and understanding as they have actually been found to be, in the course of inquiry.

Callebaut: What do you see coming as far as questions of reduction to mechanistic or physicalistic explanations (in a broad sense) are concerned?

Shapere: Science is clearly headed in the direction of reduction, if that term is properly understood. The application of modern field theories of fundamental physics to cosmology provides a framework within which one can understand the evolution of particles, atoms, chemical elements, stars, and planetary systems (1991a). One can see how the conditions existing on at least one planet—ours—led to the development of organic molecules. And although we don't, at the present time, have any good way of understanding how self-reproducing molecules evolved, most biologists agree that that is only a problem of the *incompleteness* of our picture of nature rather than an indication of its incorrectness.

Callebaut: What do you mean by that?

Shapere: Nearly all scientists expect that we probably will be able to develop a picture of how self-reproducing molecules can evolve from the simpler organic molecules that we know can be produced by all kinds of processes in primitive earth conditions. In fact the recent discoveries concerning the self-catalytic properties of RNA are a great step in this direction (Altman et al. 1987; Cech 1987). If all that works out—and as I say that is a contingent matter—then one will have shown that living organisms can develop evolutionarily through the fundamental laws of physics and chemistry, and ultimately from the conditions existing at the very beginning of the universe, and the laws thereof.

Callebaut: Does that mean that biological explanation must necessarily be physical explanation?

Shapere: Again, it's a contingent matter. Even if we do manage to show how organic life evolves, by the laws of physics, from the conditions in the first microseconds of the Big Bang, there might still be forms of explanation needed in biology which are not identical with those used in physics or chemistry. But it still might be possible to explain, in terms of our understanding of nature—including not only physics but also of planetary environment and the interaction of organic life with that environment—why it is possible and necessary to use a certain type of explanation in biology that is not found in physics. It's not only "facts" that can have explanations; the possibility of forms of explanation can have an explanation, too.

Callebaut: Prigogine, if I may bring him in here, maintains that his views on dissipative structures have no direct bearing on, say, the present controversies about neo-Darwinian evolutionary theory in biology, because his theory is just not intended to address that kind of specificity (e.g., Nicolis and Prigogine 1977).

Shapere: I don't claim that the physical theories have any *direct* bearing on that. What I want to say is that the problem of reduction is to be recast in different terms, such that one can understand not only the organic structures in the light of the prior physical laws and conditions, but also why one explains in a certain way on the biological level, in terms of the same prior physical conditions and laws. That is, one can understand a different mode of explanation as, well, you know: "The universe has developed in this way, the relation between environment and individual organisms has developed in such and such a way, and when that kind of development has taken place, explanations must be of a different sort." We can understand *why* the explanations must be of that different sort in the light of the prior conditions and the way the new conditions were developed. So I would say, yes, it does have to do very deeply with the questions of reduction. This way, you see, *you can have your cake and eat it too,* because you have reductionism in a very full sense, but you still have levels-type explanations—there is a physical type of explanation, there is the biological type of explanation.

4.2.1. *Supervenience*

Callebaut: Part of the problem of reductionism is not empirical but conceptual; it is to try to be more precise about what terms like "reduction" and "autonomy" mean. An idea that has been discussed a great deal in various branches of philosophy is *supervenience* (e.g., Kim 1978, 1982; Rosenberg 1978b, 1983b, 1985; Haugeland

1982; Kincaid 1988; Tuomela 1989; Bonevac 1991). Elliott, you rather like super-venience, because you think it enables us to make sense of the idea of autonomy as opposed to reduction. But what does "supervenience" *mean?*

Sober: Let me talk about this idea in connection with the idea of *fitness,* to give an idea of what the intended concept really is. We look at a population of zebras and we notice that one zebra is more able to survive and reproduce than the other. We ask why, and perhaps the answer is: "The first zebra has stronger legs, so it can run faster than the second." And it is in virtue of that difference between the architecture of the one's legs and the architecture of the other's legs that we say that there is a difference in fitness, that the one with the stronger legs is better able to survive and reproduce, and that is what makes it fitter. Notice that in answering the question about why one organism is fitter than the other, I have described a physical property—having to do with the structure of the leg—of the one and a physical property of the other. Suppose I change species and I look at a pair of cockroaches. I ask, "Why is one cockroach better able to survive and reproduce than the other?" The answer might be: "The first cockroach is resistant to the kind of poisons that human beings put down to kill cockroaches, and the second one is not." And so we explain why the one is fitter than the other by appealing to a physical characteristic of the first and a physical characteristic of the second.

Callebaut: What is the upshot of your comparison?

Sober: Here is the important point: The physical characteristic I appeal to in the cockroach case is entirely different from the physical characteristic I appeal to in answering the question about the zebras. The pair of zebras differ in fitness for one physical reason, the pair of cockroaches differ in fitness for an entirely different physical reason. If I went on to yet another species, say a pair of corn plants, and asked questions about fitness differences, I would undoubtedly get a third physical reason for explaining the fitness differences in that population. And so on for innumerably many pairs of organisms that I might consider.

Callebaut: Where does the idea of supervenience come in?

Sober: Suppose I ask what fitness is, what property we are talking about when in evolutionary theory we talk about fitness. The examples about cockroaches and zebras and so on show us, I think, that *fitness is not a physical property.* There is no single physical property that varies as fitness varies. Rather, what we can say is that for any given pair of organisms, there is a physical explanation of why one is fitter than the other. But *there is no general physical theory of fitness.* That is what is meant by philosophers when they say that fitness is a "supervenient" prop-erty. There is a physical explanation for every example, but there is no physical generalization that can be stated to answer the question, "What is fitness?"

Callebaut: Bill remains unmoved by the supervenience sirens' songs.

Wimsatt: All of this is true, but this doesn't show that fitness is not a physical prop-erty in at least an extended sense of that term. Fitness could be, for example, a

complex relational property, even having, for instance, the same viability and fe-
cundity components in organisms having the same generation times, related to
different lower-level physiologies and behaviors, etc., down to the atomic level.

Callebaut: Elliot, part of the theory of natural selection consists of attempts to get at
the nature of fitness. What role can the idea of supervenience play here?

Sober: It is because fitness is a supervenient property that, I think, one can draw anti-
reductionist lessons about the theory of evolution. The theory of evolution is not
going to be reducible to a physical theory, partly because one of its major concepts—the concept of fitness—is a supervenient property. There is no answer to
the question "What is fitness?" that can be given within a purely physicalistic
theory.

Wimsatt: Of course, we would both agree that fitness is not a concept in any present
or plausible future physical theory. But I think functionalists and supervenience
theorists have sometimes acted as if their special properties were not anchored in
underlying levels of organization—or tie supervenience to irreducibility, which is
a mistake. It is a mistake not in terms of the classical model of reduction, but in
terms of any scientifically adequate model of reductive explanation as "articulation
of parts" explanation (Kauffman 1972; Wimsatt 1976a).

Callebaut: Harold Kincaid (1988) makes a rather similar point, I think, in trying to
answer the question what can be the explanatory adequacy of lower-level theories
when their higher-level counterparts are irreducible: Lower-level theories *do* explain (as is shown by certain facts about causation and explanation), but there may
also be important questions about counterfactuals and laws that they cannot
answer. How about other scientific domains, Elliott?

Sober: I think this kind of anti-reductionist conclusion is not isolated to evolutionary
theory. If one looks to another kind of theory, say psychological theories, I think
it is arguable that the properties that are investigated there are also supervenient;
and that similarly, one can reach a conclusion to the effect that psychological theories will not be reducible to physical theories.

Callebaut: Can you give an example?

Sober: Psychologists investigate the nature of *memory*. If we look at human beings
and ask about the nature of memory—what enables one human being to remember
this or that fact—there is presumably a physical explanation, having to do with
the architecture of that individual's brain. But I take it that other species can have
memories too, and perhaps can even have memories without having brains that are
at all similar to the brains we have. To extend things even further, I would suggest
that it is only a matter of time before computers can be said to really have memories. Right now we have machines that are perhaps too primitive and simple to
really qualify as having the psychological property we call "memory," but I think
this is only a limitation of technology, and in a hundred years maybe we may have
such machines. If so, we will have machines that remember things. We could ask
what the physical basis is for a computer's memory and get an answer. But it won't

be the same answer we'll get for a human being, because human beings are made out of one kind of stuff, computers are made out of something completely different. Let us go back to the beginning. Psychologists will ask questions about the nature of memory. Will those questions be answered by talking about human beings and the particular architecture of their brains? One kind of question will be, but there is another that won't be. If we want to know the particular physical mechanism of memory, attending to the details of the brain will give us an answer, in principle. But if we want to ask a more general question about the nature of memory—a question about what *we*, an advanced computer, and a species other than our own might have in common that enables us all to engage in this activity we call memory—I suggest that we won't get a single physical answer. It is just as in the case of fitness: there are many physical mechanisms that can be memory mechanisms, which is to say that memory is a supervenient property, which is to say that a theory that answers the question "What is memory?" will not be reducible to any single physical theory. So for this broad kind of reason, I would suggest that psychology is also going to have a certain autonomy from physical theory.

Wimsatt: There are stronger and weaker ways of being a supervenient property, as Ron McClamrock has led me to discover. Fitness could be a supervenient property of an absolutely identical set of organisms because it is a relational property and they occupy different but functionally equivalent environments—thus the supervenience arises by quantifying over differences in the environments. In real cases, of course, the supervenience of fitness spans variations both internal and external to the organism, including some correlated changes, either of which without the other would put the organism in a different fitness class. I had always meant this by supervenience, and Rosenberg's (1978b) article, for example, entails this. It turns out that philosophers of psychology mean a stronger thing by supervenience—that the supervenience is due just to variations internal to the system.

Callebaut: By thinking about supervenience, Elliott hopes to address at least one aspect of the traditional question of reductionism versus autonomy.

Rosenberg: But supervenience is just a metaphysical solution to a reductionist's problem. It cuts no methodological ice.

Callebaut: John Collier (1988) argues that supervenience is not robust enough to deny reduction while supporting explanatory relevance. Bill, you object to the Peircean image of the ideal scientific community (cf. 4.2.3). Closely connected to that is your objection to something that philosophers do a lot: talk about what we can do "in principle." You don't think supervenience is going to be of much help.

Wimsatt: Well, it's OK as a first order characterization, but actually there isn't any supervenience in real science, and not for the reasons that most people think. For instance, the mind-body problem is formulated something like this: "In principle, once we have apocalyptic physics that can explain everything, what is going to be the place of the mind in this? What are going to be the mappings between psychological and physical predicates?" and so on. Often they imagine that psychology will also go through an evolution to apocalyptic completeness. And then, super- venience deals with the relationship between predicates in those ideal apocalyptic theories. Well, we don't have such theories and I doubt we ever will. So I think that what we need are concepts that will be useful to us now, given our limitations, given that our data sets involve a lot of noise, given that we introduce errors in our computations, and so on. That is, *what we need is a philosophy of science for real, fallible people* . . .

Callebaut: . . . for satisficers, to use Simon's term . . .

Wimsatt: . . . and that has to be built, from the beginning, on the assumption that errors will be introduced in the structures, that there will be some in the data sets we've got already, that we ignore lots of detail, and that we need to have a concep- tion of theory and of scientific activity that renders an effective way of proceeding even if there are errors in the data and the assumptions—that our laws or what passes for them are not airtight, but a little leaky.

Callebaut: What would be your alternative to supervenience, then?

Wimsatt: The relevant heuristic analogue to supervenience is Levins's (1966, 1968) notion of a "sufficient parameter," which he suggests is "a parameter which cap- tures (most of) the relevant information of a larger set of other variables, in most cases, for the purposes at hand." This vagueness and multiple qualifications are deliberate, but if we look at the profligate use of approximations and idealizations in going from a lower-level characterization to an upper-level one (even in what is called a reductionistic explanation), we will see that every purported case of su- pervenience, as well as every case of purported reduction, is really an example of sufficient parameters. Supervenience is a limiting special (or idealized) case of sufficient parameters which is never found in nature. On the other hand, the very existence of stable higher-level entities in the face of systems which are fluctuating rapidly at lower levels of organization entails the existence of sufficient parameters at every level of organization—whether or not we have satisfactory reductive ex- planations.

4.2.2. Robust and Reliable Organization

Callebaut: Although you don't take supervenience to be very relevant, you don't object to some of the conclusions derived from arguments based on superveni- ence—conclusions to the effect that there are various *levels of organization* in nature which, although being interconnected to some extent, also display a relative autonomy (Wimsatt 1974, 1976a, b, 1980a, b; cf. Campbell 1973, 1974b).

Wimsatt: The reason why levels of organization are important is that they turn out to be collections of robust or real entities that interact most strongly with other robust or real entities at that level. That is why we need to take, as it were, each level in its own terms, and study all the levels, rather than trying to reduce all of the higher levels to lower ones. Rather I see the aim of science to be to figure out what the appropriate levels are in nature and to understand phenomena by explaining them either at the same level or, when necessary, in terms of upper-level and lower-level phenomena; and to as it were "maximize" the explanations we have of everything, and to articulate the relations between different level descriptions of the same entity.

Callebaut: What do you mean when you talk about "robust" or "real" entities (Wimsatt 1974, 1976a, 1981b)?

Wimsatt: Robustness has turned out to be very important to me in a variety of ways. For instance, it has to do with the kind of scientific realist that I am (1981b): I believe that we ought to be looking for *what criteria scientists use in deciding whether to trust an entity, or claim, or whatever*—rather than wondering what happens in the long run, as traditional scientific realists did, or whether our terms really refer. So-called "Scientific Realism" is a misnomer.

Callebaut: What do you mean?

Wimsatt: It is a term invented by philosophers to describe a collection of views which are neither scientific nor realistic. Scientists may assent to it, but it bears little connection with how scientists of a realist persuasion—almost all of them!— decide what exists (but see Lloyd at 4.3.5). For scientists to trust an idea is not to give it a claim to infallibility; it is a kind of heuristic judgment . . .

Callebaut: . . . May I interrupt you for a moment? This reminds me of a similar move Glymour (1982:177) made in reaction to Putnam's "internal-realist" demur regarding reference: "Putnam is quite right that some sort of Glue is required, but why will causal glue not do instead of Metaphysical glue?"[22]

Wimsatt: . . . Right! And if you ask what criteria they use, it turns out that they use *robustness*—the detection or derivation of an entity or result in a number of independent ways. This is related to Hacking's intervention criterion ("If you can spray it, it exists"—said originally of electrons), but that's primarily because if you can spray it, you can usually also apply it with a brush, or put it on with a roller. If you've only got one way to apply it, detect it, derive it, or measure it, you can't separate out the contribution of the applicator, detector, etc. from the object or property being applied or detected. If you've got multiple ways, you can "triangulate" on the object (Campbell) and simultaneously calibrate the means of detection. It's almost magic, and should be the universal solvent for instrumentalist or skeptical worries of anything less than Cartesian proportions. (I'm not interested in those kinds of doubts; I assume we're playing a real game.)

Callebaut: Hacking (1983) was talking about electrons . . .

---

22. Glymour (1982:180) concluded that the "metaphysical realism" Putnam wants to discard "is no more than the framework presupposed by every kind of scientific realism, whether physicalist or dualist. It is a conception whose depths and difficulties are still not mapped. If there are indications from modern physics that it may be wrong, and others from epistemology that it may be excessively optimistic, there is yet, I believe no valid reason to think it incoherent." Putnam (1990:x and ch. 5) has since replied that the notion of causality at stake here is not a physicalist but a "cognitive" one, as "intentional" as the notion of reference itself.

Wimsatt: . . . but I pretty clearly changed the subject matter to paint by adding two other means of application. This isn't quite fair to Hacking because I didn't flesh out his example—you can't spray paint with an electron gun or electrons (at least not very efficiently!) with a paint gun. But the point remains that all of his interesting examples involve multiple means—whether of detection, or intervention, and he could have added derivation or measurement.

Callebaut: All of this falls right out of our concept of an object, which (whatever else it is) is something having a multiplicity of properties.

Wimsatt: But these properties, being different, will be detected in different ways. Thus robustness is a criterion for objecthood (cf. also Star and Griesemer's "boundary objects" [1989]).

Callebaut: Q.E.D.

Wimsatt: It's actually a little more complicated than I've laid out here. You need to make sure that the different means are in fact independent, and that all of the things you are detecting are one and the same; but still, it works.

Callebaut: So you are interested in what you call "local realism" . . .

Wimsatt: . . . that is, "How do we judge, now: Is this trustworthy? Is this veridical, or artifactual, or illusory, or whatever?" Not in a Cartesian sense, in which you have to guarantee the results; but in a differential sense, namely how do we decide which of the various things we posit in the world we want to put much trust in? And I think this might be parlayed into a stronger defense of realism, although I'm really not that much interested in either analyzing or defending the traditional philosopher's account.

Latour: I think Wimsatt has a network theory without naming it so. Same thing with Mary Hesse (1980). I think the connection is simply that these guys, although they are not in disagreement, don't put the weight or the solidity in one pole—either the social or reality. And if you don't have poles any more to explain the solidity of reality, you have a network theory; because then you have to *disseminate the force* (see fig. 4.1). It's a point of material resistance. I know Bill won't agree with that, because he is a realist.

Wimsatt: I have regions, determined by the intersection of the various relevant laws of nature, in a phase space[23] of alternative modes of organization of matter, where any matter there behaves in a simpler, more regular, and predictable manner. They are in effect stable states. But then things that get there will tend to stay there, and these regions will fill up with relatively stable and stably interacting objects.

Callebaut: Dawkins talks about . . .

Wimsatt: . . . "the survival of the stabler," and it's implicit in G. C. Williams; but something really clicked in the early 1970s with Levins's claim that organisms

---

23. One can conceive of a (physical) system as capable of being in any number of states. The variables used in a mathematical model represent distinct actually or potentially measurable physical magnitudes. Any particular configuration of values for these variables is called a "state" of the system. The phase space (state space, property space) is the collection of all possible configurations of the variables.

evolve in such a way as to minimize the uncertainty in their environments. The multidimensional (or Hutchinsonian) concept of the niche is formulated in ways suggestive of a phase space. So in effect, I used a phase space or dynamical systems metaphor for stability to talk about the stability and evolution of all different kinds of organization of matter, at scales from the subatomic through the biological to the cosmological.

Callebaut: In your paper on "Reductionism, levels of organization, and the mind-body problem" (1976b), you mentioned complications.

Wimsatt: There are many complications I'm leaving out. For example, in a deterministic formulation of the problem (ignoring quantum complications, or talking merely at higher levels), talking about relative degrees of regularity and predictability requires that the phase space in question be of lower dimensionality than, and thus lose information relative to, the phase space in which the dynamics are deterministic, for in that space everything is equally regular and predictable. Talking in a phase-space of reduced dimensionality is desirable for other reasons anyway, however.

Callebaut: What is the main point?

Wimsatt: There are regions in this phase space or property space where matter is stabler, and things tend in time to accumulate there—ultimately, I suppose on their way to a heat death elsewhere! And that is where levels of organization—robust collections of robust objects which interact more with one another than with things at other levels of organization—are found. I think this is a good way to think about ontology, because it naturally integrates the temporal dimension, ideas of evolution, stability, multiple detectability, regularity, and contextual dependence into our

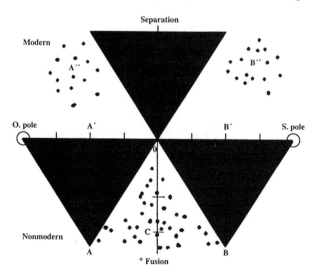

Figure 4.1. Modern philosophy polarized subject and object; "nonmodernists" like Latour prefer a network view instead.

notions of an object. I think you could get a kind of notion of *natural kind* out of it (unnatural kinds as rare, transient, or both?), and do some other things as well. But for all that, it may turn out to be nothing more than a suggestive metaphor which will break down under closer examination. I want to emphasize that although this idea uses robustness, robustness does not depend upon it in any way.

Latour: It's a nice idea, but if you want to have something which is really solid, you have to disseminate the force, not to gather it. And though the force is not isotropic, there are lots of other things. That's why I think all the philosophers who are moving away from the two poles (cf. 3.6.9) are network theorists in one way or another.

Wimsatt: To give Bruno a point, this stability depends not only upon intrinsic properties, but also upon the relational properties of all the interacting systems. But there are some intrinsic properties of systems, relative to the level of organization at which they are described. Networks are networks of interactions among things, just as things are characterized at least partially in terms of the networks they interact in.

4.2.3. *Peircean Cables versus Cartesian Chains*

Callebaut: We will take up the idea of robustness again in 5.3.2. In Wimsatt (1979:368 n. 4), you trace the misconception of the functions of reduction by some philosophers all the way back to Cartesianism.

Wimsatt: The problem with the traditional deductivist or formalist model is that it tries to make the structure of a scientific theory or conceptual schema as *certain* as possible. The way in which it does that—I agree with that, I think it is a useful thing to do—is to say: "All right, let us start with presuppositions or axioms or assumptions about which we can be absolutely certain, and then let's use the most certain means we have at our disposal for getting out our information from that." That means logic and deductions, because valid logical deductions are *truth-preserving* . . .

Callebaut: . . . if you start with true things and use only truth-preserving rules in operating on them, then you'll only get truth out.

Wimsatt: But what if there are some false premises, or if we use rules which are not truth-preserving: heuristics (Gerson 1990), or inductive principles (Holland, Holyoak, Nisbett, and Thagard 1986), or whatever? Well, classical philosophy of science has really nothing to say about that. Of course, they spend some time working on *confirmation* theory, but it is pretty much of a scandal that they weren't able to get out much, if anything, from it.

Callebaut: What, then, do we need, according to you?

Wimsatt: What we need is a model in which you have theoretical structures that may have erroneous assumptions, in which some of your data are false, that you may operate on with rules that can introduce falsehoods, that may have contradictions and so on, because those are the kinds of problems we really face (cf. Priest, Routley, and Norman 1989 on paraconsistent logic). We need to understand how you are able to work with that kind of mess and get useful information out of it. All that traditional positivism has to say about this kind of situation is the computer scientists' lament: "Garbage in, garbage out!" That is, an axiomatic system won't help you at all unless you've got everything certain. But of course that is a counterfactual assumption; we need to deal with the real case. In a way all of my work is asking: *"What kinds of tools do we have, and what kinds of constraints do we operate under as real people, as real scientists solving real problems in the real world?"* That requires a recognition of fallibility through and through.

4.2.4. *From the Piecemeal Approach to a Theory of Everything*

Callebaut: Explanation and reduction are intimately related, although the picture of how they actually interrelate tends to become more and more complex. Dudley, you distinguish the "piecemeal approach," which is concerned with the explanation of specific areas of nature, from the "compositionalist approach." What do you mean by the latter?

Shapere: By an "approach" I mean a very general view about how our experience, or nature, is to be investigated and understood. For example, there's the idea that nature is precise, and that precise explanations can and should be given for every detail of nature. In the seventeenth century, *that* approach was tied to a mathematical approach, the idea that nature is to be understood in terms of mathematics. The compositional approach is still another one (it became tied to the preceding two in the seventeenth century), and it had long been in opposition to a view that understanding of material substances comes through understanding their *perfected states,* their fulfillment . . .

Callebaut: . . . the Aristotelian tradition?

Shapere: Right, but often expressed in this particular way in much (but not all) of the alchemical tradition. As opposed to that, the compositionalist approach, which came to real prominence in the sixteenth through eighteenth centuries, says that material substances are to be understood in terms of their constituent parts, the arrangement of those parts, and whatever it is that holds those parts together. All of these—the compositionalist, the perfectionist, the mathematical, and the piecemeal approaches—are very general approaches. All four have changed. The idea of composition that is found in present-day fundamental physics is very different from the idea of "elements" found in Lavoisier. Any or all of these approaches could have been false, and were in some formulations.[24]

24. Shapere gives an example: "The universe could have been constructed in terms of perfectible substances which don't have constituent parts at all; dirt and lead as imperfected states of gold, for instance. The universe could have been constructed in such a way that precise explanation of details wouldn't have been possible. Plato in fact believed that the world of our experience—the realm of Becoming—can't be described precisely. But what we have found is that certain approaches are possible and others are not; and we have had to alter our understanding of our approaches as we learned. It does turn out, as a matter of contingent fact, to be possible to get extremely precise descriptions, in mathematical form, and to get understanding of the constituents of things. (We may yet have to give that up at the most fundamental

Callebaut: I would like to come back to your distinction between the piecemeal and compositionalist approaches.

Shapere: The piecemeal approach has to do with investigating isolated domains, on the assumption that that's possible. It need not be a compositionalist approach. There *have* been "global," as opposed to piecemeal, compositionalist approaches: the atomism of Democritus and of Newton were of that sort. Piecemeal compositionalism was a characteristic of early modern chemistry, even before Lavoisier: the elements of Becher and Stahl, for example. It really became a way of explaining the domains of physics only later. But when it did, scientists found that that approach could do what it promised to do. It even did more. For through the application of that piecemeal approach, in its compositionalist form, scientists gradually found that they could achieve unified accounts of what had originally been approached as separate, isolated domains. Today, science is really on the verge of—if I can borrow the expression used by many physicists today—a "theory of everything" (e.g., Hawking 1985; Barrow 1991).[25] Although really impressive things have been accomplished in the direction of getting such a theory, lots of problems still have to be overcome. No doubt the supertheory, if it's achieved, will change still more of our understanding, not only of nature, but of how to understand.

Callebaut: Dick, physicists and biologists obviously tend to think differently about reductionism.

Burian: I don't know if I have a fully settled view about that (1985a, b). I don't think there is going to be a unified language of science or a grand unified theory. Even if the physicists produce a successful GUT of the sort they seek, it won't be a Grand Unified Theory for all of science. It can't be, because it would not be capable of dealing with the sorts of higher-level entities whose behaviors and regularities do not depend on the behavior of the fundamental particles out of which, in some sense, the universe is composed. To the extent that the apparently hierarchical structure of our universe is taken seriously—and I believe that it is a correct reflection of the structure of our world—there are significantly different rules for the behavior of entities at different levels. Although the universe is built out of

---

levels: the division of nature into discrete, independent objects may itself turn out to be only an approximation that works on certain levels.) It is possible to give compositional explanations, even though we have to modify our understanding of what composition is, what the forces involved are, and so forth. The compositionalist approach has, for instance, been altered by the idea that something need not be a constituent in order to be separable from it: electrons aren't separate constituents of neutrons before they're released in beta decay for instance. Something can be a constituent but not be separable: quarks are 'confined' in protons and neutrons."

25. Such a theory, physicists expect, would combine our understanding of matter and force, and of matter-force and space-time, into a single unity. It would, according to Shapere, "combine a unified quantum field theory of forces and particles with the theory of space-time which is embodied in general relativity; and that unified theory would be entwined with Big Bang cosmology so that, ultimately in terms of that supertheory, we would be able to understand the history of the universe from beginning to end."

common materials, the theories appropriate to different levels are not going to be derivable from or have any such simple interrelationship with lower-level theories. Some version of supervenience might be correct in some cases, but I think there are also other cases where you have interesting type-type or token-type relations that are not supervenient,[26] where you have ontological composition all the way down of lower-level entities but behaviors that cannot be derived from those at the lower levels.

Wimsatt: The piecemeal approach goes very deep.

Callebaut: To avoid possible confusion, let me note that your usage of this expression is the more generally accepted one and broader than the meaning Dudley attributes to it.

Wimsatt: Popper talked about "piecemeal engineering" without really appreciating, I think, the importance of the idea. Simon went much deeper with his *near-decomposability* (see Simon 1981a:ch. 7, for an idea that was already extensively developed two decades earlier).[27] And Shapere is definitely right that the possibility of a piecemeal approach depends upon basic facts of nature. Abner Shimony appreciated this very early, and has to be regarded as one of the earliest prophets in the recent rise of naturalism—I remember a talk of his in 1968, where he remarked that the very separability of laws and boundary conditions in physical theory rested upon basic empirical facts of nature. More recently, I applied this approach to the analysis of levels of organization (1976a, b). And Lewontin pointed out, at least a decade before his article on "Adaption" (1978), that evolution required that it be generally the case that you could modify one thing without screwing up everything else (this is his principle of "quasi-independence"—so much for Leibnizian "functional coherence" theories in biology). He also showed

---

26. On the received view of (vertical) reduction, a one-to-one correspondence between the primitive terms (which may be entity and/or predicate terms) of the upper-level theory and the corresponding terms of the lower-level theory must obtain (this is the "condition of referential identity"; see, e.g., Schaffner 1976:618). As Fodor (1974) and others have pointed out, this requires that the statements being interconnected are true laws, i.e., are formulated by means of "natural kind" or "type" terms (cf. 6.6.1). Advocates of the strong identity theory concerning mind and brain thus claim that it is (or ought to be) possible in principle to define a one-to-one correspondence between psychological and neurophysiological natural kind terms (cf. 8.4.1). In practice, what one usually finds are at best weaker many-many, many-one, or one-many relations between upper- and lower-level theories (see, e.g., Hull 1974, on the reduction of classical or Mendelian genetics to molecular genetics). These weaker relations tend to exist between a type and a number of tokens exemplifying it, or, worse, between tokens and tokens. For the purposes at hand, "money" would be an example of a type term from economic theory—it occurs in economic "laws" (but see Rosenberg 1980, in particular ch. 6). Particular coins, or cheques that have been made payable, would be examples of tokens going with this type. According to Hull (1976b) and some others, a many-one relation from the lower- to the higher-level theory will typically make actual reduction so complicated as to make it impossible in practice, whereas a one-many relation in the same direction will make it impossible in principle (cf. Wimsatt 1979).

27. Boyd and Richerson (1987:27) trace "the conscious use of the strategy of using simple models to study complex phenomena" at least as far as Max Weber's use of "ideal types" to study human societies. The interesting relationship between Simon's and Weber's work and influence is discussed in Ostrom (1974).

that small changes in phenotype generally resulted in small changes in an organism's fitness and environmental niche (his "principle of continuity"). In two papers (1981a, b) I argued that differential selection ought to favor more nearly decomposable or more quasi-independent phenotypes, so that the evolution of or selection for quasi-independence paradoxically meant that if (or rather: to the very limited extent that) a Williamsian or Dawkinsian genic selectionist picture is true, it is true because selection at higher levels produced it! Indeed, there is here perhaps a "meta-piecemeal" approach, but—following Stu Kauffman's (1972) paper in another context—at least a very large number of different ways of decomposing the pieces. That doesn't mean that it is arbitrary of course, but just that nature is intrinsically multidimensional.

4.2.5. *From Inner Environment to Interfield*

Callebaut: The issue of intertheory connections features prominently in much of your work, Bill Bechtel. In particular, you seem to have taken quite seriously the work of Lindley Darden and Nancy Maull—who were students of Shapere and Toulmin—on "interfields" (Darden and Maull 1977; Darden 1978), which they offered as an alternative to reductionism (Maull 1977). Let us begin with your work on cell biology (1984a, 1989b), biochemistry, and some related things (1984b, 1986b, c, 1988c). Were/are these separate lines of inquiry?

Bechtel: They fall out of each other. (Cell biology is a term I shouldn't have used for the early period, it should be called "cytology.") My primary interest was how work on metabolism and nutrition developed in the nineteenth century. Cytology was relevant as people located the cell as the center of metabolic processes—Schwann's contribution. He and others developed theories about how cells incorporated material, how they received nutrition, and what their composition was. These kinds of questions were explored by others such as Bernard in the nineteenth century (Holmes 1974). I think I fell into the same trap as the actual history did.

Callebaut: Unwittingly, you were a victim of reductionist thinking?

Bechtel: Yes. Coming into the twentieth century I kept looking at how more basic level explanations were developed. Clearly physiological chemistry—later biochemistry—did look downward with successful results as it became the most progressive area of work on these problems in the twentieth century. Cytological level studies simply couldn't make the progress as there weren't the research tools at the level of the cell as there were at the lower level, where you ground up cells and did analytical and functional chemical studies. So the work I did in biochemistry in the first thirty or forty years of this century was in fact the direct successor of my earlier work. Hence I was really looking at the relation between physiological studies of metabolism and nutritional studies, which had been going on parallel paths, and the question of how they became integrated with the work on vitamins and the discovery of coenzymes and of the close compositional relationship between vitamins and coenzymes (1984b).

Callebaut: And somewhere along that road you changed your view about reduction (1983b)?

Bechtel: At least insofar as I took lower-level explanations to be more fundamental and took theory reduction to be the means for linking disciplines. As I worried about the relations between work on nutrition and metabolic studies, I realized that there weren't clear levels there, and certainly no attempt to derive a theory at one level from one at another. Hence I became attracted to Darden and Maull's notion of an interfield theory, which allows for connections between fields of research without derivations of theories. It provided a far more useful framework for describing the historical developments than the theory reduction model.[28]

Callebaut: I wonder whether your own position in general couldn't be compared to that of Bernard? I am thinking of your favored way of using and interpreting connectionism (cf. 8.2.3), of your views on the mind-brain issue, and on the philosophy of mind in general. Your position on these and other issues is quite complex and balanced; you are not just siding with the proponents of one extreme and opposing the other (e.g., Bechtel and Stiffler 1978; Bechtel and Richardson 1992).

Bechtel: In the philosophy of psychology you've got the strong West Coast, San Diego reductionists (the Churchlands) and people like Fodor (1975, 1983a) at the opposite pole. I have found neither the claim that you have to have either a totally autonomous level of description of the mind (Fodor) nor the claim that it must be reducible to a neuropsychological account or even eliminable (San Diego) to be satisfactory. It seemed to me that mentalistic accounts as found in cognitive psychology—not necessarily folk psychology—are needed because they describe real causal processes, but that these processes are grounded in neurophysiological processes and that this grounding makes a difference.

Callebaut: I would like to go back to Bernard; it seems to me that you are fascinated by him.

Bechtel: In fact Bernard was the hero I took in an early paper (1982b), in which I looked at implications of his position for psychology. I developed an analogy between those seeking autonomy for psychology and Justus Liebig, who had developed a complete model of nutrition for the animal organism without any physiological studies of the various processes. Bernard showed the mistake of that

---

28. Bechtel's interest in reduction stems from his work with Wimsatt in graduate school. At the outset he was interested in the relationship of physiology and chemistry: "I examined the path by which physiological chemistry and later biochemistry sought to explain physiological functions. I shared Wimsatt's reluctance to endorse the reductionistic scenario, and so I contrasted those who sought to reduce physiological activities to single chemical processes with those who looked for integrated chemical systems that performed physiological processes." He was also excited by developments in colloidal chemistry in the early decades of this century, which he saw as a resistance to the lower-level explanations developed by biochemists. "The colloidal chemists sought to trace some physiological phenomena to the physical structures established in the cell and not to discrete enzymes in the cell. I was attracted to the proposal that physical chemistry might explain ion gradients without requiring biochemical mechanisms like the sodium pump."

endeavor by showing that once you did understand the internal processes, it put additional constraints on the kinds of overall chemical reactions that you could postulate in the system.

Callebaut: How does this relate back to psychology?

Bechtel: The blindness of trying to do chemistry without physiology is the same kind of blindness that I see in trying to do neuroscience without psychology or psychology without neuroscience. As in many other cases, you need additional constraints that come not from the level of inquiry you're working at but from other levels of inquiry. That is quite compatible with the kind of analysis of reduction Wimsatt advances, even though I am no longer attracted to the theory reduction model as a tool for analyzing the development of science.[29]

Callebaut: And you consider interfield theories to be a step in the right direction?

Bechtel: Yes, since the fruitful connections often don't take the form of derivation but of supplying needed information. However, Darden and Maull's analysis leaves out a lot of what you need to understand science—the social dimension of the people doing science.

Callebaut: An interfield is supposed to consist of many items of various sorts in addition to theories: problems and expectations as to how to go about solving them, methods, techniques, and what have you. In Darden and Maull's view of scientific change, *problem shifts* are crucial; problems arising in one field typically occasion the import of concepts, techniques, etc. from another field.

Bechtel: But they're all intellectual items: problems . . .

Callebaut: . . . I guess that in their defense one could claim—siding with Kuhn or even Rorty here—that like justification, problem definition is at bottom social (see 5.3.4; cf. Gutting 1984).

Bechtel: You have to bring in more of what Hull (1988a, b) talks about in terms of the community of scientists; and you need tools for identifying that community and determining what maintains the structure of the community. A field has got to include the scientists, the mechanisms by which they come into a domain of work—which is going to involve the vehicles of education—the grant awarding process, the hiring process, membership in professional organizations. . . . In terms of such social networks, one can look at what kinds of techniques these scientists are using. Darden and Maull's notion of a field did include research techniques, but that needs to be highlighted much more (cf. Griesemer 1991; Clarke and Fujimura 1992). Problems—if the notion is treated relatively loosely—also help constrain the group you are working on.

Callebaut: Not only the received view of reduction has capsized. General Systems Theory, which for a while looked like an attractive alternative unity of science

29. At this juncture, Wimsatt offered the following specification: "My *inter*level reduction model is not a theory reduction model; only the quite different successional or intralevel reduction model is, and it doesn't apply to this kind of case. I hope we're in the same camp, because I regard Bechtel and Richardson's work as a natural and very rich extension of mine. I wish I could have written it!"

movement, also seems to have largely failed to deliver the goods—as a *theoretical* enterprise, that is (this is not to deny the *practical* relevance of various *specialized* systems approaches).[30] The "structural uniformities" to come out of GST have been few and rather uninteresting (Miller 1978). Maybe *connectionism,* which we'll discuss in chapter 8, will be able to do some of the things that systems theory promised at the time but was unable to deliver. I mean, systems theory really seems to be dead . . .

Bechtel: . . . it seems to me that's right . . .

Callebaut: . . . although in Europe it's not usually viewed that way.

Bechtel: I've never understood it. I once went to the Society for Systems Theory's meetings and came away totally unclear as to what I was hearing. There wasn't a precise underlying conceptual framework to work with. The thing with connectionist systems is that you can figure out how they work. You're not talking in the air; you build them, you tinker with them, and so you got a real experimental entity on your hands.

Callebaut: Actually I guess that what the defenders of systems theory would say— or rather, the systems *approaches,* because the thing certainly lacks unity—is that with the advent of self-organization and *autopoiesis* (e.g., Varela 1979; Maturana and Varela 1980; Krohn and Küppers 1989; Krohn, Küppers, and Nowotny 1990) they now have something similar to play with.

Bechtel: If you have strong enough mathematics, you're off and running. Perhaps the advantage of connectionist systems is that they're simple enough; so that those of use who lose it when we just get abstract mathematical analyses at least see the point of developing the mathematical analyses. We can experience the way the systems work; we can actually build and tinker with models and play with them long enough so that we then get some sense of what is going on. Clearly, *the mathematics is what does the work.* If you have the kind of mind that deals with mathematical arguments, you can gain much from that alone. I just don't have that kind of mind.

4.2.6. *Ephemeral Interdisciplinarity*

Callebaut: You have produced an excellent volume on interdisciplinarity (Bechtel 1986a). How did you get interested in that topic?

Bechtel: Classical reduction got you nowhere in the interdisciplinary area. What kinds of fruitful links *are* built, then, between disciplines? That's where Darden and Maull seemed on the right track. But I realized, even in the introductory paper

---

30. A founding father of the systems movement, Ludwig von Bertalanffy, described the enterprise thus: "Reality, in the modern conception, appears as a tremendous hierarchical order of organized entities, leading, in a superposition of many levels, from physical and chemical to biological and sociological systems. Unity of Science is granted, not by an utopian reduction of all sciences to physics and chemistry, but by the structural uniformities of the different levels of reality" (quoted in Oppenheim and Putnam 1958:29–30 n. 10; cf. von Bertalanffy 1968).

(1986b), how much more was involved; I have in mind the social factors. But it did not come home to me—even though I said things there—*how* critical these factors are. It has become clearer in my recent work on modern cell biology, which was an interdisciplinary project at the outset; by 1960 it ends up being a discipline (which is not at all necessary for interdisciplinary projects).

Callebaut: Was this development foreseeable?

Bechtel: By 1950 you might have guessed it, but it was not an inevitable outcome. There you see the struggles of people coming from morphology to work with people from biochemistry, from genetics, from pathology. They all have different orientations. You see the ignorance of the literature of other fields, so that people are rediscovering what other people have done.

Callebaut: Don Swanson (1990, n.d.), an information scientist at the University of Chicago who is aware of Darden and Maull's work on interfields, is currently developing the tools to study this very process—the coming together of complementary literatures, which he describes as itself a discovery process, a source of new knowledge.

Bechtel: I don't take that ignorance as a bad thing. In an evolutionary model of science (see 7.2), you would value the fact that people *don't* know what everybody else knows; *it helps* to be ignorant, it allows for new variants to get into the pool.[31] Again, that's why I am now interested in the social structures. One of the things that struck me as a question that I need to pursue more is: What are the possible arrangements for interdisciplinary work? In some cases, what looks like interdisciplinary work ends up congealing into a discipline. Often, once the research tools become available, there is a niche for a discipline. But that outcome is not inevitable. I certainly don't think cognitive science will become a discipline. In fact I argue against it—it should *not* be looked at as a discipline in the making, but ought to remain an "interdisciplinary research cluster." You have anthropology, psychology, philosophy, AI, and linguistics already there, capable of doing their own kind of work. People need to be trained in one of these disciplines extremely well. Cognitive science can integrate the activities of people in those different disciplines, but it need not ever jell into a discipline.

Callebaut: How about philosophers?

Bechtel: Maybe philosophers will become generalists, but that's, you know, what happens when you don't produce your own data. Having generalists is not going to make for a good, productive research program. There is too much richness in that interdisciplinary milieu to have a single discipline.

Callebaut: One of the founders of General Systems Theory, Kenneth Boulding, used

---

31. Compare how Latour entered the Salk Institute and thus sociology of science: "Professor Latour's knowledge of science was non-existent; his mastery of English was very poor; and he was completely unaware of the existence of the social studies of science" (Latour and Woolgar 1986:273).

to insist that people be trained well in some discipline before being allowed to get involved in interdisciplinary work.

Bechtel: There has always been a communication problem between people who have different training. I never appreciated that there could be so much of such a problem between AI and cognitive psychology until I participated in some early meetings between Larry Barsalou and Janet Kolodner, and their students. Bob McCauley and I occasionally sat in as kibitzing philosophers. It was revealing to see how often someone from AI would say something and be followed by someone from psychology who would say, "What we would say . . . " There was a whole process of *negotiation of the language* going on here. They didn't necessarily understand each other, and there was the *naïveté* of a philosopher sitting on the outside, assuming that these two groups of people had already found how to interact. Well, some people had, but they still had to learn about each other.

Callebaut: Apostel and Vanlandschoot (1988), who tried to launch a very ambitious series of interdisciplinary projects at the Free University of Brussels some years ago, give an instructive description of this negotiation process.

Bechtel: An inevitable part of working in an interdisciplinary area is the realization that *the other person is not going to do exactly what you do*—something I find mainstream philosophers have the most trouble with. When I talk to my colleagues about what's going on in psychology, they say, "Aren't they making the silly mistake *X*?" To which I say, "Well, you don't understand what they're doing, but you've already formulated the framework in which you are going to filter everything they do! And it's a framework that's totally incompatible with the enterprise these people are engaged in. You haven't got the tools . . . "

Wimsatt: Yeah, there's a real paradox here, and it's why incommensurability isn't total BS. It took me between five and eight years, starting as a philosopher, to really digest the input I was getting from Lewontin (e.g., 1963, 1969, 1974a, 1974b) and Levins (e.g., 1966, 1968). Or rather some parts of it took that long, and it was a real struggle throughout that entire period. I think it wasn't until the late seventies that I really felt that I was thinking about science like them (not as fast, you understand!). But it had a cost. I can remember Bill and Bob Richardson both telling me, three or four years after they had left Chicago and had taken the hardcore turn towards science: "We were always being asked by the younger graduate students, 'What the hell is Wimsatt up to anyways? I don't understand what he's doing or why he's doing it.'" After which they tried to explain it, and usually did—successfully.

Callebaut: And what did you say to them?

Wimsatt: I said to them, "Just you wait, you'd better write some 'real philosophy' papers now to prove you can do it, or work really hard to keep feet in both camps, because pretty soon, you're going to be in exactly the same boat." This is one reason why I don't think there is any best way to do philosophy of biology (cf.

10.3); we need a bunch of intertranslators standing in different positions, and all talking to one another. Even if two people can't communicate well directly, if things are going right, you can find translators. Elliott talks better to many mainstream philosophers than I do, and, although he does an excellent job with many scientists, there are some I know who can't understand what he's saying, or—more critically—why. We need all of us, each with our respective different perspectives.

Callebaut: Interdisciplinarity may be handled at all kinds of degrees of difference.

Bechtel: Consider for instance Ulric Neisser's approach to Gibson (see 8.3 for the background to this example). Neisser came out of a *Gestalt* tradition. Then he worked with Selfridge on letter recognition models, pandemonium models of neural networks. Subsequently he went to Cornell and wrote *Cognitive Psychology* (1967). At Cornell he starts listening to Gibson. The way he himself describes his transformation is, "This guy is a raving maniac. This guy is *saying* something. I'm going to incorporate some of what he is saying. But of course he is not going to like the rest of what I'm saying." That's still the dynamic now. Although James Gibson has gone, Eleanor [Gibson] has still that reaction to Dick: "Well, you're right here, but why don't you stop?" It was a negotiation between two approaches within psychology—or at least from Dick's side—to negotiate the language, to figure out what's relevant, to decide what's important, and then to go on with it as part, now, of *his* approach to doing psychology.

Callebaut: In his abstract for your conference on Integrating Scientific Disciplines (Georgia State University, 1984) David Hull discussed possible undesirable consequences the institutionalization of interdisciplinarity could have. A primary locus of innovation is the small, ephemeral research group. David wrote: "We specialize, compartmentalize, reduce, clump, etc. because we have to. We all work within a big picture without ever being able to stand back and see it as a whole." He was emphatic that "periodic bouts of Platonic angst to one side, most of us are capable of functioning under such circumstances. In whatever bits of time that we can snatch from our 'real' work, we should do what we can to make sure that this informal system of periodic cooperation and regrouping is allowed to continue." Bureaucratic attempts to improve education often have had counterintuitive results, and David very politely expressed the wish that bureaucrats should not be allowed to strangle research as well.[32]

Bechtel: I think attempts to organize interdisciplinarity have failed. I am most comfortable when I am appointed in a philosophy department. Even though I reach far outside of philosophy, I still know the rules of being in philosophy. My previous two appointments were in interdisciplinary programs, and they were far less comfortable because we didn't share the background of working from a discipline. Also

---

32. David Hull wrote: "All I ask of the bloodless creatures sitting on the Course and Curriculum Committee is that they not make teaching impossible. I ask no more of the organization of universities with respect to research. It would be nice if it also facilitated research, but I suppose that this is really asking for too much."

you can't really tell in advance where the fruitful links will be made. *Interdisciplinary work is not valuable in its own right.* There is a very great value to people pursuing their own paths and becoming fairly isolated from each other. Interdisciplinary endeavors are valuable when people reach the point that they can borrow usefully from what someone else is doing, that they can put together what they're doing with what someone else is doing. But one certainly does not want to impose the connections between people in advance. While it's nice to have frameworks in which interdisciplinary research can go on (the Chicago committee structure is very nice in that way), that leaves people with their own departments, with their own potential to go their own directions. Administrators, at least of American universities, have often wanted to engineer things, so that people will work together . . .

Callebaut: . . . It has been the same in Europe, you know . . .

Bechtel: You can't anticipate where fruitful connections are going to come five years from now. So by the time people are actually doing the work, it's not going to be the people that you're housed with that you are going to want to work with but the people in some other unit.

Callebaut: I know what you're talking about!

4.3 The Realism-Constructivism Debate

Callebaut: The last big issue in philosophy of science we will treat in this chapter is realism. Latour told me he thought it is "a bit of a dead issue" (and he seems to mean this quite literally—see his fable, Latour 1980), but reading and talking to philosophers one gets a rather different impression. In Kitcher's words, "Here all hell breaks loose. That issue is completely up in the air!" (cf. Hones 1991). I will venture to say provisionally that the realist claims that scientific knowledge is in some sense "about the world out there." This position, which is in line with our commonsense view of knowledge (Popper 1972:ch. 2), is rejected by instrumentalists, conventionalists, and other philosophers tending towards skepticism. Let us try to bring some order in here. How do *you* define realism, Ron?

Giere: Realism is first of all an understanding of the nature of scientific *representations:* Is science a representational activity in the sense of attempting to produce models of the world, or is it not? The realist says yes. The antirealist says no, or maybe says that scientific representations are restricted to representing limited aspects of the world, like those that can be observed with the unaided senses. Secondly, realism is a claim about scientific *judgment*. It is minimally the claim that scientists at least sometimes have a reliable basis for judging that their models do in fact correctly represent at least some deeper aspects of the world. Antirealists deny this claim.

Callebaut: It would seem that philosophy of science is going back and forth on the realism issue. Do you think it is possible to assume a "neutral" stance on it?

Giere: Carnap (e.g., 1963) took a neutral view; his official position was "It's a pseudo-question."[33] Nagel (1961) did too. He said the difference between realism and antirealism is merely linguistic. My position is that a neutral view is really a nonrealist view. It says that we can understand science without deciding whether science in fact ever produces adequate representations of the world. That, to me, could not be a satisfactory understanding of science as an activity. The question whether science produces adequate representations of the world is fundamental to an understanding of how science works. Moreover, I think the realist answer has to be right.

Callebaut: There is also Arthur Fine's NOA ("natural ontological attitude"), which he claims is neither realist nor antirealist: an "attitude that seeks to ground scientific belief in reasonable practice—and to understand that belief in those terms" (Fine 1986:171; cf. his 1984).

Giere: Fine thinks scientists are realists about some things and not about others, and that is fine with him. What he objects to is realism as a *philosophical* thesis. I have not been able to determine whether he also objects to it as an empirical thesis made by people outside the specific scientific field of interest.

Callebaut: After the demise of logical positivism, it looked for a while as though realism "was coming into its own again" (McMullin 1984:9; cf. Leplin 1984). Putnam (1975a, b; cf. also "middle-Putnam" 1978) came out for a strong sense of realism.[34] His student, Richard Boyd (1981, 1984), continues in that tradition. Within the historical school, McMullin and Shapere are also realists. Salmon (1984) also claims that a realist account of scientific theories is correct, but seemingly on grounds different from those put forward by Boyd.

Kitcher: Boyd's view, to oversimplify, is that you have to be a realist if you are to be able to make any ampliative inferences at all. In other words, the only alternative to realism is a rather extreme skepticism. Salmon, on the other hand, thinks that there is a rather specific type of argument that gives us confidence in the existence of unobservables. When we find that independent phenomena would be explicable if there were entities of a particular type, we have grounds for believing that there are entities of that type, since otherwise there would be a strange coincidence. Both these arguments have problems, but I think there is something insightful about them.

Callebaut: Someone like van Fraassen (1980, 1985) reacted to that. He claims that classical instrumentalism is wrong but that there is a version of empiricism—he dubs it "constructive empiricism"—that is correct and antirealist.

Giere: Most of the historical school is also antirealist, explicitly. Lakatos did not talk

---

33. Carnap (1950:206–7) distinguished between "internal questions" concerning the existence of certain entities within a certain framework (e.g., "Is there a piece of paper on my desk?") and "the external question of the reality of the thing world itself." The latter he took to be a question, not about reality (in which case it would be empirically decidable) but about language; a question which calls for a decision— are we interested in using a *reistic* or thing language rather than a sense-datum language? See Gochet (1986:65f.) for a discussion.

34. In the same vein: Churchland (1979) and Newton-Smith (1981).

Disentangling rationalism, realism, and relativism

Ronald N. Giere (b. Cleveland, Ohio, 1938) is professor of philosophy at the University of Minnesota and director of its Center for Philosophy of Science (1987–). He studied physics at Oberlin College and Cornell University (M.S. 1963). At Cornell he shifted to philosophy (Ph.D. 1968).

Barbara Hanawalt

Giere previously taught and did research at Indiana University, New York University's Courant Institute of Mathematical Sciences, the University of Pittsburgh, and Indiana University's Institute for Advanced Study.

Starting with Thomas Kuhn and his fellow travelers, much work in the history and sociology of science has aimed to debunk the myth of "great science as consisting of solitary geniuses along with their thoughts or experimental apparatus—the mythological Newtons, Faradays, and Einsteins of yore," as Giere (1989a:7) puts it. A good deal of Giere's philosophical efforts in the last decade aimed at showing that an insistence on the centrality of individuals in science does not imply acceptance of this myth at all. This single idea provides one clue (there are others) that enables us to make sense of Giere's enterprise, which is at once thoroughly naturalized (Giere 1973, 1984b, 1985a, 1989b), cognitivist (1985c, 1988, 1991a), evolutionary (1988, 1990), and realistic (1985c, 1992b).

Giere is often referred to as the first philosopher to have undertaken a lab study, but he himself is more modest: "Ian Hacking (1983) did it before I did." He wanted to beat the antirealists on their own ground: "An overall kind of nonrealism or antirealism I find very hard to even take seriously. I know people do it. But look at electron microscope pictures of chromosomes, of DNA—we have pretty nice pictures of DNA! The overall idea that there is an observational realm and that we should not go beyond that I think is just not tenable. Part of the reason for the laboratory study I did was simply to say: 'Look, antirealism is just *bizarre;* it is not logically *impossible,* of course, but it's *bizarre!*'" His book, *Explaining Science* (1988), exploits the idea that the issues of realism ought better be disentangled from various other issues. Giere believes that there has been a tremendous conflating of the issues of realism, relativism, and rationalism: "I think people just haven't sufficiently sorted those out. As I already suggested, I think realism is primarily an issue about the nature of scientific representations. The question of rationalism, on the other hand, is really a question about the nature of scientific *judgment:* are there autonomous, external principles of rationality by which you decide what is a good theory and what is a bad theory? The opposite of that is usually taken to be *relativism.*"

Giere's own position on the rationalism-relativism issue is that one can have theories of judgment that are not "categorical" theories: "There are reliable strategies for making judgments about the fit of models to the world. You might say these strategies define a kind of rationality, but it is purely *instrumental* rationality" (1989b). Here he endorses Herbert Simon's (1982) position. Scientific judgment is both social and rational. It is neither the application of categorical principles of rationality nor just social consensus, but something in between. Giere insists that "scientists really do judge the fit between their models and the world," and that the theory of how they make these judgments can be completely naturalistic: "Their judgments are hooked causally to the world—that is where experimentation comes in. Experiments causally connect scientists to the world in a way that helps them to make judgments about the world. That is what clever experiments are all about. This involves creating something completely artificial, to be sure. But it allows nature to provide clues as to which way it is going."

Giere (1991b:522) has pointed out an interesting policy consequence of evolutionary models of science (cf. Campbell 1986b): "If the development of science fits an evolutionary model, then science policy analysis is more like applied ecology than management science. The task is not to exploit the relevant social scientific laws by devising and enforcing rules that scientists should follow, but to design an environment conducive to optimal evolutionary development given the normal range of cognitive abilities found in the typical scientific community."

With Giere, says psychologist of science Ryan Tweney, "one has trouble sometimes understanding the extent to which his work is philosophy of science. In most respects, I see it as not being philosophy *at all.* He is engaged in the empirical study of science. Great!" Being at such an outpost, it is not too surprising that Giere has been receiving large amounts of flak lately, both from philosophers to whom a naturalized philosophy is anathema (Siegel 1989; Efron and Fisch 1991 go so far as to suggest that his account "denatures" science) and from sociologists who think his naturalistic stance does not go far enough (thus Pickering 1991 belittles his position as "philosophy naturalized a bit").

Giere has written *Understanding Scientific Reasoning* (1984a; the first edition was published in 1979), *Explaining Science: A Cognitive Approach* (1988), and numerous papers, mostly on probability and science dynamics. He is the editor (with Richard F. Westfall) of *Foundations of Scientific Method: The Nineteenth Century* (1973) and of *Cognitive Models of Science* (1992a).

The original interview was taped at the Club de la Fondation Universitaire in Brussels where Ron gave a talk to the Belgian Society for Logic and Philosophy of Science on February 25, 1989. His exchange with Karin Knorr Cetina was recorded in Ghent in 1984 (see her biographical note at 4.3.6), right after Ron and I first met.

about it, but his view is nonrealist. Laudan (1981a, 1984, 1988) was right up front, and he is an antirealist.

Kitcher: Kuhn, I think, has a very complicated position, a very interesting position actually; he is beginning to formulate quite clearly the argument that was really motivating him in the *Structure.* And there are people like Larry Sklar (1982, 1984), who believes that scientific realism is becoming popular at the cost of abandoning some of the legitimate questions that the positivists tackled quite successfully. Larry puts it in a kind of ecumenical fashion by saying that he thinks that realism actually doesn't have any more going for it than positivism.

4.3.1. *How to Approach Nature*

Callebaut: Karin, as a sociologist you tend to associate realism with the way philosophers since Aristotle have approached the problem of knowledge.

Knorr Cetina: The way scientific accounts have been traditionally analyzed by philosophers is that they have been put in relation to what one might loosely call "the World," "Nature," or "Reality." How does the scientific finding, the scientific fact, relate to the world? One answer to the question, put in this way, is the realist answer, which basically says that science represents the world as it really is.

Callebaut: The next move of philosophers is usually to point out that many of their ilk have always disagreed with *that* kind of realism. Listening to anthropologists and sociologists of science, one is sometimes given the impression that all philosophers have been realists at all times. They thus disregard the important skeptical tradition in the history of philosophy (cf. Campbell in 7.1), nineteenth-century positivism, etc. etc. The strong connection between rationalism and realism which sociologists of science suggest when criticizing philosophy of science (e.g., Collins and Pinch 1982; Hollis and Lukes 1982) is not obvious either. Rationalism and realism may be profitably disconnected (Giere 1988:7–8).

Latour: Oh yes. The French tradition since Duhem is rationalist and absolutely antirealist in many ways. That's a very old tradition, going back at least to Auguste Comte.

Callebaut: So we have a straw man here that ought not to be taken too seriously. You are, of course, aware of this, Karin.

Knorr Cetina: Philosophers like Charles Sanders Peirce have argued that scientific concepts are constituted with respect to what can be done with them in experimental research. Take the hardness of the diamond. The concept of "hardness" itself is constituted with respect to the possibility of rubbing another stone or rock against the diamond. The constructivist program has extended this idea by claiming that the information produced by science is first and foremost *the product of scientific work,* and what we should do is try to describe how scientific work produces scientific information, rather than locating the focus of the analysis between the finished scientific product and the world.

Callebaut: We will pursue this line of reasoning further on. But let us first look somewhat more closely to a couple varieties of realism that philosophers have considered. Bill, you object to some of the traditional formulations of a realist position.

Wimsatt: One way to define realism is, "Do our terms really refer to things in the world?" Another is, "Can we have a view of science in which it appears likely that we are asymptoting towards a truer and truer description of the world?" And this is usually cashed out in terms of what an ideal Peircean community of scientists would know in the end rather than as a question as to whether we are going in that direction or whatever. Well, I don't think we will ever go as far as the standard conception of the Peircean community would picture. The reason for that is that *our theories are cost/benefit structures adapted to dealing with real-world problems,* and it is not the aim of theories to capture all phenomena (cf. 9.4.1). So we'll never be going to have "theories of everything" in which you can imagine plugging in all the real details of a specific case and getting out all of the real details, predictive details of it. Rather, we'll have a bunch of modular pieces of theory of which in each case we use a special selection to apply jointly to a phenomenon. Each of these modular pieces remains a general-purpose piece of equipment that we can carry and use elsewhere. A single theory to cover any phenomenon in all of its detail (in a way, there's "thick description", even in physics) wouldn't apply anywhere else. Such theories would be both conceptually wasteful and nonexplanatory, because you would have no idea what parts of the "explanation" were generalizable. It would look like an extended narrative which could have been all coincidental.

4.3.2. *Do the Laws of Nature Lie?*

Callebaut: How would you reformulate the problem of realism, then?

Wimsatt: When we construct a scientific model or theory, we want something that applies to a variety of cases. In doing that, we are going to abstract away from a lot of the details of the situation. To make a theory, say, of the lac operon[35] that included not only how the basic mechanism works but also how it would work in the face of all possible perturbations would be to make a theory not only of the lac operon but of everything else. That is not a theory we can use. Rather, what we want is something that throws out a lot of details that can sometimes be causally relevant, but not frequently, or details that are frequently relevant but have very small effects, and so on. We want to capture *major features* of the account.

Callebaut: That view goes some way in the direction of Nancy Cartwright's "entity realism" as put forward in her book *How the Laws of Physics Lie* (1983; cf. Hacking 1983).

---

35. The *operon* (a concept proposed by Jacob and Monod in 1961) is a unit of gene expression that includes genes and elements that control their expression. Operon activity is controlled by one or more regulator genes, whose protein products interact with the control elements. For an illustration of the basic control circuit of the lactose operon see Lewin (1987:222–23).

Wimsatt: Yes, indeed, I think Nancy's book is beautiful. I recommend it a lot for her stuff on models. I think she has taken van Fraassen too seriously in her skepticism about theory, though. There is also a remarkable convergence between what I said on mechanisms (1976a) and her book and subsequent stuff on capacities—even down to a lot of details in our common attack on the centrality of laws in explanations (Cartwright 1989). We discovered this when I was a commentator on a paper of hers at the APA Pacific Division meetings in 1986. All of which means that philosophers of biology had better not ignore the work of philosophers of physics, or conversely.

Elisabeth Lloyd (University of California, Berkeley; see biographical note at 5.5.1): I'm a real fan of Nancy Cartwright. Her book gets at very fundamental assumptions in formal philosophy of science and in realism. She argues that the role of laws of nature is at a very high level of abstract theories, and she shows how they relate to various explanations that are being used and implemented in the sciences. I think that it's a very important piece of work.

Callebaut: How do you find her way of dealing with the realism issue?

Lloyd: The realism issue is not an obsession of mine. That she takes a particular stand on what counts as entity realism . . . I can see why she wants to do that. At the same time I don't find plausible at all the reason she gives for thinking that entity realism is OK and other types of realism that go further are not. On the other hand, I'm well aware of the fact that people consider van Fraassen's distinction between "observable" and "unobservable" to be untenable.

Callebaut: Don, for instance.

Campbell: We would never have achieved the marvelous sets of "facts" that a modern theory of physics has to be empirically adequate to, had *empirical adequacy* been the only goal. It's the quarreling about realist hypotheses that has generated all these elegant, demanding "facts" that a modern physical theory has to be empirically adequate to (cf. Harré 1986).

Giere: I find the supposed distinction between "entity realism" and some other kind of realism difficult to comprehend. What I think Cartwright is getting at is this. What are touted as the high-level "laws" of physics, such as Newton's laws or Schrödinger's equation, are not true of anything in the world. To get claims about the world that might be even approximately true one must idealize and add boundary conditions. There are always several incompatible ways of doing this—even starting with the same general "law." I would put this point somewhat differently. In brief, *there aren't any universal laws,* understood as true, high-level generalizations. That way of understanding these bits of science is an artifact of a mistaken view of science, a view reinforced by standard logical formalisms.

Callebaut: A reification?

Giere: Yes, a reification. I think universal generalizations play almost no role in science when you see how it actually works. This is obvious in biology. But I think it is also true in physics. What you have is general schemata for constructing specific models—that is what Newton's laws and the Schrödinger equation really are. Some of the detailed models scientists construct fit the world better than others. But universal generalizations? Who needs them? So, do the laws of physics lie? My answer to that is no. But not because they tell the truth, but because there are no laws in this sense. *There are no laws to lie!*

4.3.3. *Reality as Immanent to Science*

Callebaut: Dudley, you criticize traditional philosophy of science for having assigned a "science-transcendent" meaning to the concept of realism and for having tried to frame the realism issue in terms of that concept (e.g., Shapere 1989a, 1990).

Shapere: Philosophers have always tried to formulate their problems in terms of fundamental concepts that they really take for granted, like "realism," "entity," "exists." But the terms "reality" or "existence" are not terms that transcend science and have some kind of a priori meaning which we all understand in advance of investigating reality, so that we can formulate philosophical issues like "Does science deal with reality?"—it's not good enough. *We have to learn even what to understand by the term reality.* That is shown clearly by modern quantum field theories, where probabilistic fields become fundamental. Here the fundamental entity—I am sorry I have to use the traditional word—is a quantum field; but the quantum field is a field of probabilities. The traditional philosophical distinction in terms of which issues like realism versus nonrealism have been formulated, the very distinction between actuality and possibility, between reality and unreality, does not fit the kind of "entity" that we have in quantum field theory. In other words, the content of what we are learning even about the word "about" has to be learned in the development of science; and we are still learning what it means "to exist."

Callebaut: Concepts like "truth" and "reality," as you stress, were developed in the course of our dealing with a middle-sized world or "mesocosm" (Vollmer 1975) . . .

Shapere: . . . as were all our concepts, originally; that's something we should have learned from evolutionary theory. Whether these concepts are going to be adequate or not at other levels is itself something that we have to find out. You know, to the commonsense, middle-sized mind, it is almost a contradiction to say that what is real are mere probabilities. And yet, of course, our theories may tell us that! It is an open question as to whether our scientific theories will give us a realistic theory, even in my extended sense. They may tell us, for example, that we can know that our experience tells us that we will never be able to find out anything beyond experience. So we can know that. But that tells us that we cannot know truth in the correspondence sense. That is also a possible outcome. So what realism would mean, and whether realism, even in my extended sense, is true, is a matter that we've got to find out.

Callebaut: What do you hold of Quine's famous dictum that "To be is to be the value of a bounded variable," which is meant to tell us not what *is*, but what the theory (etc.) in which the variable occurs *says* there is; that is, the ontological commitments of a discourse (cf. Gochet 1986:69).

Shapere: My criticism of this is that scientific theories "quantify over" (to use Quine's language) idealizational and fictional concepts as well as realistic ones. Therefore what count as existence claims (in the broadest sense) in science cannot be reduced to what you quantify over, but rather presuppose a background of substantive knowledge (as opposed to purely formal considerations) that enables us to distinguish between those concepts that are "realistic" (existence-claiming) and those that are not. That seems to me a devastating criticism of the "To be is to be the value of a variable" idea.

4.3.4. *Humans Do Not Process Probabilities on Propositions*

Callebaut: Ron, Bayesians,[36] whose number is growing in philosophy of science . . .

Giere: . . . I think it is, unfortunately! . . .

Callebaut: . . . hold a much more idealized view of human rationality than you do. How would you contrast Bayesian decision theory involving subjective probabilities (e.g., Howson and Urbach 1989) with your own satisficing approach (Giere 1988)?

Giere: There are all kinds of evidence that as an empirical theory of how humans think, Bayesianism doesn't work. *Human beings do not process probabilities on propositions.* That is just not how they work. Supposing they do is not going to give you a good theory of human judgment. Satisficing might be a better theory. I don't think it's a great theory, but it is the best simple theory I know. I would be quite happy with a more elaborate theory of human judgment. I am looking around for one but haven't found it yet! But I am quite convinced that the Bayesian idea that scientists are really calculating the probabilities of the hypothesis space is crazy—they just don't do it! There is no evidence whatsoever that they do. What philosophical Bayesians will do is go normative and say "Oh no, this is not a theory of how people *actually* work. It's like logic: it tells us how they *should* think."

Callebaut: John Maynard Smith (1984), one of the pioneers of the use of game theory in evolutionary biology, has called "Sahlins' fallacy" the presumption that the organisms whose behavior the biologist interprets in terms of fitness maximi-

---

36. On the Bayesian account of inference and decision making in terms of subjective probabilities (named after the Reverend Thomas Bayes, ?–1763), "agents" or information processors like scientists have highly structured preferences, such as coherent preferences regarding the various possible outcomes of a decision situation (see, e.g., Levi 1967, 1986; Jeffrey 1983; Giere 1988:145–57). According to Giere (1988:149), there "is now overwhelming empirical evidence that no Bayesian model fits the thoughts or actions of real scientists." Contrast Anderson (1991:485), who at a more general level maintains that "many of the major characteristics of human cognition can be explained as an optimal response, in a Bayesian sense, to the informational structure in the world"; see also Cosmides and Tooby (1992a).

zation are required to actually "do the calculations in their heads."[37] What his counterargument boils down to, translated into philosophese, is that you can remain an instrumentalist about that—it's the predictions that count (cf. Heyes 1987). Playing the devil's advocate, one could claim that the Bayesian approach to science is as justified as is the game-theoretic approach to the evolution of animal and even plant populations.

Giere: No, no, no. My critique of Bayesianism is deeper than that. I am a realist both about evolution and scientific judgment. If organisms are maximizing in a way that leads to the outcomes predicted by game theory, there must be some mechanism producing those outcomes. Consciousness is irrelevant. What matters is that there be some mechanism that is producing results equivalent to the maximization calculations.

Callebaut: Something genetically wired?

Giere: The basic capability is almost certainly genetically wired. How much is learned is difficult to sort out empirically. The situation is similar with scientists. Although many scientists use the probability calculus in the context of theories such as genetics and quantum mechanics, and even in processing data, it is rare that scientists explicitly attempt to calculate probabilities of theories. And just because scientists may on other occasions act as *if* they had particular prior probabilities and are calculating posterior probabilities of hypotheses on evidence, that does not show that the scientists' actual process of evaluating hypotheses has the structure of a probability model. All the experiments on human judgment suggest that they are doing something else. Satisficing is a crude theory of what else they might be doing. I would be happy to find a better theory. Maybe connectionist models of human thinking (cf. chapter 8) will provide better models of scientific judgment. I would be pleased to see that.

4.3.5. *Saving the Phenomena from the Realist*

Callebaut: In your scheme of things, could one be a relativistic realist, then (cf. Margolis 1986)?

Giere: Only in an attenuated sense of realism. One might say, "Oh yes, scientists intend their models to capture the deeper structure of the world, and for all we know, some models might actually succeed in doing so, but there are no legitimate ways of judging whether this is so." That is van Fraassen's view. He thinks that theories may be used to refer to the hidden structure of the world, but that, *epistemologically,* no one ever has any basis for believing that they in fact do so.

Callebaut: This is related to the distinction between "observable" and "unobservable" referred to before.

Giere: According to van Fraassen, who is a true positivist, we can only believe our

---

37. As Sahlins (1976) correctly saw, that might become exceedingly complex.

models to be *empirically* adequate—to capture the observable aspects of the world.[38] This also seems to me basically the position of social constructivists like Karin Knorr Cetina.

Lloyd: The issue that van Fraassen is talking about in what I take to be the *epistemological* realism debate is about justification and warrant. Is it reasonable at all to say that someone can think that a theory is empirically adequate without being compelled to believe in the reality of the theoretical entities? It's a position which I think van Fraassen did not make clear in *The Scientific Image* (1980), but which he makes very clear in the reply to his critics in the Churchland and Hooker volume (1985). What he says there makes sense to me, and although I've never in public taken a strong position on antirealism, I feel that the realists have not addressed the key problem that van Fraassen raises for realism, which is this argument—on logical grounds—that *a theory which is more likely to be true is less likely to be explanatorily powerful.* If you take explanatory power to add something to the truth of a theory, you've got to deal with that logical problem! As far as I can tell, no one—in any of the thousands of varieties in the realist camp—has even dealt with that question. And I take that question to be very basic; I take it to be a very serious challenge to a realist position.

Callebaut: What are your problems with Ron's realism, Lisa?

Lloyd: (*Impatient.*) The pressure to take a position on the realism/antirealism issue has astonished me from the very beginning. I thought, and I still do think, that you can do a huge amount of philosophical work without taking a position on that issue. My tactic in looking at the structure of evolutionary theory was not to worry about the interpretation of the entities at stake but to actually describe the structure of the theory, learn things from that, and clarify some of the debates in a very precise way. The issue of realism as it has often been raised, I have to say, has always seemed very naive to me.

Callebaut: Is that because of your scientific training?

Lloyd: In part, I think. There is a tendency among philosophers to be realists whereas there isn't so much a tendency among scientists to be realists. I don't mean to say that scientists always say that they are fallibilists. They rather think that "this is as close as we can get until the next time we find something out." But there are lots of scientists who have a fairly instrumental approach to what they are doing anyway. That is neither here nor there. I mean, unless you are going to run your philosophy

---

38. Van Fraassen (1980:154–55) sees the conventionalist Pierre Duhem (1861–1916) as a historical ally to the extent that Duhem already tried to debunk the view—which is widely held, especially by realists—that the "explanatory power" of a theory is evidence for its truth. But he departs from Duhem to the extent that the latter may be said, in retrospect, to have "fostered that explanation-mysticism which he attacked" by granting explanatory power to metaphysics only. What van Fraassen means is this: "fifty years later, Quine having argued that there is no demarcation between science and philosophy, and the difficulties of the ametaphysical stance of the positivist-oriented philosophies having made a return to metaphysics tempting, one noticed that scientific activity does involve explanation, and Duhem's argument was deftly reversed."

by democratic means it doesn't really matter; and I understand that that is really quite a separate issue. Therefore I think that Ron's approach, in supporting his view of realism, of always going to the scientists and pointing to the physicists and saying, "Well, they're behaving in their everyday lives as if these things exist" (cf. also Wimsatt at 4.2.2 and Hacking 1983) is not relevant to the epistemological issue at stake. It seems to me as if the important debate is strictly an epistemological debate. The fact that scientists' behaviors and interpretations get put into this epistemological debate . . .

Callebaut: . . . as in the exchange between Ron and Karin below . . .

Lloyd: . . . turns it into another kind of debate, one about scientific behavior. I think both of those are interesting debates, but that the thing has been completely confused by putting the two of them together.

Callebaut: Someone who actually tried to disentangle some of these issues is Hacking (1988). He basically makes the point that someone like van Fraassen is addressing rather different issues than someone like Latour.

Lloyd: Absolutely. You see, that's why I don't like to talk about "realism." It's like the words "communist" or "Marxist" or "leftist"—you've got a zillion different varieties, people have equally ornate views about them, and you have to spend four days interviewing the person to find out what kind of realism they're actually all worked up about! It seems to me that there is a huge amount of very useful work that can be done on the content of theories without looking at the interpretations—epistemological realism, etc. I think a lot of the labor that has gone into the realism/ antirealism debate is wasted. This is why I like van Fraassen's new book on laws of nature (1989): he stays away from this debate; he is doing something new, addressing another fundamental problem in thinking about how to do philosophy of science. I like to avoid the realism/antirealism debate; I found that it has clouded issues.

4.3.6. Realism versus Constructivism in the Laboratory

Callebaut: The following exchange between Karin and Ron will help to clarify a number of issues having to do with the interpretation of what goes on in situ, in the scientific lab.

Giere: I have been most impressed by people looking at real science. The most exciting work has been the work that Karin and her colleagues have done by actually going into the laboratory and looking at how scientists do science (e.g., Knorr-Cetina 1981a; Latour and Woolgar 1979; Pickering 1984a). I might even say that they have awakened a number of people, including some philosophers, from their dogmatic slumbers. I would regard myself now not as a representative of the traditional philosophical schools that would say that we should analyze science from a totally a priori basis and understand it solely in terms of logic and other logical categories.

Callebaut: May I come in here? Karin, there are many senses in which the notion of "constructivism" has been used. How do you define "constructivism" in your own work?

Knorr Cetina: Two senses of "constructivism" are especially relevant here. The first refers to a problem shift. Philosophers, having located the problem of knowledge in the relationship between the scientific theory and reality, asked, "How is the world represented in these accounts?" (mainstream realism), or "How is the world empirically adequately represented?" (van Fraassen's type of empiricism), etc. Since we constructivists believe that the world as it is is a *consequence* rather than a *cause* of what goes on in science, we have reverted the arrow between the scientific account and the world, considering the latter as a consequence rather than a cause of the former. The focus of attention has shifted to *what goes on in science when it produces these accounts.* That has opened up a whole new area of research.

Callebaut: In this sense, constructivism is a claim about the "constructedness" of scientific findings . . .

Knorr Cetina: . . . which are not just *found,* as the notion suggests, but are *fabricated*—the Latin root of the word fact is *facere,* "to make something." When you observe scientists in the laboratory, you find processes of *negotiation* at work, processes of decision making, which influence what the scientific findings are going to look like. In a sense, the scientific finding is construed in the laboratory by virtue of the decisions and the negotiations it incorporates. Of course, these negotiations and decisions are no longer visible once the finding is published and is considered a *scientific fact.* Then the negotiations have disappeared from the surface, but they are embodied in the scientific finding (1984).

Callebaut: That is also the sense in which you think scientific findings should be deconstructed.

Knorr Cetina: We should make visible all the processes—including social ones, but *not only* social processes—which bring this scientific finding about. We claim that a lot of decision making, a lot of negotiation, a lot of interpretation—but *not only* interpretation—goes into the production of the scientific fact. Of course, if you have many decisions incorporated in such a finding, this also means that *these decisions could have been otherwise* and might have led to a different scientific "fact."

Callebaut: In other words, science and scientific change are contingent; *there could always have been alternatives* (Böhme, van den Daele, and Krohn 1972). What is the other sense in which the notion of constructivism is being used by people who have adopted your kind of approach to science?

Knorr Cetina: The other issue concerns the shift from time to space in constructivism. In philosophy, you had the conception that science is a succession of theories, theories-cum-methodological rules, or some such thing. The question then was: "How does one go from one theory (etc.) to the other?" That was a temporal account, which usually also assumed progress and commensurability. What social

"You don't have to go outside to find context."

Karin D. Knorr Cetina
(b. Graz, Austria, 1944)
is professor of sociology
at the University of Bielefeld, Germany, where she
began to teach after her
Habilitation in 1981. She
started out studying cul-
tural anthropology at the
University of Vienna, but,
finding the teaching too
historically oriented ("It
wasn't really cultural anthropology in the Ameri-
can sense"), she shifted to
sociology. She obtained

her Ph.D. from the University of Vienna in 1971. She
then went to the Institute for Advanced Studies in Vienna.
"That was a major thing for me, because the Institute had
the policy of inviting guest professors from all over the
world. I didn't have to travel around the world to get to
know many prominent (general) sociologists!"

She got involved in science studies as a result of "sort
of growing up" on the *Positivismusstreit* between Jürgen
Habermas and Hans Albert, which was very prominent in
Europe in the late sixties and early seventies and was
mainly about the status and the methodology of the social
sciences. Should the social sciences continue to emulate
the natural sciences? Knorr's answer was, "of course, no!"
She found it a very exciting debate at the time. It also had
the effect of putting her into contact with philosophers.
"The continuous disagreement I had with philosophers like
Hans Albert or, at one time, Lakatos, whom I met in Al-
pach (Austria), but also with my philosophical colleagues
at the Institute for Advanced Studies, was a constant stimulus for reflection on and for the continued interest in science."

Knorr's first empirical investigation was a study of the
social sciences, which grew out of her methodological interest (Knorr, Haller, and Zilian 1979). Then came the opportunity for doing a quantitative study of the organization
of research groups in the social sciences and the natural
sciences, which was an international project. "I felt too
confined in Vienna and wanted to get out (*blushes*), and
the project offered an opportunity to do so." In retrospect,
Knorr feels, the real effect of this project was a frustration
and a worry. "A frustration with the kind of data we collected: the correlations were always between 0.1 and 0.3;
that means you explain about 9 percent of the variance.
Doing a life-scale quantitative study is quite expensive; it
costs a lot of energy and time and man\womanpower. It is
also very unsatisfactory intellectually: you find some correlations, but it's not clear what they tell you; and there
sould be correlations that you don't find and you don't
know why—things like that. The other thing was a continuous worry. Before the actual study, I did a pilot study and
did some of the interviewing myself. I always had to de-

part from my own standardized questionnaire in order to
explain what I wanted and to accommodate someone's
frame of reference and context of opinion in order to even
make the question understandable and get an intelligible
answer. So what these little crosses you have on the questionnaire actually mean is very unclear in the light of such
experiences." All this prompted her, when she went to
Berkeley in 1976, "to do something completely different,
non-quantitative, to use a different approach, to incorporate some of the things I had read out of this frustration,
like Aaron Cicourel's *Method and Measurement in Sociology* (1964)." Knorr Cetina's *The Manufacture of Knowledge: An Essay on the Constructivist and Contextual Nature of Science* (1981a) was the first major statement of her
new position. With its emphasis on the contingency of "selections" of all sorts and on the opportunistic logic of the
"tinkerers" Knorr takes scientists to be, the book could not
pass unnoticed. It was in fact hailed as "doubtless the bold-
est and most consequential attempt to state the case" for
the microstudy of scientific cultures in the laboratory by
one of her fiercest critics, Gad Freudenthal (1984:285).

Knorr Cetina has been with the Department of Sociology at the University of Pennsylvania (1979–82), the Center for the Study of Science and Society at Virginia Tech
(1981–82), and Wesleyan University (1982–83). In 1992–
93, she is at the Institute for Advanced Studies, Princeton.
In addition to *The Manufacture of Knowledge*, she has
written *Mikrosoziale Theorien* (1989) as well as more than
fifty articles. She had co-edited six books, including, with
Roger Krohn and Richard Whitley, *The Social Process of
Scientific Investigation* (1980); with Aaron Cicourel, *Advances in Social Theory and Methodology: Toward an Integration of Micro- and Macrosociologies* (1982); with
Burkart Holzner and Herman Strasser, *The Political Realization of Social Science Knowledge and Research: Toward New Scenarios* (1982); and, with Michael Mulkay,
*Science Observed: New Perspectives on the Social Study
of Science* (1983). *Epistemic Cultures: How Scientists
Make Sense* (1993a) and *Mikrosoziologie* (1993b) are
forthcoming. She is one of the founding members of 4S
(Society for Social Studies of Science) and a member of
the editorial board of several important science studies
journals.

The main interview with Karin was recorded in Marburg, Germany, on April 21, 1990, on the margin of a
meeting of the theory section of the *Bund deutscher So-
ziologen* on "Kommunikation and Konsens." Also included in the book are materials from an exchange between Karin and Ronald Giere under the roof of the his-
toric "Pand" (where the historian of science George Sarton
lived as a student) in Ghent, Belgium, on November 16,
1984, which I arranged for a Belgian Public Radio broadcast on the occasion of the joint meetings of the 4S,
EASST (European Association for the Study of Science
and Technology), and C&C (Communication & Cognition).

studies of science have done is they have shown the *spatiality* of all kinds of things which happen if one "cuts across" the temporal dimension. There is a whole space of activities which has to be reckoned with and that has nothing to do with the sequence of theories. There was actually a discovery of spatiality in two senses. First of all, the discovery that context is not outside but *inside* the "epistemic."

Callebaut: On this score, your brand of constructivism departs from the sociology of knowledge account of, say, Edinburgh or Bath.

Knorr Cetina: Sociology of knowledge, in formulating the question as it does traditionally—"How is scientific thought conditioned by societal influences?"—continues to assume that context is something outside science and asks "How does it get in?" Whereas what one does in these studies, particularly the laboratory studies, is to show that context is always already inside. There is an epistemic culture in there, and it differs across fields. You don't have to go to class interests—even though that might be an interesting question to pursue! You don't have to go outside to find context. The second important sense of "spatiality" is that we found a specific locale to be an important ingredient in the whole thing, and that is the laboratory. The laboratory is a device, it's a specific, local space that has theoretical importance and relevance. It's a machinery, a technical device for science to produce its results.

4.3.7. *What Is the Purpose of All This Equipment?*

Giere: Having said how important I think your kind of work is, I still think that when one goes into the laboratory and looks at the *processes* by which the scientific results are created, one can find a different interpretation than the constructivists want. In particular I would be inclined to emphasize that *what is going on in the laboratory is a highly developed and organized process for interacting with the world.* And indeed I think scientific theories and models are designed to *represent* the world, and the activities of the laboratory are to actively try and *decide*—the notion of decision, I think, is a very good one: *making many decisions*—which of competing models—and there are almost always *competing* models—best represents the world, but through a process of *active interaction* with the world. This gives me something of a more traditional understanding of what is going on in science. But I think it is important to see this in the laboratory.

Knorr Cetina: One of the first results of the laboratory studies—or *surprises,* perhaps, because they went into the laboratory with that very same notion that scientists would interact with the world: basically that is what is expected from science—was that the World or Nature did not appear in the laboratory at all. What you find in the laboratory are highly *preconstructed* products, substances, which have been purified, animals which have been specially bred, plants which have been cut up and reduced to cell size and whose cell wall has been destroyed, machines and apparatus, instruments—we all know about that . . .

Callebaut: . . . Bachelard (1973) gives a masterly description of this process of pre-construction in the case of the history of chemistry.

Knorr Cetina: That is one part of the laboratory. Biologists, for example, distinguish between "wild" mutants and laboratory mutants; and they don't care about wild mutants, they are not interested in them. What they need are the mutants which have been created under controlled conditions or which they have been supplied with by industry.

Callebaut: Yet scientists, in the laboratory, interact with *something* . . .

Knorr Cetina: Scientists do not, we think, interact with the world *directly;* they interact with, for example, *what other scientists have said about the world.* The concepts in terms of which they think are taken out of the literature. The interpretations which they impose on their experimental results are interpretations that have been established by other scientists or by themselves, latching on to what other scientists have said. So it seems to us that when we really look at what is going on it is not the world which "appears" there in any pure sense, but scientists interacting with each other, with the literature, with established knowledge, with what you could possibly claim to be based upon established knowledge, extending it. You find their own interpretations; you find the arguments of other scientists against which they argue; you find the concepts of other scientists; and you find a highly artificial nature *as we built it* within the laboratory, but in very much reduced size, very much changed, incorporating already all the decisions about which I have talked before.

Callebaut: So you contest the claim that you can put it in the way in which Ron puts it: that there is something like *interaction with the world* going on in the laboratory.

Giere: Of course I agree that in the laboratory one is dealing with highly prestructured materials. Laboratory research is very carefully organized to have as much control as possible over the materials you are dealing with. So indeed you are not getting the materials in their natural situation. There *is* that kind of research; but the typical research we are talking about is indeed with highly preorganized materials. But I don't see how that means that you cannot be interacting with reality. What that *does* show, and what is very important to understand about science, is that one is constructing—I would agree, *constructing*—highly simplified models of the world, and one knows that the world is much more complex than any model you could possibly construct. But the hope is that you can construct a model that will capture various aspects of the world; and this you try to do in a controlled situation, using indeed highly purified substances and highly controlled kinds of interactions. To take the point further: We both agree that if you go into a modern laboratory, one of the most obvious features is the amount of equipment, and technology, and machines of all kinds. One wonders what could possibly be the *purpose* of all this equipment if it is not to probe and interact with nature—I agree, in very specialized and highly controlled ways. Maybe this is a slight parody of the situation; but one might say, on the constructivist view it *seems* as if maybe one could just get rid of all this machinery and just have the scientists in the laboratory interacting and talking with one another and writing articles for one another, and reading other people's articles and thereby constructing science. I mean, they could do it without all this machinery. One could wonder, why do they need all this machinery? What is it doing there if it is not serving as what they would call their "probe of reality," to put it in those slightly grandiose terms?

4.3.8. *Have You Talked to the Azande?*

Knorr Cetina: If I were Feyerabend I would now draw a comparison to the magic of the Azande and say, "Well, all belief systems have an elaborate machinery to substantiate and sustain their beliefs, haven't they?" (Feyerabend 1987; cf. Evans-Pritchard 1937). All belief systems in a way try to *externalize* their beliefs, to attribute them—to have "Nature" speak to them or have their gods speak to them, for example, through some chicken oracle—rather than to have their own opinions or their neighbors' opinions registered and taken seriously. Some of these oracles may be highly sophisticated and articulated. One could ask the question whether science is not just our magic? Now, I am not Feyerabend. He might perhaps ask this question; I won't go quite as far. (Cf. Barnes 1974 on the laboratory experiment as just another divination experiment.)

Callebaut: You also wouldn't deny being an *ontological* realist.

Knorr Cetina: All of us constructivists, I think, are what they call *ontological realists:* We believe in the existence of the material world "out there," and we believe in the fact that this material world offers resistance when we act upon it. It will resist; we can't just do everything with it. So in that sense we are all realists.

Callebaut: How about scientists' self-understanding as "truth-seekers"?

Knorr Cetina: I would agree that the purpose of what scientists think they are doing is to get a better picture of nature. That is indeed the declared purpose of their work, and also what they believe they are doing. One might call this *teleological* realism. In this sense I am a realist too; but not in the sense of *epistemological* realism, which will make the claim that in the long run the results of science approximate nature.

Callebaut: Why not?

Knorr Cetina: Because it seems to me that the big question, even assuming that there is a world out there, is how you interpret this world. You can actually see a very concrete correlate of this question in the laboratory. The biggest question for scientists working at the forefront of things is always, "What have I got there? Is this thing I measure here real? What does it mean? What *is* it actually? Is it an ion or is it not an ion? Is it noise or what is it? When can I say, 'Now I have got it and now it is real?'" So they operate in a realm of *uncertainty,* and within this uncertainty they have to interpret what is going on, what they have got and what they haven't got, where the noise starts and where the "real thing" begins. This they do very often interactively, not just by imposing their own interpretations but also by taking the opinions of other people into account and then negotiating, in a sense, the interpretation.

Callebaut: Please give an example.

Knorr Cetina: Negotiating, for example, when they can stop the measurement, at what point they've got enough data, and at what degree or position they can say, "Now it is real!" and "Now it is no longer noise!" This *interpretative flexibility,* which I think is sustained by the tremendous uncertainty they confront, prompts me to doubt that you can ever get at the real world as it really is. You can get *resistances* in the laboratory; but in order for these resistances to make sense, they have to be interpreted. The very moment you interpret them, you enter the realm of the social world, you enter the thoughts of previous scientists, of your colleagues in the field, of what you think yourself. The thing becomes contextual, perspectival, dependent, and relative to ruling paradigms, for example—to use an expression of Kuhn—and can no longer be represented in the very simple terms of scientists "getting at the thing as it really is."

Giere: I can start by agreeing with most of that. It is most indicative of how difficult it is, in fact, to do modern scientific research. That is, one does not just simply look at what is going on and immediately come up with what is the correct picture of nature. That just does not happen. Especially in the forefront of research, there may be years of great uncertainty as to what is going on and what kind of model is best; and one is constantly interpreting and reinterpreting. This is indeed so. Now the question is, "Does this mean, though, that in the end it is impossible for us to say at any point that 'Yes, the scientists have now got this part of reality right in some respects and to some degrees?'" I mean, that is the kind of claim I would like to say they have. I would like to say, *in some cases,* "Yes, we can see that." In some of my own laboratory research I find that scientists use what to many people are highly theoretical things—protons and neutrons—but they use them as research tools to do sophisticated experiments on the structure of matter about which at this time they are not very clear. But I would want to say that there is at this point no reasonable doubt—although it is *logically* possible that they are wrong—nor do *we* have any reason to doubt, that they understand pretty well the major characteristics of, say, protons and neutrons, especially as they use them in the laboratories.

Knorr Cetina: I would not contest that at all. But what I want to debate is whether the *picture* that is associated with these technical effects must necessarily be a true picture in order for the technical effect to be successful. I think you can have technological success, success in replicability, success in achieving certain technical effects, without necessarily assuming that you actually have a true picture of the world.

Giere: I think one of the most important things that scientists do is take their new knowledge and use it to construct instruments to develop later knowledge. That is one of the ways—only one—in which science can indeed progress. That is an important thing. But it seems to me that the object and the goal of this pursuit is not just to get new technology, but is indeed to use the technology to, I must say, *improve our picture of the world*. Now of course we could be mistaken. There is no guarantee that even though we have got the effect that we think we should get if one particular model were the right model, it could still be the wrong model. This is always possible. The purpose of experimental design is to try and carefully eliminate the possibility that we would get this effect if some other model were correct. That, of course, is difficult; but that seems to me the goal of design.

Callebaut: You haven't really addressed the problem of the Azande yet.

Giere: Everybody talks about the Azande, but none of us of course have ever talked to the Azande![39] But I think my moderate interpretation of what was going on with the Azande is in fact very similar to what happened here in Europe in the Middle Ages, when one held trial by ordeal; one would subject a person to a physical assault—burning or something—and then see whether they got cured in a certain time, and if they did, that was a sign that God was on their side and they were innocent. Now the Azande situation seems to be very similar, except they were a little more humane and they killed chickens instead of people. But the result, I would say, from a modern scientific standpoint is that they had a complicated way of *flipping a coin;* a chance mechanism which they built into a whole social practice. But what it was, was in effect a *chance* mechanism. The Azande, I assume— I also have not observed the Azande!—did not have what we have, which is a social practice for consciously, deliberatively investigating the reality behind certain practices. They *used* the practices, but their purpose was not to investigate further what was behind. They did not do controlled experiments on chickens. That, maybe, is the difference: *our modern science is not a modern witchcraft!*

4.3.9. *Turning the Tables on the Relativist?*

Callebaut: By relying on a kind of empiricist methodology in their own case study approach, relativists in current science studies may be running up a blind alley (cf. 3.6). Maybe it is not too farfetched to conclude that like the Cartesian philosophy of the subject, current epistemological relativism is on its way to turning into a rather sterile global skepticism.

Giere: You see, I think that without knowing it, relativists become a kind of empiricist. Of course they don't have a relativistic view about people, or tables and chairs, or what people say. "We are not going to commit ourselves to the belief that there are other people"—they would not say that.

Callebaut: They could say that science is not about chairs at all, though (Campbell and Paller 1989).

Giere: But the point is that there is still the problem of where you draw the line. I

---

39. To further complicate matters, some epistemological relativists call it to be their aim to leave the Azande "in peace" (Collins and Pinch 1982:6)!

think that is just an impossible thing to get around. One of the really interesting questions is: *How do scientists decide when something is real and when it isn't?* In quantum theory—maybe a nice case—there are some things that are really open. "Are these particles real or not?"—it is an open question. A lot of questions are. Other questions are closed. That there *are* protons is not an open question! Exactly what protons are made of—*that* is open. But there is something like a proton. The idea that you cannot use that as a resource for explaining what scientists are doing seems to me silly; like saying, "Well yeah, we are not going to assume they have amplifiers," or "We are not going to assume that they have these other kinds of detecting apparatus," or who knows what. That seems to me just wrong. But of course you can't assume the truth of the theory they are working on in order to explain what scientists are doing; because of course that is still open.

Callebaut: Dick, you have used the label "realist methodology" (Burian 1987a). What do you mean by that?

Burian: At a low level, if you look at the sociology of many scientific disciplines, you find a division between people who are "just modellers" (some scientists sometimes say that in a rather derogatory way), work-bench people, and "theoreticians." In different disciplines, the scales of values connected with this may vary. But the "just modellers" are in a certain sense very strongly nonrealists. Give them any damn set of assumptions and they'll see what follows. Their game—and I don't mean anything unflattering by that term—is simply to work out consequences. *If that were all that there was to it, it wouldn't be science.* But as a piece of a larger enterprise it is a very crucial part of science. The division of scientific labor requires us to supplement modelling with work that is closely tied to both experimental and theoretical practices. It is the interaction among these different modes of work, sometimes allocated to different disciplines, that enables us to pin down processes and entities well enough to secure reference to them.

Callebaut: And on a larger scale?

Burian: On a larger scale, my unpublished "Ontological progress" paper (n.d.) argues that there are weak controls on the progress of science—when things work right (which they do not always do). I employ a very loose analogy for that: gravitational force is 23 orders of magnitude weaker than any of the other fundamental physical forces. But if you look at the shape of the large-scale universe, astronomically, it is shaped by gravity. *A weak force is not necessarily one which does not have any shaping importance.* In that sense, it is necessary to step back from the forefront stuff which is caught up in all the *tohubohu* of negotiation, experiment, controversy, etc. When you're caught in the middle of things, you don't know quite how to do the evaluations in a narrowly epistemologically sound way. You step back from that far enough—you can do sound epistemological evaluation. The result comes out as a realist hopes it will often enough, though by no means always; as Ron Giere suggests, we are justified in claiming that there are protons even if we are not justified in saying what they are made of.

Knorr Cetina: That you can't see the place in which a boat will land in thirty or one hundred years while you are drifting in it may be right. But as the history of science shows, the fact that some entity is considered "real" or "true" after many years of research also does not mean that this "fact" will not be overturned another thirty or one hundred years later. Hence I don't see the point of Burian's argument. Furthermore, the new sociology of science is interested in the mechanisms of knowledge production, in the *cultural machinery* deployed in the sciences in regard to the generation of knowledge. These mechanisms cannot be learned from considering what emerges as accepted knowledge (and disappears as accepted knowledge) every now and then. You have to be in the boat in order to find out on what it runs and how it moves.

4.3.10. *Stepping Back from the Tohubohu*

Callebaut: You also point out that one quarrel between history-and-philosophy-of-science types and traditional philosophers of science concerns closeness to versus distance from the forefront. What, exactly, do you mean by that?

Burian: In scientific work, you are seriously in the middle of things. According to Mulkay's, Latour's, and a couple of other people's symmetry thesis, no side in a dispute can be presumed to possess the truth, and truth cannot therefore be a vehicle for decision making. One is *in mediis rebus* in evaluating ongoing work. One of the reasons why I want to step back from being too close to the forefront as a scientific realist is because it's only *after* one has learned which modalities are reliable for examining a particular domain (and what their weaknesses are) that one can tell which claims are secure and which are not. Such problems are surmountable, but only with time. There is no longer a serious question about the existence of electrons. Electrons may not be exactly what we conceive them to be. But if you had asked seventy-five years ago whether electrons exist, the answer might not have been so clear as it is now. It is going to remain typical that until the modalities for putting oneself in causal interaction with the entities are sufficiently developed, one does not know what one is dealing with.

Callebaut: In that sense, one has an ambiguity . . .

Burian: . . . about what are things and what are not—a point for Latour. At the same time, a point against Latour and also Hacking (1983) is that "intervening," in Hacking's sense, is not the only way of putting oneself in causal relations by manipulating things. If there are black holes, we are not going to manipulate them (cf. Hacking 1989). Nor, if current theory is right, are we going to manipulate isolated quarks, for we will not be able to isolate them singly. *If*—note the "ifs"—it works out that we can get the modalities for registering their causal interactions well enough, we may not *have* to intervene to know that there they are. In that sense, the proper distance can sometimes be achieved, but not while you are in the forefront of discovery.

Callebaut: What do you conclude from this as far as the realism/constructivism debate is concerned (cf. also Shapere 1992)?

Burian: If I'm right about the need to retreat from the forefront in order to get honest epistemic evaluations, work on the scale of laboratory life is not going to give you answers to the right questions if you're asking epistemic questions. In fact, in my view, in such cases it takes two to three scientific generations before the epistemic status of major theoretical entities is resolved. If this is right, piling up micro-studies to facilitate philosophical analysis of science will be a very difficult (though inevitable) job, one that requires immense labor, but one that we can and should do. A full-scale evaluation of how things went, of the long-term successes and failures of scientific research in a given domain, is not an easy task. The historico-philosophical reconstruction of the history of a major theory or program of experimental research (which reconstruction must include a lot of sociological analysis too—for instance, of the organization of experimental work and the division of labor among disciplines) is like the weaving of a gigantic tapestry. It is easy to lose the grand picture if one looks too closely at the warp and woof of the weave. On the other hand, without close weaving, there is no picture!

5

NEW ROLES AND TOOLS FOR PHILOSOPHERS OF SCIENCE

ONCE UPON A TIME, a practicing scientist (it was difficult to decide which discipline he belonged to, though) complained that "in the philosophical literature, the adequacy of proposed formalisms has too often been tested only against simple hypothetical cases, often contrived to be paradoxical, but none the less contrived" (Simon 1967:33). His purpose in pointing to *actual* applications instead "was to have examples formidable enough to demonstrate adequacy to handle real-world complexity, and not simply adequacy to handle toy situations." The strange ways of philosophers of science in dealing with "examples" have puzzled scientists and laypersons alike. It puzzled me as a student. Why would philosophers be so . . . different? The answer, it would seem, can be straightforward. Philosophers began to handle examples in a disreputable way as a part of becoming *professionals*. They were going to show the rest of the world—the scientists in particular—that they were different from *them*. There are other aspects to the professionalization gambit that deserve being looked into, like the tendency to "go meta" and the relationship between research and scholarship; which is what I propose to do as a way of easing into this chapter.

If you take away a kid's favorite toy, different things can happen, including such contradictory things as regression to a previous developmental stage and growing up faster (well, let's put it this way—I can't help being helpless myself when it comes to dealing with examples!). The naturalistic turn left philosophers bereft of their cherished menagerie of examples—white swan, black raven, unicorn—so that whether they liked it or not, they had to face the question, *What to do next?* If naturalized philosophy of science is "to fall into place as a chapter of natural science" (Quine), will the naturalized philosopher turn to using empirical methods? The bulk of this chapter delves into several answers to these questions that have been given a try recently. It is suggested that the philosopher of science can become a therapist with respect to scientific strategy or, related to that, an underlaborer helping the scientists solve theoretical questions. In that world of Specialists of ours, he (and she as well?) may still be tempted to keep abreast of The General Picture, but the *context-dependency* of everything (well, almost . . .) will have to be reckoned with as a serious killjoy.

New roles not only take new clothes but new tools as well. The "semantic" or "model-theoretic" view, which takes *structure* rather than deductive quality to be the key to scientific explanation, is promoted by many as a new tool to be put to profitable use in studying the fine-grained structure of various sorts of scientific theories. Not unlike an ordinary map, which is "a physical thing on a piece of paper" (Giere), a theory, when viewed as a model structure, does not make *statements* about anything; it is given an *interpretation* by relating it to a part of the world and making claims about the modalities of the "fit." A question of some concern is whether the new view of theories allows us to dispense with laws in the sense of universal statements of correlations of *necessity,* for empirical claims are particular.

Philosophers of science have traditionally acted as (self-proclaimed) legislators with respect to scientific method. Logical positivism was a "gross misreading of the method of the already successful sciences" (Campbell), as was—I hasten to add—Popperian "methodology." Is there a role left for methodology in the new picture of science? Most naturalists agree that if there are any norms to be found, they will have to be derived at least in part from *substantive* scientific knowledge. If all methodological rules are of the "if/then" type, as "normative naturalists" claim, should we be able to specify the goals of scientific inquiry in advance? That would seem to imply yet another essentialism. But maybe methodological claims amount to little more than rhetoric?

Gone forever, it would seem, is the relatively peaceful unanimity of philosophers in the heyday of positivism!

5.1 WHO NEEDS PHILOSOPHERS ANYWAY?

Callebaut: People outside philosophy of science who know I am doing this book often ask me if I am going to come up with something *they* will be able to understand. I try to reassure them, but they aren't always convinced. And who knows—they may be right! What I find most frustrating is that many of these people are scientists of some sort; after all, this book is supposed to be about them! David.

Hull: Realize that until about a hundred and fifty years ago there wasn't this difference. "Is it philosophy? Is it science?"—we didn't even have the word "scientist" until 1834, when William Whewell coined it in a review of a book by Mary Somerville. Obviously we had the *concept* before then, and finally it dawned on us that we needed a word for the concept. But it is a fairly recent innovation, and I think it is one which is not all that good! I wish that we did not distinguish so sharply between philosophers and scientists.

Callebaut: The professionalization of philosophy of science is of recent origin, Tom.

Nickles: Early in the century, with the important exception of a few writers such as Mach, Peirce, and Duhem, philosophy of science was an amateur pursuit of scientists, in both the good and bad senses of "amateur." Even in the 1930s, when *Philosophy of Science* was founded, the original editorial board was made up almost entirely of eminent scientists, with just four or five historians of ideas and a philosopher or two thrown in for good measure.[1] In this sense, this academic journal was *not professional of philosophy of science*. Over the years, scientists, particularly physicists, have lost interest dramatically in what philosophers are doing. Philosophers became mostly concerned with their own logical and linguistic models, their own methods and the problems that arose within their models. Philosophy of science became a second-order inquiry; it was almost as if the philosophers were studying themselves rather than science!

5.1.1. *From Scholarship to Research*

Callebaut: Professionalization also calls to mind all sorts of pressures that, in this case, would tend to isolate philosophy of science from the very sciences and scientists it was presumed to be studying. You alluded to this before, when you talked about the need to carve out one's own space in the intellectual universe (2.3.3). In addition, most scientists have moved far away from philosophical preoccupations.

Nickles: The scientists themselves felt tremendous pressures to pursue their own work and are now commonly regarded as somewhat soft-headed or tender-minded if they become too interested in historical and philosophical questions. There is a fairly sharp distinction made in science and engineering departments between "scholarship" and "research." Research—the production of new data, new instruments and techniques, or new theories for publication in scientific journals—is one thing; that might get you grants. Studying historical episodes and engaging in philosophical disputes, well, if you want to do that a little bit on your own time that's OK, but too much of it actually tends to discredit one, suggesting that one is not entirely serious about one's scientific profession.

Callebaut: The intellectual and wider cultural climate in which scientists are working has changed as well—one does not have to endorse Allan Bloom's (1987) dim view of the predicament of Western Culture to accept that much.

Nickles: Scientists, especially in the U.S. today, are not trained in terms of general intellectual culture the way they used to be. As Gerald Holton (1986) has remarked about physicists, scientists seem no longer to regard themselves as culture carriers.[2] Maybe they *think* that they are creating the new culture and that the old one

---

1. This is documented by Butts and Churchman in volume 51 (1984) of *Philosophy of Science*, in which they commemorate the journal's fiftieth anniversary.

2. Cf. Jean-Jacques Salomon's (1973) distinction between the "savants" (aristocratic scholars) of former times and the "scientifiques" (scientific workers) in our era of "technopositivism," and John Ziman's (1960) parallel between scientific "vocations" and "careers" on the one hand and the English cricket class distinction between "gentlemen" and "paid players" on the other (Ravetz 1977). Still in the same vein, Levy (1982:154) writes about Karin Knorr Cetina's choice of the title of her book, *The Manufacture of Knowledge:* "This author, I felt, is teasing us. She knows that we manufacturers of knowledge pride ourselves on being intellectuals, scientists, academics, thinkers, and so on. Manufacture has to do with a crass world of trade and commerce. It is dominated by people who are concerned with getting and spending. Those who actually do the manufacturing are blue-collar workers who are certainly not encouraged to think (creatively or critically). Our author, I mused, knows that we live in a world that, unlike our past

is of no use—that may depend on the individual. But professionalization has introduced many approaches that have pulled apart the two areas—now the two cultures! And it has contributed to a general fragmentation of intellectual life, as we are all aware. Analytical philosophy, which has characterized much philosophy in Britain and America and can now be found in many places on the Continent (certainly in Helsinki and Oslo but also in Belgium, the Netherlands, Germany, and Poland . . .) has been of little interest to intellectuals who aren't professional philosophers, because it no longer seemed to address the important old questions about human beings and their place in the universe and what life was for—that sort of thing. But people end up saying the same thing about modern music and modern arts . . .

Callebaut: . . . the postmodern patchwork . . .

Nickles: . . . and all of these disciplines, in professionalizing, have withdrawn into their own communities of experts; and they have—for some very good reasons, I must admit—regarded only those experts, only their peers, only their fellow experts in their field, as being a qualified audience—qualified to criticize and to judge them. So they just simply don't care what the rest of the world thinks. That has been as true of philosophers as of scientists. Finally, some of the things I said earlier about the retrospective nature of philosophical work versus the prospective attitude of working scientists (2.3.1) could easily be plugged in here too.

Callebaut: As professional philosophers of science, some of us tend to suppress the knowledge that not everybody is happy with the *meta*-inquiry view of philosophy of science, let alone the *meta-meta*-inquiry view Tom was just alluding to. Alternative views have been influential in other times, such as the view that philosophy of science ought to expose the *presuppositions* of scientists, or that it ought to formulate *world views* that are in some sense based on scientific theories (cf. Losee 1980:1–4). To the extent that naturalized philosophers of science see their own function as "therapeutic" in some sense (see below), this would seem to bear some resemblance to the first alternative. On the other hand, some persistent misunderstandings between scientists and philosophers are certainly due to their laboring under different conceptions of what the proper task of philosophy of science is. The clash between Gunther Stent (1986, 1987) and Alex Rosenberg (1986) in *Biology and Philosophy* (see also Thompson 1989b) would seem to be a case in point. Or, to mention another example, it has always struck me that in a social science like economics, the practitioners of the discipline have showed a marked preference for the philosophy they produced themselves, to the almost complete neglect of what was going on "outside." I guess something similar is happening in

---

worlds, depends critically on exponentially increasing dollops of scientific knowledge. She knows that what for most people has been the dalliance of affluent elites has become a pressing necessity involving so many people that it is even hard to think of them as elite. I thought the title had a certain Swiftian ring to it." Alan Chalmers has since written a book called *Science and its Fabrication* (1990).

physics, which has always been the paradigmatic science for economists. It would seem that at some point the naturalization of philosophy of science may involve a clash between research and scholarship of a new kind: *philosophical* research versus *scientific* scholarship!

Tweney: It's interesting in this context to compare philosophy of science with psychology, which latched on to logical positivism in a most incredible way. Psychologists often still argue as if only operational definitions matter in science.[3] Most are about fifty years behind the philosophers on this. It's curious too that, for all of our "physics envy," psychologists don't by and large have much of a clue about what physics is about—we have a "mythical physics" that we appeal to for justification! (Cf. Wimsatt on Max Dresden, 3.1.)

5.1.2. *Standing For a Cause Again: The Creationism Debate*

Callebaut: A curious episode in the recent history of the philosophy of science should be mentioned in the context of our discussion of the roles of the philosopher of science. I am referring to a phenomenon I have the greatest difficulty explaining to my students in Europe: so-called "creation science," and how the philosophy of science community in the U.S. a decade ago got involved in the debate over it. Philip, you were one of a handful or so people who have written a whole book— *Abusing Science* (1982) in your case—to expose all that is wrong with this "lunatic fringe."[4] How do you look back on this episode, now that the heat has dissipated? Was it worth the effort?

Kitcher: Since most of the book is taken up with details of bad creationist arguments, I hope that most of the book will become redundant soon! I think there are some points in there of interest even after the lunatic fringe has gone away, but they tend to be points about how you give a capsule presentation of evolutionary theory or of the methodology of evolutionary theory. I have been delighted by the fact that some of my philosophical colleagues think the second chapter is a nice way of introducing students to contemporary philosophy of science.[5] But the message had to be gotten out to school boards: they had to have something to which they could point and which they could use. Ordinary thoughtful people in school districts all

---

3. Campbell (1988a, cf. 1990b) considers *definitional operationism,* in which operations are regarded as defining terms in a scientific theory (e.g., IQ), to be positivism's worst gift to the social sciences. It should be noted that Hempel (1966:ch. 7) was aware of several of the reasons for the failure of definitional operationism. See also Hull in 6.1.

4. See also the biographical note on Kitcher at 3.4.2, Futuyama's *Science on Trial* (1982), Ruse's *Darwinism Defended* (1982), and the collections Godfrey and Cole (1983), LaFollette (1983), and Ruse (1988c).

5. The experience with the creation/evolution controversy convinced Kitcher that scientists and the lay community need a better picture of how science works. The example of Nagel, Hempel, Carnap, Reichenbach, and Feigl continues to deserve emulation here, he thinks: "They did the scientific community a tremendous service by formulating methodological ideas which could then be taken out by scientists and used by them, not only in internal disputes where they are relevant, but also in confronting challenges from outside." He grants that there may be a touch of nostalgia there (cf. Sarkar 1992b).

over the country were getting bamboozled, and the scientific and intellectual community owed it to those people to *do* something. So we sat down and we wrote. I think that was right. There is a lot more that can be done, like getting educational programs organized for improving high school science education, which is much more positive than "just doing down the creationist."

Callebaut: It struck you in the creation/evolution controversy that extraordinarily intelligent scientists like Steve Gould were floundering; there is a hint at that in the motivation for the project of your "New Consensus" institute, which you were so kind to let me read.

Kitcher: They were floundering because they knew they couldn't really rely on the simple and persistent ideas that they'd gotten from a philosophy or history of science course, or just from people talking to them about the methodology of science in general.

Callebaut: In a way this can be understood as implying a contradiction: To the extent that these people have gotten philosophical ideas, it must have been from the kind of people you mentioned (see note 5)—or am I mistaken?

Kitcher: What they didn't have was a general picture they trusted. Very bright scientists I know were constantly saying, "What can I read in philosophy of science? We know that logical empiricism faces severe problems. We know about Kuhn and that it is very provocative, that people have reservations about this. What is it that philosophers of science believe these days? Give us a basis from which we can go out and reply to the challenges!" They had an acquaintance with the philosophical ideas of the previous generation. But, in a way, they lacked commitment to a set of philosophical ideas. So there was an uneasy attempt to mix together divergent points of view that didn't always fit with one another . . .

Callebaut: . . . A kind of Popper-Kuhn mix, which is rather odd . . .

Kitcher: . . . Exactly right! You can see this in the writings of a number of biologists: you can see their sympathies for Popper as he is read in the popular literature, and their sympathies for Kuhn. One thing that was very clear to me is that many practicing scientists today have a kind of second-hand understanding of people like Popper and Kuhn. They don't know the sophisticated discussions of the opening chapters of *The Logic of Scientific Discovery*—a very sophisticated book!

Callebaut: Pat, I was rather surprised when I read class hours on your door for "philosophy of religion." What is happening there?

Churchland: I have quite a lot of fun. I announce that it's going to be a skeptical course and this is essential since I do not want fundamentalists of any kind down my neck. We go through the major arguments for believing in God and discuss the weaknesses. We do the argument from evil, and that pretty much sinks God. Given the argument from design, we have a lengthy debate on evolution and creationism. Philip Kitcher and Duane Gish debated this issue in Minnesota, and there is a tape of the debate. The students get very involved in this. You see, for most students it's the very first time they've ever confronted these issues, which *matter* to them.

I do a little bit on mysticism—what people think it is and why it is epistemologically goofy. And we look at contemporary superstitions: alleged paranormal phenomena, claims about homeopathic medicine, astrology, auras and acupuncture—
some of them may have some value, others not. The approach is always, "Let's
look at the data." I have a good time in the philosophy of religion, but I prefer to
be straightforward, not precious.

Callebaut: Michael, the creationism controversy was quite important for you.

Ruse: It involved me being a witness for the American Civil Liberties Union in the
Arkansas creation trial in 1981, along with people like Gould and Francisco Ayala, Langdon Gilkey (the theologian), and others.

Callebaut: Why you? Why philosophers at all?

Ruse: A number of people realized that there were philosophical issues that came up
here, particularly about the nature of science; and Popper's name was being
banded about by the creationists and others. Why me rather than others? By that
time I had debated Gish and Henry Morris, and I felt that I had got a lot of
biological knowledge at my fingertips, and I had got the philosophy, but in particular I also had the historical background. I felt and still feel that many of the
creationist arguments were arguments that Darwin's critics had brought up and
that had been run a hundred years ago—the nature of the fossil record, the age of
the earth and those sorts of issues, not to mention concentrating on the fossils
rather than the biogeographical distribution.

Callebaut: The reason I'm raising this whole issue is not the creationism stuff per
se—being European, I can comfortably say, "I don't have time for that!" But I
was and am still intrigued by the fact that many of your fellow philosophers of
science and of biology in particular refused to have anything to do with it. How
come?

Ruse: They felt that they didn't have enough expertise or that if they were going to
appear in court, they wouldn't necessarily make a particularly good showing.
Some felt—this came out in particular afterwards from some of my critics like
Larry Laudan—that philosophy has no place in court. A philosopher is concerned
with a disinterested inquiry into the truth; and a fight is a matter of winning and
losing, not right and wrong. The court system in North America is an advocacy
system; you try to beat the other side. A lot of the people felt, I think, that if one
was to appear as a witness, one would in some way be degrading one's subject
and prostituting oneself, even if one wasn't going to make any money out of it.

Callebaut: You felt differently.

Ruse: I think, to give them credit, that some of them had serious doubts about
whether or not a *philosophical* case could be made against creationism and for
evolution. They had serious doubts about whether one could demarcate between
science and religion in *any* way, but certainly in any simplistic enough way to
make the case in court. But I didn't then and I don't now have those sorts of
doubts. I mean, I'm happy to agree that there are grey areas; but I really do believe

there is a difference between science and religion, and I think the things are separate.

Callebaut: Some philosophers—Laudan (1972), Quinn (1984), Burian (1986b) to a lesser extent—have criticized you for having used "psychological" rather than "logical" arguments, among other things. How do you react to this sort of criticism?

Ruse: This brings up the naturalistic turn. For me, the psychological arguments and the logical arguments aren't separable. I don't draw the same line of demarcation, perhaps, as they do. So there we are! I want to say something else, though, and that is a personal thing. I, for a long time, was worried about being an academic philosopher. I'd often wished that I'd done something else with my life, like being a doctor. Although there are many faults with the medical profession, at least one was helping people in a genuine and tangible way. And being a doctor is a life of action. Whereas being a university professor . . .

Callebaut: How about teaching?

Ruse: It is true, of course, that you do a lot of teaching—you help people there. But that had never been the side which appealed to me. I think I'm a good teacher, but what turns me on is the research. And I'd always had a lot of doubts about the real value of doing these things.

Callebaut: Incompletely socialized into the profession.

Ruse: I still do about a lot of academic philosophy, but I'm not convinced of the virtues of pure research. Perhaps it's OK for some, but it was never quite that for me. That's why I was interested in getting into the biology. I think I can say, without sounding pretentious, that I've got fairly strong moral urges. I have a Quaker background. So, often quite explicitly I felt a real sense of dissatisfaction about what I was doing, in the sense of the worthwhileness of my positions. So taking on creationism for me was personally a very satisfying thing, because I really did see creationism as a moral evil. I don't think the creationists are. The creationists I know—many of them I think are wrong, but I think them sincere.

Callebaut: You still talk to them?

Ruse: Oh yes! I don't think they're *evil people,* like Hitler, but I think creationism is a moral evil. What is really a moral evil is forcing it on young people in schools! Taking on the creationists was also psychologically satisfying, because it involved me in a job of action. When one went to court, one had to stay on top, and one had to use one's wits. I spent time going to New York to pre-trial conferences and things like that. I found that quite thrilling. It was all a question of taking philosophy and being able to use it in a concrete way, in what I felt unambiguously was a good way. And I felt very good about that.

Callebaut: The critics.

Ruse: What happened, as you know, is a number of people started to criticize me afterwards: my fellow philosophers, Larry Laudan, Phil Quinn, Dick Burian, Ernan McMullin. The first time they did this, I felt a bit sore and a bit hurt, as you might imagine. But then, very quickly, my mind changed. I think they are wrong, I think I am right. But I enjoy a good fight. Therefore, I put these things together in *But is it Science* (1988c), where I deal with all of these issues. I should add that, as is appropriate, I give my critics more space than myself. I see the strength of philosophy as opposed to creationism in that one can have debate and differ about these things.

Callebaut: So you've rather enjoyed your philosophical critics taking you on.

Ruse: I really think that people like Larry Laudan and Philip Quinn and Dick Burian got their heads in their asses—quote me on this—when it comes to some of their reasons against taking the sort of stand that I did. I'm not arguing with them about my particular *philosophical* position; but I think that if they were able to come up with a strong viable alternative—which they haven't—I'd take them more seriously than I do. So I've enjoyed fighting with them on that! I think, to a person—I'd include Dick Burian here, who does generally take real science seriously—I don't think they've read the creationist literature, so they don't know how bad it is!

Callebaut: Philip Kitcher did read it.

Ruse: Yes.

Callebaut: You may want to say something about his *Abusing Science*.

Ruse: Well, I think his book is very good. It's certainly better than mine. I'm a great admirer of Philip's book. I suspect that Philip would have taken the role that I have taken. I was just there earlier than he was. He has always been supportive on that one; we differed on other things. I have nothing but praise for that book. I wish I'd written it myself! It's an excellent book.

Callebaut: The creationism debate was referred to in Philip Kitcher and Wade Savage's motivation for the "New Consensus in the Philosophy of Science" institute at the Minnesota Center. Do you see this coming: a new role for the philosopher as someone involved in public controversies—we've also heard about philosophers assisting lawyers in their firms with their analytic skills, at least in the U.S.—as someone who can do for the scientists what the scientists can't or won't do for themselves?

Ruse: I don't know that I particularly want a new consensus in philosophy of science. I think life a lot more fun when there are differences! In fact, I'm a great believer in "dialectical" processes at that level. (*Pauses.*) But what about philosophers being involved in public issues? Yeah, I see this as a role for the philosopher of science. You hear a lot today about "universities under pressure"; governments are expecting us to do things that *they* want rather than pure research. Of course, academics generally rear back from this, particularly people in the arts. I personally think there is something to be said for this governmental expectation. I don't think all research should be immediately oriented on some immediate payoff, but I do think there is a place—and a need—for the philosopher of science to apply his or her expertise. In fact, just recently, with David Hull and others, I've been on a committee of the National Academy of Sciences, putting together a little booklet on fraud in science (Committee on the Conduct of Science 1989); and we've been on meetings doing that. There is something where expertise can be used to help on that. So I agree with that.

Callebaut: Philip and Dick Burian want to say something about Michael's participation in the creation/evolution controversy.

Kitcher: Michael did something that required great personal courage, and his testimony was practically successful. It's quite correct to say that his remarks about philosophy of science were simplified. Even a highly intelligent man like Judge Overton couldn't have appreciated an intense presentation of recent ideas in philosophy of science. While I sympathize with those who feel that Michael's presentation somehow sold philosophy short because it was too simple, it seems to me that they should remember that a court of law isn't a seminar room. Michael had a job to perform for the good of science and science education in the U.S. In my judgment his responsibilities to those constituencies were primary, and he discharged them excellently.

Burian: Creationism raises two entirely different issues, a metascience issue and a political issue. I'm not sure that I want to quarrel with Ruse about the political issue. He got up front and took a stand. In one fashion or another, I admire him for taking it. I don't mind confessing to having my head up my ass—if that is what he wants to say—in terms of dealing with issues like that, politically. There is a second issue which I think is entirely different: is there a clear demarcation criterion of the sort that he asks for? The answer to that, I am still firmly convinced, is *no*. It's not because we cannot separate science from non-science, but because it depends on the practices of the individuals and not on the logic of the dispute. To that extent, I will stand firm on the kind of position I was advocating. There is room for both of those, and I admire what Michael did in a public context, but disagree with it in a science studies setting.

5.2 UNDERLABORERS

Callebaut: Let us flee from these muddy waters to the quiet safety of our study rooms. Bill Wimsatt, you are definitely not the kind of philosopher who sees scientists making "silly mistakes" all the time (cf. Bechtel, 4.2.6). Your work remains close to the science. How, on your view, are a science and the philosophy of that science to be related? You have written that in addition to their traditional role of contributing to conceptual clarification, philosophers can also be useful as "therapists with respect to scientific strategy" (Wimsatt 1976a:673). What did you mean by that?

Wimsatt: You can analyze conceptual problems and to try to sort things out, clarify

alternatives, refine concepts, make recommendations, and so on. That is pretty
close to the traditional function of philosophers, the only possible difference being
that you're somewhat of a *participant* if you're doing that with scientists.[6]

Callebaut: You see a continuum between relatively applied philosophy of science
and theoretical work in the sciences.

Wimsatt: There is an intermediary ground where it is hard to tell which you're doing.
Maybe I've always wanted to be a scientist, and I've delighted in working in that
ground. In fact I'm even more delighted when I can do something that might be
mistaken as science than when I can do something that looks like straight philosophy.

Callebaut: You've done almost nothing that looks like straight philosophy, to put it
mildly. And you enjoy the borderline work that you do very much.

Wimsatt: I can give a rationale for it. I would not recommend that we no longer do
traditional philosophy. There are a variety of different perspectives on any given
problem, and we need the whole continuum. I've been told a number of times that
I think more like a biologist, and I probably do. Really *what I'm doing is what I
love*.

Callebaut: And what you love is closer to the sciences than most philosophers go.

Wimsatt: But that is changing. There are more and more philosophers who are in the
sciences up to their ears. Chicago has an unusual collection of people all of that
persuasion. As a matter of fact we have trouble getting someone to teach a standard
introductory philosophy of science course. Everyone feels that if you do that,
you're going to spend a lot of time talking about the positivists, and nobody really
wants to! When we get a new person coming in the department, they teach it for a
couple of years until they can pass it on to someone else.

Callebaut: Philip, like John Locke, you compare the work of the philosopher to that
of an "underlaborer."[7] What are the tasks of the philosopher according to the "conservative naturalist" you are?

Kitcher: One thing is to help with the resolution of theoretical problems in particular
areas of science. Scientists sometimes get themselves into difficulties because they
either use concepts which are inadequate, or fail to make certain kinds of distinctions, or get muddled up with various methodological canons. Elliott Sober's
(1984a) book on the nature of selection is a fine example, especially the last chapter—with which, by the way, I now disagree (Sterelny and Kitcher 1988). It is

---

6. Wimsatt emphasizes that "really good scientists of a reflective bent like Lewontin, Simon, or Feynman do it too—some of them quite a lot." And these, he boasts, "are just samples; I could name twenty
others in five minutes. But having philosophers do it adds an additional perspective that is salutary."

7. "The commonwealth of learning is not at this time without master-builders, whose mighty designs,
in advancing the sciences, will leave lasting monuments to the admiration of posterity: but everyone must
not hope to be a Boyle, or a Sydenham; and in an age that produces such masters as the great Huygenius
and the incomparable Mr Newton, with some others of that strain, it is ambitious enough to be employed
as an under-labourer in clearing the ground a little, and removing some of the rubbish that lies in the way
to knowledge" (Locke [1690] 1959:14).

absolutely superb for getting all kinds of things straight, and it may help, I think, evolutionary biologists. Some work in philosophy of psychology has been enormously helpful to practicing psychologists too, I think (e.g., Fodor 1975; Dennett 1987).

Callebaut: Physicists seem so much more self-confident.

Kitcher: It is true that they feel that they really don't need any help from philosophers.

Nickles: The situation in physics is rather curious. One finds a great deal of work in philosophy of physics by people who are very competent.[8] But it does not seem to me that the physicists are terribly interested in what they are doing. It's not that the physicists think their work is bad; it's just that most of them don't think it's really physics. Problems that philosophers think ought to be fundamental problems of physics as a description of what the world is like and ought to command specialities of their own (for example, tests of the quantum theory, nonlocality, and separability) are treated by most physicists as peripheral, a possible sideline to more serious research. This divergence in problem appraisal has to do with the philosophers' retrospective versus the scientists' forward-looking orientation. While philosophers are preoccupied with foundations[9] and constant testing (especially with theories they don't like, such as quantum theory), the physicists want to "get on with it" and do something new. There are always important exceptions, of course, such as Abner Shimony, who is highly respected in both fields.

Kitcher: Many of the physicists who are working on the Bell inequalities would concede that people like Shimony and Fine have really done a lot that is quite valuable for them; and I suspect that there are people working in space-time who would say the same about Malament, Earman, Friedman, and Stein (see, e.g., their papers in Earman, Glymour, and Stachel 1977).

Callebaut: David, like Bill Wimsatt, you see the philosopher's task as continuous with that of the scientist.

Hull: If you're going to do philosophy of biology, a lot of things you are going to be talking about will be indistinguishable from what theoretical biologists do. Those activities merge so imperceptibly into one another that if you spend much time in deciding, "Am I doing philosophy of science or theoretical science?" you are wasting your time, because the two merge together at the foundations. If you're capable of doing good philosophy of biology, you should be able to do good theoretical biology. Younger philosophers of biology take for granted that they must thoroughly understand some area of biology and even contribute to it. Maybe not the more contingent, empirical sort of biology, but the foundations. I do not see how, honestly, you can do the one without the other.

Rosenberg: I am very much a student of Willard Van Orman Quine's. That is, I never

8. People like Howard Stein (1983), Arthur Fine (1986), John Earman (1987), Michael Friedman (1983), and Clark Glymour (1977) have written on quantum theory, space-time, and relativity theory and are very knowledgeable in those fields.

9. In a rather different sense than the "foundational epistemology" discussed in chapters 1, 2, 3, and 7.

studied with him but I was brought up philosophically during his ascendancy in American philosophy. I learned those lessons so well that I think that, as Quine once said, "Philosophy of physics is philosophy enough." What he meant was that the questions of philosophy are continuous with the questions of science; that philosophical problems are just very high-order theoretical problems. They're relatively unpractical, they're extremely theoretical; but the difference between a philosophical problem and a theoretical problem in natural science is only a matter of degree, not a difference of kind.[10]

Brandon: I was a student of Quine's in the sense that I actually studied with him during my graduate school years. As my work became more and more biological, he asked, "Shouldn't you be in a Ph.D. program in biology rather than philosophy?" So it is not clear to me that he took this position of his very seriously.

5.2.1. *Intuition and Judgment*

Callebaut: Do you think of yourself more as a social theorist or as a philosopher, Jon?

Elster: I don't really think of myself as a philosopher of science or as a philosopher of the social sciences. I am not sure that I have any kind of well thought out conception of what philosophy of the social sciences ought to be. I think that is mainly because the social sciences and the natural sciences are in many respects very different. I am talking about differences in their stage of development, the consensus that has been achieved, the robustness of the general theories that have been constructed. It seems to me that many of the things philosophers of science do with respect to physics or biology simply are not feasible with respect to the social sciences. For instance, to study the logical structure of theories is certainly feasible in physics and, I guess, in biology; but there are no general theories comparable in scope, rigor, or precision in the social sciences.

Callebaut: Of course, that is in itself a philosophy of social science claim . . .

Elster: . . . and it needs to be defended. But I tend to think that all the attempts, without exception, at producing general theories in the social sciences have been unsuccessful. I think that what exciting, good social science is about is small and medium-sized case studies that allow *mechanisms* to be transferred to other cases.

Callebaut: Social scientists—sociologists in particular—are notorious for spending much of their time discussing foundational and methodological issues.

10. Thus the methodological problem of whether we ought to adopt hypothetico-deductivism (cf. 4.1) is for Rosenberg a question on a par with the question whether we ought to calibrate our *pH* meters with distilled water: "There is a good theoretical explanation for why we ought to calibrate our *pH* meters with distilled water. If it is indeed the case that we ought to adopt hypothetico-deductivism, then there has got to be a good theoretical explanation for that fact, not an a priori philosophical theory. So philosophy is continuous with science, and therefore philosophy doesn't deal with a special kind of problem that bears some relation to scientific problems. It deals with scientific problems; perhaps not truthfully, perhaps very general theoretical problems, but there isn't any difference between these two activities. We just don't have laboratories."

Elster: Many of the attempts to create a "professional" philosophy of social science have been premature. I personally tend to rarely read such journals. I think it is much more important to know what social scientists are doing than to keep abreast with what philosophers of social science think about what social scientists are doing. In fact, I guess I think of myself more as a social theorist than as a philosopher of social science.[11] I also think that *intuition* is—I guess unfortunately— much more important in social science than in the natural sciences . . .

Callebaut: . . . or than in mathematics, for that matter—where it may be less mysterious than is often thought (Putnam 1979) . . .

Elster: . . . Intuition and judgment still seem to make the social sciences to some extent into an art, not a science in the more rigorous sense of the term. These are all very general statements; but they express at least my reluctance to just think of myself as a philosopher of the social sciences. Now to the extent that I do philosophy of the social sciences, or just to the extent that I do social theory, I suppose it would be natural to include me among analytical philosophers or analytical social scientists, although that label of course depends heavily not just on what it includes but also on what it excludes.[12]

5.2.2. Eclecticism as the Response to the New Question de Méthode

Callebaut: If naturalized epistemology is "simply to fall into place as a chapter . . . of natural science" (Quine 1969:82),[13] then this is bound to have methodological consequences. How can a philosopher investigate a science empirically, David? By turning into an empirical sociologist or a cognitive psychologist?[14]

---

11. Elster qualifies this statement immediately: "I think that the two tend to go together, in the following sense: *Given the highly precarious or undeveloped state of the social sciences, any social theorist has to be very self-conscious about what he is doing.* He has to be aware of the highly artificial nature of many of the assumptions that are used in the social sciences; he has to be aware of the importance of *intuition* in social science." Looking back on his scholarly career, the anthropologist Claude Lévi-Strauss recently issued a similar caveat (Lévi-Strauss and Eribon 1990).

12. Elster also finds some of the oppositions that have been set up, e.g., between analytical and "Continental" thinking (Marxism, hermeneutics, structuralism . . .), quite artificial: "I think that Marxist social scientists should embrace the general principles of analytical (philosophy of) social science, such as *methodological individualism, rational choice theory,* and the like." Elster also maintains—and this may come as a surprise to some people—that hermeneutics, if properly conceived, is essentially a subvariety of intentional explanation: "There is nothing particularly mysterious or non-analytical about hermeneutic philosophy. I guess I should add one qualification. I think that to the extent that hermeneutics takes as its object the study of works of art, that is quite a separate kind of undertaking that does not really fit in with the general aims of science. In my opinion the difference is simply that studies of work of art, literary criticism, and similar esthetics disciplines do not have the pretension to *explain* anything. It is the explanatory pretension that constitutes the essence of a science."

13. Most naturalized philosophers of science and a growing number of naturalizing epistemologists grant that individual (as opposed to social) psychology alone is too narrow, because epistemology is quintessentially *social* (Goldman 1987, 1991; Fuller 1988, 1992). But there is considerable disagreement as to the nature of "the social" (cf. ch. 8, n. 5).

14. See Heyes (1989, 1991) on the difficulty of identifying the special skills of the naturalized epistemologist (as well as of some others).

"A professional philosophy of social science would be premature."

Jon Elster (b. Oslo, 1940) is research director at the Institute for Social Research, Oslo (1984–), and adjunct professor in the Department of Economics of the University of Oslo (1989–).  From 1984 to 1991 he was professor of political science and philosophy at the University of Chicago. He has also taught at the University of California, Berkeley, Stanford University, Caltech, the Université Paris VIII (Vincennes), the École des Hautes Études en Sciences Sociales, and Oxford University (All Souls College).

Elster wrote a thesis, Prise de conscience dans la *Phénoménologie de l'Esprit* de Hegel, for his M.A. in philosophy, which he obtained from the University of Oslo in 1966. In 1968 he went to Paris to study with Jean Hyppolite, who died before they met. Elster, who was "pensionnaire étranger" at the École Normale Supérieure, preferred the historical sociology of Raymond Aron (who agreed to be his thesis supervisor) and the rigorous formulations of Marx's economic theory in the "capital controversy" to the Althusserian Marxism which was dominant among Rive Gauche intellectuals at that time. He completed his thesis, Production et reproduction: essai sur Marx, in 1971 and became docteur ès lettres in sociology from the Université de Paris V the following year. His thesis laid out the program for his subsequent work in that its emphasis was already on rational choice theory, microfoundations, and the philosophy of explanation—the main methodological commitments characterizing Elster's outlook on society and social theory, which in the sixties and seventies were unacceptable for Marxists but are now being endorsed by a growing number of social theorists and even 'political animals.'

Elster—his name, as he likes to emphasize, means "magpie" in some languages—is the author of eleven books and about a hundred articles and reviews. He has also edited or co-edited seven volumes, including *The Multiple Self* (1986b), *Rational Choice* (1986c), and, with John Roemer, *Interpersonal Comparisons of Well-Being* (1991). His first book, *Leibniz et la formation de l'esprit capitaliste* (1975), was an attempt to understand the polymath in the light of the transformation the European economy was undergoing at the time, that is, a (quite sophisticated) case study in externalist historiography. In his second book, *Logic and Society* (1978), Elster applied modal logic to sociological problems and theories, in order to get a grip on, among other things, the elusive Marxist notion of "social contradictions." *Ulysses and the Sirens* (1979), *Sour Grapes* (1983b), and *Solomonic Judgments* (1989c) are studies in rationality and irrationality, with an emphasis on preference formation and the scope and limits of character planning (e.g., limits due to weakness of will). For instance, against the neoclassical economist's wisdom that preferring the present over the future is like preferring apples over oranges, and that "de gustibus non est disputandum," Elster maintains that a person who takes his future states as given, rather than something to be created, is fundamentally irrational. Or, to take another example, the "sour grapes" in Aesop's fable of the fox—who decided that what he couldn't have he didn't really want in the first place (and the opposite, "the grass is always greener on the other side of the fence" phenomenon)—are "perverse" desires, because they are not formed on the basis of the merits of what is being desired. Nonetheless they are interesting desires in that they are paradoxical: "sour grapes" may reduce the pain of failing to get what we once wanted, but also stop us trying to get something we want that is within reach. *Explaining Technical Change* (1983a) is both an exposition of Elster's views of scientific explanation (which are discussed at some length in this chapter) and a case study of the problem of technical innovation.

In *Making Sense of Marx* (1985) and the shorter book *An Introduction to Karl Marx* (1986a) Elster returns to the subject of his dissertation work, "because I became aware that the intellectual atmosphere was changing" (Elster 1985:xiv): "Above all, the publication of G. A. Cohen's *Karl Marx's Theory of History* (1978) came as a revelation. Overnight it changed the standards of rigor and clarity that were required to write on Marx and Marxism. Also I discovered that other colleagues in various countries were engaged in similar work." A small group of people applying the methods and standards of analytical philosophy that included Cohen, Elster, John Roemer, and Philippe Van Parijs, began to meet regularly in 1979. As one reviewer suggested, *Making Sense of Marx* might more accurately have been called "Making Mincemeat of Marx," given that it is a "comprehensive demolition of the best-loved parts of Marx's oeuvre" (Ryan 1991:19). *The Cement of Society* (1989a), a study of social order inspired by the Swedish system of collective bargaining in which "everything is up for grabs," challenges the economist's assumption that the identity of the players of the game (the actors) and the rules of the game are fixed. *Nuts and Bolts for the Social Sciences* (1989b) is an easily readable nutshell formulation of Elster's views. With *Psychologie politique* (1990), Elster returns to the language of his "normalien" years.

Elster is a fellow of the American Academy of Arts and Sciences (1988–) and a member of the Academia Europaea (1989–).

The original interview, for Belgian Public Radio (during which Jon complained about a "lack of focus" as he found it difficult to imagine what sort of audience he was talking to), was taped at the student radio station WHPK in the Reynolds Tower at the University of Chicago on June 5, 1985, some time after I had started sitting in on Jon's political science class.

Hull: If you worry about those things—"Am I doing psychology? Am I doing sociology? Is this epistemology?"—you're done for. What you have to do is try to figure out science using any technique that looks good. Right now it is fashionable to put down citation analysis. I think you can learn things by citation analysis; because we tend to approach any subject matter with all sorts of preconceived ideas, and the hardest thing is to undo those preconceptions. There are preconceptions with citation analysis, but they are different preconceptions than the ones I've got. (Cf. Wimsatt on robustness, 4.2.2.) So when I use citation analysis, I will every once in a while notice something that I would not have noticed on my own.

Callebaut: So you just use every possible technique that you happen to come across if you feel that it is pertinent to do so?

Hull: Yes, but given my training, I have not been able to use all the techniques that I think would help. For example, although I have read what psychologists have said on the subject of discovery, confirmation biases, etc., I have not been able to make much use of the techniques devised by psychologists. I did send a questionnaire to about twenty of the scientists whom I was studying to see which psychological types are liable to found movements and redirect the course of science, but that is all. I was in no position to give the MMPI to several hundred scientists, even if they would hold still for such an intrusion. That is why I find Ron Giere's (1988) analysis of science in terms of cognitive psychology so valuable. It covers one of my most exposed areas. My work is a combination of history, sociology, philosophy, and evolutionary biology. Now am I sure that my mix is the right mix? No! But it is the only mix that I know enough about to use. You've got to use what you know, and I am using that combination. If it turns out to be the wrong combination, I will fail; if it turns out to be the right combination, I will succeed. But there is no way to know in advance which is which. My work is very sociological. I am not tremendously well trained in sociology; but what I have done is every time I needed to know something, I went to learn it.

Callebaut: Karin, you also think that eclecticism can be beneficial, don't you?

Knorr Cetina: It depends on what you mean by eclecticism. Take physics. Physicists shift from one language game to another, and this shift is an important resource for their endeavor. For example, they use Feynman diagrams, and that's a thing in itself. They can do certain things with Feynman diagrams, but not others. They use vector calculation and matrix algebra. They use probability theory, which is a different thing again. They also use the language of electronics and the language of informatics a lot. If they cannot solve the thing in terms of particles, they move to waves. So there are lots of different languages they can draw on in going from one to another. I don't object to eclecticism if it means that. If you can draw on different kinds of resources and put them together for your own purpose, there is nothing wrong with that!

Callebaut: Are you implying now that there are no incommensurability problems!?

Knorr Cetina: Again, it depends on what you mean by incommensurability. It is quite obvious that, in practice, people switch. Now you can say, to cut it relatively short, "It's a jump." But people know *how* to jump. You don't have to solve the incommensurability problem and you don't have to say they translate one thing into the other. I'm not sure that they translate a lot, although they may at times— it's an unsolved issue anyway. Even if you say that there can be no translation and that there is no commensurability—and I quite appreciate Quine's work on translation (e.g., Quine 1960; cf. 7.1.5)—*people know how to jump!* And as long as they do that, they can draw on different resources. I find nothing wrong with that. Of course, in a specific case it always depends on what you combine, and how you combine it, and what you achieve by doing this. It's not an automatic consequence that the result is beneficial. But it can be.

Callebaut: David, the first group of scientists you have studied in detail were the numerical taxonomists (Hull 1967). Why did you pick those?

Hull: That was an accident. As I was completing graduate school, numerical taxonomists were devising methods for using computers to classify plants and animals. Because I had spent two years running IBM machines in the army, I was in a position to involve myself in the dispute over numerical taxonomy. If I had not had the background I did, I might have been scared off. When the cladists came along, I had to learn cladistic analysis, but I was able to learn it as they developed it. In general, those of us who study science must learn the relevant science or we cannot even begin to do our work. Initially, this may seem like a tall order, but science is inherently interesting. Once you get into a particular area of science, you are dragged along whether you like it or not. Even those bitten by the discourse analysis bug find it difficult to ignore the content of science.

Callebaut: How did you get involved in the sociology of science?

Hull: When I began to realize that the social relations among scientists are more important than I had been led to believe, it was too late in my career to go back to school and study sociology formally. I have considerable admiration for Alex Rosenberg returning to graduate school to take courses in molecular biology. Can you imagine what it would be like to be graded again, perhaps getting a B? But I did read the literature in the sociology of science, go to 4S meetings, talk with sociologists, argue with them; in fact I think I am on their governing board now or something. If you want to understand a subject, you've got to get involved with those working in that subject. Just reading is not enough. And this makes genuine interdisciplinary work difficult. I got involved with taxonomists, I got involved with sociologists, and don't forget I am in a department of philosophy. If I don't keep involved with philosophers, I am done for. I have to get grants. I go to a lot of meetings, and each group has its own mores. I had to learn how to behave in ways appropriate for each group. I certainly was not able to be spontaneous and genuine all the time!

5.3 DISCOVERY IS EVERYWHERE

Callebaut: Positivists à la Reichenbach and Popperians made a rough and ready distinction (following Herschel) between *context of discovery* and *context of justification* which allowed them not to bother about the "pragmatics"—the "details" and "contingencies"—of ongoing research. They were only interested in justification (confirmation and testing issues) and in the logical reconstruction of the "final research report," not in the investigative pathways that led to those results. One obvious symptom of the naturalization of philosophy of science is the renewed interest in research as a *discovery* process. About a decade ago, Tom, you brought many relevant lines of thinking about this topic together in two books (Nickles 1980a, b; cf. 1977). Where do you stand on these kinds of issues today?

Nickles: I treat discovery in a more general way than do many of the "friends of discovery" (that term I owe to Gary Gutting) in that I adopt a *problem-solving approach,* as opposed to a directly truth-seeking, theory-oriented approach. Philosophers in the past have tended to talk about *theories* as the units of discourse. I prefer to follow Kuhn on this point: the *solved problem* is the primary unit of scientific achievement—the most useful kind of success—and scientific activity consists mainly of problem finding and problem solving. Since discovery mainly involves search or construction tasks, most scientific activity is discovery in a broad sense. So when philosophers shun this topic, they ignore most of what scientists do, most of scientific cognition, most of the human learning process. Any sort of search, generation, or construction falls into the discovery ballpark for me, including recognition and articulation of problems themselves. "Where do problems come from?" is a very important question that even the "friends" have neglected.[15]

Callebaut: Part of your inspiration, as you acknowledge, comes from Simon's *information processing approach* to AI (see 8.1). When you agree with Doug Lenat that problems and discovery are everywhere, do you mean that discovery is relevant even to justification?

Nickles: Absolutely! And in all sorts of ways, assuming that you mean the justificatory aspects of the *process* of inquiry. Suppose one already has a decent-looking theory or hypothesis "on the table," as the Popperians like to say.

Callebaut: . . . a theory deemed worthy of "pursuit," in Laudan's (1977:108–14) lingo.

Nickles: One is now faced with the task of criticizing and testing it. Well, here are problems arising immediately: search problems—one has to search for *novel predictions,* for various kinds of *errors,* for *consistency* with various kinds of *constraints,* and so forth. So here we have several search tasks arising immediately,

---

15. Nickles (1976, 1978, 1980c, 1981, 1988).

even within the hypothetical- deductive viewpoint. I find it highly ironic that Popperians, for example, who put so much emphasis on novel prediction, gloss over the crucial discovery tasks involved in the very process of searching for novel predictions.

Callebaut: Another dimension related to justification in which discovery is deeply implicated, according to you, is *heuristic appraisal* (Nickles 1990a).

Nickles: This goes back to the point that scientists are forward-looking. They are constantly evaluating current proposals, projects, hypotheses, and so forth, heuristically: not simply on the basis of their track record to date, but on the basis of what outstanding problems they seem to be able to handle in the future—their *problem-solving potential*. That is, I think, a terribly important dimension of justification that philosophers have horribly neglected.

Callebaut: "Final" justification is not to be had anyway—neither in (philosophy of) science nor elsewhere. Don?

Campbell: We have certainly naturalized justification. If we insist upon apodictic justification and that all things that we admit to be knowledge are also completely justified and true, then we aren't going to have *any* knowledge. The skeptics point to a very real problem. It's not just a simple category mistake; it's not just confusing deductive and inductive logic. As Laudan's great essay against convergent realism (1981a) shows, the non-provenness of our very best accepted theories has been a real problem again and again. Similarly, the skeptics were right in saying that we are not going to be able to get certain knowledge in science, or in ethics for that matter. Their excellent self-critical philosophy has shown that such kinds of "proven" objective norms are not available to us.

5.3.1. *The Mystery of the Great Logical Inversion*

Callebaut: You also talk about *discoverability* (or *generalizability*) in this context (Nickles 1983, 1984a, b, 1985, 1987b). What do you mean by that horrible word?

Nickles: If someone can think of a prettier word, I wish they would tell me! Discoverability has to do with the way in which *scientists are constantly reconstructing their previous work,* again not always so much with the intention of tidying it up logically so that it has a neat foundation to make the philosophers happy, but in order to have a solid platform from which to leap into the future. It is a kind of justification in which one reasons *to* the theory or claim in question, rather than *from* it to its testable consequences.[16]

Callebaut: Is that a critique of the familiar forms of deductivist or "consequentialist" philosophy of science?

Nickles: Yes, it certainly is. Almost all philosophy of science, almost all theory of

---

16. In other words, Nickles specifies, "a discoverability argument is an *idealized* discovery argument; because by the time scientists know enough to construct such an argument, they know far more than the poor wretches who made the original breakthroughs. This is one instance of a bootstrap process."

justification is "consequentialist" in the sense that people consider *all* supporting evidence of theories to be the results of tests of the theory. The vast majority of philosophers, I believe, regard virtually all empirical evidence for theories as resulting from empirical tests. But yet if you look back to an older tradition associated with people like Isaac Newton, you find a quite different use of empirical information in science, namely, to reason *to* a scientific claim. Unfortunately, this tradition has been somewhat tainted because many of these earlier writers—whom I would call "generativists" as opposed to consequentialists—were also inductivists; that is, they talked of somehow inferring theories from facts. Popper and many other people have rightly attacked inductivism as a general methodology on many grounds.

Callebaut: Do you think, then, that it is possible to drive a wedge between induction and the generation of ideas?

Nickles: A generative methodology is not necessarily inductive. Inductive methodologies are only one specific variety of generative methodologies. In some recent inquiries I have made into nineteenth-century history of methodology, so far as I can tell there has not been one single methodologist who has justified the great methodological turn that took place in the nineteenth century among philosophers, from a constructivist, generativist methodology to a consequentialist one (1987a, c). This change is more apparent in philosophical writings than in science, I claim. I call it the *great logical inversion,* because instead of reasoning from data and other constraints to the theory, we are now reasoning from the theory to the data and other constraints as tests. So far as I can see, the problem of explaining how that came about is unsolved. No one has given sufficient logical reasons for abandoning whole hog the generative approach for a purely consequentialist viewpoint.[17]

5.3.2. Robust Bootstrapping

Callebaut: How does your account of discovery—including its justificatory aspects—relate to Clark Glymour's (1980) "bootstrap confirmation"?

Nickles: Glymour's idea is that one part of a theory can be used to help test another part of the same theory. This is a step in the direction that Wimsatt and I want to go, but I use the bootstrap idea in a much broader way. There are all kinds of bootstrap mechanisms used in science and other inquiries. A familiar, still hypothetico-deductive one is that introducing an hypothesis to get predictions and otherwise guide research can help, in favorable instances, to produce the very evidence

---

17. Nickles regrets that many "otherwise enlightened souls" still take for granted the empiricist dogma that all evidence is consequential. He agrees that predictive testing is important, but denies that it is *all*-important. Putnam (1962:216) said that justification in science doesn't go just one way but in *any* direction that may be handy. Nickles goes one step further: "If it proceeds in many directions at once, so much the better, for then we get a kind of *robustness*." He thinks this "sort of an *engineering conception* of justification" is a powerful antidote to the usual sort of linear and "Euclidian" models of justification: "reasoning from axioms to theorems, corollaries and so forth—the kind of thing we all studied in school."

that eventually confirms the hypothesis. Discoverability takes this a step further: the hypothesis can eventually become linked into a network as a *derived conclusion*—not only as a premise. Thereby it becomes less conjectural. "Linear" empiricists tend to ignore this dialectical, self-transforming feature of scientific development. Ultimately a bootstrap epistemology becomes nonlinear in allowing a broadly circular kind of reasoning through the nodes of the network, a kind of *virtuous circularity.*

Callebaut: That always raises eyebrows (cf. Vollmer 1975; Walton 1985).

Nickles: To be sure! Nonetheless, I think we need to wean ourselves from the idea that the only possible sort of epistemic support is the old Euclidean model of justification used in foundational epistemology. It's not just a matter of rejecting absolute foundations—the self-justifying premises—but of discarding the whole model. Let's try some new models that permit *mutual support* and even *self-support,* in a stronger sense than Glymour's, when you look at it as a process in time! A move may vindicate itself by "working" in the right sort of way. Here we have Bacon's old image of a temporary scaffolding or the one about using something as a ladder to reach a higher platform, whence one can throw the ladder away.

Callebaut: What about mutual support?

Nickles: It occurs all the time in nature. The members of a building or truss bridge are mutually supporting. They hang together, and they could *not* "hang" separately. Or we might draw an analogy to space-time theory. In a local context reasoning will be linear, because if you use circular reasoning involving only a few claims, you're going to have a *vicious* circularity; you are short-cutting the reasoning process, so to speak. We may call that "short circularity." *In local contexts one is Euclidean,* so to speak. But more globally, there are no founding assumptions; *everything is justified in terms of everything else,* and there is an elaborate crisscrossing of interderivations and so forth that makes the whole thing—one hopes—hold together in a coherent manner. So globally one can no longer be a Euclidian; one has a virtuous circularity; one has to go over to a sort of *spherical* conception of justification.

5.3.3. *The Invasion of the Social and the Locus of Control*

Callebaut: A nonlinear view of justification neatly ties in with a thoroughly socialized conception of science, which in Tom's case is clearly influenced by the recent developments in anthropology and sociology of science. Now the shift in orientation in people's work from a traditional, purely intellectual approach to one that aims to encompass "the social" has become so *natural* (if I may say so) overnight that it may take place almost surreptitiously. Therefore, before resuming the discussion with Tom, I would like to focus for a moment on the "social conversion experience" of one naturalistic philosopher. Bill Bechtel, when we discussed Darden and Maull's interfield theory (4.2.5), you noted that at bottom, it still excludes the social. What made *you* realize that "the social stuff" is inescapable?

Bechtel: The case I've been working on most recently. It wasn't from reading the sociologists. I previously found reading sociology of science very hard to do and not to answer any questions I had at the time. Examining the work of scientists, however, I came to realize what proportion of their time was devoted to social activities and how important those were in terms of determining the intellectual content of their work. Such things as which scientist would respond to which other and what experiment someone would do were affected. They might do something, say, because David Green had said this outrageous thing and they wanted to nail David Green. I needed the sociologists' resources here. I still haven't moved that far into the tools they have given us. So I arrived at an emphasis on social factors on my own way, but then turned back to read someone like Latour, or Mulkay, and said, "Now I understand *something* of what they're saying!" Now I turn to them for credibility, since there aren't many philosophers I can turn to for credibility on this issue! Whether I understand *them* right I don't know . . .

Callebaut: . . . It shouldn't matter, according to Derrida. (*Bashful.*) I'm sorry! . . .

Bechtel: . . . But they're now saying something *to me* and giving me insight in how to think further about the problems I'm working on. Even though I still can't understand the French notion of *discourse analysis,* I certainly do see that there's more complexity to a scientific text than merely conforming to the styles prescribed in a scientific tradition. Even *that,* in fact, is an important social constraint: you realize that scientists are writing in a particular way because that's the only style that's acceptable to get their ideas across. They're fitting into a channel. And you can't take what fits into the channel as reflective of the way they understood problems to begin with. Also, I came to realize that there is a dynamic that involves who else is in the community and that this influences how one scientist uses words to establish something. As I came to recognize the importance of the social dimension, I was suddenly like a kid in a toy shop: so much richness was around me all the time, but I had no appreciation for it because it just seemed totally irrelevant to what I was after all interested in as a philosopher. Now it's swimming around me! The risk, of course, is that at this point I'm not very rigorous; I'm just exploring wildly because there is so much there.

Callebaut: But you must have been aware of, say, the clashes between Bloor (1981) and Laudan (1981a, b; 1982a; cf. his 1982b on Collins).

Bechtel: I never got very much into that debate. I read a little of it and found both sides unattractive. There seemed to be no justification for dismissing social causes just because scientists' reasoning fit a rational pattern. That issue never consumed me. When I was doing *Integrating Scientific Disciplines* (1986a), there were various pieces of work in sociology that I found relevant for my understanding of a discipline, but it still didn't hit me in the way it's hitting me now. It wasn't going to be a future project. Now I find myself really wandering onto their turf and saying, "I really have got to develop intellectual tools to deal with the social aspects of science."

Callebaut: What do you reply to the colleague in your department who feels that "this is no longer philosophy?"

Bechtel: I am now beginning to understand Don Campbell's notion of ERISS: "epistemologically relevant internalist sociology of science"—although I don't think I will ever understand that whole expression.[18] (*Laughter.*) I see sociology of science as epistemologically relevant; if I'm going to understand how the cognitive content of a science develops, I have to embed it in the social context of science. At the same time, I'm coming much more to think of cognition itself as embodied in the physical body that we have. I can't think of cognitive processes simply as abstract symbol manipulations.

Callebaut: Social systems, psychic systems, and organic systems (to use Luhmann's 1990a terminology) can all be knowledge-carrying vehicles (cf. Campbell 1979b). Actually, it seems to me that in your critique of Popper's "World Three" view of scientific knowledge (Bechtel and Richardson 1983), the idea of *embodiment* was already there.

Bechtel: Sometimes ideas are there and I don't appreciate them. It was probably then beginning to filter through to me. But it's become very clear to me now that you can't cut cognition off from the environment, that you can't cut off the cognitive work of scientists from the social embodiment of scientists.

Callebaut: One distinction I have found tremendously useful in trying to make sense of what has been going on in science studies is the distinction between *contextual* (or "ecological") models of science and *cognitive* models (De Mey 1982). In a contextual model, the border between a system and its environment is defined by an outside observer. In a cognitive model it is defined by the system itself. Mertonian sociology of science, but also Edinburgh and Bath, would belong to the former category; the views of Kuhn and, I guess, Latour, to the latter.

Bechtel: There is certainly something of that last stage that I find right. In the book with Bob (Bechtel and Richardson 1991), one of the things we articulate is that the first decision in trying to develop a mechanistic explanation is to find a *locus of control*. I'm not sure any more that expression is the right one, but what we mean is: the system that can be looked at as the primary locus of causal interactions. This does not deny that it receives inputs from the environment and sends exports out into the environment. But we can still isolate the system and examine the causal processes within that produce the phenomena of interest.[19] Is the cognitive system

---

18. Campbell (1981b); cf. his (1986b), (1988b), and Callebaut and Pinxten (1987).

19. An example of a locus of control would be the cell as the unit of respiration, where the rate of respiration is controlled by external factors such as availability of oxygen, but the processes of respiration are largely going on within the cell. (Cf. also Star 1989.)

a locus of control? Behaviorists would say "No," cognitivists have said "Yes." Making a boundary is an important move (see also Fisher 1990); without one, you cannot begin to develop a mechanistic explanation. Once you reach the boundary the rest of your analysis can be directed to the internal workings of the system. What I think has happened in my own case is that I might have taken the cognitive system as the locus of control without fully appreciating how much transaction goes on at the boundary. I think this turns to a Simonesque kind of point: that interface might be much more crucial than I recognized at the outset. That's often what happens as someone develops an internal model: they realize later that it's sensitive to all kinds of different environmental factors!

5.3.4. *From Global Skepticism to Local and Social Justification*

Callebaut: Back to justification. You make a daring claim for a philosopher there, Tom: that in a sense all justification—and hence all rationality—is at bottom social (Nickles 1987b, 1989b, c).

Nickles: It sounds daring, but in a way it's trivial. At bottom what else is there? Justification comes down to addressing human critics. The naturalistic turn dispenses with the idea that we are trying to justify ourselves before God or before Plato's heaven or maybe Popper's Third World. "There's nobody here but us chickens," and if your fellow chickens don't object to something, it has passed a minimum level of justification. You have not violated any constraint if the community thinks it's worth enforcing. As Rorty (1979) puts this minimalist view of rationality, it is ultimately a matter of *what the critical community lets you get away with.* This is philosophically liberating, I think, but so far it is just a slogan that would have to be fleshed out. For example, "ultimately" and "at bottom" signal that several other things are going on. If we look only at the bottom, the view becomes too reductivist. "Rationality" has so many different meanings—some of them nonnaturalistic—and operates at so many different levels, that I can understand why some people just avoid the word altogether.

Callebaut: In science, in contrast with some other endeavors, the communities are of highly trained specialists. Does that make a difference?

Nickles: One fact about scientific specialist communities is that they are usually quite small. At the subspecialty level, we're talking about maybe a dozen or two people. At the journal publication level, the referees will be drawn from a somewhat larger pool. The important philosophical implication is that justification as it really operates in ongoing inquiry is quite *local.* The arguments and moves that make a difference, that *cause* investigators to behave one way rather than another, are typically quite local. Philosophers have arrived at this conclusion by a different route in arguing that the *global* kinds of skepticism of Descartes and other foundationist philosophers can be ignored; that only local skepticism, based on specific reasons for doubting particular claims, count as genuine skepticism. Only local skepticism makes a difference, since global skepticism undercuts all the competitors more or less equally—assuming, of course, that foundationism fails.[20] (See also the box at 2.3.)

5.4 A GENERAL PICTURE OF THE SCIENTIFIC ENTERPRISE: PROS AND CONS

Callebaut: After three decades of historicist sensitization to uniqueness of detail and context, many philosophers of science still believe they can make themselves useful by asking questions about "the Nature of Nature," "the Life of Life," "the Knowledge of Knowledge," or "the Becoming of Becoming."[21] This seems to me to require some explanation. Philip, you have expressed your concern about the lack of philosophical commitment of many people, including scientists, before. In two of your books you also gave nutshell formulations of your methodological views. And David, you haven't shied away from making general ontological claims, if I may say so.

Kitcher: It would have been, I think, futile to wade into the sociobiology debate with some very highly articulated, detailed methodology which my fellow philosophers could have chipped away; that would have undermined the enterprise. So both in the creationism book (1982) and in the sociobiology book (1985a), what I tried to do was distill from my own general view that part which I think is necessary for the subsequent detailed analysis of particulars, and which at the same time is likely to be broadly shared.

Hull: Every once in a while, after you have been immersed in these detailed empirical issues, you can step back and ask more general questions, such as "What are the general features of functional systems?" But to do that you need to have a broad knowledge of a variety of functional systems. Too often, when philosophers think of functional systems, they think of vertebrates—hearts beating and all that (for an exception, see Horan 1989). Vertebrates *are* functional systems, but we are only one sort of functional system.

Callebaut: You yourself have been working on the general structure of selection processes (cf. 7.2).

20. Shapere, who distinguished local and global skepticism in his (1984) book, puts it this way: "Descartes set philosophers worrying about the possibility that a person might be dreaming, or that a demon might be deceiving him. Some of his philosophical heirs still talk about the possibility that we might be brains in a vat, deceived in all our experiences." (For critical assessments, see Hofstadter and Dennett 1981:pt. 3, and Putnam 1981.) The trouble with such suppositions, Shapere maintains, is that "that kind of consideration can be raised against *any* statement whatever, any belief whatever, with equal weight." Whether I believe it is raining outside or that it is not, such arguments apply equally to both the proposition and its denial. Universal skeptical doubts ("philosophical doubts"), Shapere concludes, "have only the function of reminding us that we can never be certain." They are not reasons for doubt, but "merely reminders that reasons for doubt might, in principle, always arise, though they need not."

21. I am referring here to the titles of four published books in a longer series that together will constitute an influential French philosopher's magnum opus, *La méthode* (Morin 1977–91).

Hull: I think that to do a decent job, you've got to study real selection processes and how they operate. Not because inductively, you can just generalize from particulars to universals. But I think a deep knowledge of a wide variety of examples really helps to come up with general analyses. Now I don't know whether this counts as philosophy or as science; but I don't think it makes a difference.

Callebaut: A general picture of the scientific enterprise—for whom, Philip?

Kitcher: Not just for the practicing scientist. It seems to me that understanding this important facet of Western culture is intrinsically valuable. There is an instrumental purpose as well—enabling people in the broader community better to appreciate science—and an educational purpose—helping us to see how best to communicate to the people entering the discipline what is really going on.

Callebaut: In your own work—as far as this aspect of it is concerned—you seem to be moving in the direction of an analysis of the scientific enterprise in microeconomic terms.

Kitcher: I have a social epistemology which stresses connections both to models in population biology and to models in microeconomics. In one of a trilogy of papers (1990b) I work out a model of social decision making in a very artificial scientific context. I've also been working on problems of authority and deference (1992b) and on the problem of assigning scientific credit (in 1993), using the same kind of apparatus. I think we can develop for the study of the growth of scientific knowledge something that stands to evolutionary epistemology as population genetics stands to evolutionary theory. I start from various conceptions of what the agent is like and try to understand how agents might interact with one another in the pursuit of the cognitive and practical goals that they have, where some of these goals are directed towards the attainment of accurate representations of the world and some are quite mundane things such as trying to advance their careers or to seek credit. It is, I think, a general enterprise that philosophers ought to engage in: to try to understand how particular kinds of cognitive agents, situated in particular sets of social arrangements, might proceed to maximize, from the community point of view, the chances of attaining various kinds of goals—which might be epistemic or non-epistemic—and from the individual point of view, the individual's chances of attaining his or her goals. This general study seems to me what social epistemology and social decision making are all about.[22] I regard this trio of papers really as the beginnings of a much bigger venture. They will be combined and extended in a book (1993).

---

22. Kitcher grants that his approach bears some relationships to Hooker, Leach, and McClellen's (1978) application of decision theory to epistemological and social issues ("I think I have achieved a more general formulation"); to some of the work that Isaac Levi did in *Gambling With Truth* (1967); to Tom Schelling's work (e.g., 1978) on the micro-motives of individual agents in social activity and social interchange; and to work undertaken by Nelson and Winter, Michael Ghiselin, and Nic Rescher. "There are also connections with the kind of stuff Howard Margolis (1982, 1987) is interested in, or Elster. Alvin Goldman and I have had some exchanges on this; we are in many ways like minds."

5.4.1. *The Unfathomable Goals of Inquiry*

Callebaut: Why would philosophers be especially equipped to formulate general pictures of science? Also, if they want to be more than just the scientists' mouthpiece and to assume a critical stance toward certain developments in the sciences, what should be their vantage point? (Remember: no foundations!)

Kitcher: One thing that is clearly possible on my view is that the philosopher or epistemologist has a vantage point from which he can criticize what the scientific community does. One can say that a particular development or research program is faulty because it fails to abide by the rules for maximizing one's chances for attaining these goals.

Callebaut: Doesn't that presuppose that the philosopher has some rather explicit picture of what the goals of inquiry might be?

Kitcher: It is true that there is no sort of special epistemological act of intuition in which one *sees* what the goals of inquiry are. One gets this through the analysis of the practice of science past and through one's reflection (*hesitates*), one's intuition about human rationality and what rational human activity is. One tries to do the best one can to systematize these facets of human life. The philosopher's knowledge of the goals of inquiry is fallible.

Callebaut: So your claim is really that the philosopher can serve the scientific community in principle as—*fallible—critic.*

Kitcher: Exactly. The view I adopt about the ends of inquiry is somewhat like Chomsky's view about grammar. On Chomsky's old view (I am not *au fait* with what is going on in his thinking in the recent past) there are these principles of universal grammar, and a sort of innate knowledge that we all have. Whether epistemological vocabularies are appropriate here or whether we want to talk in some other terms—about some hard-wired dispositions of some kind[23]—is not the issue here. But there is something that all human beings have: a capacity or form of innate knowledge that enables them to learn the natural language as they do. We don't have explicit knowledge of this. It is a painful and hard task for linguists to figure this out; they can get it wrong. In fact in the foreseeable future they probably will get it wrong.

Callebaut: What is the analogy?

Kitcher: I want to claim that something like this is at work in human knowledge; that there are special epistemic goals that in a certain sense we all tacitly know; and that the philosopher's task is to make these kinds of things explicit. Just as Chomsky tries to make explicit the knowledge he takes us to have of the principles of universal grammar by studying natural languages and their structural intricacies, by looking at language development, etc., so it seems to me the philosopher tries

---

23. See, e.g., Fodor (1975); the exchange on "Must beliefs be sentences?" between Brian Loar, Jerry Fodor, and Gilbert Harman in *PSA 1982*; Churchland (1986a, b) on the "sentential paradigm"; Lycan (1990:pt. 5) on the "language of thought" hypothesis; or Pinker and Bloom (1990).

to figure out what these built-in goals of inquiry are by taking a look at the practice and the history of the sciences.

Callebaut: So far you have only referred to these goals in the abstract without spelling out any of their content—or would that be too tall an order?

Kitcher: This is very hard. What I want to say—and this is where my neo-neo-Kantianism comes in (1983, 1988a, d, 1989)—is that there is an ideal of inquiry towards which we aim, which is the *fitting together of the phenomena of the world into unified patterns* (as far as we can do that). Science attempts to produce a practice, in my technical sense (Kitcher 1983:ch. 7; cf. 3.4.2), in which we achieve *the maximum of conformity between stimuli and scientific description,* and simultaneously *the maximum of unification.*

Callebaut: Now these are two different dimensions . . .

Kitcher: . . . and there will have to be tradeoffs. I want to say that the traditional problems of the philosophy of science about such things as natural kinds, causal dependencies, explanations, correctness, and so forth, come down to facets of the character of this ultimate system. I want to say that all of the old concerns about what science is trying to do—find causes, give explanations, make adequate predictions—have to be understood from the perspective of an ideal system, conceived of in its most rudimentary form as something which involves a tradeoff between what one might loosely call "accuracy of prediction" and "unifiedness." I hope this gives you at least something.[24]

5.4.2. *Irréduction*

Callebaut: Bruno disagrees *carrément.*

Latour: Philosophy is exactly the opposite of what is said by the Americans. The Americans say, "If you don't have philosophy, you just have case studies, and no unification of the empirical matter. You need generalization." They identify philosophy with generalization. My impression—and that is exactly the opposite—is that if you *don't* have philosophy, you have a unification of the history by a point of view which might be sociology, "force," "networks" (since you kindly mentioned them), "allies," or economic metaphors; maybe it will be logic, or whatever—one of them will win and unify the others. You need philosophy to *undo* that, to protect the networks—their scarcity, their heterogeneity, their historicity,

---

24. Kitcher is adamant that drawing the general picture continues to make sense: "You asked me some time back about the relationship between my work on mathematics and that on biology (in 3.4.2); now it may be possible for me to give a somewhat clear answer. I have just been sketching my general approach to the growth of knowledge; and it is really hard to say much more without getting into all kinds of odd aspects of my philosophical views. That general approach is really the view that is in my philosophy of mathematics book (1983) and that is subsequently developed in the mathematical naturalism paper (1988d). A stripped-down version is in the creationism book (1982) for the purposes of criticism; in the sociobiology book (1985a) it is minimal. (*Bursts out:*) The view is in just about everything I have written in the philosophy of biology or in any particular scientific theory! My current task is to try to bring this out and give a full-dress version of it, so that people can see how all this stuff really fits together (1993). I am very excited about it, as you can probably tell! *I don't believe in demarcation!*"

their locality. *Saying that has nothing to do with relativism;* because relativism is just one possible way of arguing against universality. We have grown out of that now. So the only way to respect the heterogeneity and the locality is, on the contrary, to do *a lot* of philosophy. But philosophy is not unifying factors; philosophy is protecting against one factor's hegemony.

Callebaut: You have written: "maybe philosophy frees the sciences from part of their violence."

Latour: Did I say that?

Callebaut: You wrote that in your paper on Michel Serres' philosophy (Latour 1988b:96).

Latour: Oh yes. That's the same argument. My argument would be Serres' argument: that you need philosophy as a *reservoir* for all the new sciences to grow and as a *protection* against the hegemony of the present sciences (cf. Serres 1992).

Callebaut: It's an ecological metaphor.

Latour: He has this beautiful metaphor of the primitive forest: Science is always the extraction of one variety, and it has an extreme productivity, like a pure line of maize or corn. But you need a wide variety there to protect it, to renew and rejuvenate the pure lines. Science is very good at producing pure lines; philosophy is good at protecting impure lines. So it's exactly the opposite of the argument that we need philosophy to unify the sciences. No, no: we need philosophy to protect ourselves, *including* from the sort of theory you mention, that sociologists of science *like us* produce.

Callebaut: Could you be more specific?

Latour: Like this absurd idea—often attributed to me—that winning in science has little or nothing to do with rationality but has everything to do with having more allies. This is an absurdity, of course, because you don't know what "more allies" are! That you could have a unification of all the case studies from this point of view would be as much a catastrophe as if it were "Darwinism," or "Reason," or whatever. If anything, I prefer "reason" to "allies," because that at least has a more cultivated past! Now the whole question is: how do you redifferentiate all these different types of networks, etc.? That's the thing we are working on now. And I say that with some hesitation, because it is in a way against my own *irréduction* (Latour 1988a:pt. 2). An *irréduction* has the big advantage that the heterogeneity is back; but not for long, for it immediately sneaks in the mind of the writer—and the reader, of course—that *force* is the unifying factor. Then you get another hegemony and another universality. Even though the whole argument of network building is that networks build their own references, their own measuring instruments of time and space, their own value meters . . .

Callebaut: . . . Your version of incommensurability.

Latour: So you have no outside criteria to unify the forces. But nevertheless, "force" immediately becomes a unifying factor and you are back where you were first, that is, you have a *reduction* instead of an *irréduction*. Philosophy, in a sense, is a protection against that. I moved a lot on that by doing this piece on Steven Shapin and Simon Schaffer (Latour 1990a). Because I think that's the difference between modernism and postmodernism. We are not postmodern, and these people are postmodern. (*Laughter.*)

Callebaut: A comment, Philip?

Kitcher: There's a marvelous remark that Russell makes in his *A History of Western Philosophy* (1946), where he compares Nietzsche to the mad Lear who says "I shall do such things/ I know not what they are, but they shall be/ The terror of the earth." I often get the same feeling when I read Latour. It's wonderfully entertaining, but the talk of a new program in which philosophy of science has withered (Why?) and traditional philosophy (What traditional philosophy?) enjoys a new relationship (What new relationship?) with empirical studies seems to me like emphatic advertising from a master salesman. Bruno is right to claim that philosophers like me want to give a general account of various facets of science. One way to read him is as simply denying that generality is possible here. But why should that be? Why can't we have theoretical ecology as well as natural histories, lovingly done? There's no real argument in Latour, just rhetorical flourish and, I'm afraid, serious misunderstandings of lots of views. However, I do enjoy talking to him, and I hope that, in time, we can clear some of our disagreements up.

5.5 THE SEMANTIC VIEW OF THEORIES: A NEW TOOL FOR THE TRADE

Callebaut: If philosophers are going to assume new roles, they will need new tools. One such tool, a number of them believe, is a relatively new account of theories that is called variously the non-statement view, structuralist view, semantic view, or model-theoretic view (see already Adams 1959). On this view, theories are not collections of statements, but rather are "extralinguistic entities which may be described or characterized by a number of different linguistic formulations" (Suppe 1977:221). Jim Griesemer (1984:103), among others, has pointed out that the success of the semantic approach is related to the naturalization of philosophy of science.[25] By focusing on "models for a theory rather than theories in themselves," the semantic view contributes to the trend "away from logical reconstruction of theoretical structures in science toward description of the actual structures employed by scientists."

Kitcher: There is a sense in which the semantic conception is founded on the famous remark of Patrick Suppes that "mathematics, not metamathematics is the right way for philosophy of science to go" (cf. Suppes 1968, 1979).

---

25. On the other hand, there is considerable continuity between the logical-empiricist and structuralist approaches to science. Kuhn, for one (e.g., 1987:21n), has welcomed the latter.

5.5.1. *Specifying Blank Slots with Particulars*

Callebaut: Lisa, you are a visible proponent of the semantic view of theories (Lloyd 1983, 1984, 1986a, 1987a, b, 1988, 1989). Was your teacher van Fraassen (1970, 1980) responsible for your adopting that approach?

Lloyd: No! I'll tell you what happened. When I was reading Darwin in my first year of graduate school, I thought what he was doing in a particular explanation was developing a *plausible scenario* for that explanation. He had a general outline for an explanation that he filled in with different components. For instance, I saw evolutionary explanations as types of models which were not completely specified. There was a blank slot for selection pressure, for the environment, etc. Each of these specific claims could be checked separately. It's a more complex, less linear picture of what was happening.

Callebaut: So you thought that the way to understand Darwinian explanations was to see them as a blank sort of model which was then specified with particulars . . .

Lloyd: . . . and had a particular *structure*. The key was in the structure, not in the deductive quality of the explanation. Once you looked at the structure, you could see very clearly why Darwin was interested in all these different kinds of evidence . . .

Callebaut: . . . "variety of evidence" seems to be a thread that connects much of your work . . .

Lloyd: . . . and I wrote up a draft of a paper. At the same time I was taking a course on quantum mechanics from van Fraassen, who was a visiting professor at Princeton at the time. I first became overwhelmed by this course. I went to him and said, "I'm interested in this thing in Darwinian theory. Can I show this to you?" I showed him the paper. He very nicely took me out to lunch and sat me down and said, you know, "I'm very interested in this paper; you might like to have a look at my new book (van Fraassen 1980)"! (*Laughs.*) I had basically just regenerated a very primitive version of the semantic view! It must have been a bizarre experience for him. I want to make something really clear: He did not push it on me in any way. What happened was a very bizarre kind of separate generation of it from the biological and historical material on Darwin. *I got it from Darwin against the interpretation offered by Ruse!* That's how I actually got started. You can imagine my delight when I found out that this intuitive idea I had—very undeveloped, very vague—was all developed by a philosopher who had all this fancy logical equipment to go with it. I thought, "Here's this big guy and he has everything figured out! This is wonderful, this is a genuine option." And people had not, to my knowledge, applied it. I rewrote the paper somewhat in light of what I could get from his mechanism. But basically it was the same paper; I could just refer to his stuff.

Callebaut: And *Philosophy of Science* accepted it (Lloyd 1983)—congratulations!

"Don't underestimate the arrogance of philosophers!"

Elisabeth Lloyd (b. Morristown, New Jersey, 1956) is associate professor of philosophy at the University of California, Berkeley. She has been interested in biology and evolution ever since she was a child: "My parents were intellectuals; they had lots of books around, and I spent time reading biology. When I was a kid, my mother would buy me any books that I wanted in biology. And I spent a lot of time with animals and thinking about animals."

In 1974, when she went to Queen's University in Kingston, Ontario, Lloyd wanted to take a biology course. "The academic advisor said that first level biology—Biology 101—was a very difficult course, and that I perhaps should consider not taking it, and why don't I go into geography—which I did (I was very intimidated)." Then she dropped out of college; she played electric lead and slide guitar in a country rock band. When she went back to studying at the University of Colorado in Boulder in 1976, she wanted to become a doctor, and she actually became a biology major. But she also took political science for a distribution requirement. "I fell in love with political theory. So I decided to be a double major—political science and biology." Then she got bored with the biology. "We weren't doing anything except rote memorization. When I would ask questions—about evolution in particular—they would just give me very unsatisfactory answers and tell me to go talk to the philosophy department, which I thought was a joke. In a scientific family, philosophy is the furthest thing from something legitimate. So I didn't go to the philosophy department."

Then—this is, in Lloyd's own words, "a very weird story"—she took a course in the honors program, on human nature, which is a very important part of political philosophy in Anglo-American culture. "I found out that the man who was teaching it, Gary Stahl, was a philosopher! I felt that this was a very damaging thing against him, but I was already in love with the course anyway. So I continued taking it, and then he coaxed me into taking some more philosophy courses." By that time it was clear to Lloyd that she didn't want to be a political science or a biology major. She continued her philosophical education by studying . . . phenomenology. "We read Husserl and *Being and Time* (Heidegger 1962). I was also studying with Horst Mewes, a student of Hannah Arendt. I became interested in Heidegger's views about technology. Also, no matter where I looked in political theory in the contemporary scene, the role of expert knowledge had taken over a huge part of actual political decision making." So Lloyd became interested in policy making on science and technology, and in the role of scientific knowledge in policy making.

She was on her way to becoming a hard-core continental philosopher at that point: "I was also taking a course in intellectual history with David Gross, a Marxist, and I became very interested in German romantic philosophy. At that point I left college and moved to Germany to learn German, so I could read Marx and Hegel and Heidegger *auf deutsch.*" Lloyd returned to obtain her B.A., *summa cum laude* in Science and Political Theory, in 1980. Then, although she was not a philosophy major, she went to Princeton to graduate school in History and Philosophy of Science, a subject she had discovered in her last year by reading Feyerabend's *Against Method* (1975) in a course on the history of modern philosophy. "I liked his ideas a great deal because he got at the political nature of the generation of knowledge, including scientific knowledge. At the same time it went against my strong scientific training." Then she read Kuhn's *Structure* and thought it was much more along the lines of combining the kinds of interests that she had. So she decided this was worth studying. "Besides it was only a lark, because I was still pre-med, and I intended to go to medical school after I went for a year to graduate school in philosophy just for the hell of it. Right?" Lloyd spent her first semester at Princeton studying metalogic and . . . Spinoza.

Why did she like Kuhn and Feyerabend? "I fell in love with philosophy of science through history of science. I learnt a lot of history of science from Kuhn actually. I liked both Kuhn and Feyerabend because they were pulling the rug out from under a very strict positivist (although I didn't realize that at the time), progressivist picture of the growth of scientific knowledge, and because they brought in the elements of the social into the picture of science." Was she aware of the conflicts between these two authors? "No. I'm interested in science, society, and culture, and how human beings act in a scientific context. That was enough for me. It was a sense of reaching an oasis rather than a comparison of one oasis against another."

Lloyd defended her Ph.D. thesis, A Semantic Approach to the Structure and Confirmation of Evolutionary Theory, at Princeton in 1984 (cf. Lloyd 1988). She took her first job at the University of California, San Diego, in 1984. From there she moved to Berkeley. Lloyd, who has spent time at Harvard, working with Gould and Lewontin, is amazed by "the levels of misunderstandings in the sociobiology debate." For instance, she has been asked whether she had been vetted—got a clearance—politically: "whether I had been shown, you know, to be radical, liberal, have the correct political positions to be in Lewontin's lab. I laughed, but I soon realized that that person was quite serious. People tell me the most amazing things about Lewontin that have virtually no correspondence with the person that I know."

Lloyd is the co-editor, with Evelyn Fox Keller, of *Key Words in Evolutionary Biology* (1992). A new book, *All About Eve: Bias in Evolutionary Explanations of Women's Sexuality,* is in progress.

The original interview was recorded at my home in Liège on December 11, 1989, on the occasion of a talk Lisa gave to the Belgian Society for Logic and Philosophy of Science.

Lloyd: People think van Fraassen put me on a leash and said, "I've done physics; you go do evolutionary biology!" But that is just fantasy on their part. Take Jonathan Hodge, for instance: In the course of criticizing my work for being ignorant about the history (see below), he proceeds to make up a history about how I came to write that paper. It was interesting to me that a historian insisting that philosophy must be adequate to the history proceeded to make up an entire history of how it came to be that it was inadequate to the history.

Callebaut: Continue to give us your story *wie es gewesen*.

Lloyd: I already knew some genetics. The mathematical models in genetics were immediately transformable into the semantic view. It became immediately clear to me that the semantic view would be a beautiful tool to analyze *any* mathematical theory, including population genetics, and worth a try. The only things that were out there to analyze population genetics . . .

Callebaut: Mary Williams's (1970) axiomatization of evolutionary theory . . .

Lloyd: . . . is interesting and shows some important things about the structure of evolutionary theory in that it was presented as a standard of what an axiomatization would look like and made it clear on what level the theory was actually operating. I don't see what I'm doing as being opposed to what Williams was and is doing. Her analysis is on a different kind of level. I wanted to do something that dealt with the details, the actual conflicts in population genetics. Michael Ruse's approaches to population genetics are very sweeping, very high level, and you don't get into any of the debates that the geneticists are actually having. So I wanted to have a philosophical tool that would go in and be capable of being used in a context to analyze the differences, the actual structure, the multiplicity, the non-unification, the non-uniformity of that theory.

Callebaut: At that time there were various versions of the semantic view around.

Lloyd: Well, I was a graduate student at Princeton and I was having van Fraassen as a tutor for the semantic view. It's not the simplest view to understand; and his was my real first exposure to it. I understood what he was saying better than what anybody else was saying. He is one of the originators, and I didn't see any harm. His state space version[26] of it was much more natural to the actual biology than

---

26. In general, a structure "presented" by a theory, where the theory is understood as intended to represent the empirical phenomena within its "scope" (Suppe 1977:223f.; Griesemer 1984), is a model of the theory if it satisfies the theorems of the theory. In a semantic definition, the set of sentences that are *theorems* of the theory is not defined by interpreting a set of axioms (as in the statement view), but through directly defining the class of structures. A *model* here consists of the following elements: (1) a set of entities that are operated upon; (2) a state space (cf. ch. 4, n. 23) such that the location of an entity in the space specifies its state; (3) a set of transformation laws that state how the entities change state, or at least predict the location of the entities in state space; and (4) a set of parameters that fit into the laws to determine the quantitative changes in state. These elements must fit together in a "dynamically sufficient" model so that the state variables are unambiguously transformed by the rules of transformation in the state space (see Lewontin 1991). Under the semantic view, theory structures are considered phase spaces of configurations imposed on them in accordance with the laws of the theory; therefore, they are themselves seen as constitutive of the theory (Lloyd 1988:19).

was the set-theoretic version,[27] because the biology was directly transferable into his view. The set-theoretic view is a long distance from the way actual scientists speak and the way the stuff is presented. I thought it was better to use the state space version of the semantic view.

Callebaut: So you were aware of the various alternatives.

Lloyd: Sure. When van Fraassen told me about the semantic view, he told me to read people who had worked on it. I read Suppe (1979, cf. 1988), I read Suppes (1967). He didn't say, "Only read me!"—he would never do that.

Callebaut: We'll look at your application to the units of selection controversy in the next chapter. Right now I want to ask you to react to Jonathan Hodge's criticism that your analysis of Darwin's support for his theory (Lloyd 1983) is "an incongruous conjunction of disparate traditions, the one Darwin belonged to and the one Lloyd herself descends from through van Fraassen" (Hodge 1989:177). Hodge does not mince his words: "Lloyd's attempt to shift Darwin from his tradition to her own is, unknowingly, an attempt in effect to change the past, something even God is traditionally unable to do. Lloyd cannot be blamed, therefore, for failing." What Hodge suggests is that your adoption of van-Fraassian "anti-causal-realist empiricism" blinds you to Darwin's own vindication of causal explanation. Now who is right as to the descriptive accuracy of the semantic approach: the "optimist" Griesemer (cf. above) or the "pessimist" Hodge?

Lloyd: Hodge criticizes my view for being completely ahistorical, because I don't focus on the role of Herschel's *vera causa* principle in Darwin. I think he is right in the sense that if we were to understand completely what Darwin's picture was and to understand all the things—philosophically—that went into his argument, we would have to include an examination of the role of *vera causa*. On the other hand, explaining things in terms of *vera causa* is also insufficient for the details of understanding what Darwin had in his mind in terms of the structure of his own theory. Hodge has done a beautiful job of analyzing *vera causa*. If he thinks that is the only thing that anyone can ever say about Darwin, that's quite unfair. So what I had to say about the structure of Darwin's theory I will still stand by. I think my paper offers what it is intended to offer as a—whatever—25-page paper in *Philosophy of Science*. It is not intended to be a definitive history of the philosophical content of Darwin's theory.

Callebaut: More generally, Hodge suggests you are doing a "rational reconstruction" that does violence to the history.

Lloyd: My answer to him is: "You do violence to the history by oversimplifying it and not seeing some other parts of the structure of what was going on for Darwin." *Rational reconstruction*—I have a real problem with such a label. There is intelligent rational reconstruction—stuff that is very close and faithful to some aspect of

27. In the set-theoretic version of the semantic view, a set-theoretic predicate is defined to specify a certain mathematical structure. The set of models of the theory is identical to the extension of this predicate. See in particular Suppes (1967), Sneed (1971), and Stegmüller (1976, 1979).

the history, though not necessarily to all of the history—and there is stupid rational reconstruction—stuff that is uninformed by the history. I think that what is being done now, by students that are coming out, is very faithful, very close to some period of the science. That is what makes it good, because it's careful and it does not assume some vast philosophical project or domain and vast philosophical theory that is then imposed on the actual events as they are interpreted. Rather people just go in and muck around and see what happens. That's what I think is the most fruitful approach: it is this mucking around and seeing what happens.

5.5.2. *The Model-Theoretic Version*

Callebaut: Ron, in my experience it took your presentation of the semantic view (in Giere [1979] 1984a; cf. 1983 and 1988) for many people to understand what it was really all about.

Giere: I no longer refer to the "semantic view," partly because many people don't like the word "semantic." My favorite currently is *model-theoretic view* of theories, because scientists talk about models and people are comfortable with the terminology. So it is the model-theoretical view of theories, referring to a family of models. The term is actually van Fraassen's.

Callebaut: Is it possible to give a nutshell characterization of your present interpretation of the model-theoretic view?

Giere: There are two things going on. Let me use an analogy. You draw maps of the world; you *create* the maps. Now a map does not say anything; a map makes no statements. It is a physical thing on a piece of paper. You create a model, which is a structure, and then you interpret it. But it does not by itself say anything; it is just a picture or a map. The way you give it empirical content is to go out in the world and say, "Aha, that part of the world is like my map! My map represents it, my map is a good map! It represents that part of the world!" And then you say how and why it fits. There are two separate things: there are maps and there are the claims you make using the map.

Callebaut: It may not be superfluous to stress the difference between the kind of "fit" you are talking about and the traditional philosophical theory of truth as "some form of correspondence between belief and fact" (Bertrand Russell), which is most naively interpreted as "correct reflection of reality in thought," as my old Marxist-Leninist dictionary of philosophy (Rosenthal and Yudin 1967:461) wanted to teach me. There is nothing mysterious—nothing "mind-like" or whatever—about your maps, however (compare and contrast Sinha 1988 on the "materiality of representation" and the social dimension of cognition; also Star n.d.). As you say, they are physical things—they are spatiotemporally located. You use such maps to make claims about other physical things—(other) parts of the world. Moreover, a map, being a representation, is no replica.[28]

---

28. Martin Rudwick (1985:454) dwells on this point in his book on geology: "Bookish people with no practical experience of mapping often assume that a map is an unproblematic replica of reality, or merely

Giere: Notice that on this view, there is *no necessity for talking about laws* (cf. van Fraassen 1989:ch. 8, "What if There are no Laws? A Manifesto"). In a way, all empirical claims are particular. They are saying, "Here is some real system—do you see it out there? It fits my model, and my model fits it. There are some other systems, and it fits those too." That is how you do science! Whereas you don't say: "What we're looking for is a generalization, a universal statement." Forget it. That is the difference: You create models and you apply them to the world. This is a view that you can actually find in Toulmin (1972). He talks about "concepts." As near as I can tell, for Toulmin a concept is what I would call a model, and it applies to the world or not.

Lloyd: Ron's (1988) version is wonderful. He makes it very clear; he has succeeded in making it clear where van Fraassen has not made it as clear. The only part that I disagree with Ron about is his commitment to realism (see 4.3).

Callebaut: Helen, the logical-empiricist picture of science is hierarchical like the old Cartesian picture (Longino 1990; cf. Wimsatt in 4.2.2), with an empirical basis replacing the rationalists' intuitions and, of course, a consequentialist rather than a generativist interpretation attached to it (cf. Nickles at 5.3). Let us resuscitate the good old white swans for the occasion: I see one, two, fifty white swans. When I'm going to make the jump—let's not bother about the exact modalities of that jump now—to "All swans are white," the only *evidence* I have for that claim . . .

Longino: . . . is just those fifty white swans.

Callebaut: *Is there any way to go beyond that?* "All swans are white" must have nomic necessity attached to it in order to be a law of nature and not just an empirical generalization (cf. Hull in 3.5.2).

Longino: That has been an obsession of philosophy of science.

Callebaut: Would you be prepared to say that the ascription of necessity is just a relic of our religious history (cf. Keller 1985; Longino 1990, and 7.1.1)?

Longino: I think I'd be quite sympathetic to that view. Given everything else that I think, I'd be quite comfortable to think that notion of "nomic necessity" really is a kind of theological, religious inheritance that we don't need. I don't see why we should need it for science.

Callebaut: Ron, how do you reconcile your model-theoretic view with your realism?

Giere: Van Fraassen, as in most things, is very clear about this. He says explicitly that the semantic view of theories is neutral with regard to realism and nonrealism. He develops it in an antirealistic way; but he says explicitly that the view is neutral. I think there are two separate issues here. There is a picture of what theories are and then the further question of how deeply we interpret them, as it were.

---

a miniaturized version of what one would see from the air. Those who make intensive use of cartography know on the contrary that any map is a pervasively conventional representation. . . . It makes no sense to talk of ever achieving a uniquely 'perfect' representation . . . since different kinds of maps are designed for different uses, and there is no limit to the further representations that may be needed for new and unforeseen purposes."

Lloyd: This conflation has actually affected the reception of my work. Several people have said to me, "I really don't think your analysis of population genetics and of evolutionary theory is going to work, because it is committed to antirealism!" I point out that it is not committed to antirealism at all. The fact that the semantic view is tied up with antirealism to these people means that they do not understand the semantic view. The semantic view of theory structure says that whether you are a realist or antirealist is irrelevant to the issue of explaining theory structure. It is one of the strengths of the semantic view that you could be a realist or an antirealist. So the rejection of, for instance, my work on the basis that I am an antirealist and that my analysis is necessarily antirealist is a fundamental misunderstanding of what's going on.

Giere: I think the issues are separable. Now as a matter of fact, what has happened is that the semantic views have been associated with various kinds of nonrealism and antirealism. I think that is not an essential feature. I really do want a realist interpretation, and I don't see that there is a contradiction at all.

Callebaut: Nor does Rudwick. He thinks that the analogy of mapping "yields a way of retaining the constructivists' insistence on the social processes that went into the making of a piece of scientific knowledge such as the Devonian, while also accommodating the realists' insistence that the real natural world of rocks and fossils, and the real history of the earth that originally brought them into existence, had a more than marginal effect on that claimed knowledge" (Rudwick 1985:454).

Giere: It may be a minority view, but I don't see any internal problem with it.

Callebaut: Philip, you and Lisa Lloyd have both studied in "the same small place in New Jersey" (Hodge), yet you favor different metatheories. What difference, in the end, does the semantic view make, according to you?

Kitcher: The honest answer is, *I don't know.* I used to believe that the semantic view was good for solving certain kinds of epistemological problems, that the traditional syntactic view was helpful in addressing others, and that neither of these could tackle some important issues. Hence I introduced this view of theories as sets of problem-solving strategies, which I've used in certain contexts in philosophy of biology. Lisa has argued with me for several years, suggesting that everything I want to say can be translated into the idiom of the semantic view. I've been inclined to disagree on the ground that theories, on my account, are primarily *explanatory* devices, and I don't see that the semantic view has an account of explanation (apart from van Fraassen's, which, I believe, trivializes it). However, Lisa has challenged me to show that there really is a difference, and, since I've never had the time to work it all out, I should probably be more cautious than I have been. Perhaps, though, I can issue a counter-challenge: when the semantic conception shows how it can capture the idea that what primarily distinguishes various biological theories is their ordering of the phenomena for explanatory purposes, then I'll concede that there's no interesting difference.

5.6 A ROLE FOR METHODOLOGY AFTER ALL?

Callebaut: The final issue we will discuss here is the future of methodology—which is traditionally regarded as an important component of the philosophy of science—in a genuinely naturalistic approach. Bill Lycan (1985:137) has observed with respect to "justification," "warrant," and "rationality"—key notions of epistemology—that "their place or ground in the closed causal order we call *nature* is not at all clear"—the reason being, of course, that such notions are "as fully evaluative as their moral counterparts." Methodology would seem to be in a similar if not identical situation.[29] Put bluntly: doesn't naturalism lead straight to methodological *nihilism,* as epitomized in Feyerabend's (1975) infamous dictum, "Anything goes"?[30] In terms of your image of science, Dudley, how can one go beyond the purely descriptive or even tautological? How do you avoid having to say, "This is what science has been reasoning like (or using reasons like), so this is what we must accept as rational"? What I'm asking really concerns Hume's problem of going from "is" to "ought."

Shapere: Here's a capsule summary of the history of particle physics since 1968. A certain form of quantum field theory, gauge theory, has been found to be enormously successful in quantum electrodynamics (cf. ch. 3, n. 16). It turned out that a theory could be constructed which unified that area with another, the weak interactions. Furthermore it was found that gauge theories in general remove certain difficulties in older theories of those areas—they are "renormalizable." The immense successes of the gauge idea led to the normative principle that that is the sort of theory we *ought* to be looking for in other areas. This was tried and found successful in the area of the strong interactions; and it points the way to possible unification with the electromagnetic-weak theory. The normative principle was strengthened further— *"that was the way we ought to proceed."*

Callebaut: So, not unlike John Dewey, you take normative principles to be testable hypotheses based on generalization (cf. Sackman 1967:530: "ethics and values may be treated as hypotheses, as provisional and subject to the continual test of human experience, and to reformulation and improvement in response to changing conditions") . . .

Shapere: . . . but going beyond the mere descriptive in saying—and this may be what you mean by "guiding principles" too[31]—that, because of their successes of

29. See the discussion of nomicity in 5.5.2.

30. Or as Quincy Rortabender, the imaginary relativist portrayed by Laudan (1990a:135), puts it: "Now that we know that philosophy is not prior to, nor more sure than, science—and that, above all, is what the naturalistic epistemologist is committed to—we can see that philosophy can provide no firmer grounding for science than science can provide for itself. And that is why normative epistemology has become wholly gratuitous."

31. This conversation was held some time after the conference on Testing Theories of Scientific Change (Blacksburg, Virginia, 1986), where there was a lot of talk about "guiding assumptions" that were sup-

the sorts mentioned (and other sorts), we have reasons to think they may be applicable more generally. On the basis of considerations like those, which are factual, descriptive, we go to the normative principle, "We ought to try to construct our theories in this form!" For example, "We ought to try to understand chemical substances in a compositional way" became a guiding principle in nineteenth-century physics. This has not only the force of what kind of thing we ought to do, but also of what kind of thing we ought *not* to do!

Burian: One needs to be in touch with the science. Let me show it loosely by analogy. In the 1960s, I had long discussions with a dear friend who eventually drank himself to death, Henry Aiken, about the status of ethics. We drew interesting parallels between ethics and meta-ethics on one hand, science and philosophy of science on the other. He had a funny little slogan (which can be taken too far, of course): "We don't need *meta*-ethics, we need *better* ethics!" Part of what he had to say about that was that the very standards you employ in meta-ethics depend on what you count as good, right, an obligation, and so on.

Callebaut: A similar point is made by Bob Richards (10.1).

Burian: What you count as evidence is going to depend on what is available in the laboratory in ways that are quite critical. How you evaluate scientific models must be ultimately tied to the resources available that are appropriate to the subject matter. In that sense (and here I take a stand on a major dispute in the philosophy of science), I am suspicious of the more abstract attempts to give armchair standards about the evaluation of theoretical claims or evidential relationships that do not touch down intimately to the substantive knowledge of the disciplines that are at stake. I do not anticipate that there will be much more than empty form and content in common between the epistemological evaluation of theories in physics and those in molecular biology, say. In molecular biology, my own view is that it's *mechanisms* all the way down; in physics, there are *symmetry principles* and things that are of an entirely different character. The evidence relations that are involved are very different in the two cases. In that sense I think one is driven to a very close look at the science in a way that does not allow a very general epistemology of science.

Shapere: You have three alternatives. First, you can say that the normative principles and their changes are wholly divorced from factual changes, but that your choices of normative principles are just fads. Second, you can agree that they are divorced from factual changes, but say that they have some other basis—that philosophy of science can make normative recommendations to science, which are presumably above the level of substantive considerations. I take the third alternative, that *there is a connection between what science learns and how it thinks it ought to go about learning in the future*. The first alternative is the way of relativism. I see many

---

posed to be heuristic—a normative feature—*and* referring to substantive features of science as well (Laudan, Laudan, and Donovan 1988).

recent philosophers moving to the second alternative when they want to subscribe to the existence of normative principles in science. But that's the way of old philosophy, and it is a way that has always failed.

Burian: If relativity theory is right, any knowledge claim that depends on signals that proceed at speeds faster than the speed of light must be dismissed. That becomes a normative claim. *If there isn't normative content of that sort in our substantive scientific knowledge, then it's not functioning fully.* If there is, then the norms are to be sought at least in part in that substance.

Callebaut: Traditionally, such normative principles were taken to go beyond purely nomic generalizing. Wanting to have some kind of *epistemic necessity* accrue to them is, I think, one of the things which could mark off a more traditional epistemological approach from a thoroughly naturalized one.

Shapere: I can't agree with that introduction of epistemic necessity. I want to see how it is, given that human beings do not have the power to find any kind of a priori or necessary principles—ethical or normative as well as substantive—that we have managed to come to certain beliefs, and to certain ideas even of what counts as a reason in science; how we could have built up, over the course of coming into interaction with the world of our experience or nature, the entire range of our present beliefs, including normative, rule-like, as well as substantive. This also includes our goals. Many philosophers of science—McMullin (e.g., 1983) is an example—say that at least we have to specify the goals of science in advance and independently of science, because all methodological rules are of the "if/then" type.

Callebaut: Larry Laudan says something similar ("methodological rules are . . . hypothetical imperatives"), and insists that these rules are therefore "generally contingent claims about matters of fact," which will likely be applicable in some but not in all possible worlds (Laudan 1987b:349–50; cf. 1987a, c, 1990b; contrast Siegel 1990).

Shapere: The "if" part has to do with the goals of science: if you want to achieve such-and-such a goal, then they are all hypothetical imperatives. That does not leave sufficient scope, I think, for analyses of how we get the goals and how the goals of inquiry have *in fact* changed over the history of science.

Callebaut: Philosophers like Popper or Quine, who hold that *ought* statements cannot be derived from *is* statements, have been led to believe that no conceivable facts of the matter could have any bearing on methodology (Laudan 1987b:349). What would distinguish a naturalized philosophy of science without normative pretense—say, Quine's (1969a) "descriptive epistemology"—from a purely descriptive and interpretive history and sociology of science? Why have philosophers of science at all (cf. Latour)?

Lloyd: I had a talk with Bruno about this very subject. I have a problem with the way "normative" is frequently used. (*Gets nervous.*) Typically, normative philosophy of science goes in and says what's good science, what's bad science, congeals the "essence" of what good science is, makes it abstract and goes back to the scientists

and philosophers and says: "This is what it means to be doing good science!" I
take that to be something like a description of what normative philosophy of science would do.

Callebaut: How about, say, Bill Wimsatt's work on the biases of reductionistic research strategies?

Lloyd: Do you call what Bill does normative?

Callebaut: In a way, yes.

Lloyd: OK, that's a different sense of "normative," then.

Callebaut: Don Campbell would say, "hypothetically normative" (1986b). If your
goal is X, you can learn from history or current scientific practice that Y is not an
appropriate way to do it, whereas Z has shown to be an (the?) efficient means.[32]

Lloyd: Both Don and Bill are actually good examples. I don't have any quarrel with
that kind of normativity. What the logical positivists meant by "normative" or what
Popper intended is by comparison completely beyond the pale.[33]

Callebaut: We can *learn* from history—even Feyerabend accepts that much.[34] Yet
few hard-line relativists, especially in SSK, are prepared to grant that.

Lloyd: The kind of normativity that's in Wimsatt's or in my work now is much more
constrained, specific, meant to be informed by the historical circumstances, by
actual science at actual times. It is meant to be an analysis of the "map" of the
situation, so that if you find yourself in similar terrain later, it may do some good
that someone has mapped the territory in some way. That is work philosophers can
do. It is normative, but in an extremely constrained way. That kind of normativity
is taken by many traditional philosophers of science to be piddling, descriptive,
getting lost in the trees, losing sight of the forest—this whole kind of thing. I think
that's a mistake.

5.6.1. *Limits of Internalization*

Callebaut: Do you take *complete internalization* (cf. 2.7) to be an ideal of rationality
in science, Dudley? Has *that* become a normative principle?

---

32. In Laudan's view (1987b:349), a methodology is "a theory (in the quite literal sense) about how to
conduct inquiry." Once again, things look different depending on whether you approach this issue from an
optimization or a satisficing perspective. In Philip Quinn's (1987:355–56) terms, "If a methodological rule
is not acceptable unless we have grounds (presumably, good grounds) for believing that doing what it
prescribes is more likely to realize the goal it specifies than doing any alternative course of action open to
us [this is one of Laudan's requirements—W.C.], then I fear that very few, if any, methodological rules
will be acceptable. After all, when we contemplate some action that is a means to one of our ends, we
typically do not even know what all the alternatives open to us are. We usually have little or no idea of how
likely each of those alternatives is to lead to the realization of our goal." Or, put differently: "We can watch
no contest of the theories we have so painfully struggled to formulate, with those no one has proposed. So
our selection may well be the best of a bad lot" (van Fraassen 1989:143).

33. The issue discussed in this section is treated in more detail in the symposium on "Normative
Naturalism" (with contributions by Gerald Doppelt, Jarrett Leplin, Alex Rosenberg, and Larry Laudan) in
the March 1990 issue of *Philosophy of Science* (vol. 57).

34. Feyerabend (1981:322) insists on a more moderate interpretation of "anything goes" than one
usually encounters. With Ernst Mach, he sees no place for a philosophy of science, but only for (scientific)
research (*Forschung*) and "historically illustrated rules of thumb for future researchers."

Shapere: If by "complete" you mean *everything,* including ethics, for example, that's a matter for investigation. An evolutionary framework of thinking would suggest that it's true. One of my present projects consists of asking about the relations between factual beliefs and normative-ethical principles. Ethical principles presuppose certain factual beliefs, just as epistemic ones do—say, about the way people are. If those beliefs are overturned, they at least pull the rug out from under the ethical principle. The point of the naturalistic fallacy is that the factual beliefs can't prove or disprove the ethical belief. Nevertheless, the grounds of support are removed, and we find in the history of ethics that ethical beliefs often are rejected in the light of such factual findings—and that suggests that there is some sense in which the factual beliefs support the ethical conclusion.

Callebaut: This reminds me of Abraham Edel's (1970) "existential perspective" of an ethical theory (cf. also Richards's evolutionary ethics at 10.1). "Ought" implies "can?"

Shapere: That at least has some relevance to the issue. But the question you were asking was, "Can ethical beliefs—as an example—be internalized into science?" My present answer is that not everything can. (I don't promise to see it that way next year!) Right now my idea is that although the factual presuppositions of ethics, if overturned, can lead, and have led, to changes in ethical beliefs, nevertheless I see no way in which one can say that science *positively* supports ethical conclusions (as opposed to furnishing grounds for rejecting some of them). But the issue needs more looking into.

5.6.2. *Methodology as Part of the Rhetoric of Science*

Giere: I don't have a whole lot of role for methodology in the traditional sense. I mean, I have a theory that scientists make judgments and they have strategies for making judgments (Giere 1988:ch. 6; also 1992b).

Callebaut: We'll discuss that issue in chapter 8. But will you agree that scientists can and do learn from the past?

Giere: Oh yes, they do learn from the past. But now we can't be careful enough about how much that's content-dependent. I mean, in a scientific field, most of what scientists do is really tightly bound up with a particular content. They try a certain kind of model. Why do they try that? Well, one reason is that this kind of model is a successor of past models; another reason is scientists try the models they know best.

Callebaut: Like Dudley, you would reject the "levels view" of science: "the content of science changes, but methodology stays . . ."

Giere: What does stay the same is something like the human scientist who has certain capabilities and certain strategies—that is stable. But the interesting things are the more specific methodologies that develop—methodology with a small *m*—like techniques for experimentation. These are all pretty close to the science. New instruments—those kinds of methodologies. That is really important. But they are all close to the subject matter. High-level methodologies? Actually, what I think happens is that appeals to grand things like simplicity, fruitfulness, and all this stuff, that is part of the rhetoric of science.

Callebaut: We don't know that much about simplicity anyway, to take just that one. The simplicity/complexity issue has only been scratched at the surface, philosophically speaking, and then usually by people who are rather peripheral in the profession like Simon (1973a, 1977b, 1981) or Pattee (1973) (see also Morin 1985 on Gaston Bachelard; Bougnoux et al. 1990; Fogelman Soulié 1991).

Giere: But that is not what is going on in science. I think that is a kind of rhetorical resource. People say, "This is simpler!" and others say, "This is more beautiful and elegant!" I don't think those notions have any content. I think it really is rhetoric.

Callebaut: Maybe the study of rhetoric (Bazerman 1988; Prelli 1989; Gross 1990; Dear 1991; Pera and Shea 1991; Howe and Lyne 1992) can itself help the normative dimension of science studies (Fuller 1993; cf. Shapere 1991b). At any rate, (some) philosophers' nostalgia for "an overarching normative perspective that addresses the ends of science, in addition to its means: a perspective unafraid to suggest how science might change and improve" (Fuller) will have to be reckoned with. Even the odd Marxist sociologist of science begins to be tempted again by "recovering and expanding the normative" (Lynch and Fuhrman 1991). But it is certainly premature to warn the remaining Popperians all over the world that the specter of Marxism is rising again (in *theory,* that is).

5.6.3. *More Methodology, Please!*

Callebaut: Don Campbell has noted that the rejection of logical positivism left the theory of science in disarray. "Under some interpretations," he worried, "it has undermined our determination to be scientific and our faith that validity and truth are rational and reasonable goals." So far, there is nothing unusual about his assessment. But then he went on to observe that this is a rather paradoxical twist: "What we should have learned instead is that logical positivism was a gross misreading of the method of the already successful sciences." He concluded that "properly interpreted, the dethronement of logical positivism should have led to an *increase* in methodological concern rather than its abandonment" (Campbell 1984:316). A comment, Bill?

Wimsatt: I think that's perfectly explicable. If you're not going to do armchair biology, psychology, or physics, or whatever, you've got to learn the territory before you can make methodological pronouncements. When positivism fell from power, we threw away the maps of the court cartographers, knowing that they were incorrect. But we still had to use them for heuristic guidance because we hadn't yet had time to do our own surveys. I bet that if you look at the history of almost any of us who started twenty to twenty-five years ago, as well as most of the people who started since, you see initially fairly "conservative" papers which may involve new material, but relatively minor epistemological, ontological, or methodological changes. After a few years (I would say that it took me between five and ten years from when I started the post-doc with Lewontin) you begin to assimilate and incorporate the values of those you work with sufficiently to be able to transform them to your own tasks rather than drawing just on your philosophical tradition, and then you see rather broader and usually more radical methodological pronouncements. I think that, particularly in the last five years, we have seen a broader exploration of divergent methodological proposals than at any comparable period in this century. Look at the recent books by Hull (1988a), Giere (1988), Thagard (1988), and Margolis (1987): philosophers going biological, psychological, AI, and sociological *simultaneously* for their methodology. Or if you want to see something really provocatively radical in outlook, look at the papers by Star and Griesemer (1989) and Griesemer (1991) on museums and material models.

PART TWO

Doing It

6

PHILOSOPHY OF BIOLOGY

THE METASTUDY OF BIOLOGY is doing well. In April 1990, the President of the newly founded International Society for the History, Philosophy, and Social Studies of Biology (ISHPSSB) could announce that "we are now official, with tax free status and an honest-to-goodness charter from the state of Virginia" (Maienschein 1990:1). If you think there is nothing special about that, you should realize that only a decade ago many philosophers made no secret of what they thought of philosophy of biology— "is it any good?" (cf. Hull 1979:421). From the Scientific Revolution until just yester-year, philosophers of science neglected to "do justice to the living world as well as to the physical one" (Mayr 1988:v).

One philosopher who was more than "instrumental" (as they say) in bringing about the "much needed change" (Mayr) is Marjorie Grene, who is now in her eighties.[1] Says Burian: "In the philosophy of biology, she was in some sense the grandmother of us all." Grene had an undergraduate desire to be a biologist, but was not able to "manage it with her hands" in the laboratory. "It was always a science that interested her, a science which didn't require a complex mathematical manipulation, and a science very close to the kinds of issues she was concerned about in pure philosophy: *human nature,* and the settings of humans *in the world*—as biological beings—which she feels very strongly can't be anything like the Cartesian kind of being, which she criticizes." Grene came to philosophy of biology from an entirely different background and with an entirely different perspective than any of the standard Americans would have had, having worked with people like Adolf Portmann and Michael Polanyi, to mention only those two. Hers certainly was an unusual scholarly career: "After she studied—get this combination!—some with Carnap, some with Heidegger, some with Jaspers, she married David Grene, who was at the University of Chicago. When he and, I guess, the Provost at the University of Chicago had some disagreement, the Provost took revenge by not allowing Marjorie to have a position there. She had studied at Radcliffe, which was no help getting jobs. After finding a couple of small things in Illinois, and David Grene being an Irishman, she ended up being a dirt farmer in

---

1. See the special issue of *Synthese,* July 1992, in honor of Grene's eightieth birthday. It includes two articles on her philosophy of biology: Burian (1992) and Eldredge (1992).

Ireland for about fifteen years. She worked her way back in, first by continental philosophy, and by some incidental teaching in Britain." She came to entertain some unorthodox views about evolutionary theory (Grene 1959)—she represented a way of looking at evolutionary theory that was in some disagreement with the evolutionary synthesis—which may help explain Mayr's reluctance to acknowledge her crucial role in bringing about an autonomous philosophy of biology. She also came to see, partly under Polanyi's tutelage, the necessity of setting philosophy of science in the context of real knowledge of the science. And, Burian hastens to add, "she worked her way into a much different understanding, after fifteen or twenty years, of the synthetic theory than she originally had."

Wimsatt remembers Grene taking the developmental biologist Kauffman and him under her wing at the Philosophy of Science Meetings in Boston, in 1970: "We've been friends ever since—though we haven't always agreed about trends in the discipline—and she is a fierce friend in every sense of the word." Wimsatt also recalls consulting with her about her summer institute on "Biological and Social Foundations of Human Nature" at Colorado College in 1977 "on which scientists would be best to get—we agreed about 95 percent" and then planning "on how to put it to the institute for *philosophical* studies that we wanted mostly *scientists* for the faculty!" They ended up with Gould, Kauffman, and Levins. Wimsatt wondered, "How many biologists would give their eye-teeth to be at this?" There were social scientists, too, but that was less successful. "It was a stimulating summer!"

Many of the people at Colorado College were carry-overs to Grene's second institute, with Burian, at Cornell University in 1982, which, in retrospect, turns out to have been crucial sociologically, if not intellectually (see the biographical note on Burian, 10.6.1). "Marjorie is often seen as a crusty and somewhat harsh critic. But it's always in the spirit of deeply caring about people and about the subject matter" (Burian).

This chapter begins with a reflection on the question what it takes to become a good philosopher of biology. The next section is the first one of five devoted to different aspects of evolutionary biology. After a survey of anticipations of Darwin's theory of evolution by natural selection, evolutionary theory is interpreted as a theory of forces, and evolution as "heritable variation in fitness" (Lewontin). Fitness being a relational property, the next question to tackle is how one should go about formulating the relationship between organism and environment, including, in humans, the sociocultural environment. The section, "The Social Construction of Genes and Ecosystems," analyzes the gene-centered stance characteristic of current evolutionary thinking. We then focus on the hot potato of current philosophy of biology, the *units of selection* controversy, starting from G. C. Williams's case against group selection. Sober's causal analysis is contrasted with Wimsatt's analysis of variance (ANOVA) approach. The multilevel view of evolution favored by all our discussants fits nicely with Hull's conception of biological species as (quasi) individuals and of the evolution of lineages.

6.1 The View from Within

Callebaut: How does one become a good philosopher of biology, David?

Hull: Philosophers of science can do many things, but one of the most important is to learn some area of science really well and deal with those problems that scientists have not been trained to deal with. Scientists are trained to handle empirical issues, but they falter sometimes when it comes to conceptual problems. For example, numerical taxonomists were in the midst of rediscovering operationism. I wrote a paper pointing out that these same issues had arisen in physics and psychology (Hull 1968). Here were the problems that arose, here were the solutions that were proposed, take your pick! I think having the history of operationism laid out for them in terms that they could understand helped. Sober (1988a) is doing the same thing right now with his book on parsimony. Brandon, Lloyd, and lots of others are doing the same for fitness in evolutionary biology. So I think one role for a philosopher of science is . . .

Callebaut: . . . therapeutic?

Hull: Yeah, therapeutic.

Callebaut: Apart from that, you also think philosophers of biology can usefully contribute to the formulation of a general picture of science (cf. 7.2).

Brandon: David is to a large extent responsible for the view that almost all philosophers of biology give lip service to, but I think most of us really take seriously: that *you really need to know the subject matter before you can do it.* I think that is an enormously important contribution. He is of course not the only one who did this, but I really see him as the person who more than anyone else made it *unacceptable* to work in this area without knowing biology.

Lloyd: I do have a couple of strong feelings about the field of philosophy of biology. One is that philosophers must become conversant in the technical details of the field.

Callebaut: Is that feasible? I mean, how many people will be able to get involved in this?

Lloyd: If it turns out to be as difficult as philosophy of physics, that would be just about right! You don't become a successful philosopher of physics because you have taken Physics 101 in engineering school! Yet overall, that has been the standard in philosophy of biology. I am not talking now about *real* philosophers of biology like David, but about the impressions of general philosophers of science, who don't know biology, when they talk about it. The increase in technical expertise in the field I see as inevitable. You know, without actually knowing the science, we might as well just all go home, join an orchestra or whatever!

Callebaut: You spend some time in the first chapter of your book, *The Structure and Confirmation of Evolutionary Theory* (1988), whipping various philosophers for saying silly things about evolutionary theory—Sellars especially.

Lloyd: He was an easy target on this issue. A lot of the silly things philosophers have said about evolutionary biology—that it is not confirmable, not predictive . . .

Callebaut: . . . not falsifiable (cf. Ruse 1977 on Popper) . . .

Lloyd: . . . yadi yadi yada, blah blah blah—those silly statements were made out of sheer ignorance. A familiarity and respect for the complexity and the actual content of the theory is the best cure for this kind of silliness. Some people apparently thought that what they learnt from opening up a biology textbook for fifteen minutes one day was going to give them the goods to discuss the content of evolutionary theory. It is a professional disgrace that this happened, and one of the main things that I tried to do in my book was get into the details of the theories. Maybe the theory does not have some of the mathematical complexities of quantum mechanics, but is not as simple and easy as *they* understood it to be!

Callebaut: Maybe many "philosophers of science" of previous generations were interested in science only very indirectly, only to the extent that it allowed them to address issues more or less extraneous to science: morality, politics, and what have you. Think of Popper's "critical community," for instance (Ravetz 1980).

Lloyd: But don't underestimate the arrogance of philosophers!

Callebaut: It would seem, Alex, that if there is one field within present-day philosophy of science, broadly speaking, where there is a real symbiosis with the science, it is biology.

Rosenberg: It is no surprise: biology is the fastest moving of all the natural sciences right now, except for high-energy physics. And it has got the most conceptual problems. Therefore it is the one in which the philosopher has, recently at least, had the most influence. In addition, biology is in sufficiently primitive a state compared to physics that very many intelligent philosophers can learn enough biology to get right into the subject without devoting half a life time, the way they would have to in high-energy physics—which also explains why philosophy of biology has become such an active field: many philosophers of science have realized that this is an area in which a small investment will be repaid very quickly.

Callebaut: Why do philosophy of biology rather than something else, Bob?

Brandon: Philosophy of biology is a really interesting area of philosophy of science because there is a bit more consensus about the sorts of problems that are of interest than in other areas of philosophy. I think the reason for that is that it is a very small discipline, and there is a high degree of inbreeding in the discipline.

Callebaut: Part of the relative coherence within the philosophy of biology may be traced back, perhaps, to the influence of some founding fathers. Michael, the publication of your first book, *The Philosophy of Biology* (Ruse 1973) more or less coincided with that of David's *Philosophy of Biological Science* (Hull 1974). Some people say that yours came out more on the logical-empiricist side, whereas David's showed more of a historicist persuasion. Alex Rosenberg, for instance, makes that kind of opposition.

Ruse: There's no question about that. Nevertheless, I would perhaps put it a little bit differently: Although we were both publishing books which were intended not so much as serious research books but rather as introductions to the field, my editor, Stephan Körner (who in fact was my old professor) probably had a lighter hand than David's. I always saw David's book as intending to be an introduction to the field; and for that reason he couldn't and wouldn't—and wasn't perhaps allowed to—bring his own philosophical overarching thesis to the fore. I always read David's book as one where when you came to the end, although you knew where the author stood on particular issues, you didn't altogether have a feel of what his overall Weltanschauung was. Whereas in mine, I was allowed to put my own position forward very strongly.

Callebaut: Alex, could you say something more specific, as a "second-generation" philosopher of biology, about the "failures" you told me you found in Hull and Ruse's books?

Rosenberg: Oh yes. They are different kinds of failures in each one. It seemed to me that there were particular arguments that Ruse made about logical-empiricist applications that were defective in one respect or another, that he wasn't as sensitive to what was going on in biology, particularly molecular biology, as he might. And I thought that Hull wasn't enough of an empiricist, and that you can make a much more empiricist or positivist case for the relationship of biology to the physical sciences than Hull made in his book.

Callebaut: We will return to the issue of the autonomy vs. "provincialism" of biology—your term[2]—below (6.5.6).

Rosenberg: While Ruse was eager to make such an argument that biology was very much like the physical sciences—his aim was to show that the positivist picture applied to biology—it seemed to me that he didn't see the difficulties in the positivist picture well enough, and that vitiated parts of the argument. On the other hand, Hull wasn't enough of a positivist. I could sort of go at both sides here, and I did at first; I wrote papers on both of these books.

Lloyd: The part of *The Darwinian Revolution* (Ruse 1979a) I like is the naturalizing part. What I didn't like—even as a completely naive student—was that it seemed to be a Procrustean bed into which Darwinian theory was being put. The historians persuaded me of this, because I tried to defend it, for other reasons. So I thought that in some sense it wasn't sufficiently naturalized. And then the question left to me was, "Is there something philosophy can do that does not violate the history,

---

2. "The provincialist holds that at best biology is a province of physical science, a dependency that can advance only by applying the methods of physical science and, nowadays especially, the methods of physical and organic chemistry. According to this doctrine, biological findings and theories must be not merely compatible with those of physics but must actively cohere with its theoretical achievements. To do this it must, at a minimum, surrender some of its aims and methods, adopt others in the light of what we know about how physics proceeds, and import from the physical sciences its research programs, theories, laws, and concepts. Provincialists hold that the differences between biology and physics are either inconsequential matters of investigative tactics or reflect deficiencies in biology that must be extirpated." (Rosenberg 1985:16).

that is more sensitive to the actual historical context?" From the very beginning I had a sort of unreflective commitment to a naturalized philosophy of science.

6.2 GETTING A GRIP ON EVOLUTIONARY THEORY

Callebaut: We are now ready to move on to substantial matters. The philosophy of biology has been primarily concerned with evolutionary biology. Charles Darwin is, of course, the patron saint of that discipline. Elliott, let us talk about the Darwin you picture in your book, *The Nature of Selection* (1984a), which was received very favorably and is now widely used as a textbook. Darwin was not the first biologist to talk about evolution, nor the first to think about the role of selection in influencing the composition of a population. What *was* his original contribution, then?

Sober: What was really novel in Darwin was bringing these two ideas together— evolution on the one hand, natural selection on the other—and advancing the theory that the diversity of the living world we observe was mainly created by a process he called "evolution by natural selection."

Callebaut: Kitcher (1985b) uses Darwin as an illustration of his theory of explanation as unification. What influenced Darwin in formulating the idea of natural selection?

Sober: Anticipations are found in a number of places. A well-known source, one that Darwin cites in his autobiography as having catalyzed his thought process, is Malthus's *Essay on Population* ([1798] 1989). Malthus describes the situation of the poor in England at the end of the eighteenth century—how increasing population will deplete the food supply and thereby create increased mortality. A selection process will then take place, eliminating the excess population. So it is a kind of *equilibration process:* an excess of population corrects itself and brings the number of individuals back to some equilibrium value. Malthus understood this process in a very pessimistic way. His recommendation to poor people was that they should stop procreating so much. Malthus, like many political thinkers of his time, did not think of society as having to change to accommodate people; he thought of people as having to change to try to accommodate some sort of inevitability in the social arrangement, an idea which had many conservative implications for the politics of the time.

Callebaut: Now Darwin is known for having been a progressivist (see, e.g., Desmond and Moore 1991).

Sober: Darwin took this idea of a selection process but he changed it in one very important way (cf. 6.2.3). Malthus's idea was essentially static; it had the effect of keeping population the same. There is always a return to the same point; nothing much changes. Greater numbers return to lesser numbers, but the nature of the individuals present really did not change. Whereas Darwin had the idea that this selection process could be *progressive,* could improve the level of adaptedness and fitness found in the population. New individuals, new variants would appear; organisms would die, the fitter would survive on average; and so a population would be transformed and improved. So Malthus's static and pessimistic idea was in Darwin's hands transformed into a dynamic and in fact optimistic picture of how life may evolve.

Callebaut: Another influence on Darwin you discuss in your book are the Scottish economists, in particular Adam Smith.

Sober: Darwin, like many of the intellectuals of his time, was widely read in political theory; and we know from his copious notebooks that he read McCulloch and Adam Smith.[3] Now Smith also had an idea of competition and of a selection process. He thought of individuals in an economy as competing against each other, with the aim of advancing their self-interest. Smith thought of this competition process as having the effect of improving society. In the twentieth century, we are very familiar with the idea that capitalist competition can sometimes have a deleterious effect on the social welfare. In the eighteenth century intellectuals didn't have this idea so much. There were a few people who did; but for the most part, the Scottish economists had an optimistic view of laissez faire capitalism. Again we see here the idea of a selection process: individual entrepreneurs compete against each other; the fitter ones will be more successful, the ones that are less fit will go bankrupt. And as if by an "invisible hand," the wealth of nations would be increased.

Callebaut: What did Darwin take from this idea of competition?

Sober: Both the idea of a selection process as weeding out the fitter from the less fit and the idea that improvement is somehow built into the nature of competition. What Darwin mainly failed to see here is the same thing that Adam Smith and the other economists of his day mainly failed to see: the way in which competition can sometimes be deleterious. Economists have told us about the *prisoners' dilemma,* about the *tragedy of the commons.* These are situations which can exist in economic and biological contexts in which the following can happen: Each individual acts only for his own welfare, and everyone ends up worse off than they would have ended up if everyone acted against his own welfare. It is sort of a paradoxical twist, something that one would not anticipate until one saw the details of the kinds of situation; but nevertheless we know that these are possible and even actual in the world we live in.[4]

3. Hodge, Richards, and other historians of evolutionary biology have come to doubt the evidence for and the importance of this influence (cf. 7.3.2).

4. The example of a prisoner's dilemma originally used by A. W. Tucker can be described as follows: "The police present two captured suspects with the following alternatives: if either confesses while the other does not, he will receive two years in prison while the other will receive 12; if neither confesses, each will receive 4 years in prison; if both confess, each will receive 10 years in prison. Each prisoner reasons as follows: 'I am better off confessing no matter what the other one does: supposing he confesses, then if I don't confess I receive 12 years while if I do confess I receive 10, which is better; and supposing he doesn't confess, then if I don't confess I receive 4 years, while if I do confess, I receive 2 years, which

Callebaut: You also mention Darwin's acquaintance with the work of animal and plant breeders in Britain in the eighteenth century, that is, about how his theory was influenced by ideas concerning *artificial selection*.

Sober: The practice of artificial selection gave Darwin a kind of touchstone for his idea of natural selection. The longhorn cattle of England were totally made over by a very rigorous and carefully conducted program of artificial selection, and species after species of domesticated plants were made over as well. Darwin read this literature and discovered several very important things, not just about what happens in artificial selection, but about what happens in the wild without human intervention. The first thing he noticed was that the animals and plants in a species are not the same. If you look at a species undergoing artificial selection by a breeder, you see that there is an immense amount of *variation* among the organisms. Those of us who do not look carefully at organisms might think all cattle, or all corn plants, are all basically the same. Not so. And this is very important for selection to work, because . . .

Callebaut: . . . it can only work when there is variation to select upon. What else did Darwin learn from looking at this literature?

Sober: The second lesson he gathered was how impressive the changes were that could be made by a process of selection. If we take a population of corn plants and start to select which shall breed and which shall not, which shall be allowed to produce seed and which shall not, the question arises: how much change can you get in, say, the yield of the corn plant, in its height, in its color, in whatever characteristic you are interested in? There is no way of foreseeing this in advance; you have to look at the experiments, at the actual activity of artificial selection, to see how large a change can be brought about by selection. What Darwin thought he saw in the literature was that artificial selection can make over a population, can radically change its characteristics.

Callebaut: Did he also actually observe things changing?

Sober: *Darwin never observed a new species being created by natural selection.* What he observed was a given species being modified by human intervention. But he reasoned by extrapolation: If human beings in their activity as breeders can change a population that much in the comparatively short span of human history,

---

is better. So in either case, I am better off confessing.' Since each reasons in this way, each confesses, and so each receives 10 years in prison; whereas if each had not confessed, each would have received only 4 years in prison. So each would have been better off if neither had confessed. Although they realized this as they were deciding what to do, nevertheless, each also realized how he was better off confessing whatever the other did, and so each confessed himself." (Nozick 1981:542). The expression "tragedy of the commons" (Garrett Hardin) refers to situations of the following sort: each individual shepherd in a community, if rational, will be inclined to add sheep to his herd, which is grazing on communal grounds; as a result, the commons will be overgrazed, and eventually their net productivity will decline, to the detriment of all shepherds (Hardin 1968; Hardin and Baden 1977; Bajema 1991). Cf. also the "perverse effects" of social action discussed by the sociologist Raymond Boudon (1977).

how much greater will the changes be that nature can make when it selects over the many, many times longer period of natural history? Human beings can produce impressive changes in a short length of time; nature, acting on a much longer time frame, can produce much more radical and dramatic changes—changes large enough to count as the origin of new species. In artificial selection, you do not see the origin of new species, at least not at the time when Darwin was writing. But he reasoned from artificial selection to the origin of species by a kind of creative extrapolation.

6.2.1. *A Theory of Forces*

Callebaut: One of the original things you do in your book is you attempt to reconstruct evolutionary theory as "a theory of forces."

Sober: Evolutionary theory is not a simple, single, monolithic idea. It is a very diverse, heterogeneous collection of questions, approaches, disciplines; it is really quite a messy area of inquiry (cf. 1981c). In my book I try to provide a few road maps to understand the organization of this thing. One of the main organizing ideas is to think of evolutionary theory as, in part, a theory of forces. It tries to explain change of a certain kind: "evolution." In doing so, it tries to describe the different *possible* forces that can produce change, and how they will work. One can think of this theory of forces as organized in a certain hierarchy.

Callebaut: How would you describe this hierarchy of possible causes of change?

Sober: First one asks what will happen to a population if no evolutionary forces act on it. This might seem to be a trivial question—the answer being, "Nothing will change." But the *details* of how this is to be stated are in fact quite subtle, and at least the initial part of what I call evolutionary theory is given by a law in genetics called the Hardy-Weinberg law, which describes what will happen in a population when no evolutionary forces act on it.[5]

Callebaut: Evolutionary theory as you describe it becomes really interesting in the next stage, where one asks of given evolutionary forces what kinds of change they produce.

Sober: Here there are several. It is important not to confuse evolution with natural

---

5. If the frequencies of two alleles a_1 and a_2 on the same locus are p and q, respectively, and they occur in a Mendelian population within which sexual breeding occurs at random, the frequencies of the diploid genotypes can be written as

$(p + q)^2 = 1$, that is $p^2 + 2pq + q^2 = 1$

where p^2 is the frequency of a_1a_1 individuals (a_1 homozygotes), $2pq$ is the frequency of a_1a_2 individuals (heterozygotes), and q^2 is the frequency of a_2a_2 individuals (a_2 homozygotes). This result holds generation after generation. "Thus sexual reproduction allows individuals to produce offspring with a diversity of genotypes, all similar to but different from its own. Yet the process does not alter the frequencies of the genes; it does not cause evolution" (E. O. Wilson 1975:64). For a lucid discussion of the role of alternative "null hypotheses" in evolutionary biology—i.e., roughly, hypotheses about the probability that a certain kind of result is the result of chance rather than of a causal influence one wants to investigate—see Beatty (1987).

selection. Natural selection is one possible—obviously quite important—cause of evolution, but it is not the only one. Besides selection we have mutations, migration from one population to the other, and several other things that are properly called "forces," that is, causes of change. So the second stage of the theory of forces will describe the consequences of what we might call *single* forces: What happens when selection is the only force at work? What happens when mutation is the only force at work? What happens when migration is the only force at work? and so on for the others.

Callebaut: And next you combine forces.

Sober: Yes, the third stage occurs when one will try to describe models in which the joint activity of, say, selection and migration, or mutation and migration, are taken into account, and so on. You can see what the next stage will be: you will take three at a time; and the stage after that, four at a time, five at a time and so on, until all are simultaneously incorporated into an encompassing model. This progression makes for more and more complicated theories.

Wimsatt: In fact the complexity increases so fast (if you don't make lots of simplifying assumptions) that there is no single general set of equations that covers them all in any meaningful way. *Drosophila* are probably segregating at 2,000 loci. And if we assume that there are just 2 alleles at each of these loci (as many as 60 are known), we have 32,000 genotypes! Our most complex models, while more realistic, are inestimably simpler than nature.[6]

Sober: In nature, populations typically experience a multiplicity of evolutionary influences. Rarely if ever does it occur that only a single evolutionary force is acting on a population.

Callebaut: The question to ask next is, of course, what the *actual* causes of change are—or have been—in specific evolutionary processes.

Sober: Yes. We look at the living world; we see diversity, we see organisms that used to exist but are now extinct, and we ask why? It is in this context that many of the most important controversies now under way in evolutionary theory take place; e.g., controversies about whether group selection or individual selection has been important; questions about chance and random genetic drift as opposed to selection.

6.2.2. Heritable Variation in Fitness

Callebaut: We'll tackle some of these questions below. Is my impression correct that there is a quasi consensus, among present-day philosophers of biology, as to the *definition* of evolution by means of natural selection? I am thinking of Lewontin's (1970; cf. 1968) analysis, of course.

Sober: What Darwin realized and subsequent biologists have elaborated was that under natural selection, evolution will occur if three conditions are satisfied. First

---

6. This is why biologists look for qualitative and robust rather than for detailed quantitative models (cf. 4.2.2).

of all, you have to have *variation* in the population. Organisms have to be different from each other. Secondly, they have to be able to transmit their characteristics from parents to offspring; you have to have *heritability*. And thirdly, for selection to be the cause it has to be the case that the organisms differ in ways that make a difference for their *chances of surviving and reproducing*. If we add these three conditions together, we can say that the theory of natural selection basically asserts that *heritable variation in fitness* will produce evolution.

Callebaut: There is a set of questions that have occupied evolutionary theorists quite substantially about that threefold requirement—heritability, variation, and the existence of fitness differences. Could you summarize some of these issues?

Sober: Let us start with heritability. Darwin took it as a given that organisms are such that parents resemble offspring. One can ask why they do so, and what the mechanism is by which this heritability—this resemblance—exists. Darwin had a theory of inheritance: the theory of *pangenesis*. His idea was that hereditary particles migrate from each part of the body to the reproductive organs. It was one among many nineteenth-century theories of inheritance that turned out to be wrong. The theory that we now think (and have thought for a long time) to be basically correct is Mendel's theory. Mendel, in a sense, asked a deeper question about one of the assumptions in Darwin's idea of natural selection. Darwin simply *assumed* that there was heritability. Mendel asked the question about the *mechanism* that gives us heritability.

Callebaut: We'll discuss some of the historiographical aspects of the discovery of Mendel's laws in another context (7.2.2). "Neo-Darwinism" is the label for the theory of evolution by natural selection that incorporated this mechanism of heritability. What about the other requirements in the theory of natural selection—the existence of variation and of fitness differences?

Sober: Again one might ask, "Why is there variation in nature?" Again, Darwin did not have an adequate theory of this; he simply assumed at the outset that organisms are various. What makes them various? Well, we know now things that Darwin didn't know: about the existence and nature of mutations, about genetic recombination—indeed a whole set of disciplines in evolutionary theory focus on understanding the origin and maintenance of variation in populations. The third component, the existence of fitness differences, was something that Darwin *had* something to say about, unlike heritability and variation. It was a problem that he addressed in a very detailed way. He did not just assume that there *are* fitness differences; he asked *why*. He looked at the details of the relationship of organisms to their environment and to other organisms, and he tried to figure out what it is about those relationships that creates fitness differences.

6.2.3. *The Nature of Fitness*

Callebaut: In your book you talk about two ways of thinking about fitness, which, as we have seen before (4.2.1), cannot be regarded as a *physical* property.

Sober: One can think of fitness differences in terms of their *consequences*, that is, in terms of the way a population will change its composition in virtue of the existence of fitness differences. But evolutionary theory also has a great deal to say about the *sources* of fitness differences. It tells us a great deal about which environmental circumstances will produce fitness differences. For example, there is the theory of *mimicry* which tells us in what circumstances certain kinds of visual appearance in an organism will produce a fitness advantage. There are many models of this sort which give us, in a sense, an *engineering analysis* of how organisms can be built, which explains why some organisms are fitter than others.

Callebaut: More generally, what seems to be at stake here is the very definition of evolution in biology. Henry, you have pointed to the great variety of conceptions of evolution in the literature: biologists have talked about the *relatedness* of creatures, about *descent with modification,* about *changes in gene frequencies* in a breeding population, about *natural selection.* Help me out, Henry!

Henry C. Plotkin (University College London; see biographical note at 9.1.2): If you press biologists a bit on natural selection, they will say: "Well, some animals are reproductively more competent than others; there is a *differential reproductive competence,* and that is what we mean by evolution." At any rate, if you were to look at the community of biologists worldwide, my guess is that at least 95 percent of them would be taken up by one or other of these definitions. Biologists usually talk about evolution as a *consequence* of something. I would think that if we understand much more—and we know very very little—about *the process of evolution itself,* we might find that it is a very rich device, much richer than has ever been thought of it, certainly in the nineteenth century.

Callebaut: We will pursue that line of thinking about evolution as the *cause* rather than as the consequence of something further on (9.4). Why, in the end, do you think the tautology line of argument is misguided, Elliott?

Sober: It simply misdescribes what fitness means, and even after it is corrected by bringing the concept of probability in, it still neglects this very rich source of information we have about fitness that is different from a mere description of what consequences fitness differences have. The claim that the fitter organisms are more probably the ones who survive and reproduce describes the consequences; but fitness is also understood in evolutionary theory in terms of its sources.[7]

6.3 THE MISSING HALF PANCAKE: THE ELUSIVE ENVIRONMENT

Callebaut: Fitness—and I am now referring to the "engineering" aspect Elliott just talked about—is a *relational* matter. It has roughly to do, on the one hand, with the fit between the properties of organisms and the properties of the environment these organisms live in and, on the other, with how the organisms compare with

---

7. For further discussion of the notions of evolution and natural selection, see Sober (1987, 1989b), Maynard Smith (1987b), and Shimony (1989a, b).

respect to these properties. It is a truism, then, I guess, to say that the characteristics of organisms depend on both their *inheritance* and their *surroundings*. It would seem to be quite natural, therefore, for biologists to try to describe the relationship between genetics and the environment in quantitative terms.

Richard Levins (Harvard University; see biographical note at 9.2.4): People will ask questions like, "What proportion of a character is determined by the genotype?" or "What proportion of it is determined by the environment?" Or they will use formulations such as, "The genes create the potential, and the environment the realization." This kind of analysis is used also in epidemiology in a peculiar way, where if you ask people "What is the cause of tuberculosis?" you'll very often get the answer that the "cause" is the bacteria and the "influencing circumstances" are the social conditions—nutrition or environment.

Callebaut: What is wrong with that?

Levins: You could just as well take the same statement and turn it around backwards, and say that the *cause* of tuberculosis is poverty and malnutrition and the *conditioning factor* is the presence of the bacteria. This symmetry of interaction is disguised when people give priority to one or another of the components, and they do this in the relations of heredity and development, of organism and environment.

6.3.1. *Genes Do Not Code for Traits*

Callebaut: This reminds me of one of my university teachers who, in a course in historical critique, insisted that it was pointless, methodologically, to try to distinguish between the "real causes" of some historical event and circumstances that had been "merely conducive" to it, as historians of previous generations used to do all the time. I take it that this would be pretty much the same point as the one you are making.

Levins: The reality is that the *genes do not code for traits*. What the genes do is mediate the production of proteins, or influence when the production of protein begins or is terminated. These proteins then act with other materials in the cells, giving rise to various chemical reactions, and through a long and tedious chain of causation give rise to the characteristics that we are able to observe. The fact, therefore, that different proteins are produced in different individuals and that, in that sense, some things are "determined" by the genes, does not mean that the character is an inevitable consequence of the presence of a particular gene.

Callebaut: An example should make this clearer.

Levins: Suppose that we decided that all individuals with a given blood type shall be sent to school and those without that blood type shall not. If this program were carried out, and accepting that the appearance of the appropriate protein in the blood cell is very much determined by the genes, what will happen then after a while is that people carrying this particular protein will be educated, and that those who do not carry that protein will not be. If this program is carried out rigorously, then *we have created a hereditary characteristic!* In fact, by all the ordinary tests of heritability, this is a hereditary characteristic. The obtaining of a graduate degree will then be a genetic trait. It fits into the rigorist definition of heritability. Nevertheless it is certainly *reversible*. We can change that; we can decide that it is the other blood type that will be sent to school, and you'll have a complete reverse of effect. So one of the fundamental things to recognize is: *the statement that something is genetically determined in no way means that it is any less environmentally determined.* The partitioning of an effect into an "environmental component" and a "hereditary component" does not really make sense. That's part of an atomistic, reductionist view of the world.

6.3.2. *The Social Environment*

Callebaut: But how should one go, then, about formulating the relationship between organism and environment?

Levins: In general, when we are confronted with opposite and mutually exclusive categories like "organism" and "environment," the first thing to ask about is, "Do they in fact interpenetrate?" That is, is there a sense in which the environment is determined by the organism *and* the organism by the environment? We have found in our explorations a number of ways in which there *is* a very intimate interpenetration of organism and environment.

Callebaut: Organisms to some extent *select their own environment.* Conrad Waddington used the concept of "exploitive system" to refer to this feature of life. The exploitive system is "the capacity of animals to select, out of the range open to them, the particular environments in which they will pass their life, and thus to have an influence on the type of natural selective pressure to which they will be subjected" (Waddington 1975:273; cf. Plotkin 1987a).

Levins: Animals can move to a place that is warmer or colder, they can nest on the ground or up in the trees. And even those organisms such as plants, which themselves do not get up and walk and choose an environment, determine the environment in which their offspring will grow by the kinds of seed: whether the seed will stick to the fur of animals, will pass through the gut of a bird, will float down a river. So organisms in one way or another—consciously or unconsciously, through their anatomy or their physiology—select the habitat within which they will develop. Secondly, by living in a habitat, they change that habitat. *Organisms modify the environment around them.* At the surface of the skin of a human being, there is a thin film of air which is different from the atmospheric air, which has a higher concentration of carbon dioxide, of urea, of moisture, and will have a different temperature. And this film is a boundary between ourselves and the rest of the meteorological conditions. At the surface of a plant leaf, similarly there will be a layer which is rich in oxygen, poor in carbon dioxide, high in moisture. This is an environment perhaps a millimeter thick. It determines which insects, which fungi can develop in the plant leaf.

Callebaut: So the organisms modify the environment around them in the microscopic scale . . .

Levins: . . . the chemical environment around the roots of a plant is very different from that in the general soil.

Callebaut: What happens in the macroscopic scale?

Levins: Forests collectively modify the habitat, creating the water regime of a forest which fluctuates much less in response to the rainfall fluctuations than would a desert environment. The friction of the top of the forest changes the behavior of wind. So organisms determine properties of their surroundings, individually in the small, collectively in the large scale. They determine the abundance of their resources. They modify the environment in ways that make the habitat more congenial for their survival or less congenial.

Callebaut: At another, deeper level, organisms really *define* what their environment is, according to you (Levins 1968, 1975) and Lewontin (1978, 1982b) (see also Levins and Lewontin 1985).

Levins: When you ask, "What is the environment of an organism in a particular realm?" you have to ask, "What organism is it?" A mosquito coming into a lecture hall will be responding to gradients of temperature, of urea, of ammonia, of the clues that indicate that blood meals are available in some locations in that room and absent in others. The mosquito is totally indifferent to whether the bearer of the blood supply has an advanced degree or not, to the race of the person or what their economic condition is. People sitting in that hall are responding to other things, not particularly responding to the gradients of carbon dioxide or urea, but may be very much aware of who is the boss and who the underling, how your own actions will affect your future of employment. The environment becomes a social environment of cooperation, of conflict, of rank and hierarchy. A tiger coming into that same habitat may be interested in very different things—in the availability or nonavailability of meat. A lizard chasing insects responds to anything the right size that moves. It does not care what kind of insect it is; it does not recognize different species of insects. If its eats something it does not like, it will spit it out and start all over again. So the food world of the lizard is a spectrum of different sizes of moving things. A bird is smarter and at that same place sees insects of different species, recognizing their color, making it possible for insects to fool the birds sometimes by adopting the color of a distasteful insect.

Callebaut: So organisms define their own environments, determining what is relevant about their surroundings.

Levins: When anything impinges on the organism from the outside, it gets transformed as it penetrates through the organism. What may be an unusual substance at the outside—a new dish—goes through the gut as sugars and carbohydrates and elements of protein. A new pesticide reaching the outside of the insect may be experienced inside the insect as one of the old challengers—a loss of oxidative capacity. Hot weather, impinging in the insect on the outside, may limit the distribution of that insect, not because it's too hot but because the heat keeps it from feeding; and therefore the insect may experience a hot environment as a hungry place. So the insect's own biology transforms an external signal along pathways that are already there in the previous evolution. Furthermore, *every part of the organism is environment for the other parts of that same organism.* In the same way that you can talk about the evolution of algae in hot springs to elevated temperature, you can talk about the evolution of the chemistry of the liver to constant temperature conditions in spite of the fact that the skin surface has to deal with changing temperature conditions.

Callebaut: All of these are aspects of the interpenetration of the organism and environment at an individual level.

Levins: A similar analysis could be performed at the level of whole populations, and such aspects of the surroundings as "Is the environment an uncertain environment or a predictable environment?" become important. But *the uncertainty or predictability is very much dependent upon how the organism relates to its environment.*

Callebaut: What do you mean by that?

Levins: A mammal lives very intensely; it has a high metabolic rate, it uses up its food rapidly, it has to feed often. This means that the food environment of a mammal, or of a small bird, will be uncertain if it can't be sure of feeding at least every day. A hummingbird has to be able to feed many times in a day in order to have a stable food supply. A scorpion in that same place comes out and feeds, and if it doesn't find anything it can go down underground or under a rock and hide. Its environment is a secure one, provided that on a time scale of weeks or months on the average it can get enough food. So by averaging the environment—the food supply—over a long period of time, the scorpion is able to take fluctuating conditions and turn them into a predictable and reliable environment.

Callebaut: Moral?

Levins: What this means is that the statistical structure of the environment as well as the particular physical characters of that environment are very much determined by the organism living in those surroundings. In a socially structured population, the different social relations also are both part of the environment and a product of the interactions of the organisms in that environment in a very intimate interpenetration. So that it really doesn't make sense, in most cases, to talk about "Is something determined by environment or by heredity?" or "Is something more environmental or less environmental?" Crude measurements of relative importance might be useful in plant breeding experiments, but in fact for understanding of evolution and for understanding ecology the important questions are always, "What of the nature of the interpenetrations among categories that we define as separate?" rather than the pulling them apart and assigning of relative weights.

Callebaut: Dick Lewontin, you assign special importance to the social environment in the case of social species in general and of man in particular.

Richard C. Lewontin (Harvard University; see biographical note at 9.2.4): One of the most important interpenetrations that we have to deal with is *the way in which the social environment transforms the properties of individuals.* We do not understand collectivities by simply looking at some isolated properties of individuals, but the properties of the individuals themselves are changed by the social environment in which they occur. (Cf. Richard Levins's models of the regulation of blood sugar in fig. 6.1.)

Callebaut: This is another aspect of your anti-reductionism. When culture is added to sociality, things change even more.

Lewontin: One of the most interesting cases is the way in which characteristics that we assign to animals have been transformed in humans under the same name into totally different meanings. Let us take for example *eating;* in the theory of evolution, the nutrition of an animal is considered after all an extremely important part of the adaptation of the animal. Yet what we call "eating" in human society has become in many senses disjoined from the process of nutrition: bringing calories into us and keeping us alive and reproducing. When we invite a friend to dinner, we are usually not really inviting that friend to increase his or her caloric intake; we're inviting that person to a social interaction. When we say "Come over and have a drink!" we are not trying to adjust the electrolyte balance of that person; we are engaging in a social interaction, which may not involve drinking anything at all in the end. When we have a breakfast together at a business meeting, the purpose is to have a business meeting. Eating has undergone an extraordinary number of transformations. One of the most interesting historical transformations as an example is a cold lunch that was prepared by the Israelites when they were to leave Egypt after being released. That cold lunch became incorporated into Jewish religious ritual as a ritual supper, called the Passover Supper. That particular religious ritual became transformed into a particular historical example of the Passover Supper, called the Last Supper, in Christian religion. And that act of nutrition at the Last Supper has been converted into a purely symbolic act, the taking of Communion, which is carried out by hundreds of millions of Christians, without any caloric intake or nutritional value whatsoever.

Callebaut: Sex would be another nice example, I assume?

Lewontin: The same is true for sex. When sociobiologists think of sex, they think of something that has to do with the rate of reproduction of the species. But in fact, in human populations sex has become almost completely disentangled from reproduction. Given modern technology, for example, there is no absolute and necessary relationship between sex and reproduction, between copulation and reproduction, between copulation and giving birth, between having a child and having had copulation in the first place. It is now entirely possible to bear a child without ever having to copulate. Most sex is involved in pleasure, in dominance, in aggression, in cruelty, in giving pain, in forming bonds, in a whole variety of different kinds of human relations, not in the simple reproductive act. So again, what human

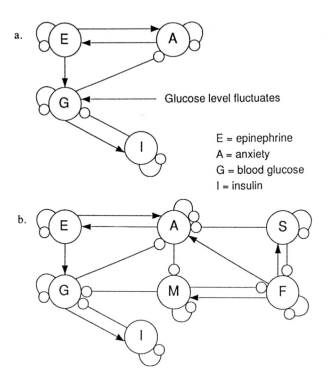

a. The variables are: E = epinephrine, G = glucose in the blood, I = insulin, A = anxiety.

Glucose level fluctuates

E = epinephrine
A = anxiety
G = blood glucose
I = insulin

Figure 6.1. Physiological and expanded model of the regulation of blood sugar.

a. The variables are: E = epinephrine, G = glucose in the blood, I = insulin, A = anxiety. The links between them show the direction of direct effects. → is a positive effect, —○, a negative effect. Thus epinephrine increases glucose through the conversion of liver glycogen, glucose releases insulin, insulin removes glucose, glucose releases the anxiety symptoms of hypoglycemia, and epinephrine and anxiety increase each other. All of the variables are self-damping. In the absence of input from other variables, they return toward some equilibrium. The arrow to glucose without a source represents an external input, in this case perhaps from eating. Other sugar regulating pathways are omitted. An analysis of this graph (Puccia and Levins 1985) shows that an increase in input to glucose reduces anxiety and epinephrine. But the effect on glucose level is ambiguous. If the positive feedback loop linking epinephrine and anxiety is weaker than the product of the self-damping terms of epinephrine and anxiety, the commonsense expectation of increased glucose and insulin will occur. But if the positive loop is stronger we get the counterintuitive result that adding glucose reduces the glucose and insulin levels. The negative loop G —○ A → E

→ G, if strong enough, can cause the system to oscillate.

b. Expanded model. The physiological network shown in (a) and shared by all people, is embedded in a larger network in which the individual is an industrial worker. The negative input to glucose is identified as coming from the metabolic rate M that reflects work intensity. The negative link from anxiety to M is a possible response of the worker who reduces the intensity of effort when the hypoglycemic anxiety is perceived. The foreman observes this slacking off of effort and acts to order an increase in work intensity (F → M) which increases anxiety (F → A). But this is a well-organized union shop. In response to the foreman's actions, the shop steward and workmates' solidarity S acts to stop the foreman's harrassment (F ○—S) and reduce anxiety (A ○—S).

Model (a) is more universal and is already complex enough to have interesting and unexpected properties, but the input M —○ G is arbitrary. The physiological network of (a) is arbitrary. The physiological network of (a) is embedded in larger social networks that differ among people, open up possibilities of richer dynamics, and would therefore be a preferred object for the management of blood sugar.

society has done is to transform the meaning of individual acts which seemed so simple in the nonhuman case.

6.3.3. *Culture Transcending Biology*

Callebaut: You insist that it is even possible for the sociocultural level to actually "negate" the purely biological. Could you explain this?

Lewontin: That is another important transformation, and one which is very important for our understanding not only of sociobiology but of evolution in general: The social milieu—the social level—may actually *negate* individual properties of organisms when those organisms are regarded in isolation. Examples come immediately to mind. When I go to Europe, I fly there. Now no person can fly by flapping his or her arms and legs. I mean, that was tried by da Vinci and it didn't work. Nor can ten people fly if they all simultaneously flap their arms and legs. Yet ten people can fly, indeed hundreds of people can fly, in an object—an airplane—which is the creation of social forces. The airplane, the pilot, the fuel, the technology, the airports, the radios, and so on are all consequences of social interaction. And it is very important to understand that it is not society that flies; it is the individual who flies; that is, *a constraint on the individual has been destroyed by the process of social interaction.* Over and over again, social interactions, then, have the property of negating—of turning on their heads—individual constraints.

Callebaut: And you think that is extremely important to understand when sociobiological or indeed evolutionary explanations of human properties are considered.

Lewontin: Any theory of the evolution of human life which begins with what are said to be individual biological constraints on individuals, and tries to create a picture of society as the sum of those constraints, misses what is really essential about the social environment, which is that in moving from the individual to the social level we actually change the properties of objects at the lower level. This whole problem of levels of explanation, of levels of evolution, of levels of action, is one of the deepest ones with which we have to deal in our understanding not only of sociobiology but of evolution in general.

Callebaut: Playing the devil's advocate, I could point out that the examples of transformations of characteristics of people by their social environment we just heard apply only to behaviors and things external, but not to things that are entrenched in people's physiology or anatomy. Could you produce instances of that as well?

Levins: Human posture would be an example of the latter. Human posture is not simply the result of our ancestors' coming down from the trees and walking erect; it is very often an indicator of class. And this is known very well by actors, especially in street theatre where a character has to be indicated very quickly, and where in the acting workshops actors are taught how to walk like a peasant, or a lord, or somebody from the working class. Now this means that the system of mechanical stresses within the body will be very much determined by your social position, not because you've inherited them along with your land, but because the social position imposes a kind of posture. This is also true in women, where the binding of feet, or the use of high heels, or a whole series of patterns of clothing and posture are designed either to accentuate or to minimize sexual properties.

Callebaut: OK, this is one example.

Levins: When we examine the workings of the nervous system, it has become increasingly apparent to medicine that the distinction between the autonomic nervous system and the cerebral nervous system really breaks down. The autonomic nervous system is not that totally autonomous. The things entering us through our consciousness are transformed through the endocrines and the autonomic system into physiological responses: the stress responses which are recognized as so important in health. What happens therefore is that aspects of the environment which may be occurring far away—say, news about the threat of war—entering into our consciousness can impinge upon the endocrines, the autonomous nervous system, the degree of muscular tone, the concentrations of sugar. So our whole physiology, by way of this mediation, is transformed by social surroundings, even operating at a very long distance and mediated by signals of very low levels of energy.

Callebaut: What does all this imply for the study of evolution?

Levins: It means that we cannot conceive of human evolution as if each stage in evolution "adds on a layer"—the layer that we acquired when animals first emerged from the sea, the layer of the early vertebrates, the layer of mammals, the layer of primates, the layer of humanity . . . As if each of these layers remains intact and then a thin veneer of subsequent evolution is superimposed on top! Quite the contrary, what happens is that later stages in our evolution transform the structures and processes also of earlier stages . . .

Callebaut: . . . what Campbell (1974b) calls "downward causation" . . .

Levins: . . . and this means that evolution is a much more dynamic and innovative process than the conservatism of the sociobiological outlook would lead us to believe.

6.3.4. *The Importance of Frequency-dependent Selection*

Callebaut: Until recently, Lewontin's (1978) paper, "Adaptation," was about the only thing one could refer to when talking about the environment (cf. Brandon 1978a, b; Conrad 1983). Now you have written a whole book on the notion of the environment (Brandon 1990). How did that come about?

Brandon: I wrote that abstract[8] thinking that I could solve the problems I was worried about in a paper. I've been thinking about the problems ever since, on and on. This is really what I am most interested in and what I really want to work on. The problem, I think, is absolutely central because any "adaptationist" explanation— any explanation of something being an adaptation—implicitly is saying something about the environment, because it is an adaptation *to* some environment. And

8. At the time the interview was conducted I had only seen an abstract (Brandon 1984) of the book.

selection occurs *within* some environment. So it seems to me that it is one of the central concepts of all evolutionary thought. I found it really remarkable that no one says anything about what the environment is. People sort of wave their hands and take it for granted that we all know what the environment is . . .

Callebaut: . . . the exception confirming the rule being a book written by the chemist and sociologist L. J. Henderson early in this century, *The Fitness of the Environment* ([1913] 1958).

Brandon: So I've been led to this just from my general interest in adaptation and how we explain adaptations by means of evolutionary theory. I've come to see that there is a big gap in my understanding of all this, and that is: What exactly is environment? How do we individuate environments? How can we say that these organisms are living in one environment and these organisms are living in another environment?

Callebaut: The easiest and, I guess, an utterly naive approach would be to say: "The environment consists of those factors that are exogenous to the evolving population." I mean, that would be the standard way of approaching the problem by, say, a systems person.

Brandon: It would be nice if you could say that, because then you could study it independently. That's the way people typically think about it. But unfortunately that won't work. If the environment is that thing which determines fitness values, if fitness is a relationship between phenotype and environment, then the environment is obviously going to *include* the evolving population. This is most obvious in cases of frequency-dependent selection. If fitness is frequency-dependent, then it is dependent on the frequency of one's conspecific. But now the question arises: "The frequency of which group of organisms?" How do we determine if it is the entire species or not?

Callebaut: Typically biologists have talked about "demes": *local* populations or *local* breeding groups.

Brandon: There are very interesting and, although simple, I think revolutionary points that the biologist D. S. Wilson makes in a number of papers and in his book, *The Natural Selection of Populations and Communities* (1975, 1977, 1980, 1983; cf. Wilson and Colwell 1981). What he argued is that demes by and large are not—I am using my own terminology—*selectively homogeneous* . . .

Callebaut: . . . meaning?

Brandon: The selective pressures experienced by this organism at *this* end of the deme may be entirely different from *that*. When you think of this in terms of frequency-dependent selection, the relative frequencies of the types this guy runs into may be quite different from the types that guy runs into. So it is something like fitness being frequency-dependent because there is one type that is altruistic and another type that is selfish—then you benefit by having altruistic neighbors, but that is a local benefit. Altruism dispenses the goods very locally. The group that the biologist distinguishes in terms of local breeding may not be the group that is determined by these ecological factors. So *genetical or genealogical divisions of nature need not and usually don't*—and that is an empirical assertion—*correspond to these ecological divisions*. This is a point that people have just ignored, just haven't thought about; but I think it is an extremely important point.

Callebaut: Why would thinking about frequency-dependent selection be a good example of this?

Brandon: We say that the environment includes the types surrounding the individual. In order to identify the environment we need to know what population we are talking about, what group we're talking about. But how do we know that? Well, it is the group that is inhabiting their common environment.

Callebaut: Now that seems to be circular . . .

Brandon: . . . and it *is* circular. So the question is, how do we break out of that circle, how can we find some way of putting a wager in there so that we have some independent means of doing this? One way of doing this is trying to divide the differences in fitness—*actualized* fitness, that is, actual reproductive success. We can explain some of that in terms of different environments and some of that in terms of different character states, different phenotypic states. Essentially we are doing an ANOVA here (see Lewontin 1974b). We are trying to partition that variance into variance within a given environment, where we can identify the differences as differences in the phenotypes of the organism, versus differences in environments. That is a problem that we can get a handle on.

Callebaut: ANOVA has a number of problems coming up.

Brandon: Absolutely. It can be misleading. It is only a tool, it is not definitive of the processes of nature, and so on and so forth; but it is a very useful tool. And so it is the sort of tool I would use in cases like that.

Callebaut: You are also getting involved in real experimental work.

Brandon: The other thing that I am exploring now—and in fact I am doing some experimental work with Janis Antonovics—we are using a method that goes back a long time, in botany at least, called the *phytometer method*. Essentially what you are doing is you're letting the plant tell you how it perceives the environment. The techniques that one might use—we have done some of this already—would be to take a bunch of clonings of genetically identical individuals and plant them on some transect; and then you could observe differences in growth rates, eventual height or something like that. That is a way of telling how the plant perceives the environment, not how we as humans perceive the environment. Differences we might perceive may be unimportant to the plants; differences we cannot see may be very important to the plants. We have a lot of data on two species of grasses that Janis has already worked on.

Callebaut: What do you expect to learn from these experiments (cf. Antonovics, Ellstrand, and Brandon 1988)?

Brandon: The main point is that we look at the evolutionary significance of the relation between deme size and size of what I call "selectively homogeneous populations." We've got two species for which we already have enough data. One species is a species where the deme is larger than the selectively homogeneous populations, and of the other species the reverse is true: the selectively homogeneous populations are larger than the demes. What we try to do is a reciprocal transplant of cloned individuals, to try to observe differences in micro-adaptation. We expect that the species where the selectively homogeneous population is larger than the deme will show a greater micro-adaptation than the reverse.

Callebaut: So you are looking for an experimental confirmation of your theoretical argument for this?

Brandon: It would be very nice to experimentally demonstrate this, because then we could show that here is something that evolutionary biologists have heretofore entirely ignored. It is going to have important evolutionary consequences. In particular, it is going to serve as *a constraint on adaptation*. That is, if demes cover lots of different selective environments, then that is going to serve as a constraint on adaptation. Put intuitively: If selection is going on here for one phenotype, and here for another phenotype, and here for a third, and here for a fourth, yet breeding occurs within the entire group, then that selection is going to be ineffective, because the effects of selection within each selective environment will be swamped by the effects of selection in other environments. This is sometimes something that organisms have no control over; and as such it serves as a constraint on adaptation. Another thing is—Darwin felt this and I think this too—that one of the real tests of evolutionary theory is not to see how well you can explain adaptations but to see how you can explain things that are *nonadaptive*. That is how you are going to answer the question, "Why is not perfect adaptation totally ubiquitous in the biosphere?"

Callebaut: The problem raised in Gould and Lewontin's paper (1979) against "Panglossian" adaptationism.

Brandon: This, I think, will serve as a nice and probably very important evolutionary principle and an important evolutionary constraint on adaptation. If that is so, then of course it means that the concept of environment is extraordinarily important. You're going to have to be able to individuate environments, because you need to know in any given case whether you have got a deme that is selectively homogeneous—living in one environment—or whether there are multiple environments within that deme.

6.4 THE SOCIAL CONSTRUCTION OF GENES AND ECOSYSTEMS

6.4.1. *A Primer on the Secret of Life?*

Callebaut: Full-fledged reductionism as defended by the positivists may be considered passé by a growing number of philosophers of science (4.2), but it is still very much an ideology that appeals to many—probably the majority—of practitioners of the "hard" sciences. Let us take a look at the situation in biology.

Levins: The sciences exploring the nature of our natural world have always worked at a number of different levels, treating objects on the scale of everyday life, smaller objects, objects that are very much larger.

Callebaut: The "micro-reductionist" approach to the complexity of anorganic nature and life is to imagine that somehow within a complex system, there is a smaller object which is fundamental to determining its properties (Oppenheim and Putnam 1958).

Levins: At different times this fundamental object would have different names. At one time, the atom was introduced as the fundamental unit of matter. Then the nucleus was identified, and subatomic particles. Particle physics is going further and further down, identifying smaller and smaller units. Biologists also were interested in seeing whether they could find a fundamental unit, and decided on the *gene*. The gene was regarded to be a fundamental unit both because it was small and because it was presumably unchanging. What was very attractive to Weismann and to others in the last century is that evolution was after all only an appearance. The organisms may change, but their fundamental building blocks are the same; and it was easy for them to comprehend evolution as the reassortment of fundamental units which themselves were not changing. That is, *the key to the understanding of change was still non-change . . .*

Callebaut: . . . the shadow of Plato and Aristotle.[9]

Levins: As a result of this, when the structure of the DNA molecule was worked out in the 1950s, much of genetics became focused on the physical structure of the gene, the chemical structure. Attention was diverted away from problems which were of great importance to biologists previously: how do characteristics develop in organisms? Now this problem was displaced. It became a trivial problem. They answered: "The characteristic is coded, the gene produces RNA, produces protein, produces a trait, and everything else around that is either helping it or interfering." In a similar way, it was assumed that modern biology is the biology of the smallest unit, of the molecules; that in principle, we can deduce the properties of the individual organism from its genetic code; and that when we knew the properties of the individual organism, we understood populations and ecosystems.

Callebaut: All this had and still has important practical significance in the development of the biological disciplines (cf. already Commoner 1961).

Levins: Universities deciding to become modern tended to neglect their museum collections, to decide that biogeography, evolution, comparative anatomy were obsolete fields; and *to go modern meant to go molecular.* After a short period of intense euphoria in which it was announced in the *New York Times* every week that "the secret of life has been unlocked at Berkeley," it was realized that the knowl-

---

9. "Only the immutable world of ideas was traditionally recognized as 'illuminated by the sun of the intelligible,' to use Plato's expression" (Prigogine and Stengers 1984:7). As to Aristotle, although he was "fully aware of the qualitative multiplicity of change in nature," the only type of change that was allowed to transpire in classical dynamics was motion: the relative displacement of material bodies (ibid., 62).

edge of genetic makeup, of the DNA molecule, of the structure of chromosomes, was important biological knowledge but didn't answer the questions at other levels. And in practice, therefore, the distinct levels of investigation recognized as the different disciplines of biology have survived in spite of the attempts to reduce it all, to dismiss it as the epiphenomenon of fundamental units.

Callebaut: Which view of life do you propose as an alternative to the reductionist picture?

Levins: What we recognize is that the causation which takes place in the hierarchies of complex systems operates in both directions. While it is certainly true that the properties of an organism depend on the makeup of its cells, it is also true that the makeup of the cells is a consequence of the evolution of the organism. And so an important field of investigation in understanding the complexity of the living world is an examination of the nature of the causation as it moves from one level to another—the relative autonomy of levels and their interconnection. But this is violating the reductionist program which seeks the fundamental unit and then reifies that into a controller for the whole of living nature.

6.4.2. Genes Cannot Replicate Themselves

Callebaut: There is more to be said about the gene.

Lewontin: Nothing illustrates more the errors of reduction and reductionism than the way in which people talk about genes. Any textbook or popular lecture on genetics will say: "The gene is a self-reproducing unit that determines a particular trait in an organism." That description of genes as self-reproducing units which determine the organism contains two fundamental biological untruths: *the gene is not self-replicating,* and *it does not determine anything.* I heard an eminent biologist at an important meeting of evolutionists say that if he had a large enough computer and could put the DNA sequence of an organism into the computer, the computer could "compute" the organism. Now that simply is not true. Organisms don't even compute themselves from their own DNA. The organism is the consequence of the unique interaction between what it has inherited and the environment in which it is developing (cf. Changeux 1985; Edelman 1988a, b), which is even more complex because the environment is itself changed in the consequence of the development of the organism.

Callebaut: Yet the dogma that the gene is self-replicating appears in every genetics textbook.

Lewontin: The fact of the matter is that no piece of DNA can replicate itself. DNA is made by a very complicated system of enzymatic machinery; namely, genes are made by proteins. It *is* true that which particular protein is made as opposed to some other protein is based on information contained in a stretch of DNA; but DNA cannot make itself. The peculiarity, then, is that in some sense, *DNA makes proteins but proteins make DNA.* What we inherit is not simply naked DNA. We inherit from our parents a complex microscopic structure of enzymes and energy-rich compounds, together with our DNA. Without those pre-existing enzymes and structures, the DNA would simply lie there inert. That is to say, the historical process of the passage of life from generation to generation is not only an unbroken passage of nucleic acid, of DNA, from one generation to generation, but the unbroken passage of the entire productive machinery of new proteins and new DNA, which *cannot* be broken.

6.4.3. *Who Is the Boss around Here?*

Callebaut: You undertake a Marxist analysis of what is going on in genetics, interpreting features of science as part and parcel of the "superstructure" and in terms of the socioeconomic "infrastructure" of Western late capitalism (Levins and Lewontin 1985).

Lewontin: There is embedded in this notion that DNA is somehow the fundamental master molecule which self-reproduces itself a broader ideological prejudice, which is the prejudice that *mental labor is superior to physical labor.* The DNA molecule is somehow the blueprint, the plan; and the physical labor is merely undifferentiated work that puts that plan to action. It's much the same as the error made when we say that Napoleon conquered Europe, or Cheops built the Great Pyramid. Cheops did not build anything; Egyptian slaves built the Great Pyramid, and French soldiers conquered Europe. But we always give the primacy to Napoleon and Cheops, because built so deeply into the ideology of our society is this notion that the plan is superior to the mere undifferentiated labor power that actually produces the object. Nowhere is that seen better in biology than in the whole metaphor of the "blueprint" and the "assembly line" of DNA and the undifferentiated labor of the enzymes.

Levins: The gene as the fundamental unit of the individual organism is paralleled to some extent in the search for the fundamental unit of ecological systems. And looking around for a justification outside of biology—looking at physics—ecologists have come up with the notion of *energy* (e.g., E. P. Odum 1971; H. T. Odum 1971, 1982). So energy has played the same role in some of the discussions about ecology. An ecosystem was reduced to a system for the transformations of energy—a flow of energy. The notion of energy as the fundamental thing to look at as the universal medium of exchange is clearly brought into biology by analogy with economic exchange, where, to the surprise of people in the early age of capitalism, objects that are physically unrelated to each other can nevertheless be transformed into each other. Lead can in fact be turned into gold through the minerals' exchange. So the notion was generalized and there was a universal medium of exchange, of value, which allows any objects and services to be transformed into each other; and that measure—the Gross National Product, or somebody's net worth—becomes a fundamental property of the system. There was a hope, therefore, that the same could be said for ecological systems; that we could ignore all the complexity of interacting species, the heterogeneity of populations, the complexities of competition and symbiosis, of mutualism and predation, and reduce everything to a single medium of ecological exchange, which was designated "energy." However, we see this is something coming not from nature but from social experience that becomes transformed into common sense.

Callebaut: Peter Taylor (1985) has made a excellent analysis of this transfer in the case of ecology in his Ph.D. thesis.[10]

Levins: This is a recurrent theme in understanding the history of science. The content of science does not come simply from nature. It is not that the observation of organisms leads us necessarily to the notion that the fundamental unit is the gene, or that looking at ecosystems means that energy is the fundamental unit. What happens is that people interpret the natural world in terms of their own social experience. And the social experience, then, dictates a particular way of looking at the world in contemporary science, which is after all a product of the bourgeois revolution. This way of looking at the world bears a striking similarity to the everyday experience of bourgeois life.

Callebaut: You extend this interpretation to the reductionistic research strategy itself.

Levins: The atomistic view, in which the whole is seen as a summation of the properties of individual parts, is parallel to the atomistic view of individuals in a capitalist economy. The notion of a universal medium of exchange is related to the experience in the market place. The search for stability in ecosystems—the talk about a "balance of nature" (Peters 1976), the evolution of stability—is a consequence of the growing preoccupation of capitalist society with law and order (see, e.g., Cook 1980). As against an earlier concern with progress in the Victorian period, which led to concern with progressive evolution, biologists have become increasingly concerned with the questions of stability and of harmony. The notion of "harmony" as a unifying principle is an older one, but it becomes particularly important at the present time when the harmony of social life is being disrupted and when the reestablishment of harmony becomes a major goal of social planning; so that it seems common sense to talk about "the return to equilibrium," "the balance of nature," "the reestablishment of harmony." In a world in which everyday experience is increasingly hierarchicalized—complex networks of bosses and sub-bosses—the way of understanding the world is to ask the question, "Who is the boss around here?" And so people have looked at ecosystems and asked, "What determines the properties of the ecosystem? Are it the herbivores, or are it the predators?"—looking again for a ruling factor.

Callebaut: How do you propose to analyze ecosystems instead?

Levins: We would suggest that the way to understand an ecosystem is neither the search for who is in charge—picking out a dominant factor—nor the *statistical kind of democracy* which says that everything contributes a certain proportion and

10. See also Salthe (1985), Eldredge (1985, 1986), Doolittle (1988), and Dyke (1988), on the need for a multilevel analysis of complex biological systems.

ANOVA tells you how much; but rather the recognition that within this complex network, the role of each component changes depending upon the others, and that an overriding influence on some processes will be located in one part of the system, on other processes in other parts of the system. So it makes more sense to look at a diffuse and fluctuating hierarchy determining the properties of systems, of systems being determined partially at their own level, and partially through their relations with other levels of analysis (cf. 4.2.2).

6.4.4. *Against Hyperreflexivity*

Callebaut: You and Lewontin have argued at various points in this discussion that the prevailing views you are criticizing seem plausible only in the context of a particular ideological situation.

Levins: The everyday experience of capitalist life generalized into universal laws; the ideology of capitalist society which becomes the common sense which indicates to investigators what are the interesting problems and how you would go about looking at them; the importance of taking something and cutting it up into pieces and assigning relative weights—a standard investigative procedure—we have attributed to a reductionist view of the world and a capitalist view of the world.

Callebaut: The common countermove against externalist explanations of the content of science is a *tu quoque* argument: the thesis of social determinism must also be determined. Bob Richards (1987:13), for one, actually believes "extreme externalism" can be "terminally infected" this way. What do you say to that?

Levins: It may be objected that our views are also very much influenced by the surroundings in which we work, by our own social environment, our own social goals, and our own political ideology; and that's certainly true. That is, the polemics around biological determinism, around how to develop an ecologically viable agriculture, about how to develop science in developing countries, these are the questions—important political questions—which influence us not only in the sense of bringing us to particular contemporary polemics, but also in guiding our thinking about the problems of nature. My own experience has been that an examination of the organism-environmental relationship, which led us to a deeper understanding of the evolutionary process, was in part motivated by the clearly distorted way in which organism and environment were counterposed to each other in debates around biological determinism.

Callebaut: Hilary and Steven Rose (1980), who are in many ways your political *compagnons de route,* in reacting to Feyerabend's analysis of science and to the Strong Programme of the Edinburgh sociologists of science, warned against an "oversocialized conception of science."

Levins: The way in which we proceed with our scientific investigation—in a sense the way in which we attain a certain transient degree of objectivity in the scientific process—is by *being aware of the social determinants of our thinking,* rather than pretending that we are value-free and operating outside of such determination. Science is not an alternative to ideology; it is one of the particular *expressions* of ideology. And one thing we can do about it is to be aware of that determination and examine what are the questions that it leads us to examine, what are the kinds of problems that it poses to us, how does it determine what constitutes a satisfactory solution? In that way, the development of a deeper understanding of nature is not only the underpinnings for ideological struggle, for struggle for a more just world, but is also a consequence of that struggle. That is, *philosophers, in trying to change the world, also come to understand it better.*

Callebaut: Karin, how does a full-time social constructivist—if I may call you that—assess this sort of Marxist analysis?

Knorr Cetina: Many of them seem to base their argument on some sort of homology, or even something stronger, between what goes on in society and in science.

Callebaut: It's almost a *reflection* (cf. 5.5.2).

Knorr Cetina: Sohn-Rethel's (1970, 1975) analysis of Galileo's concept of inertial motion is the paradigm case for that. I think some of the analyses are based on a very crude picture of natural-scientific concepts. For example, the point about atomism in physics strikes you as immediately true. But when you actually look at the conception of "particles" in particle physics, then you see that these particles are not simple point-like entities that are stable over time and then can be captured by a simple "the whole is the sum of its parts" atomism. For example, there are the particle/wave dualities; particles frequently are considered in terms of waves; particles can change into each other at higher energies; particles constantly decay into other particles . . . These are very fluid notions; it's not even clear completely what a particle is or should be. It can be many things.

Callebaut: But physics is certainly much more sophisticated than, say, Howard Odum's ecology (1971)?

Knorr Cetina: It could be; I don't know enough about ecology. But pictures that always consider the natural sciences to be "atomistic" and "nonlinear" or some such are very stereotypical and in many ways completely wrong. And their picture of society is too crude too, I believe.

Levins: But our view is not externalist nor social determinist. Rather we see science as developing at the interface of social practice and nature, and the explanation of science is internal to that larger domain. It is only when "science" is defined narrowly that questions of interests and beliefs can be construed as external. This is also the case with regard to other issues; for instance, while conservatives take an internalist (heredity) view to the nature/nurture discussion and liberals lean toward an externalist (environmentalist) view, Marxists look at the interpenetration of organism and environment and see explanation as internal to that larger whole.

6.5 THE UNITS-OF-SELECTION CONTROVERSY

Callebaut: The issue we are turning to next is certainly the biggest one in current philosophy of biology in sheer terms of investment in man- and womanpower. The problem of what the units of selection are—whether the group is the unit of selection, or the individual, or still something else, whether selection pertains to more than one sort of unit simultaneously, and related things—that problem has worried evolutionary thinkers ever since Darwin, Elliott.

Sober: Darwin and Wallace argued about this. Very often, biologists either on one side or the other of this question have come at the units of selection question in a monolithic way: they have tried to give an answer to that question which works for all populations and all examples. So people have argued that the organism is always the unit of selection, it's the only unit of selection; or that the gene is the single unit of selection in all cases.

Callebaut: In its most recent phase, we can trace this polarization from a seminal book by George Williams, *Adaptation and Natural Selection* (1966), which claimed to give proponents of group selection (most notably Wynne-Edwards 1962) short shrift.[11]

Sober: One of the most interesting conceptual parts of Williams's argument against group selection is his distinction between what he calls "group adaptations" and "fortuitous group benefits." Let me give you an example from Williams's book that brings out the distinction he had in mind. Imagine a herd of deer which is preyed upon by some predator. He says a bear, so imagine a bear that eats deer occasionally. Suppose some deer are faster and some are slower, and that being fast confers a fitness difference, because you can run away from bears more successfully if you are fast. There is a selective advantage to individuals in being fast, and the typical process of individual selection—Darwinian selection—will lead the population to be such that the fast individuals reproduce and the slow individuals do not on average reproduce as well. So over time we might expect a net increase in the average level of speed found among the organisms. At the beginning of the process you have some fast and some slow. At the end of the process, after selection has run its course, one might expect a much higher level of speed.

Callebaut: What does Williams conclude from this?

Sober: Williams's point about this example is that a group of deer might be less likely to go extinct if it is faster, because a faster herd is less likely to get all eaten up than a slow one. So it is actually good for the group that the individuals in it are fast. But Williams describes this as a *fortuitous benefit:* It is an accidental benefit that the group receives by this process; it is not a group adaptation. The reason that it is not a group adaptation for Williams is that the process was one of *individual* selection alone. The reason the average level of speed increased in the herd we are considering was that some individuals were fitter and some were less fit.

Callebaut: An *individual selection process* had the consequence of producing a fortuitous *group benefit* . . .

Sober: . . . but at no point in the process were some groups more successful than

11. "For the next decade, group selection rivaled Lamar[c]kianism as the most thoroughly repudiated idea in evolutionary theory" (D. S. Wilson 1983:159). See also Wimsatt's (1970) illuminating review of Williams's book.

"I do not think there is a defense for idealism or a coherence theory of truth."

Elliott Sober (b. Baltimore, 1948) is the Hans Reichenbach Professor of Philosophy at the University of Wisconsin-Madison (1989–).  He studied philosophy and education at the University of Pennsylvania (B.A. 1969, M.S. in Education 1970), before going to Cambridge University as a research student (1970–72) and to Harvard University, where he defended his Ph.D. thesis in 1974. He joined the Madison faculty in 1974.

His first book, based on his doctoral dissertation, was on the nature of simplicity or parsimony as it features in scientific activity in general (Sober 1975). Subsequently, however, "for reasons that are perhaps psychological and ultimately ineffable," he has taken questions at a less general level. His interest shifted to specifically biological and psychological questions. Thus the main goal of his book, *Reconstructing the Past* (1988a)—for which he received the 1991 Lakatos Award in Philosphy of Science—is to try to understand the concept of parsimony as it is used in a *certain* part of evolutionary theory, where biologists look at a set of species and try to figure out which are more closely related to which, on the basis of their similarities and differences. "Biologists will tell you that human beings and chimpanzees are more closely related to each other than either is to tigers. The question that has really embroiled scientists in this area for about twenty years is: 'What are the methods that one can reasonably use in reconstructing historical relations of that sort?' One of the methods that has been advocated by biologists is called parsimony." Sober thinks that the agenda that has been set for philosophers of science of his generation is "to try to find out to what extent generalizations can be made about the nature of explanation, of confirmation, and of scientific theories, that are *wide* generalizations. One can think of the problem as deciding *where on a certain continuum of possibilities one wishes to situate oneself,* with positivism and historicism at the extremes."

In his philosophy of biology Sober not only departs from the positivists but also from Popper. It is well known that the slogan "survival of the fittest," which Darwin adopted from Herbert Spencer, has occasioned the criticism that evolutionary theory is in fact a mere tautology—an empty definition. The criticism, which has also been voiced by a number of influential but ill-informed philosophers, including Popper, is often put in the form of a string of questions and answers: "Who survive? The fit. Who are the fit? Those who survive." In this form, Sober admits, the theory looks entirely circular and empty indeed, and therefore untestable. He agrees that the nature of fitness is itself one of the most perplexing philosophical problems about the theory of evolution, but he thinks it is a problem that has received an adequate solution: In point of fact it is a *mistake* within evolutionary theory to define fitness as actual survivorship. It is not part of the theory that any organism that survives has to be fitter than any organism that fails to survive. And the crucial way of seeing this is to realize that fitness is a *probability concept.* What is in fact true is that fitter organisms have a higher probability of surviving than less fit ones. But that is not the same as saying that the fitter ones always survive and the less fit ones always die." An example would be a characteristic in a population that enhances an organism's resistance to a disease—say, malaria. What will happen in a population to individuals with each of the two characteristics, malaria resistance and lack of it? "In saying that 'malaria resistance is the fitter of the two characteristics,' what we mean is that the chance of surviving among individuals who have the resistance is greater than the chance of surviving among those who do not. Now this does not mean that all of the malaria-resistant individuals survive and all of the malaria-vulnerable individuals die. It does not even imply that the frequency has to be greater among the former than among the latter. All that it implies is a probability."[1] Once we realize this "we see that at least one of the criticisms of the theory—as being vacuous—just rests on a mistaken understanding of fitness. The criticism says: 'Fitness, by definition, means survival.' In fact that is not what the theory says at all. What fitness means is *probability* of survival. In addition, there is a great deal more that the theory of natural selection tells us about fitness than simply the remark that fitness is probability of surviving and reproducing. It also gives us a wealth of information about *why* individuals are fitter or less fit."

In addition to the aforementioned books and about a hundred papers, Sober has written *The Nature of Selection: Evolutionary Theory in Philosophical Focus* (1984a); *Core Questions in Philosophy* (1990), a book which consists of introductory texts accompanied by readings on the philosophy of religion, epistemology, philosophy of mind, and ethics; and *Philosophy of Biology* (1991a). He has edited an anthology, *Conceptual Issues in Evolutionary Biology* (1984b), which, despite Sober's wish, expressed in the preface, that it may "soon be rendered obsolete," continues to be used widely by both biologists and philosophers of science.

The original interview with Elliott was recorded for Belgian Public Radio in a cool and dark WHA 970 Wisconsin Public Radio station on the hot Spring day of June 7, 1985.

---

1. In distinguishing probability from actual frequency, Sober makes a move that is standard in discussions of chance: "Consider a fair coin that is tossed exactly three times. Since it is fair, its probability of landing heads when it is tossed is 0.5. But because I have chosen to toss the coin an odd number of times, there is no way the actual frequency of heads in the run of trials will be 0.5. So in this case probability is not identified with actual frequency."

other groups, which is what it would take for there to be a group selection process. So Williams's starting point was this: When we look at nature and notice characteristics that are good for the group that have them, we shouldn't conclude just from this that these are group adaptations. We can't conclude that, because they might not have been produced by a process of group selection. Whenever we see a group benefit, we have to ask the question, "Is this really a group adaptation or is it simply a fortuitous group benefit?" And the burden of Williams's book and of subsequent pieces of work by people similarly oriented was to show that group benefits in nature are almost always fortuitous, that they are not the product of group selection.

6.5.1. *Causal Analysis of the Units of Selection*

Callebaut: The view of the units of selection that you yourself develop in your book (Sober 1984a) is not "monolithic" (as you call Williams's) but hierarchical; it is in fact a *multilevel view* (cf. also Richardson 1982; Vrba and Eldredge 1984; Vrba and Gould 1986; Grene 1987, 1988; Doolittle 1988). How would you summarize it?

Sober: First of all, my view incorporates the idea that in any given context, there may be many units of selection operating at once. In a given biological context it might not be simply the genes, or simply the organisms, or simply the groups, but all three or only some of them. And in fact it can vary from context to context what the unit of selection really is; it will depend upon the specific biological facts of the example what the units of selection are in that case.

Callebaut: The main tool you are using to develop your multilevel view seems to be an analysis of *causality.* Could you illustrate this?

Sober: In Williams's deer example, the reason there was no group selection is to be given in terms of the kind of causal process that produced the change. We look at the selection process and ask ourselves, "What characteristic was selected for?" The answer was: "Individual deer were selected for being fast." Notice that the characteristic that caused individuals to benefit was a characteristic simply having to do with their own physical structure. . .

Callebaut: . . . that is, when we disregard that *relative speed*—as opposed to absolute speed—is what is really at stake here (cf. also Wimsatt on fitness being a relational property at 4.2.1).

Sober: There was no selection for belonging to a group of a certain kind. If the causal factor that had produced the evolution in that case had been "belonging to a group in which individuals are fast," *that* would have been group selection. But the way Williams tells the story, that is not the cause, and so the answer is not group selection but individual selection. So the main entry that I take into the problem is to look for the causal processes that are at work.

Callebaut: How does one find out about them?

Sober: By asking which characteristics are selected for and then analyzing the kind of causal processes. One will typically come up with the analysis that it is not a single causal process that is at work but that there are several causal processes under way simultaneously in most cases of natural selection; the reason being that there are many characteristics, at different levels, that influence the survival and reproduction of organisms.

Callebaut: And your conclusion is that one should not get a single answer to the question, "What is *the* unit of selection?"

Sober: There is no such thing as *the* unit of selection.

Callebaut: This is very much in line with what Lewontin—with whom you have spent time working at Harvard—told me: that the chief philosophical point one has to understand in discussing the relationship between individuals and groups is the direction of causation. If individuals were prior to social organization and if social organization was to be seen as a (one-way) consequence of individual behaviors and properties, Lewontin went on, the classic and orthodox evolutionary view that evolution consists in the selection of individuals that have traits that leave more offspring would be a sufficient explanation of the organizational properties of collectivities, of species, of populations, of societies. But if, to the contrary, the individual properties that are manifested are a *consequence* of the collection of individuals into collectivities—Levins's and Lewontin's own view—then the Darwinian view of selection of individual traits will be insufficient. Dick, you have some qualms about Elliott's analysis.

Burian: If Elliott's causal analysis were really causal, I'd be more persuaded by it. It pretends to be causal. The problem, at one level, is that we genuinely don't have a good account of causation (cf. 4.1). So this is not to fault Elliott strictly; but I don't think his account of the causal role of units of selection is at all right. Fundamentally, the reason is this (and there's a way of parsing it in the paper by Mayo and Gilinsky [1987] on units of selection): *If* groups are to be units of selection, then group interactions must cause differential group reproduction and survival. And if they cause differential *individual* reproduction and survival, one would have a special case of context-dependent variation in *individual* fitness. Until we can straighten out what counts as the proper way of separating causal interactions at a group level from causal interactions which impinge on individuals but involve complicated contexts, we haven't got the job done. And Elliott, I think, fails to make the necessary distinctions to get that particular job done. To that extent, I don't think he has reached the end of the road for solving these kinds of problems. It's a problem which Marjorie Grene, Norman Gilinsky, and I right now are working on.

6.5.2. *Additive and Nonadditive Variance in Fitness*

Callebaut: Bill, one way in which your approach to the units of selection problem departs from Elliott's causal analysis is that you stress the importance of the additivity of variance in fitness. What is meant by "variance" here?

Wimsatt: It is simply the statistical notion of variance, but additive variance is important because of the role it plays in what Wright used to call "Fisher's not so fundamental theorem of natural selection," which in one of many formulations says that the rate of evolutionary change is proportional to the additive variance in fitness (Fisher 1930).

Callebaut: You define a unit of selection as "any entity for which there is heritable context-independent variance in fitness among entities at that level which does not appear as heritable context-independent variance in fitness (and thus, for which the variance in fitness is context-dependent) at any lower level of organization" (Wimsatt 1981a:144). You shouldn't expect our average reader to understand this mouthful!

Wimsatt: Actually, as I argue in my paper on robustness and reliability (1981b), there are close connections between additivity, heritability, and context-independence. They turn out to be different criteria for the same thing, which could be called heritability if that weren't already the name for a genetic statistic which doesn't mean exactly the same thing. The question is this: What is the correlation between fitness in parents and offspring? If a gene is one of a number of alleles at different loci, each of which make additive contributions to fitness, its contribution to fitness will be the same against all genetic backgrounds at those loci, since it will always add its own increment (or decrement) to whatever fitness contributions the others make—there will be no changes in sign or magnitude of its contribution depending on what other genes are present. In this sense, its contribution to fitness is independent of the genetic context (again at those loci).

Callebaut: Why should this be important?

Wimsatt: Well, with diploidy, sex, segregation, and recombination the genetic contexts of individual genes are constantly changing, so saying that their fitness contribution is independent of genetic context is to say that it is invariant over those contexts, and will be inherited with that gene. The more common case—immeasurably so—is for epistatic fitness interactions, where the fitness contribution of a given gene depends upon which particular combination of genes are found at other loci, sometimes being an increment, sometimes a decrement of different magnitudes. In this case, additivity, heritability, and context-independence fall together. The gene may be inherited perfectly well, but in a new genetic context, there may be no correlation between its fitness contribution in the parent or parents and in the offspring.

Callebaut: And what is the crux of your own contribution?

Wimsatt: The thing that I realized was that these components of fitness which were epistatic/context-dependent/not heritable at a given level of aggregation of the genome might, if larger contexts were fused, produce components of variance in fitness, which would be additive, context-independent, or heritable at higher levels, although still epistatic at the lower levels. There is no necessity for them to be so (or even for there to exist meaningful higher-level aggregations), but if there were, there would be additive variance in fitness at the higher level which could determine the rate of evolutionary change according to an analogue of Fisher's fundamental theorem operating at that level. This would be a component of fitness which could not drive evolution at the lower level since it was not additive or heritable at that level.

Callebaut: Sober (1984a:317) has criticized your ANOVA for "not [being] sensitive to the sort of counterfactual considerations to which an analysis of causation must attend." Even if it were reformulated so as to represent fitness values that would obtain in certain counterfactual circumstances, he goes on, the ANOVA "would still be neither necessary nor sufficient as an analysis of causation." A reaction?

Wimsatt: There are several differences in our analyses which don't make a difference. Elliott allows me counterfactuals and subjunctive conditionals but thinks it has nothing to do with causation? That's strange, since most other philosophers do. His original "counterexample" only satisfied one but not both of my conditions, as I explained in my units of selection paper (1981a). Furthermore, it's not intended as an analysis of causation per se, but of a unit of selection. I can imagine a situation where we would say that selection was not occurring at the higher level, although in that circumstance, given my analysis, we would have to say that we had drift-mimicking higher-level selection, just as it can mimic lower-level selection. But most importantly, it was intended to provide an operational criterion for determining when selection was acting at a given level of organization, as opposed to operating only at lower levels, and the longer drift persists in the "right" direction, the less likely it seems that it is not causation. Elliott wants to imagine that we are already given the causal forces at the upper level. If so, fine, then upper-level selection will meet my criterion. And if not, then the upper-level forces will not have succeeded in producing any change, but will have cancelled in some way diagnoseable with my criterion. I think Jim Griesemer (Griesemer and Wade 1988) and Lisa Lloyd (1988, 1989) have also clarified things considerably, including in my own mind—Jim by pointing out that Elliott and I were trying to accomplish different things and thus managed to talk past each other, and Lisa by showing that indeed and after all my (or our—she has done a lot to clarify and extend it) analysis produced the same results as Elliott's in any case where either analysis permitted an answer—and that Elliott's analysis was mute in the face of his own counterexamples, just as ours was.

6.5.3. *Words about Models*

Callebaut: Lisa, you have criticized Elliott's treatment of units of selection at several occasions (Lloyd 1986a, b, 1988, 1989), and vindicated Bill's approach instead. You have specifically blamed Elliott for conflating the question of the *theoretical definition* of units of selection with the question of how to determine the *empirical accuracy* of claims, thus throwing out the baby with the bathwater. You suggested that by confounding the theoretical description of a unit of selection with the *testability* of that description, Sober has in fact rejected the structural and conceptual underpinnings of his own definition. In retrospect, your judgment has become somewhat milder, I think.

Lloyd: In spite of my criticism, I think Elliott was onto something very serious and major when he was trying to analyze the causal structure and the representation of the units of selection as causes. Other people were onto some of these issues, but he did help clarify some of the structure of that problem.

Callebaut: Please spell out your diagnosis.

Lloyd: The units of selection issue is a mess! If philosophers can contribute anything at all, it ought to be conceptual clarification. A lot of people have contributed to this clarification: Brandon, Hull, Sober, Burian, Sandy Mitchell . . . This is one of the things we can do. I'll tell you what my problem was and give Elliott the credit he deserves. Elliott, I think, was not using a quantitative approach to look at the structure of the theory. As a result, he missed what was going on with the biologists. I understand why he missed it: they weren't exactly clear about what they were doing when they *talked* about it. But when they *modeled* it they were clear. Michael Wade (1976, 1977, 1978) was clear.[12] Wimsatt looked at the models; Sober looked at the words. That's my diagnosis. I mean, I made this up, but I think this is what went astray.

Callebaut: Please be more specific.

Lloyd: Elliott's criticisms of the use of ANOVA in those models was not appropriate to the way the biologists were doing their science. Elliott's criticisms are generally correct—*if* that is what the biologists were doing. But what I got worked up about at the time is that that is *not* what the biologists were doing. Wimsatt saw what the biologists were doing and tried to represent it in a way which I thought basically got it right, although it was not as clear as it could be.

Callebaut: So you went in and tried to bolster Wimsatt's intuition about what was going on in the actual models.

Lloyd: Now Elliott didn't miss what should be going on. Elliott *recreated* what should have been going on and then said that the biologists weren't doing it. He is not necessarily wrong in his actual conclusions; the only point on which I actually disagree with him is the claim that that is not what the biologists were doing, and that the biologists were completely missing the boat.

Callebaut: Did it take the semantic view for you to see that?

Lloyd: A formal approach does not give it to you automatically. It is a mistake to think the semantic view is a machine that you grind out results with. I have to think, I have to read; the semantic view gives me a tool to use. My quest in

---

12. M. Wade (1978) criticized the then current group selection models for making assumptions (e.g., of a single locus state space, or that the variation between groups arises from either drift or the founder effect) that weakened the effectiveness of group selection. He proposed models representing selection on quantitative characters and assuming a polygenic mode of inheritance instead. Wade and his associates D. M. Craig and D. E. McCauley claim to have demonstrated group selection in the laboratory.

describing theory structure in the context of evolutionary theory is to see evolutionary theory as a cluster of interrelated, linked types of models, rather than as an axiomatic system or set of statements of some kind. The latter has been Kitcher's approach, which was very linguistically oriented, although he has been moving away from that. What I try to do is describe those models, using certain sets of descriptors.

Callebaut: I beg your pardon?

Lloyd: Types of descriptions. You describe the variables of the models—what kinds of state space the models are in (cf. 5.5.1). In genetics, it is often genotype or gene frequencies. Then you describe the parameters of the models, where other information is put in as fixed, like mutation rate or migration rate. And finally you describe the laws—not Laws—of nature: equations by which the system changes; or a co-existence type of law which says "This can happen and that can't happen." It can also be a law of transformation, where you have a system changing over time and you just describe its trajectory over time. Using those basic descriptors I went through evolutionary theory and abstracted out what the commitments were and what the actual models looked like. And I tried to describe a number of different model types that are used in evolutionary theory.

Callebaut: You have to choose at what level of abstraction you are going to work.

Lloyd: There are different ways that you can describe model types, different axes of abstraction that you can use. One way is to say, "A model type is all of the genotype frequency models versus all of the gene frequency models"—that's a different state space, a different set of variables. Another way is to look at stochastic versus deterministic models: where is the stochasticity appearing in the model? Yet another way is to look at structured population models. Sewall Wright's influence, especially in the U.S., has led to the development of some very sophisticated population genetical models about structured populations (Wright 1986). Many of the models now in population genetics (Lewontin 1974a, 1985; Roughgarden 1979) are structured population models, where you have an additional set of descriptors which says what the groupings of the different entities are. That stuff doesn't appear in basic Biology 101, which is limited to Fisherian type population genetics models and selection models. All of those different types of models are important components in evolutionary biology today, and this set of descriptors allows me and anybody else to go in and analyze and describe in a very precise way how the models relate to one another and how they differ. For instance, in optimality theory and in many quantitative genetics models a (fitness) parameter links an entire set of models to another entire set of models. It's that structure that I wanted to analyze in my book.

Callebaut: What came out, for instance?

Lloyd: I looked at the species selection debate and asked, "What kinds of models are these? How do they relate to group selection models? How do they relate to individual selection models?" I looked at genic selection debates and discovered a very strong difference between Williams's (1966) empirical claim about the applicability of gene-level models and Dawkins's (1982a) purely conceptual point. This isn't something that was immediately obvious to people looking at genic selection. I actually believe that philosophers have been easily misled by the use of specific terms. The basic idea of the semantic view is that you analyze the *meaning structure* of what is being presented; it's supposed to be a very basic kind of description. Through this I try to sort out various issues, like "What is the relation between kin selection and group selection? What are the problems with genic selection models? What can they represent? What can they not represent? What is the relation of species selection to individual selection models?"

Callebaut: And you offer a definition of a unit of selection . . .

Lloyd: . . . which is meant to be simply a general description of what the biologists are doing in their population genetics models. The intention is to get away from the language problems that have come into that debate—naming things that are very much abstracted from the actual mathematical models—and instead to look at the models, see why biologists think they are group selection or individual selection models, and see what a unit of selection actually is, structurally, in those models.

6.5.4. *The Evolutionists' Boston Tea Party*

Callebaut: What we have here is, on one level, a debate among philosophers of biology; on another, a debate among evolutionary biologists. You know both groups of people equally well. Now if you're talking to the scientists rather than to the philosophers, what do you say? For instance, what would you say when they were to ask *you* how important you think group selection is?

Lloyd: My basic answer to that question is that what Dawkins and Williams and Maynard Smith mean by group selection is different from what Wade means by it. This is the philosophical contribution. I don't know the answer empirically, and I shouldn't, because (*ironical now*) I'm only a philosopher! Now the answer to "How important is group selection?" What a philosopher can do, and what I've done, is to go in and say "When they call something group selection and reject group selection, that is because they have *this* in their minds!"

Callebaut: In your book (Lloyd 1988), you dismiss people who reject the efficacy of Wade-type group selection (see also Walton 1991).

Lloyd: In the book I had a particular focus. A new paper (Lloyd n.d.b) basically says that what happened was a sort of mistaken communication. I have a relatively sociological analysis of the situation in which Maynard Smith, Williams, and Dawkins became fixated on asking one type of question, about group-level or species-level *adaptation*. They reject group selection as being efficacious because they are looking only for a group-level adaptation by group selection.

Callebaut: Maynard Smith (e.g., 1976) also reacts to David Wilson's views on group selection.[13]

Lloyd: I know. Now that is not what Wade means by group selection. It may be what David Wilson means, but it was never what Wade or Sewall Wright meant.[14] This is a complete mischaracterization of the Americans, who have a broader question about the evolutionary process in general. Claiming that group adaptation by group selection is group selection is a misnomer. That's not group selection; that's *group adaptation by group selection*. Group selection as an *activity* is something Wright, Wade, and the whole Wright school—Uyenoyama, also Feldman—were modeling. The self-righteous sorts of claims of Maynard Smith (e.g. 1978) . . . Mind you, he has taken it back; he admitted that he had muddied the waters on the group selection debate by sticking to a definition that was not defensible (Maynard Smith 1987a).

Callebaut: Can you rest on your oars, now?

Lloyd: No! The distinction has not made its way into the general consciousness in the evolutionary community; just no way. I mean, even if Maynard Smith recognized that he had gotten group selection wrong, he did not generalize it to species selection. As a matter of fact, Dawkins did. He has a small paper which is basically on macroevolution by group-level properties at the species level (Dawkins 1989). This is funny, because it is very much like the paper that I just finished with Steve Gould on this general subject (Lloyd and Gould n.d.). Anyway, when you get into the words—philosophers tend to get too hung up with the words—and don't do the conceptual work on the models, you can run into a lot of trouble. Elliott, Robert Brandon, Dick Burian, Lewontin—they've all done really important work in looking at the concept of adaptation. But I didn't see in any of it this idea that there are all these different notions of adaptation going on up there.

13. In *intrademic group selection* (IGS) or "structured deme" models of group selection, as identified and advocated by D. S. Wilson (e.g. 1977, 1980, 1983), subgroups that vary in their genetic composition exist within a (single) population. Interactions among individual organisms in these subgroups result in viability differences and differential contribution of the parts to the mating pool. The subgroups usually dissolve after selection. Groups are described on the basis of density or allele frequency, i.e., group-level traits; group-level fitnesses are calculated from these traits. There must be heritable phenotypic variance in group fitness. Rita Colwell and Wilson (Colwell 1981; Wilson and Colwell 1981) claim that biased sex ratios illustrate the operation of group selection in nature.

14. Wright's *shifting balance theory* (Wright 1984; Provine 1986; also Schull 1990) exemplifies the traditional approach to groups. Here gene frequencies are altered by differential growth and expansion of partially isolated local populations that last for a number of generations. The primary difference between Wright's "interdemic" selection models and classical, Fisherian models is that species are not panmictic but subdivided into numerous local populations, "among which diffusion is sufficiently restricted to permit much differentiation, although not so much as to prevent spreading of a superior combination of genes" (Wright 1980, quoted in Lloyd 1988:51). Demes that contain particularly adaptive combinations of genes are favored by group selection. The continued operation of the shifting balance process, which involves a joint action of all evolutionary forces (mutation and immigration pressures, mass selection, random drift of all sorts, and interdemic selection), is the principal basis of evolution.

Callebaut: And these different notions have particular consequences for the notion of selection level.

Lloyd: Elliott actually points this out in a specific example in his book, where he's looking at Williams's treatment of Waddington, which I found really beautiful and clear (Sober 1984a:199f.).

Callebaut: In your new paper (Lloyd n.d.b), you try to be more even-handed about this whole issue than in the book. Now you say, "Look, there are different questions they want to ask."

Lloyd: "Group selection" can mean any one of four different things, at least—and any combination of those four also! I divide questions about units of selection as follows: (1) Is it an *interactor,* that is, an entity that has a trait and interacts with its environment via the trait in such a way that the entity's expected survival and reproductive success is determined partly by this interaction? (Cf. Hull 1988d.) I include inheritance in that. (2) Is it a *replicator* in Hull's sense—"an entity that passes on its structure largely intact through successive replications"? (3) Is it the *owner of an adaptation* (which entity does the adapting or has an adaptation)? And (4) is it a (long-term) *beneficiary* of evolution by selection, as, say, the surviving alleles are in Dawkins's view? Of course, as far as I can tell, molecules are the only long-term beneficiaries. So he comes up with a particular answer. Now you can combine those questions. Maynard Smith (1976) combines the questions "Is a group an interactor?" and "Is it an owner of adaptation?" together; that is one question to him. Wade's question is about interactors, not about owners of adaptation. Dawkins's main question is about a beneficiary; everything else gets redefined according to that. If it is a beneficiary, it can be the owner of an adaptation, and it is the replicator. Interactors don't appear; that's what he takes pride in, and I think that's what the deadly flaw in his theory is! Because he does not answer the interactor question. So where is the overlap between Dawkins and Wade? *There isn't any!*

Callebaut: So you go to Dawkins . . .

Lloyd: . . . and I say, "There is no contact being made. You have not addressed any of the questions that have been answered by the Americans; therefore you should not say anything about it!" I use this taxonomy to divide up the camps. And I predict that certain disagreements will appear, based on the camp. I think it makes everybody look more sensible than they would otherwise, and less wrong, because if you tackle different questions, of course you can get different answers. That, I would say, is work that a philosopher can do: conceptual work about the content of a theory which can have possibly normative consequences.

Callebaut: Dick, you have qualms once again.

Burian: At this point people like Lisa Lloyd and Elisabeth Vrba—it isn't just philosophers by any means—are trying to substitute ANOVA measures of statistical correlation for causal claims. Until we have straightened out how to separate causal interactions at the higher level, that is, among higher-level entities, from the influence of groupings of various kinds on lower-level individuals, we won't have it right. Lisa Lloyd's draft of a paper with Steve Gould begs all of the interesting questions in this way: Is variability a group-level feature? Which can count as causally relevant differences among groups in the right way? Those are the sorts of questions that have to be answered to solve the units and levels of selection controversies. They aren't done yet. There is fancy mathematics for some of it, there are low-level conceptual problems for some of it, but none of them come out clear.

6.5.5. *The Importance of the Replicator/Interactor Distinction*

Brandon: The more I think about it, the more convinced I am that the replicator/interactor distinction is of tremendous importance; and if you want to understand these issues having to do with the controversies over levels of selection, you have to consistently make that distinction (Brandon 1985c, 1988, 1990). And a lot of people, including Bill Wimsatt, fail to do that. I've explained this to Bill; I think he agrees with me. He hasn't written anything about this recently, but he claimed to agree with me. I am not sure.

Callebaut: What is your problem with Bill's treatment of units of selection, Bob?

Brandon: My problem with it is that in some cases it will count replicators as the unit of selection and in other cases it will count interactors as a unit of selection. So it is a hierarchy that includes both replicators and interactors. But they are quite different sorts of entities (Brandon 1985c). They do different things in the evolutionary process. I think that is a big mistake. His discussion of unit of selection—and I think this is true of the treatment by a number of other people too—often draws on considerations that are appropriate to interactors and sometimes draws on considerations that are appropriate to replicators, and he conflates the two. I think that is a serious flaw in his approach.

Callebaut: You distinguish between "levels" and "units" of selection (Brandon 1982, 1988). You think selection can occur at a certain level only if that level is the level of an interactor.

Brandon: Selection is essentially a process that involves interaction of something with its environment, so *by definition* it has got to be an interactor. But in standard cases of individual selection, Wimsatt's conclusion is that the genome—not the organism but the genome—is the unit of selection. But we've changed the assumptions. And he is led to that conclusion by certain genetic facts: epistasis, gene linkage, and so on. But if you have a system where the genetic facts are different—say, no gene linkages, no epistasis, no interactive effects of any sort—then his definition of unit of selection would say, "The unit of selection is the individual gene," even though the selection was in fact occurring at the level of the individual phenotype. I think that is a serious flaw with that approach. The same is true of Sober's approach, which I think is more flawed. Wimsatt's got a lot of good ideas in his approach, and I think it just needs to be modified in ways that require it to apply to the level of interactors. Sober's, I think, also can place the two, but has as some other problems as well.

6.5.6. Provincials?

Callebaut: There have been a number of claims that evolutionary theory is different in essential structure from physical theory. You profoundly disagree, Lisa.

Lloyd: Van Fraassen and others used the semantic view to analyze the structure of physical theories; I use the semantic view to analyze the structure of evolutionary theory. The kinds of descriptions that we end up with are not different in kind. One can say, "That is because you're using the same tool." I don't believe that. Somebody would need to show me that I'm missing something essential in evolutionary theory to substantiate that kind of claim. I think this shows that the scientific theories in these different domains have some very basic, "essential" similarities.

Callebaut: So you are saying that when looked at by someone who is familiar enough with the science . . .

Lloyd: . . . these differences do not appear! So if there are general philosophical claims I would like to make that I think follow from the work I did in my book— and this is to be complemented also by the work of Paul Thompson, John Beatty, Jim Griesemer, and other people who have worked on the semantic view—it is, first, that biological theory is not different in kind from physical theory; second, that biological explanations are not different in kind from physical explanations. And the similarity in the actual structures is extremely strong. Opponents need to do some work to support the claim that biological theory is different in kind. And third, philosophers need to look very, very closely at the actual content of the theory, and not just make it up or ignore it in order to get a full-fledged kind of picture of what is going on!

6.6 WHAT EVOLVES? RECONSIDERING THE METAPHYSICS OF EVOLUTION

Callebaut: There are people like Bruno Latour who dislike evolutionary epistemology ("It's no big deal"), but do think there are many interesting things to be found out about *evolutionary ontology,* to which the units of selection debate could be said to belong (cf. Griesemer 1984). David, for a decade and a half now, you have been arguing that the "metaphysics of evolution" ought to be reformulated (Hull 1976a, 1978, 1981, 1989). What do you mean by that?

Hull: From Darwin to the present, there has always been something jarring about biological evolution. I do not mean religious objections and all these other kinds of objections, but people who were really serious about understanding evolution kept having problems conceptualizing it. The picture kept flipping, and that is because there was a fundamental tension that people hadn't noticed. In the West, from the beginning of time, basic to the way of viewing the world of philosophers-scientists was a distinction between classes or "natural kinds" and individuals or particulars.

6.6.1. *Don't Turn Your Gold Ring into Lead!*

Callebaut: What do you mean by "natural kinds" in this context?

Hull: Natural kinds are the things that are eternal and immutable, preferably with sharp boundaries. Eternal, that means like God: they always were and always will be. It does *not* mean that they are always exemplified. But given the nature of subatomic particles, when we work out that structure, the periodic table is going to emerge, even though some of the elements do not exist, or do not exist for very long. That means that they are always there, in the sense that there is always a slot for them; but they may or may not be exemplified at any one time. When I say that a natural kind is eternal, I mean it in that sense. The first few seconds after the Big Bang there was no gold, there were no heavy elements at all. But later on, in various parts of the universe, they came into existence. And they were gold, independent of their histories. Just their structure tells you that this is gold. Those are the sorts of terms that function in laws of nature. When you want to find eternal, immutable regularities, you have got to refer to natural kinds; because the things that you are talking about are the things that are eternal and immutable.

Callebaut: What about exemplifications, instances of natural kinds?

Hull: Well, an instance of gold is—you have got a gold ring on?

Callebaut: No! (*Laughter*)

Hull: If you did, it would be an instance of gold. But we know you can take that instance and you can melt it down and merge it with others. There are all sorts of things you can do with instances of gold. These particulars or individuals we tend to take for granted. They are not important. They come and go; they are transitory; you do not win a Nobel Prize for discovering a new instance. Scientists tend to concentrate on natural kinds and immutable regularities.

Callebaut: Could you be more concrete?

Hull: The three most common examples of natural kinds from the Greeks to the present are geometric figures, biological species, and the physical elements. Geometric figures: why is triangularity eternal? Well, given the basic axioms of geometry, there are always slots—there are always going to be triangles, squares, etc. Why are they immutable? Well, what would it be like for triangularity to evolve into rectilinearity? I do not know what that means. I can take a wire triangle and reshape it into a square. But that is just an instance. And when you are really talking about the fundamental makeup of the universe, "one part changing into another" just doesn't make sense. It's the same with the physical elements. I can transmute lead into gold. Alchemists were always trying. It turns out you *can* do it, it just costs more than mining it. But you could take a sample of lead and, with

"The distinction between science and philosophy is not all that good!"

David L. Hull (b. Burnside, Ilinois, 1935) never tells the person sitting next to him on the plane that he is a philosopher, because he does not like the questions about the Meaning of Life (Hull 1988c:34). Nevertheless he is Dressler Professor in the Humanities at Northwestern University, which he joined in 1984. He studied philosophy at Illinois Wesleyan University (B.A.) and got his Ph.D.

Larry Lapidus

in the Department of History and Philosophy of Science at Indiana University in 1964. Before going to Northwestern he taught at the University of Wisconsin-Milwaukee for twenty years, and was a visiting professor at Indiana University, the University of Chicago, the University of Illinois at Chicago Circle, and UCLA.

When he was 21, Jean Piaget (some of whose ideas Hull, probably under Toulmin's influence, does not like, but that is not the issue here), wrote his Rousseauesque novel *Recherche,* in which all the major themes of his mature work are foreshadowed (Piaget 1918). To my knowledge Hull has never written a novel—he paints. But if one is looking for a capsule formulation of his philosophical research program, the title of the second paper he published, "The effects of essentialism on taxonomy: Two thousand years of stasis" (1965), says it all. Hull feels strongly that notwithstanding the "Darwinian revolution," the philosophical, nay, metaphysical *portée* of Darwin's discovery is still often incompletely and wrongly understood, particularly so in the humanities and the social sciences. Platonism also survives in the "natural kinds" scientists must posit in order to formulate laws; if it is not relegated to its appropriate station, it may block our inquiry into evolving processes. For almost three decades now, Hull has been trying to drive home that message to biologists, social scientists, and philosophers alike. He is unique in having been the President of the Society for Systematic Zoology as well as of the PSA and the recently founded HPSSB. (He is also a Fellow of the AAAS.)

Says Hull: "People believed that species evolve and they viewed them as natural kinds, when the two claims are incompatible. And they did not really see the incompatibility; because by the time of Darwin, science and philosophy had begun to diverge. Philosophers knew what natural kinds were supposed to be, but did not understand evolution very well, whereas biologists understood evolution fairly well, but of course they did not have the faintest idea about what natural kinds are." This contrast became even more marked as the chasm between philosophy and science grew. Not until fairly recently have evolutionary

biologists asked what it means to say that *species evolve.* "What have they got to be like if they evolve? They cannot be natural kinds. The metaphysical category that has the appropriate characteristic is called 'individual' or 'particular'. Individuals are the kind of things that come and go; they split, merge, cease to exist, etc. Species belong in a different metaphysical category than classes. First Mike Ghiselin, then me, and now most of the philosophers of biology have come to see the point of this distinction.[1] One reason why evolutionary theory seems so peculiar is that one of the chief notions—species—has been viewed inappropriately. *Science as a Process* (1988a) extends this view to the metalevel of science itself.

Ernst Mayr had been talking about "population thinking" for a long time (1959, 1983), and Darwin was being read through Mayr's spectacles.[2] Hull considers him an ally: "Mayr came at it from a different direction. He said: 'You people, you logicians' (next to a philosopher, the worst thing Mayr can call you is a logician) 'you think essentialistically; you think that all the members of a species are fundamentally the same. No! The whole point of evolution is: *each one is different.*[3] And these differences cannot be explained by simple logical, statistical terms like means and modes. It is a very peculiar sort of difference.' Mayr dubbed this way of thinking 'population thinking'. Although he has some reservations about terming species 'individuals' (Mayr 1988:349f.), the distinction he was making was pretty much the same distinction that Ghiselin was trying to make in his early writings. Different words, the same distinctions."

Hull has published extensively on taxonomy, evolutionary biology, and science studies (he was involved in the empirical refutation of "Planck's principle," according to which paradigms only disappear when the old guys die, to put it bluntly [Hull et al. 1978]). His *Darwin and his Critics* (1973) and *Philosophy of Biological Science* (1974) are musts for anyone with a general interest in the philosophy of biology.

The original interview was taped in David's house in Chicago on June 5, 1985, some time after I had met him at the occasion of a talk (on Piaget, I am afraid) he and Micky Forbes had organized for me at Northwestern.

---

1. Ghiselin (1974b, 1981); Hull (1976a, 1978, 1981, and 1989); Brandon and Burian (1984); Sober (1984b). Ruse is one dissenter; he believes that "the doctrine of species as individuals is bad logic, bad biology, and bad philosophy" (1986c:146), and even sees vitalism raising its ugly head again.

2. Cf. Grene (1990:237): "It is an article of faith in evolutionary biology that Darwin shifted biological theory and biological research from a static, evil, 'metaphysical' kind of thinking called 'essentialism' or 'typology' to a dynamic, good, scientific conceptual style called 'population thinking'." See also Sober (1984a) on "natural state" explanations.

3. Popper (1976:17f.) conceives of 'essentialism' and 'antiessentialism' rather differently.

tons of money, transmute it into a sample of gold. Lead and gold are untouched; you have not transmuted lead *as a natural kind* into gold *as a natural kind.*

Callebaut: How does this spill over to biology?

Hull: People a hundred and fifty years ago viewed biological species in the same way: they were eternal and immutable. They did not mean you would always have organisms exemplifying them. You could have all dodos die out. But *they could come back into existence.* On this view, conservationists are worrying about something that is unnecessary! On this view conservationists are treating extinction as if it were forever. Well, it is. But Darwin's theory of evolution did not just question an empirical belief that people had—that species did not evolve. It questioned the fundamental way they viewed species. They viewed them as natural kinds. *If* they are natural kinds, they *cannot* evolve by definition. It is not just a contingent claim. If they evolve, they cannot be natural kinds.

Callebaut: You now characterize species as "quasi individuals."

Hull: If one makes the notion of a class or a set broad enough, then everything counts as a set, including Mount Rushmore, but if the notion of a class is given any determinant meaning at all, species cannot count as classes. They have none of the relevant characteristics. But I am not so sure that the only alternative to classes are individuals (cf. Guyot 1987; Sober 1991b:291). There are various sorts of individuals, and organisms are the commonest sort.[15] In fact, usually when a biologist says "individual," he means "organism." But organisms are special sorts of individuals. For example, they have a program that in an important sense determines their possible life histories, their reaction norm (that is, given a particular genome, a range of phenomena can result depending on the environment).[16]

Callebaut: How about other things that are equally individuals without having such a program for change?

Hull: There are other things that have all the characteristics of individuals but can change indefinitely, and they do not have a program for their change. I think species are like that: there is no program; whatever can count as a program for a species, it changes each generation. So there is no long-term sequence in anything that can count as a program for a species; and species can change indefinitely.

15. "If species are individuals, then: (1) their names are proper, (2) there cannot be instances of them, (3) they do not have defining properties (intensions), (4) their constituent organisms are parts, not members" (Ghiselin 1974b:536). Against the Ghiselin-Hull view, Caplan (1981) points out that "reification and the multiplication of entities beyond necessity have costs, too."

16. There are drastic differences among species concerning the precision of the inherited information (contained in the DNA) and the extent to which the individual can benefit from experience. Following Mayr ("The evolution of living systems" [1964], in Mayr 1976:17–25), the behavior program of an organism whose behavior is unlearned, innate, instinctive is called "closed," that of an organism with a great capacity to benefit from experience and to learn, i.e., to add information to its behavior program, is called "open." As to reaction norms, sometimes "the claim that an organism that lacks a trait nevertheless possesses the capacity for such a trait makes sense"; such organisms "have what it takes" to exhibit any one of various "character states" depending on the environments they confront (Hull, "On human nature" [1987], in Hull 1989:17). Reaction norms are further discussed in 9.2.3.

Most of them sooner or later go extinct; but they change. Those are the respects in which they differ from organisms. But there are individual atoms—individuals that have even fewer of the characteristics of organisms—"individual" being used here in a generic sense. Because there are so many different kinds of individuals, it is important to know what kind you are talking about.

Callebaut: As an aside, I must admit it took me quite a while before I realized that my own attempt, in my first publication (Apostel and Callebaut 1978), to "temporalize" the class notion in order to apply it in demography, was bound to fail. In another line of your work, you have extended this sort of analysis from the biological to the realm of socioculture (Hull 1982, 1983, 1985a, 1987, 1988a, b, d, 1989:pts. 5 and 6).

Hull: When we switch over to sociocultural change, I think societies, social groups in general, are more like species than like organisms. The early functionalist sociologists and anthropologists used organisms as an analogue for societies, but *societies do not have any programs.*[17] Organisms are the wrong analogue. I think the appropriate analogue, if you want to understand social change, is a population or a species.

Callebaut: These and related analogues will be scrutinized in the next chapter on evolutionary epistemology and in chapter 9.

6.6.2. Genealogy and Similarity

Callebaut: What makes two organisms similar, on your view?

Hull: What makes two organisms belong to the same species? It is that they are part of the same genealogical web. It has to do with who mates with whom and gives rise to whom. Mating is inherently a spatiotemporal relation. Maybe two organisms do not have to come in contact, but at least their propagules do. Classes have no spatiotemporal restrictions; anywhere, anywhen they have the appropriate characteristics. Organisms, because of the mating and descent relations, tend to be similar. It is the mating that produces the similarity. But the two do not always go together. You can have things that are fairly nonsimilar mate successfully, and you can have things that are very similar not be able to mate. When the two do not go together, you have to decide which is primary. Given the mechanisms of the selection process, what is primary is the mating relation; what is secondary—not irrelevant but secondary—are the similarities. You are selected *because* of the characteristics you have. You interact with your environment successfully or not. But whether or not you get to pass on these characteristics depends upon mating. Mating also produces the similarity.

Callebaut: So it is the genetic continuity that is primary.

Hull: That is the basic consideration. Genic change, so the story goes, also promotes

17. *Pace* Marx. According to Hull (e.g., 1982, 1988a), there is no sociocultural equivalent of the biological genotype/phenotype distinction, but others, including Wimsatt, disagree (see 9.6).

cohesiveness. It does two things which seem mutually contradictory. It promotes *cohesiveness:* it keeps the various parts of the species from diverging too much from each other. But it also promotes genetic *heterogeneity:* it produces lots of alternative possible combinations of genes. It is harder to be cohesive when you are different; it is easier to be cohesive when you are the same. So you have the same process doing both. Given this characterization of species, species should not have absolutely sharp boundaries. At any one time, you should have some borderline cases. That is typical of individuals. You should not and you need not have absolutely discrete beginnings and endings in time. Species can kind of ease into existence.

Callebaut: Wouldn't punctuated equilibria present an exception to this?

Hull: No, in punctuated speciation, we are talking about fifty generations, not an instant in time! Punctuation is not just a comma; it is a comma stretching over a long time! Species ease into existence; they also ease out of existence. You have one species gradually splitting—from our perspective—into two. You can even have some merging. We forget that in plant species, species merge. Organisms also merge. Not cats and dogs; but there are lots of organisms where two organisms can merge into one. There is a kind of amoeba such that if two come across a meal that is too big for each of them to engulf, they merge so that they become big enough to engulf it. Species are this sort of thing.

Callebaut: You insist on the utility of constructing a more general evolutionary theory—a point which will be discussed in the next chapter. You also insist that this may require giving up some of our central intuitions about what evolution constitutes (e.g., Hull 1982).

Hull: One of the problems is our relative size. We can see organisms close to our own size but not those a good deal smaller. Hence, when we think of evolution, we think of vertebrates, large insects, etc., not blue-green algae, slime molds, and nematodes. In point of fact, most organisms are too small for us to see. Conversely, species are too big and last too long for us to see. If we are to understand evolution, we have to conceptualize the relevant entities in ways appropriate to the evolutionary process even at the expense of ordinary perceptions.

Callebaut: Michael Ruse talks about "a sort of *anthropic principle* in evolutionary biology" in this respect: because we're part of the evolutionary process ourselves, we are not and maybe cannot be entirely disinterested observers.

6.6.3. *Reference Classes versus Particulars*

Callebaut: If evolutionary theory is to contain any law-like statements (cf. 3.5.2), you need *general reference classes* in addition to particulars. Which categories must one construct first, in your view?

Hull: There is no temporal order. The better you understand the particulars, the more likely you are to come up with a good reference class. The better you understand the reference class, the better you are at individuating particulars. It is a feedback between the two; a feedback between the "just-so stories" (cf. 9.2.5) and the process of theory formulation. You cannot get one all done and then start on the other. We do both of them in tandem. On this topic I disagree with many scientists, including pheneticists and pattern cladists.

Callebaut: Couldn't one accept your variety of nominalism—if I am allowed to use that technical term from traditional philosophy—and yet maintain that *epistemologically* many explanations are reductions of "the unfamiliar" to "the familiar" or some other pragmatic category which ultimately amounts to something more *general?*

Hull: From an epistemological perspective, I may look like a nominalist, but ontologically I am a realist. We have to have *some* access to the world. We happen to do it primarily through sight, secondarily through hearing; any combination would do. Light rays that are visible are no more fundamental, ontologically, than other light rays. Epistemologically yes: we can see red, blue, etc. But it does not make any difference as long as we can interact with the world *somehow.* To say that we must read the accidents of our evolution—the fact that we can see the light rays we can see—*into* the world, I think, is totally contrary to science. I tend to dismiss the epistemological considerations because *we already know we can do it.* No one has ever seen a pion. No one has ever seen an awful lot of things that physicists talk about. Yet we know a lot about them. If genealogies are scientifically important, we will find some way to get at them!

Callebaut: Evolutionary epistemologists of the Austrian school of Konrad Lorenz and Rupert Riedl make a great deal out of the idea of an "economy of expectations" as originally developed by Ernst Mach (see in particular Riedl 1984; on Mach, cf. Musil 1982). Reacting against the empiricism inherent in the view that the correction of momentary perceptions is what is really important in the process of cognition,[18] Popper (1987) writes that "our fundamental knowledge is like the very antennae we point in all directions."[19] In his example: When I am in this lecture room, to know that I am in Vienna, in Austria, in this lecture room etc. is more fundamental than to have a momentary perception of, say, a friend in the room. Popper quite explicitly associates his own conception of evolutionary epistemology with the view that the general is primordial and the particular is derived.[20]

---

18. Riedl (1978:254–55) gives the following example: "Supposing [*sic*] a bone is gradually uncovered during an excavation. We make predictions (as Cuvier is known to have done in the lecture hall). And as the uncovering continues we find agreement with expectation, or we are disappointed (i.e., surprised), we correct ourselves and predict again. Finally, when the specimen has been uncovered completely, it has already been allotted to a position in the system of similarities. The cause of this similarity, however, will be explained by a mechanism which is not accessible on the basis of this fossil bone. And this mechanism will lead us to expect identical replications on the basis of totally different experiences."

19. "Das Wissen, das fundamentale Wissen ist so etwas wie die Fühler selbst, die wir ausstrecken nach allen Richtungen" (Popper 1987:33).

20. "Wir gehen also, sagen wir, von meiner Auffassung der Evolutionären Erkenntnistheorie aus. Gehen wir aus vom ganz allgemeinen Wissen und kommen dann auch zu gewissen ganz speziellen Dingen,

Hull: I'm afraid I can't make much sense of all of this. If these people mean that those distinctions necessary for our survival up until now will turn out to be adequate for the future, I disagree. Since we are here, our ancestors must have made enough of the distinctions necessary for survival, but we may well go extinct and one cause might be the failure of those distinctions that seem natural to us.

Callebaut: I was just trying to make the point that there are good reasons for supposing that in many cases, people will start with classes and then later, when it is necessary, introduce distinctions.

Hull: No! (*Impatient.*) People may well start every generation with serious, deep-seated conceptual predispositions, many of these predispositions lead to serious misconceptions (Hull 1965, 1989). To the extent that we are disposed to these misconceptions, we will have to undo them every generation, and at just the wrong time, we may fail to make the necessary transition. Perhaps we are all born with the predisposition to conceptualize the world essentialistically. We see essential kinds everywhere, but most of these natural kinds are not kinds at all. Genuine natural kinds are hard to come by. If evolutionary epistemology is supposed to make me confident of the future, it fails. I get really depressed thinking that generation after generation, teachers will have to undo common conceptions if their students are to understand the world in which we live.

Callebaut: OK, I won't push this point.

Hull: Do you think we are guaranteed to exist forever?!

Callebaut: No![21]

---

wie etwa, daß ich dort oben gewisse Leute sehe, die ich kenne u.s.w." (Popper 1987:34). Cf. also Hayek (1972) on "the primacy of the abstract," which he takes to be largely synonymous with the "primacy of the general." In the discussion following Hayek's paper, Ludwig von Bertalanffy points out that Hayek's view "coincides with what biologists, psychologists and system theorists call progressive differentiation," and cites von Baer's law, the developmental psychology of Piaget or H. Werner, and the first, "holophrastic" stage of language development as examples (Hayek 1972:327). Conrad Waddington also insists that "in the ordinary processes of development in the biological world we do always go from the general to the specific" (Hayek 1972:328).

21. Applications to cognitive psychology of the view that species are individuals are explored in Ghiselin (1981).

7

EVOLUTIONARY EPISTEMOLOGY

EVOLUTIONARY EPISTEMOLOGY ("EE" henceforth in this chapter) is the attempt to explain animal and human cognition, including science, in a Darwinian fashion. Although as old as Darwinism itself, EE remains very controversial. Its more radical philosophical critics dismiss any endeavor to understand Reason or Science by an appeal to evolutionary biology out of hand. Van Fraassen (1985:258–63) thus includes EE in his "catalog of horrible examples" of attempts to justify science by using a "recipe for disaster," downgrading it as "the fastest and most facile" gambit. It beats Marjorie Grene: "why anybody should bother any further with a topic that was always pretty damn silly."[1] Thomas Nagel (1986:81) proposes to "take the development of the human intellect as a probable counterexample to the law that natural selection explains everything, instead of forcing it under the law with improbable speculations unsupported by evidence."[2] And according to evolutionary biologist Richard Lewontin (1989:229), "we know essentially nothing about the evolution of our cognitive capabilities and there is a strong possibility that we will never know much about it." The least we can conclude from such instances is that there is something about EE that leaves few people indifferent. A main objective of this chapter will be to find out what that something is.

It will be useful to stress at the outset that *two quite different programs*—one more scientific, the other more philosophical—*are labeled "EE."* The first program investigates the evolutionary basis of the perceptual and cognitive apparatus of living systems; it is the province of philosophers but also of cognitive ethologists, neurologists, psychologists, etc. Campbell (1987b) calls this "biological EE," Bradie (1986) calls it "EEM," with the M standing for "mechanisms." I will sometimes refer to this program as "bio-epistemology." The second program is an attempt to apply to science and

---

1. Personal communication, March 16, 1986. For all that, Grene espouses her own brand of "zoöepistemology" (Miller 1992).

2. Nagel (1986:81) grants that he can offer no alternative explanation (!), but persists nonetheless: "One should not assume that the truth about this matter has already been conceived of—or hold onto the view just because no one can come up with a better alternative. Belief isn't like action. One doesn't have to believe anything, and to believe nothing is not to believe something."

scientific change—often by means of analogy—concepts and models that were originally developed in evolutionary biology. In this chapter this second program will usually be referred to as the "evolutionary account of science"; Bradie calls it EET (T for "theories"). These two programs within EE broadly speaking must not be confused for reasons that will gradually become clear. Right now it will suffice to note with Ruse that there are "many people like myself who hate the analogy, but love the literal position" (see his 1986a and 1989b).

Although I am myself rather firmly committed to EE in both of the aforementioned senses, I have attempted to give the subject a balanced treatment. It is for the reader to judge if I have reached my aim or not. I should like to add one thing, though: If you finish this chapter skeptical of EE or some part of it, realize that "if we are to take the variety of ways in which the evolutionary perspective has infused and been used in biology as a model, the directions in which EE has gone so far have exploited but a tiny fraction of the resources available to such a perspective" (Wimsatt).

We ease into a discussion of Campbell's project of an "evolutionary theory of scientific belief validation"—the EET equivalent of his slogan, "Passing the justificatory buck to evolution"—by first considering why some people, although sympathetic to the EE enterprise at some level, hesitate to endorse the EE program. Is there anything properly *epistemological* about this agenda, and in which senses is EE naturalistic? Next, Hull's analysis in section 6.6 of species as quasi-individuals is extended and detailed here in terms of both the social and the cognitive aspects of science, which confronts us with the question whether or not the intentionality of human beings signifies an important difference between biological and cultural evolution. The general picture of science exemplified by Hull and by Richards's more historiographically oriented EE are contested by Latour, who regards them as the apex of a rationalistic approach that has shown its untenability. The chapter ends on an attempt to avoid the Platonism inherent in Popper's "Three Worlds" model of EE.

7.1 Is It Epistemology?

Callebaut: Is the pessimism of wholesale detractors of EE obvious and inescapable, Bob? Couldn't one be quite modest about one's approach and just claim to be exploring the potential of an analogy that has delivered the goods in other contexts (Grafen 1989)? My own motivation for taking EE seriously, as far as I am aware, seems to be a stubborn insistence that society and culture, including science, must ultimately be much more like *life* itself (cf. Miller's 1978 "living systems") than resemble an inanimate Newtonian system.[3]

Brandon: Absolutely. You see, that is one of the reasons I think studying cultural

---

3. If this is correct, EE may eventually have to emancipate itself from the strictures of extant population genetics (Depew and Weber 1985; Salthe 1989). Cf. also Rosenberg (1988d) on the unwieldiness of a deterministic theory of natural selection.

evolution is going to shed considerable light on this whole topic. Maybe ten years from now we will be much better able to do EE, because we will have a better biological basis.

Callebaut: Maybe we'll get to a point one day where biology will benefit from our understanding of culture (a point Plotkin has insisted on).

Brandon: I would say that I am very sympathetic to the approach but also critical. I am always skeptical about details when I actually *see* the approach in some form or other. I am skeptical because I think people are making too much of *this* similarity or that they ignore *this* dissimilarity or misunderstand *this* aspect of the biological side. I would consider myself an evolutionary epistemologist in principle.

Burian: Toulmin was a colleague of mine at Brandeis when he was writing the first volume of *Human Understanding* (1972; cf. 1967). One of the things that struck me as just dead wrong about that book was that its footing in evolutionary theory—as a footing for the analogy to the evolution of conceptual systems—was seriously wrong. When I applied to the American Council of Learned Societies for a Study Fellowship, the immediate stimulus was the attempt to get a proper footing, to examine attempts like Toulmin's, to draw an analogy with evolutionary theory. I have not really come back seriously to that—although I have some ideas about it—because I have become so intrigued with the philosophy of biology and the biology itself.

Callebaut: People like Joseph Losee (1977), L. Jonathan Cohen (1973, 1987), Paul Thagard (1988:ch. 6) or Carla Kary (1982) have argued that the evolutionary analogy is no good. Hull wants to get beyond analogies. Analogies and metaphors somehow continue to look suspect to most people, although without them science might not exist.[4]

Bechtel: I am not at all bothered by the idea that EE is only an analogy. In fact, I would prefer to stay with the language of analogy rather than "common process" (cf. 9.2), because one thing you do expect from an analogy is dissimilarities. And I'm not sure that if you come up to the level of common process, we're going to get much more than is in Lewontin's (1970) paper—the level at which we can identify mechanisms of variation, selection, and retention (Bechtel 1988d, 1990; cf. Griesemer 1988, 1992). But it may be that some of what David is developing will turn out to provide a more substantive common analysis. I would be delighted if his strategy were to succeed. I simply have my doubts.

7.1.1. *Passing the Justificatory Buck to Evolution*

Callebaut: Philosophers often ask about EE, "Is this philosophy, and how does it relate to traditional philosophical inquiry?" The answer is far from obvious. I am thinking both of what a critic of EE like van Fraassen claims and what a major

4. Black (1962); Hesse (1966); Schlanger (1971); Holland et al. (1986); but see Knorr (1980).

evolutionary analyst of science has said: that "if EE were a genuine epistemological theory, I would not be in the least interested in it" (Hull 1982:224). Don
(Campbell 1956, 1959,[5] 1960, 1974a), you are the grand . . .

Campbell (*interrupts*): . . . I want to continue to view EE as something that has
sprung up all over for a hundred years or more, and that everybody independently
invents and recognizes when other people have independently invented it. So in
no way do I want to be portrayed as a "founder" or "gate-keeper" of EE.

Callebaut: You are a bridge-builder as well.

Campbell: In contemporary Anglo-American philosophy, the epistemological formula is *justified true belief:* knowledge equals justified true belief. Biological EE
explains how our eyes, brains, inference structures, and the like have evolved and
have been selected for *referential competence.* The most frequent adoption of any
part of this first EE program comes from epistemologists in the JTB tradition who
use biological evolution literally to provide justification for beliefs which are generated by visual perception, for example.

Callebaut: Modern justificatory theory in mainstream philosophy does not demand
of a justification that it produce *certainty.*

Campbell: No. It must be reliable, or it must be the last court of appeal beyond
which it makes no sense to ask, "How do you know that?" But it does not need to
produce certainty.[6] EE enthusiasts advocating EET apply evolutionary analogues
in the history of scientific beliefs, in which they see a selective retention of competing scientific theories, as Popper said in one or two sentences in his 1934 *Logik
der Forschung* and its 1959 English translation. Popper (1972), Hull (1988a),
Wimsatt, and I are examples of evolutionary epistemologists in both the EEM and
the EET sense.[7] As to bridge building, Bonnie Paller and I (Campbell and Paller
1989) extend this bridge almost all the way to fallibilist perceptual foundationalists
such as Chisholm (1982) as well as to perception-privileging coherentists (vs.
foundationalists) such as Lehrer (1974) and Goldman (1986).

Callebaut: Finally, to further complicate things, you have recently started talking
about . . .

Campbell: . . . Campbellian dogmatic selection theory, which overlaps with both
EEM and EET but is not epitomized by either. I now require not only a selection
theory to explain how eyes came to be evolved from light-sensitive pigments—the
EEM agenda—but also a selection theory for *every operation* of the eye (1990a,

---

5. Quine (1969:90) cites this paper in what may be the first use of the label "EE."

6. Alvin Goldman (1986)—insofar as he will admit to be an evolutionary epistemologist—is perhaps
the best exemplar of this mainstream view, Campbell thinks. "He, Quine (1969b), Shimony (e.g., 1971;
cf. Shimony and Nails 1987), Rescher (1977, 1990, 1991), and Giere (1985a, 1990), to name a few others,
limit their evolutionary references to 'passing the justificatory buck to biological evolution.' All this group
fails to endorse or explicitly rejects 'analogical' EE (Bradie's 'EET') as well as what I now prefer to call
'selection-theory epistemology.'"

7. See also Ackermann (1970, 1986) and Georgescu-Roegen (1971:ch. 1: "Science: A Brief Evolutionary Analysis").

c), or more precisely, a "selective reflection epistemology" in which the referent—among other possible selectors—has, by differential reflection, selected the belief from other possible beliefs.

Callebaut: The radar or sonar analogy may be easier to understand.

Campbell: You have a blind sweep, sending out of potential locomotions—beams—and a selective reflection. You have broadcast variation and selective reflection. The eye itself is not the source of what is being reflected for vision. In my teaching, long ago, I had a little prosthesis distributed by a South Dakota Rotary Club for blind people. It was a single photo cell that generated a tone (it had an ear plug). When you moved the photo cell around, you got a different pitch according to the brightness. If there was a sharp boundary, you would get a sharp rise in tone. So you could just rub that boundary with your single photo cell; *you actually got to see the blind variation* in vision: broad sweeps until you got a sharp contour signal, then narrow sweeps recrossing the contour to reconfirm and explore the contour.

Callebaut: How about the real eye (cf. Winckelgren 1992 on illusory contours)?

Campbell: We know that there is ocular nystagmus, this vibratory motion going on in the eye. It's similar to a tactile contour. If you leave your hand still on it, your hand will cease to perceive the contour. But if you rub it, that line is continually perceived. So too with the ocular nystagmus, you are rubbing the contours in the visual field. So you need a selection theory to explain the competence of the eye *in operation* as well (1956). The competence of creative thought in operation is still more vicarious (1959, 1990a).[8] This is not epitomized in either of the other EEs. In a paper on Grover Maxwell and van Fraassen (Paller and Campbell 1990) I begin to be clear on this. And in "Epistemological roles for selection theory" (Campbell 1990a) I look back on these early studies, inspired by Ashby, and on this universal dogma. That's the only way to go; it's a tedious, improbable process; but there is no rival (*laughter*)! Nobody has put up another way except clairvoyance. I'm now claiming that I was always a selection theorist.

Callebaut: Talk about Whig history!

Campbell: Only Whig relabeling. For my inspirer W. Ross Ashby (1952) and for Popper (1963), trial-and-error learning was as much the model as was natural selection. Edelman (1974) has added the acquired immunity process.[9] My "Per-

---

8. The vicarious character of the evolution of science was emphasized by Popper (1972:70) in his now famous comparison of the amoeba and Einstein: "Although both make use of the method of trial and error elimination, the amoeba dislikes to err while Einstein is intrigued by it: he consciously searches for his errors in the hope of learning by their discovery and elimination." Scientists—and humans generally—can have their theories die in their stead, and fortunately so! Campbell (e.g., 1974a) agrees with Popper, but insists that vicarious selection operates at very many levels; cf. 7.1.5.

9. Selectionist (as opposed to instructionist) theories of antibody formation were proposed by Niels Jerne and Sir Macfarlane Burnet in the 1950s. The earlier view assumed that the antigen works as a negative template for the formation of the antibody (instruction from *without*). On the selectionist view which has come to replace it, the information which enables the antibody to recognize the antigen is of

ception as substitute trial and error" (1956) and my "Blind variation and selective retention in creative thought as in other knowledge processes" (1960) are a part of the selectionist agenda rather than an analogue to biological evolution. It is an abstract model of which biological evolution is one implementation, but which cultural evolution, with cross-lineage borrowing (1965; cf. 9.5.2), implements as well, just as does trial-and-error learning and computerized problem solving.[10]

Callebaut: Agreed. The recent work on genetic algorithms—"evolving" computer programs—could also be mentioned in this respect (Holland 1992a, b). But now one could have independent reasons to advocate a *bio-epistemology* of a generic type . . .

Campbell: . . . That's clear, I do that too . . .

Callebaut: . . . leaving open whether blind-variation-and-selective-retention (BVSR) is going to be central or even present in it or not (I am envisaging the *logical* possibility of a nonselectionist theory of evolution here rather than what is most *plausible*). Eve-Marie Engels (1989:13) makes a similar point in her recent book on German-Austrian EE.[11] So you claim to do that as well. Still you criticize Bob Richards and David Hull for their use of "biological specifics." Ron, can you agree with Don that the naturalized philosopher of science is less ambitious than his foundationist predecessors in making no pretense of answering the skeptic?

Giere: I think you have to change your ways of looking at it. We know there is no way to answer the skeptic on his own terms. That is a losing game. At that point you can say, "Look, this is a stupid game—to try and provide an ultimate justification, to find a noncircular foundation for knowledge." At that point you just attempt to change people's ways and say, "Look, this is an impossible task!" Another way is, of course, to say, "You don't need it! We have a good scientific

---

internal origin (conveyed by the genetic code); the specialized cells which produce the antibodies are *selected* (with the help of the invading environment). (See, e.g., Popper 1975; Darden and Cain 1989.) In the 1970s, it became apparent that antibodies belong with certain other immune system molecules in a single evolutionary entity, the "immunoglobulin family." Edelman (e.g., 1974, 1988a) elaborated the selectionist approach to immunology in considerable detail, hypothesizing, for instance, that the immune system evolved from a cell adhesion system that was essential to the nervous system as well. His discovery of CAM's (cell-adhesion molecules), which are proteins that mediate interactions between cells in the embryo, resulted from his interest in how cells interact in a developing embryo to yield an organism. Edelman's "topobiology" (1988a, 1989) refers to the "place-dependent" nature of these interactions. More recently he has begun to lay the groundwork for a molecular embryology linking the form and function of embryonic tissues to evolution and genetics. (For a critical discussion, see, e.g., Chernyak and Tauber 1991.) Although there are a number of obvious and less obvious links, the work of Edelman described in this note must be distinguished from his "neural Darwinism," which will be briefly discussed in section 9.2.2.

10. Cf. Darden and Cain (1989); Griesemer (1988, 1992); and—critical of EE—Piattelli-Palmarini (1986).

11. Compare Bradie's (1989:407–8) criticism of the "selectionist biases" of extant evolutionary epistemology.

"Cousins to the amoeba, how could we know for certain?"

Donald Thomas Campbell (b. Grass Lake, Michigan, 1916) is University Professor at Lehigh University, Bethlehem, Pennsylvania (1982–). A social psychologist by training (he studied with Edward Tolman and Egon Brunswik) who was later influenced by W. Ross Ashby's cybernetics, he previously taught at the University of Chicago (1950–53), Northwestern University (1953–79), and Syracuse University

(1979–82). It is an impossible task here even to begin to do justice to the stature of Don Campbell as a scientist and humanist, one of only a handful of generalists (in the noblest sense of the word) on the postwar intellectual scene in the U.S. Known among philosophers and students of science for his evolutionary epistemology, he is at least as influential among psychologists and social scientists for his contributions to their fields, which include the method of *quasi-experimentation* (Cook and Campbell 1979). Countless are the scholars who are indebted and grateful to him. But I am happy to refer the reader to Campbell's autobiographical "Perspective on a scholarly career" (1981a), which many people consider unparalleled as an exercise in intellectual honesty, if for nothing else.

Here I rather want to share with my readers some of Campbell's reminiscences concerning his dealings with the two philosophers whose thinking he finds most congenial to his own evolutionary and naturalistic Weltanschauung, Quine and Popper. Readers should keep in mind throughout that Campbell's capacity for self-effacement is considerable.

On his first meeting with Quine: "In 1966–67, while at the Center for Advanced Study in the Behavioral Sciences, I made a six-week tour to the East and visited Quine. He took me to lunch in the Society of Fellows place. The economist Leontief was at a nearby table and joined in the conversation.[1] As part of my image of this long, two-hour-or-so discussion, Quine, for whatever reason, treated me as a fellow intellectual on the basis of very little evidence. What could I have given him at that time? My paper on "Blind variation and selective retention in creative thought as in other knowledge processes" (1960). My first face-to-face contact was warm and enthusiastic." In 1977, Campbell would return to Harvard to teach "Psych. & Soc. Rel. 2730. Seminar: Descriptive Epistemology: Psychological, Sociological, and Evolution-

ary"—his William James Lectures, which would soon make him famous among philosophers. In Ann Arbor in 1980 it would become clear to Campbell that Quine was "only an evolutionary epistemologist in the sense of Bradie's EEM." At another meeting, in Columbia in 1983, Campbell was much impressed by how knowledgeable Quine was about Latin American folklore.

On one of his meetings with Popper, when Campbell was on his way to Kenya to do fieldwork with his colleague at Northwestern, the anthropologist and political scientist Robert E. LeVine in 1964: "We spent some time in London to get over jet lag. I visited Popper in Penn. We got along so well in conversation that he persuaded me not to go back to London but directly to Heathrow from his home by taxi, in return for which he gave me a book by Jarvie on anthropology." In 1968–69, Campbell had a Fullbright grant as a psychologist at Oxford (where he and Stephen Toulmin co-taught a course on "Evolutionary Epistemology") and saw Popper several times at his home. At the end of the year Popper invited him to lecture at his seminar (cf. 3.3.1). Why did Campbell (1988c) come to Popper's defense against Baigrie (1988)? "Baigrie tried to distinguish between Popper's reputation and mine. His specific criticisms, I felt, were more appropriate criticisms of me than of Popper. But because of that, I discuss differences with Popper." Isn't Popper's dualism an issue that divides them? "The only mind-brain dualism which I accept," Campbell wrote in his reply to Baigrie (1988c:375) "is purely epistemological, not at all ontological. I don't quite know where Popper stands on this, although I recognize that he is nearer to my position than is Eccles."

Campbell, who calls himself an "opportunistic animal," unabashedly admits that he wants to get a hearing from philosophers. "This has led me, in the last decade, to cite Quine where I could cite both of them or could cite Popper. The great indebtedness in my recent papers that comes across is to Quine. It's absolutely sincere, but it's selective."

The oldest material incorporated in this book is from a radio interview that was recorded at the house of my friends Katrien Raes and Karel Van Keymeulen in Ghent on November 16, 1984. Don was in Ghent to attend two conferences, the Evolutionary Epistemology (ERISS II) Conference and the George Sarton Centennial, which was a joint meeting of 4S, EASST (the European Association for the Study of Science and Technology), and the Ghent-based interdisciplinary group Communication & Cognition. I was able to have several more discussions with Don at Lehigh University and at his home in Bethlehem, Pennsylvania, in 1985 (when he was so kind as to invite me over from Chicago to give a talk to Lehigh faculty) and again in 1986. An extended interview, in which Don's wife and Lehigh colleague Barbara Frankel and my New York friend Linnda Caporael, a social psychologist at Rennselaer Polytechnic, also participated, was recorded in Don's office at Lehigh University on May 23, 1990.

---

1. Campbell's mentioning this made me remember that, in his presidential address to the American Economic Association, Leontief (1971) in a way urged the professionals of the dismal science to go naturalistic too.

explanation for how science got off the ground." That is the ground level of the evolutionary view. "Come on, we know how to get around in the world of crude objects, because we are just fancy apes, and we can do what apes can do. And that's enough to get off." The interesting questions now start there. Now we investigate how we create science. But the alternate question, "How do we know anything?"—what you have to do is convince people that it is a crazy question, in a historical deconstruction, "How did/do we get into thinking we should answer that question?" and say, "See, it was a mistake!"

Callebaut: Maybe some day people will become aware that one can say a similar thing about "Hume's problem"—the impossibility of induction.

Giere: Yes, that's exactly the same. I think Hume's problem was a blind alley, and you could see how people got into it, because science was in competition with religion. Religion claimed to have a really very solid basis for its views. People who decided to attack religion in favor of science wanted to have something just as good. So they really wanted to have an ultimate justification, because of the competition with religion, I think. And now we don't need it, and we realize you can't get it anyhow. It is not possible. But you don't need it; the need is gone. It has been gone for a hundred years. Yet people keep working on the problem. The real problem is just to make the problem look ridiculous to people; to stop working on it. There is no answer. If you get drawn in to trying to give it an answer, you are lost again.

7.1.2. *Descartes as a Brain Physiologist*

Callebaut: Within the general move of a naturalized epistemology, you like to go back, Don—it seems to be a bit of a paradox—to that great hero of the Enlightenment, Descartes-the-foundationist.

Campbell: Descartes, for me, was the major brain physiologist of his day. He saw that the neural tubes had to reach up to the brain, and that the brain—with more neural tubes—had to reach to the neural tubes that activated the muscles. He was a great mechanist. For animals, he was just like La Mettrie ([1747] 1960) a century later; the animal is a machine, and he did the neurology of that. For an epistemologist, the mechanization of visual perception confronted a normal clairvoyant realism that we are all born with. So it was disturbing to learn, for example, that for sensation of heat the neural tubes were not carrying heat, or for sensation of cold the neural tubes were not carrying cold, but that instead the neural tubes were carrying the same fluid, with the same pulse, for both heat and cold. (These are not, as far as I am aware, examples that Descartes used. Nor does he cite the "phantom limbs" of amputees. But no doubt he was aware of them.) This general mechanization showed that within the brain there were arbitrary signals that were not in themselves conveying truth in an uncorruptible manner; but there were many, many things that could go wrong. So Descartes' taking the mechanics of vision as an object in the world to be studied realistically led him to a radical skepticism, which he epitomized by saying: "We have no rules for knowing whether we are dreaming or awake." Now, stepping back from that radical skepticism, he turned to God's good will, and said that a good God would not have given us eyes that regularly deceived us (1987b; Campbell and Paller 1989).

Callebaut: And according to you, the bio-evolutionary epistemologists are doing a similar thing?

Campbell: They are passing the justificatory buck to biological evolution. They are saying: "We are justified in this 'Justified True Belief' formula; we are justified in trusting beliefs generated, in part at least, by visual perception; because had we had eyes that regularly deceived us, we would not have survived." They have the same program as Descartes. They know about the many ways—the argument from illusions—in which this knowing process can go wrong; but its general dependability Descartes justified by trusting Providence, and they justify by trusting in natural selection.

Callebaut: For you this sets the requirement that you must somehow include biological evolution in your epistemological analysis.

Campbell: If you're going to be a conscientious epistemologist, you'll have to explain how a process of biological evolution could have produced *validity*. Here Darwin becomes a major epistemologist, because what he gave to us was not the notion of biological evolution per se, but rather an *explanation* of how such marvelously designed anatomical organs and processes could have evolved without a designer. In our modern cybernetic generation, people like Ashby (1952) and others have taken that epistemological core of natural selection and said: "There is a recipe whereby order can emerge without foresight or prescience or clairvoyance." That we can epitomize as "variation-and-selective-retention." If there is systematic selection, then the variations or the thought trials or the mutations do not have to "know" in advance; they are winnowed. Similarly, beliefs do not gain their validity by being inspired by prescience. Rather, beliefs can be tried out and selected by interaction with the referent of the belief.

7.1.3. An Evolutionary Theory of Scientific Belief Validation

Callebaut: There is a certain ambiguity here: "variation-and-selective-retention" can be understood in the full-fledged biological sense, as a material process involving genotypes and phenotypes and what have you, or as a purely formal "algorithm," as you sometimes call it. I emphasize this because EE may—and has been— conceived of either biologically or as an attempt to apply the formal evolutionary or population genetics framework (as described in 6.2 and 6.5) to science, without assuming any biological basis whatsoever.[12] So if you say you want to extend the agenda that you just described to scientific truths, what do you mean, precisely?

---

12. It is rather confusing to lump together this material/formal distinction with the distinction between EEM and EET, respectively; yet this is what people usually do. What actually obtains is a 2 x 2 matrix of types of evolutionary epistemologies.

Campbell: We cannot extend it at the level of biological selection. If you examine any part of science, you find that it is heavily dependent upon scientists trusting each other. And whereas for the human eye and its competence we can say, "God/ natural selection would not have given us eyes that regularly deceived us," we cannot, either as old-fashioned Cartesian providentialists or as modern biologists, say, "God/natural selection would have given us only trustworthy fellow scientists." Sociobiology explains why this is so. For the ant, or the termite, one *can* say: "God/natural selection would not have given an ant untrustworthy scouts." But the ant is in a different situation, from a sociobiological point of view, than are we. For the ant, all of the cooperators are sterile; they are not in genetic competition with each other. But for us human beings, and all of the vertebrates, the cooperators are also competitors for inclusive fitness, for who is going to have the most great-grandchildren. This means that many times, the messages transmitted and the belief assertations are designed to *manipulate* the listener rather than to provide a description of the world which one would use for one's own action (cf. Dawkins 1986). I've gone into this in most detail in my paper on "Selection theory and the sociology of scientific validity" (1987b).

Callebaut: What consequences does this have?

Campbell: This means that if science is to produce validity, and if we are to use this Ashby-like epistemological device to explain how knowledge can increase, we are dependent upon a social system that makes scientists, in practice, *usually* honest about their scientific communications. Now they may lie to their wives or their mistresses, or in the case of Crick or Watson they may steal other people's spectrographic photographs (Watson 1968), but the social system of science was such that no one ever suspected them of faking their data. And it is that social system which enforces honesty in communicating the data from the laboratory that is the precious secret of scientific validity, and it is one that we should worry about losing, especially in an area such as the social sciences in which there is the pressure for publication in an area where nobody can check your work anyway (1986a). David Hull's sociology of science (7.2), although I have some reservations about it from the epistemological viewpoint, is excellent on how selfish scientists produce collective validity in biology.

7.1.4. *The Role of Nature Herself*

Callebaut: You sympathize with the relativistic and constructivist approaches to science.[13] Nevertheless you insist on being a sort of epistemological realist.

Campbell: I am a "hypothetical realist." I've always distinguished my position from a direct or "clairvoyant" realism, and I'm now diminishing the distance between me and antirealists like Laudan and van Fraassen. Since they are also admirers of science, my hypothetical realism and their epistemological antirealism are not all that different. On the other hand, both of them should probably be called "ontolog-

---

13. E.g., Campbell (1986a, b, 1987a, b, 1988a, b); Campbell and Paller (1989).

ical relativists," if Quine (1969a) hadn't used that expression for another purpose. Both, but particularly van Fraassen, are more nearly naive realists about ordinary perception than I am, by the way. In any event, if you're getting down to technicalities, I really want to side with the skeptics there.

Callebaut: What about "selection by Nature herself"?

Campbell: My focus on how perceptual and cognitive processes could acquire any validity at all leads me to emphasize *external selection* for either the animal or for the scientific community—the role of selection by Nature herself—in the final structure. But this should never be allowed to lead to neglect of the point that there are many, many, many other selective forces going on which have nothing to do with validity directly, but which provide the structure in which valid beliefs, or valid mutations, could be collected. One way in which these have been spoken of is through *internal selection.* There is a tremendous selection in evolutionary biology for mutations that will continue to fit in with the other parts of the already well-working machine (cf. 9.1.1). And this, in a sense, has to take priority over mutations that enable the whole machine to fit better with the environment. This is one of the valid points in the Gould and Eldredge punctuated equilibrium theory (cf. ch. 2, n. 4): once an organism has become a well-working, internally related machine that fits the environment moderately well, it may remain stable, while the environment changes, over long periods of time; because the mutations are continually being weeded out by the requirement of fitting in and not disrupting the already ongoing machine (1986a).

Callebaut: What would be the equivalent of this in the case of, say, a scientific theory or paradigm?

Campbell: A theory which works marvelously well, like Newton's, is going to be retained, even while anomalies accumulate; because the effective theory is so precious in generating almost correct predictions that it would be a shame to give it up in the absence of something more adequate. And to be a useful theory that generates predictions, it cannot fit the environment like the sand which we are using for a metal mold fits the master. It instead has to retain its internal structure which enables it to generate predictions. Here too, we have internal selection, which Kuhn has referred to as "normal science," and which Piaget has referred to as the phase of *assimilation,* in which the child's behavioral logic fits his world so well that he is willing to treat the world as being appropriately described by it until he can get another, internally coherent logic that fits—*accommodates*—it better, adapting still better to the environment.[14] But we can never have this unstructured adaptation which the sand-casting image or, within biology, the "bean bag" genetics model might lead us to expect. So that the requirement of internal structure, the requirement of having an internal logic or machinery that enables us to gener-

---

14. Ros (1991) points out that (hidden) realistic assumptions underlie Piaget's view of the causes of a shift from one cognitive stage to the next.

ate actions or predictions, keeps organisms and scientific theories from fitting the environment perfectly.

Callebaut: What gives in terms of your notion of conflicting goals (Longino 1990:32–37), Helen?

Longino: I would express that as "one or the other of the goals of science becoming paramount at different moments in the history of a particular research program (or whatever)." These oppositions (assimilation/accommodation and normal/revolutionary) are more psychological or sociological ways of describing how one satisfies or meets these goals. But the goals themselves, I would argue, are incompatible, just as one can't assimilate and accommodate at the same time and one can't have normal science and revolutionary science at the same time. So there is a similarity. The Kuhnian language is talking about processes that we can understand as determined by these different goals which are incompatible. So when accuracy or truth is paramount, we'll find work more like revolutionary science; and when the explanatory goal is paramount, work will look more like what Kuhn describes as "normal science."

7.1.5. *Selecting the Vicarious Selectors*

Callebaut: Let us focus on selection processes next.

Linnda Caporael: Are all your selectors (e.g., Campbell 1974a) themselves the outcome of selection processes?[15]

Campbell: As a part of my dogmatic program, which is utterly unjustified—just a leap of faith—I see all order as a result of selection; crystals, say. All systems that are really systematic are a result of selection. Where I infer to "selective systems," I usually should be referring to "stable selective hodge-podges." There is no *necessary* system in the physical environment to which an organism is being adapted. If it is being adapted to the period of the tides, or to night and day, it is being adapted to something systematic. But if it's being adapted to the cold on a permanent basis, as in the polar region, that doesn't have to have any system, but it has to have stability. Many things animals are adapting to are not necessarily systematic.

Caporael: They have to be stable . . .

Campbell: . . . otherwise there is nothing to adapt to. There is something about the concept of ecological niche that I need to do more work on. If we think of the ecological niche of the Asiatic wolf and the marsupial wolf, which is being a predator to large herbivores, to some extent that ecological niche has been put together by the wolf types. It isn't something *out there* in nature. It has an organized character to it, but that is being constructed by the adapted organism. In my paper on downward causation (1974b) and in other papers on cultural evolution (1965, 1972, 1983), I speak of the multiple independent invention of the syndrome of

---

15. Linnda Caporael took part in this conversation (cf. the biographical note on Campbell at 7.1.1).

stored food stuffs and the division-of-labor society in termites, ants, bees, and in a dozen isolated city states. I claim that there is selection by the laws of sociology (*laughs*), which is not there in the physical environment but is clearly put together by the animal that's moving into this niche. It is being edited by something, and there are limited ways of entering into that niche. But it has an odd Platonic status: it's not a system, it's the *possibility of a system* that's lying there. That's a hedged answer on that.

Callebaut: For creative thought, vision, and the like, the vicarious selectors are clearly the products of selection themselves.

Campbell: I still don't do better than the example of the salamander's leg: When it regenerates when broken off, does it regenerate until it reaches the ground? No! It regenerates until an internal vicarious monitor for leg length is completed. But if this regularly led to regenerated legs that were, say, too long, then an external selection would select mutations that adjusted that internal selector. So the internal selector is a vicarious representative of the external selector. Likewise, for Poincaré, mathematical beauty was a vicarious selector for mathematical truth. More generally, creative thought and vision are vicarious selectors, although vision is much less vicarious than creative thought. Even feeling solids involves vicarious criteria: you probably could create optical illusions for them, etc. Clearly, the vicarious selectors are all products of selection and are tuned—just like you readjust a thermostat—by whole-animal natural selection. Is that in the ball park of your questions?

Caporael: Yes. So there is this set of selectors that are vicarious selectors and there are other selectors. But then we can talk about stability in the larger order that is not itself selected but does the selecting. Now that seems to me in some sense to give you "foundational grounding." Imagine yourself going back; you get to a point where the only thing that is doing the selection—let's just talk about living systems—for these most primitive of DNA fooling around in the soup is going to be the stability in the external environment, where that real referent is. That real referent is *there;* there is no vicarious selection involved.

Campbell: That's what I call "external selection"—the selection that produces quick fit to the immediate environment (1987a). Note that trial-and-error learning does not involve direct contact with the environment in that you use vicarious criteria for "edible," "poisonous," etc.

Callebaut: Things become really exciting when you start looking at how one vicarious selection might impinge on another; vision on language, for example.

Campbell: I give vision a very important role in making language possible, with recognition of Quine's "gavagai" problem (Campbell and Paller 1989). And that odd paper of mine, "Ostensive instances and entitativity in language learning" (1973) . . . Ever since it's been out, when I've been teaching skeptical students, it has done more to make them feel that the real world plays a role in beliefs, because of the argument that I give there that the real world "edits" the kinds of things that can become a *lingua franca,* a way of transmitting. I attribute to Quine—although he probably didn't use those very words—the position than you cannot teach a language by telephone. I also want to maintain—*pace* Cassirer, Sapir, Whorf, and now Harry Collins (1985)—that you cannot generate a plausible model of language learning if you are going to suppose that the visual field will not be entified until you have the linguistic categories to do so (Campbell 1989; also Bloor 1989). The animal reification of perceptual data as external entities is a precursor. Any animal with distance receptors—including a fly (*laughter*)—has evolved that precursor to language.

7.1.6. *The Ideology of the Scientific Revolution*

Callebaut: How does EE, which emphasizes blind variation, to be sure, but also selection and retention—*blind* retention, as you insist—fit in with the ideology of the Scientific Revolution, which was fundamentally antitraditional?

Campbell: This is a problem, but perhaps a productive problem; and it gives us new things to work on in tying in our evolutionary theory of science with our theory of knowledge, sociology of knowledge, which will include religious and other beliefs.

Callebaut: You like to link this to your work on the evolution of moral systems (Campbell 1972, 1975, 1979a, b, 1983).

Campbell: Unlike most sociobiologists, it has seemed to me that the moral preachings that we find in the great religions, in the highly complex societies, and in the early city states with full-time division of labor are best understood as efforts to curb the selfishness and nepotistic selfishness which this genetic competition among the cooperators has instilled. So that we can see the seven deadly sins and the Ten Commandments as set in opposition to a biological human nature which is best explained by this genetic competition among the cooperators. Cultural evolution of metaphysical world systems produced inhibitory moralities, selected for optimizing group or collective coordination by preaching against the biologically-based selfish tendencies. Now, my speculation is that at the time of the Scientific Revolution and still today, cultural evolution in the area of metaphysical beliefs, of beliefs about those aspects of the physical world which could not be seen with the naked eye, had been completely coopted by this culturally useful adaptive function of supporting group solidarity. So that if the Scientific Revolution was to be able to discover things like magnetic forces, electricity, and the like—physical realities which are normally unobservable—it was necessary to be antitraditional as far as the larger, nonscientific community was concerned, and to create a new tradition within science in which the new scientific metaphysical beliefs were to be tested against short-range effects that were visible. Thus, the scientific evolution of competent theory of electricity with metaphysical speculations about invisible fields and fluids was combined with a continuing reference back to the electric sparks which were visible and were obviously under the control of the experimenter or the demonstrator (1986b). So the ideology of the Scientific Revolution says: for the variation and selective retention of beliefs of this subcommunity, this subculture, we have these new rules: "Take nothing from traditional authority!" "Feel free to disagree, even with scientific authority!" and "Limit your persuasion of fellow scientists to what you can demonstrate visually and what you can show by simple logic!"

7.1.7. *Evolutionary Naturalism, Vintage 1992*

Callebaut: How does a "friend of discovery" assess EE, Tom?

Nickles: I have great respect for Donald Campbell, who on the discovery point has often been misrepresented by friends of discovery, including myself on previous occasions (e.g., the Introduction to Nickles 1980a). I hope I don't misrepresent him here! What bothered the friends is that they read his early papers on scientific creativity as saying that there is nothing but BVSR. Not only is there no logic of discovery, there is not even anything to say about heuristics since that would pre-suppose clairvoyance. Campbell's account looked like Herschel's "law of higgledy piggledy," applied now to scientific research instead of biological evolution. It looked like the very opposite of what Herbert Simon and the AI people were saying. In fact, I think Campbell's early papers are more open to that sort of "reductivist" reading than his later pieces. Today I no longer think there is any necessary incompatibility with Simon. Campbell has a lot to say about heuristics as themselves products of BVSR, although the emphasis of the friends of discovery has been different.

Callebaut: Please elaborate.

Nickles: The friends have stressed how heuristics can "rationally" cut down search and thereby direct inquiry, while Campbell has stressed the fact that once one has exhausted the resources of available heuristics and reached the outer frontier of knowledge, one can only proceed blindly. Now I certainly agree that there is no prerecognition. But I am not completely clear what the implications for an account of scientific research are, because there is a lot of ambiguity in speaking about the *frontier* of knowledge. Most work "at the frontier" can still make use of some fairly reliable information, laboratory practices, and even heuristics and hence is not completely blind. The frontier is pushed back from the inside, so to speak. Scientists never completely leave the cocoon of "established" practices.

Callebaut: But at this bottom level, Campbell does seem to deny that there is any reasonable account of discovery to be given.

Nickles: Well, yes, but the question is whether scientists, working within the rich sets of constraints of their problems and specialty communities, are ever com-pletely at that bottom level, whatever that means. After all, you cannot even have a problem without having *some* constraints on what would count as an adequate solution (Nickles 1983). For that matter, you cannot have even BVSR without having *some* criteria of selection, and human investigators actively working on problems are more or less aware of these criteria in advance and can use them to guide their search, up to a point. I don't think Don would disagree here at all. I just reject the tendency of some evolutionary epistemologists—perhaps Popper is one—to be too reductivist, to treat modern scientists as if they were in a "state of nature" (cf. Rawls 1971) or "epistemically original condition" of total naiveté or absolûte, Cartesian skepticism. Modern scientists are never in such a fix; and the more mature and richly endowed the science, the less they are merely stabbing in the dark.

Callebaut: So do you see yourself as an evolutionary epistemologist or not?

Nickles: In a broad sense, yes. I am quite sympathetic to Don's overall enterprise. Moreover, he differs from most evolutionary epistemologists in the attention he gives to social dimensions of knowledge-seeking communities. What I have less sympathy for is the biological reductionist attempts of some other writers to make EE into a kind of *epistemobiology,* analogous to sociobiology. And speaking of "socio," Werner, I am even leery of the label *"naturalistic* turn," especially when you and many other people speak of historicists as having made the naturalistic turn.

Callebaut: The conflicts between naturalism and historicism will be taken up again later (10.4). Don wants to react.

Campbell: Those who emphasize the importance of scientific creativity but reject a selectionist model for scientific creativity are making scientific creativity foundational and unexplained. My vicarious BVSR processes are explicitly a theory of heuristics, not a denial of them. In my paper on creative thought (1960), I explicitly discuss Newell, Shaw, and Simon's heuristics, arguing that heuristics are a product of BVSR, that they do not guarantee truth, that they inhibit roundabout solutions; but that insofar as they represent an approximate mapping of the world, they short-cut. And in the "Three introductory dogmas," in that very paper, all I ask is that within the heuristics, when you get beyond what they are not assuming insofar as they explore a more limited possibility space, they explore it without clairvoyance, without prescience, and with variation. I take the AI community as moving more and more to recognize that there is no real difference there.

Wimsatt: There are very close conceptual connections between heuristics, Campbell's vicarious selectors, and biological adaptations. I have argued since 1980 that they all meet a list of crucial properties which—most recently, in Griesemer and Wimsatt (1989)—has grown to five important characteristics of heuristics. These include not guaranteeing a solution, being cost-effective relative to available substitute algorithms or heuristics, having systematic biases, transforming the problem to which they are applied, and being purpose-relative. These properties explain why they are there, when they break down, and how they can be analyzed in terms of how they break down.

Callebaut: Don, you probably want to be exempted from Tom's charge of biologism as well.

Campbell: Tom knows that I want to include a social system for science that lets "the way the world is" be at least one of the selectors of scientific beliefs. He will applaud my criticism of recent evolutionary models of science as being epistemologically vacuous, and also, I hope, my shift to a selection-theory epistemology. If one sees the history of knowing as the historical selection of presumptions, one can recognize the extremely social and presumptive nature of science without adopting Rorty's (1979) or Nelson Goodman's (1978) ontological nihilism.

Callebaut: Dudley, your philosophy of science seems to me to be in several respects a naturalized theory of science (cf. Giere 1985b) as well as an evolutionary one. Yet in other respects, you clearly disagree with many of the views that have been put forward recently under these labels.

Shapere: It *is* a naturalized theory, in that it maintains that all our beliefs, including our methods and goals, develop within, or at least from, the context of struggling to deal with the world around us, and even to figure out what "dealing with" means. This is not something that we have determined on the basis of a priori reasoning; it is something we have learned. It is a conclusion from the best beliefs we have attained, in a broad range of scientific areas. In this sense, science provides both the object of our inquiry as philosophers of science and the framework within which that inquiry is to be conducted. This conception of "naturalized" philosophy is very different from Quine's. For him, a naturalized epistemology is one according to which psychology is the key science in understanding inquiry. That seems to me to be a hangover from the days when "reasoning" was thought of in terms of something separate from the content of scientific belief—then it was seen as something formal, purely logical or methodological, only now it's psychological, but still the rules of thinking, which don't depend at all on what's thought about. But as we can see from the analysis I've given of the solar neutrino experiment (Shapere 1982), what is relevant to inquiry or knowledge about a particular problem or domain will be what we have found to be relevant to that problem or domain, and not something general like psychology—which we know little enough about anyway!

Campbell: I am happy to put in a brief comment on Shapere's comments on Quine. I find most accessible Quine's essay "The nature of natural knowledge" (1975) rather than his better-known "Epistemology naturalized" (1969b). In the "Natural knowledge" paper, Quine clearly states that epistemology is to be done as a part of science; and that is what Shapere is saying should be done too. Shapere is clearly passing the justificatory burden to evolutionary theory and not to psychology; so I think that essay would minimize the distance he sees between his own point of view and Quine's. I, myself an unqualified admirer of Quine, have chided him on the epistemological vacuousness of his references to psychology in "Epistemology naturalized," where he fails to say *what kind* of a psychology would be relevant (Campbell 1986a, 1990a).

Shapere: No, Don, I am not passing the justificatory burden of science to evolutionary theory. Justification (and, I might add, problem formulation and theory construction, that is, "discovery") is a matter of reasons taken from the science most directly relevant to the situation or problem being dealt with. A clear example is given in my paper, "The concept of observation in science and philosophy" (1982), where it is shown that the reasoning in that case is purely physical and astronomical. Evolution plays a role only in showing how that concept of reasoning could have arisen. As to Quine, for all his statements about the role of "science" in epistemology, in "On the nature of natural knowledge" or elsewhere, he has never spoken in more than generalities, not bringing out the implications of this remark other than to place the burden of justification on psychology. And that latter is the doctrine for which he is known in this connection.

7.2 THE EVOLUTIONARY APPROACH TO SCIENCE

7.2.1. *What Scientists Agree About*

Callebaut: David, you have applied biological-evolutionary ideas both to the conceptual and social sides of science. Do you have a preference for treating one of these facets before the other?

Hull: I think social systems are easier to conceive of as individuals than conceptual systems. I also think in point of fact that the cohesiveness of conceptual systems is promoted as much, if not more, by social cohesiveness than by conceptual cohesiveness. So it would be better if we treated social cohesiveness first and then conceptual cohesiveness.

Callebaut: One way to characterize your work in this area would be to say that—very much in the vein of Don Campbell—you are trying to define social mechanisms which will allow people who disagree with each other in other respects to cooperate in the scientific enterprise (Hull, "Altruism in science: A sociobiological model of cooperative behaviour among scientists" [1978], reprinted in Hull 1989:243–62; also 1988a, b).

Hull: It is easy to cooperate when everybody agrees. But when everybody agrees, there is no evolution. If conceptual change is to occur by a selection process, you've got to have variation. But when you have variation in ideas, it is hard to cooperate. What mechanisms enable scientists—and I am sure it works for other groups too, but all I study is scientists—who disagree with each other on particular points to still cooperate with one another? *Social cohesiveness.* And the social cohesiveness allows you to continue producing conceptual variations and testing them. I have been studying the mechanisms for social cohesion. One of them is *pretending that you don't disagree:* emphasizing your areas of agreement and not mentioning your areas of disagreement; and when disagreements *do* come up, saying, "Well, he is crazy here, that's his private lunacy; otherwise he really has a good head and agrees with me; hence, he has to be right!" *Agreement is the operational criterion for truth in science.*

Callebaut: The way you formulate it, it sounds a bit like this is how it has *got* to be for a variation-and-selection approach to be feasible. But do you have any independent evidence for the correctness of your view?

Hull: In part, this view was forced on me from empirical investigation. I was studying groups of scientists who claimed to agree about fundamentals. I tried to find out what these fundamentals were. Each person could tell me what they were. The trouble is that no two ever gave the same list. When they did give the same list, when I went past the words to see what the words meant, I discovered that there was plenty of disagreement. When I mentioned this disagreement, the scientists I was interviewing would get really angry, because it was threatening the social cohesiveness of the scientific communities, and they had to cooperate.

Shapere: I simply can't agree that "investigations show that scientists disagree on everything," even slightly. On the contrary, I would say that they take for granted a lot of propositions that they use as background information. I think no working physicist will vary in his interpretation of a statement like, "All electrons must have exactly the same charge."

Callebaut: But the evolutionary epistemologists I'm talking about would argue that without variation, the scientific process would be killed off. Variation is not only all pervasive; having it is how science *should* be. So isn't the construction of a "transpersonal unity" in your concepts (cf. Tennant 1988) a kind of Platonism, or even a falsification of what goes on?

Shapere: I'm not saying at all that science should be understood as monolithic, as a one-view "reified" system. Where there is ignorance, there must be variation. But the idea of variation mustn't be elevated to some kind of basic feature of science, on some misconceived analogy with biology. There can also be agreement, and there is an amazing amount of that.[16] We must recognize variations, and include them in our account. But we want to understand them, the rationale (where there is one) behind them, and their function in science. We want to understand the major directions that are open, the problems with them, the alternative approaches that are open in the light of the problems, and so forth. Are there reasons why a certain scientist's "idiosyncratic" variations are justified or promising? Certainly, variation is necessary to assure the investigation of alternative possibilities; but that doesn't mean that there aren't boundaries, that "anything goes." And sometimes there is variation outside the general range of agreement; but even there, there are boundaries, reasons that distinguish the speculative but still scientific from the mere crackpot variations. Again, we might always be wrong; but that's part of the risk we face because of our human epistemic circumstances.

Callebaut: In Kuhn's (original) view, which implied a form of idealism and tended toward a coherence theory of truth, consensus formation plays a central role in

16. Shapere adds: "Concepts of species may be 'constructions,' but if we're going to admit that there's such a thing as knowledge, the arbitrariness that comes with taking a biological analogy seriously must be rejected."

science (cf. Lehrer and Wagner 1981). EE questions this. What moral can be drawn from lab studies in this respect, Karin?

Knorr Cetina: One has to get away from a conception of consensus formation as essentially a process of *argumentation,* as a sort of mentalistic process that involves the discussion of many scientists among each other. One has to see it much more as a *material* process. Presumably, the whole question dissolves into a variety of issues, like the closure of research through the publication of scientific papers, or the decision formats that are used in the laboratory, or the choice of investments that are made in the laboratory into the right kinds of questions that have a greater chance of acceptance, or the creation, even, of consensual groupings in scientific papers through rhetorical means. That's all part of it. It's a mixture of intentional pursuits, by scientists and laboratories, of acceptance advantages and persuasion advantages, and of, on the other hand, an emergent stabilization of certain kinds of research results which are taken over because they help *you* in the laboratory. Even such things as the periodic kind of self-summation ("What do we know in the field?") which takes place in high energy physics belong here.

Callebaut: Where does the "sociobiological mechanism," as you have called it, come in here, David?

Hull: The sociobiological mechanism comes in with the question: *How do you get competitors to cooperate?* In biological evolution, you find the greatest cooperation among eusocial organisms. It is easy for eusocial organisms to cooperate because most of them are neuters. They cannot pass on their genes directly; all they can do is contribute to their sexual kin passing on their genes. Science isn't like that: there are very few neuters in science.

Callebaut: To avoid misunderstanding, I emphasize that we are discussing *intellectual* not sexual competition among scientists here.

Hull: The problem among social organisms is that they must cooperate with their sexual competitors. How is that possible? The usual explanation, the clearest one, is kin selection; the second clearest is reciprocal altruism. Kin selection is genuinely biological. As far as I can see, reciprocal altruism is just what sociologists have been talking about all the time, but with an assumed biological basis which may or may not exist. But I found that the kin selection model worked pretty well for scientists: no scientist can do all he needs to do by himself, he needs other scientists; he must have conceptual kin.

Callebaut: Conflict and cooperation in science are unpacked in your analysis: scientists (can) serve various functions for each other.

Hull: Let us say you are producing a set of ideas and you want *support.* By citing another scientist—preferably a well-placed one—as supporting your views, you give your own views some substance. But of course, then you also imply that this guy is worth referring to; you also give him credit (cf. Bourdieu 1975; Williams and Law 1980; Fuller 1991a). So there is mutual support when mutual citation is positive, when you say "As so-and-so says . . . , as so-and-so says . . ." That is the kind of credit scientists give each other. To get support you have to give *credit*. That is the conceptual part of the mechanism for cohesiveness—that we trade credit for support. People have always wondered why priority disputes are so common in science, and why scientists insist upon getting credit—"Why didn't I get the credit I was due?" They think of that as somehow "unscientific" or "not nice" or "unethical." "*Truth is truth*. Why should it make any difference who thought it up?" Well, this is the mechanism that has evolved in science, that allows scientists to cooperate with their competitors.

Callebaut: How does this functional analysis of the behavior of scientists extend to "marginal" phenomena such as plagiary and fraud (cf. Hull 1985b; Committee on the Conduct of Science 1989; for historical and current examples see Broad and Wade 1982)?

Hull: Scientists want as much credit as they can get. One way to get credit is to steal another scientist's ideas. I don't think that conscious stealing is all that prevalent in science, but in the midst of doing research, it is difficult to keep track of who did what, who thought of what first, etc. In the midst of the hubbub of the lab, a graduate student mentions a finding to a more senior scientist. Later the scientist thinks that he thought of it, but the graduate student remembers. Such hassles are common in science. They threaten the cohesiveness of research groups. As common as complaints about plagiary are, they are not punished as severely as fraud. Once again, it is difficult to distinguish fraud from sloppiness and honest error. But when fraud can be demonstrated, it is punished very severely. Plagiary, error, and fraud are punished in science, but these punishments vary in their severity. Fraud is considered the cardinal sin, errors less serious, plagiary less serious still.

Callebaut: How do you explain this difference in severity of "punishment"?

Hull: My explanation for this difference is that in cases of plagiary immediate harm is done only to the person from whom ownership was stolen, while errors, whether intentional or not, do immediate harm to everyone who uses the mistaken result. It is easy enough to make errors without introducing them consciously into the literature. This is why fraud is considered more serious than simple errors, but they do equal damage to anyone using them.

7.2.2. *Threats to the Social System of Science*

Callebaut: Don, you like to point out that David is doing more than just describing the scientific enterprise.

Campbell: My own sociology of scientific validity (1986a, 1988b, 1990a) is contingent on a guess about human nature, and so is David's. His description of the social processes of science make it plausible that these selfish, dishonest, greedy human scientists operating in the right kind of social system will produce improved competence of reference in their beliefs. That is *contingently* normative. He would agree with me that this is a *fragile* social system, hard to implement if you're a young scientist from a Third World country, hard to implement under some totalitarian regimes unless you're working in the war industry, etc. David is not only *describing* science as a process, but also saying, "When science works this way, improved belief will be produced." He is not proving that you should *want* to improve belief; he is contingent in describing human nature and the society that we live in.

Longino: The social system of science does require openness of communication. That is threatened in a number of ways (Longino 1986). It is threatened by the privatization of research which we find in biotechnology, where a certain amount of applied and even basic research becomes private property in order to support various initiatives in biotechnology. You don't want to share it, because that threatens your potential profits.

Callebaut: Friends of mine—a couple—work for two different biotech research institutions; they're not supposed to talk about their respective work to one another.

Longino: Do they have tape recorders attached to them so that this can be monitored? That must be crazy-making! That's one kind of interruption of open communication. Another interruption is of the sort imposed most notably by the Reagan administration during the 1980s, when scientists from different nationalities were prohibited access to scientific information developed in the U.S. They were prohibited—even though they were working in the U.S.—from attending conferences on materials science, for example; they were prohibited from participating in certain computer science conferences and from getting access to data that was produced in those contexts. That seems to me a blatant violation and interruption of the scientific process and of the openness necessary for science and for what Hull calls "checking" or I call "criticism," both of which are essential to the purported objectivity of science. It's a political interference which has very severe epistemological implications.

Callebaut: Are you more optimistic about the future?

Longino: Well, we're all right now in this euphoric haze over what's happening. I think it is not going to last very long. I'm sure that we will just fall into sorts of new configurations of global oppositions. I am not hopeful that peace is going to break out all over. And the Bush administration is clearly continuing some of these restrictive practices.

7.2.3. *Speciating Cladists*

Callebaut: One of the groups of numerical taxonomists you have been studying for quite a while, David, the cladists (the term "cladistic" was formed on the Greek word for "branch"), was in the process of speciating into two groups (Hull 1988a).

Hull: The cladists are breaking down into phylogenetic cladists and pattern cladists. I first noticed it about ten years ago. When the cladists first formed an organization and had their first meeting, I gave the after dinner talk. It was supposed to be funny and anecdotal. In it I mentioned that it looked like the cladists were speciating into two groups. People afterward complained, "No no no no!" Well, after a year or two, the split got so wide that they could not pretend it did not exist. And then it was *my* fault: I had caused it! By noticing it I had caused it! I think they are exaggerating my power: there is no way that a single person can cause something like that. But the social cohesion was breaking down, and with it the conceptual cohesion. Neither is the sole cause or effect. There is interplay: the social cohesion was probably slightly more important than the conceptual cohesion, but both were breaking down. In short, social groups and species split up in much the same way.

Callebaut: Another example you have analyzed is Mendelian genetics.

Hull: Mendelian genetics provides a good example of the contrast between treating conceptual systems as classes and as historical entities. If Mendel's laws are considered classes, then anyone who came up with 3:1 ratios in their study of inheritance and explained them in terms of pairs of contrasting factors discovered Mendel's laws. All that is necessary is that these various formulations are similar enough to each other. Hence, dozens of people both before and after Mendel discovered Mendel's laws—including Darwin.

Callebaut: Many people, including people who like to think of themselves as radical epistemological relativists, like to think that is literally true.

Hull: There is a sense in which that is true; but that sense is not the sense which is operative in science. If you really want to understand how conceptual systems evolve in science, you have got to treat them as if their origins matter. They matter in two respects. They matter because credit is involved, and one of the mechanisms that allows scientists to cooperate, I repeat, is giving credit to one another. But it is also because of *successive generations*. For a scientist, his or her immediate past experience will determine his future. That someone thought something a hundred years ago and you do not know about it makes no difference. You find out, let us say, afterwards. For anyone who publishes some great idea, immediately there will be people who go back and find precursors. Unappreciated precursors are curiosities, of antiquarian interest only. They make no difference if the scientists involved had not read these people, or read someone who did. Unless the past ideas feed into the present, they do not matter.

Callebaut: You have illustrated this in Darwin's case: Darwin's "discovery" of evolution by natural selection.

Hull: Darwin *did* read. At the relevant time he had not read Lamarck, because his French was not so good, but he had read Lyell; and in Lyell is a good explication of Lamarck, so that Darwin had contact with Lamarck's ideas. The effect of Lamarck's ideas was not to cause him to accept evolution. In fact, just the opposite: Lamarck's views counted against him becoming an evolutionist. But what they did do was at least allow him to see what the obvious weaknesses were so that he was able to protect himself. Lamarck counts not as a precursor—it's not that Lamarck's ideas caused Darwin to have similar ideas—but Lyell's reaction to Lamarck caused Darwin to have the ideas. So it is Lyell, a creationist, who is one of Darwin's appropriate "precursors." A guy named Patrick Matthew, it turns out, published a paper which had natural selection cold (1831). No one noticed it, certainly Darwin did not; *Patrick Matthew is irrelevant*. From the point of view of ideas as classes, Matthew's ideas are part of Darwin's class of natural selection, but from the point of view of the ongoing process of science, Matthew is irrelevant (cf. Hull 1973, 1985a).

Callebaut: How is one to individuate ideas?

Hull: If the class notion is irrelevant, then it obviously is not the one you should be using. You've got to organize ideas as if they form spatiotemporal lineages, otherwise nothing makes sense. In the case of why Mendel is given credit (cf. Olby 1979; Brannigan 1979), we do not have the history down too well because things happened in a period of a couple of weeks and it is very hard to date who read what when. De Vries *may* have read Mendel in time for Mendel's papers to have influenced his ideas. If he did, then I say: Mendel is relevant. If he did not, then Mendel is not relevant with one exception. There were three people who could claim priority. One of them was not William Bateson. But William Bateson was the father of Mendelism. He was not going to be one of the precursors. What is he going to do? With these three people all claiming priority, you are bound to have a big, nasty priority dispute. If Bateson gives credit to Mendel, then there is no priority and there is nothing to argue about. I think the main reason why Bateson and the two lesser rediscoverers were so quick to give Mendel credit was so that it would not go to de Vries. And the reason why de Vries was *not* willing to give Mendel credit was that de Vries had the greatest legitimate right to have genetics named deVriesian genetics.

Callebaut: How does your lineage approach deal with the ubiquitous phenomenon of *multiple discovery* (Lamb and Easton 1984)?

Hull: In a sense, *by definition* there cannot be multiple discoveries, because multiple discoveries mean they are independent: if they are independent, they are not the same discovery. So in a sense it is defined away. But it does give a different problem: What do you do?

Callebaut: Exactly.

Hull: Well, if the people do join in the same research program, then, though they might have had independent origins, they merge into the same lineage. If they do not, then they are separate. It is always interesting to find which ones merge and which ones don't; because you never could predict at the time who is going to merge. Darwin and Wallace merged! They did not let their relationship degenerate into a fight over priority. Darwin at the time was already a major scientist. He said to this obscure guy, "Why don't we do this together?" And the obscure guy said: "Great!" Who could want something more? And they cooperated all their lives.

7.2.4. *Intentionality Doesn't Make Much of a Difference—Or Does It?*

Callebaut: A classical objection to EE is that whereas biological evolution proceeds "blindly," scientists act intentionally.

Hull: That is the toughest problem, in part because you cannot talk about intentionality without getting into really deep, complicated philosophical issues. *What is wrong with EE is that it is 99 percent philosophy and only 1 percent science.* I would like to rectify that proportion. I think I can get you to see how the problem can be circumvented. Let us say you distinguish biological evolution from conceptual evolution in terms of *mode of transmission.* In biological evolution, the transmission is genetic, and in conceptual evolution the transmission is conceptual—words or what have you. Then you can also distinguish those changes which some intentional agent *attempted* to bring about and those that were not.

Callebaut: With respect to the latter, Elster (1983a:20) talks about both "sub-intentional" and "supra-intentional" causality. Sub-intentional causality is "involved in the mental operations that are not governed by will or intention, but—metaphorically speaking—take place 'behind the back' of the individual concerned." Supra-individual causality refers to the *non-intended* causal interaction between intentional actors, as opposed to the strategic interaction studied by game theory.

Hull: Let us look at biological evolution, natural selection. No organism prior to human beings knew anything about evolution; so they could not be trying to evolve. Among *Homo sapiens,* a minuscule percentage of people have understood evolution and believed it. Most people today do not believe in evolution. Of those of us who believe in evolution, what percentage are trying to change how we evolve? Very few. The Population Zero movement has attempted to control our evolution, but from the point of view of the biological evolution of *Homo sapiens,* you can ignore intentionality. Yes, we are intentional; some of us do understand evolution. Yes, we would like to keep ourselves from going extinct.

Callebaut: So it is an in principle possibility . . .

Hull: . . . that we can ignore in practice. It makes no difference. However, there is one area where *Homo sapiens* has had an effect: we have domesticated plants and animals. For most of the time, we did not understand genetics; we did not know what we were doing. We wanted better wheat, we wanted better cows; but we did not really understand what we were doing. Artificial selection, is that biological or social? The transmission is genes; so in that sense it is biological. Yet intentionality played a role: we were trying to get more milk. Has artificial selection suddenly become a matter for social scientists and philosophers? No! It is still straightforward biological, even though intentionality plays a role; which means that *intentionality doesn't make a difference.* We can keep it in the area of biology and forget about intentionality.

Callebaut: Notwithstanding all this, you will grant that science is one of the most intentional activities there is?

Hull: Certainly. If scientists did not try to solve problems, they *would not* solve them. But I think that the general nature of the system can be worked out holding the issue of intentionality off to the side for a while—for a long while, as far as I am concerned. I think it introduces incredible complexities that probably are not going

to make much of a difference. And I am not saying that people are not intentional—they are. All scientists are trying to make big discoveries; only one in a
thousand does. So yes, it is a difference; but it is not that much of a difference.
Leave it out and you introduce bias, but not that much bias.

7.2.5. *Let's Keep Science and Politics Separate*

Callebaut: David wants to circumvent—not to deny—human intentionality, and one
can see why. Other people have qualms here. Helen, why do you insist on keeping
the intentionality in our picture of science?

Longino: *I insist on intentionality for political reasons.*

Callebaut: Great opening line!

Longino: I think that any form of democracy—whether we're talking about liberal
democracy or democratic socialism or whatever—only makes sense as an ideal if
we think of human beings as having the capacity to make decisions about their
lives based on what they believe about themselves and the world. My claim of a
right to participate as an equal in the governance of my community is grounded in
this capacity, and makes no sense if I lack such a capacity. Democracy makes sense
as an ideal only if intentionality is in some important and meaningful sense a
characteristic of human beings. That's why I would insist on intentionality.

Callebaut: Philip, you have reinterpreted David's evolutionary model in terms of a
credit economy (Kitcher 1993).

Kitcher: I see David as having an interest in a general view of scientific change in
evolutionary terms. I want to figure out the kinds of microprocesses that lie behind
that. That's why I think what I'm doing is an extension of some of the things that
David has to say, even though I disagree with some of the details.

Callebaut: Helen, you think that a concern for truth has to be operative in the scientific community. On one reading of David's view and Philip's interpretation of it,
that would seem to be something one can dispense with . . .

Hull: . . . but not on your life.

Longino: When I was talking there, it was probably the philosopher reconstructing
science, asking "What has to be the case for this to work?" Let me use truth as an
example here; there can be other values across communities that make the discourse in one relevant to the discourse in another. I'm saying there have to be some
values that are shared in order for criticism to be relevant. When I say that a
concern for truth has to be operative in the community, or that truth has to be a
value across communities, it's not that individuals have to care; it's not that they
have to be motivated by the search for truth, but that the activity of their community has to have as a goal the production of accounts of the natural world that—in
some sense of "true"—are true. Some individuals as we know them may be motivated by a desire for a Nobel prize; but it's the goal of the community activity
that matters. It's at that level that the goal has to operate. And I think the same
thing is true in Hull. He talks at the level of individuals; individuals have to use

other individuals' results in building their views. That's where the concern for accuracy . . .

Callebaut: . . . and trust . . .

Longino: . . . comes in. Why is it important that other people's results be accurate if you are going to use them? Because of the goal of the overall activity, even though I am just engaged in it in order to win the Nobel prize. But the goal of the overall activity is some description of the natural world.

Hull: It is important that a large segment of any one scientific community be committed to finding the "truth," sophisticated philosophers to one side. However, the system is so organized that even those scientists who are not so committed can also contribute. Helen's response is right on target.

Callebaut: It would seem to be only a small step from Hull's "innocent" view about the relative irrelevance of intentionality in science to the muddied waters of a Hayekian view of political institutions and societal institutions in general as the essentially non-intended consequences of human action (Hayek 1967; O'Driscoll 1977). Marx, in his day, thought he had an elegant solution for this problem— roughly: (only) in a communist society are social relations so transparent as to allow for effective goal-directed collective action. But game theorists, among others, have made plausible that the problem of non-intended consequences is probably to stay with us forever (e.g., Boudon 1977; Elster 1978, 1983a).

Longino: It is of course the case that the outcome of many of our political actions is not what we intended. This is treading into problems in political theory and issues concerning the appropriate size of political units: what size of political units can we have and retain any meaningful form of self-governance? How far can our collective intentions be effective? They can be effective in a certain short run, but in a longer run we don't have complete control of the effects of our actions, because we don't know how they are going to interact with the actions of other groups and so on. But we engage in what we hope are corrective measures when the consequences of our original actions deviate from our intentions.

Hull: It all depends on what you mean by intentionality. Scientists *have* to be intentional agents. Enough scientists *have* to intend to discover the truth. However, what particular beliefs they have about nature are not so relevant. Most hunches turn out to be mistaken. For all the scientists who intend to make a great breakthrough, only a few do. A very high percentage of athletes training for the Olympics intend to win medals. It is important that they do. Only a small percentage actually do.

7.2.6. Evolution or History? The Sociologist Strikes Back

Callebaut: In David's book (Hull 1988a), there is a certain tension, I think, between entities—individuals—which are *historical entities* and what he calls "general reference classes." The elements in Mendeleev's table may be taken as natural kinds, and you can, of course, do a whole lot with them. Now that is just a fiction, but a very useful one indeed. Whereas in the case of social science . . . Alex Rosenberg would say that it is almost impossible to do social science, precisely because of the lack of natural kind terms (see 3.3; Rosenberg [1992b] applies his generally pessimistic view to Hull's account of science). It's a pity, but there isn't much which can be said in a scientifically meaningful way about something as particular as humans . . . In some cases—chemistry, physics—the fiction of natural kinds works; in others it doesn't. On such a view, you have two layers in your science: (1) the particular historical objects, which are somehow linked to (2) the level of the natural kinds.

Latour: Hull mixes up a lot of things, so it's difficult to talk sensibly about his approach. He's using a big metaphor of biology; and biology is entirely uncertain about what is the history of a species. How can you use the obscure history of species to enlighten the obscure science and history? It's absolutely unbelievable to me. I read the whole book. It's exactly "conservative naturalism."[17] I mean, it's just a way of accepting a bit of history without doing the history.[18] It's only a history of ideas; the only difference is that now, ideas are endowed with a sort of reproductive mechanism, so that it imitates, in some sense, the genes. And then it can be hooked up on very classical sorts of model building and very classic ideas armchair philosophers of science have about the making of science. But if you read Hull's book, there is no history; there are only ideas, ideas fighting with ideas. No people. Or when you do have people, the ideas remain the same; it is only anecdotes.

Callebaut: I saw a lot of resemblances—and of course also many differences—between *Science as a Process* (Hull 1988a) and *Science in Action* (Latour 1987). Hull's social-conceptual divide seems to me to be still very Mertonian. But on the other hand, he has this metaphor, "A scientist is but a text's way of producing another text." When you take the idea of the replicator/interactor distinction seriously, as for instance Wimsatt and Griesemer have done (cf. the exchange with Richards in 7.3.2), it provides you with a new tool, and you can get all kinds of new things out of it. For instance, you can now consider the scientist as part of the environment of "her" or "his" ideas. I agree with you that we may have to go even more to the microlevel than David has done (Tennant 1988; cf. also Rosenberg [1992b] on memes as tokens not types).[19]

17. "Conservative naturalism" is Kitcher's label for his own views. In the interview, Latour often refers to Kitcher (they are colleagues at the University of California, San Diego).

18. While completing the bibliography for this book, the following sentence from a paper by a professional historian of science caught my eye (the paper is on the role of "myth" in science and in the history of science): "Latour and Woolgar, in *Laboratory Life,* point to a possible fruitful direction, although there is no historical dimension in their work" (Abir-Am 1982b:285).

19. Similarly, the social factors typically invoked by SSK advocates as having causal influence on the content of science may not be sufficiently fine grained to match the details of their supposed effects either. For my money, Slezak (1989a:589) was right on target with his rhetorical question: "Did Gödel's 'incompleteness' theorem arise from lacunae in the Viennese social order of 1930?"

Latour: There is a common principle, which is probably very interesting, between Darwinian evolution and the history of science. But you have to try to *discover* it. Elisabeth Lloyd or Bill Wimsatt are much more interesting in that respect. Of course it's not the *vulgate*—a vulgarized account of the evolution of species— which is then used as a metaphor to understand historical progress in science. If you want to compare Darwinism and the history of science, you should not black-box Darwinism at all, nor should you black-box history; you should leave them both open, complex, and try to see what common philosophical problems they raise. I think that is what Lloyd does. But not if you take a very, very reactionary way of accepting a bit of history without paying the price. That's exactly, I think, the situation of Hull's book.

Callebaut: You're not suggesting that of all people, David has got his biology wrong, are you?

Latour: No, his biology might be right, but what does "right" mean? There are as many Darwinisms as there are Darwinians. It's very much like Kitcher. They want a bit of history, but not too much; because what they want to save is ideas, and models, and the universality of science. They have to pay the price of what? They are ready to shift from individual to collective. That can't be avoided; now they know they have to talk about collectives of scientists, and collectives of scientists fighting with one another. That's what they've taken from sociology. But they want to stop after that. After that, Kitcher wants to be able to build models, like econ-omists or game theorists do; and Hull wants to do evolutionary biology. But they don't want history—history meaning localities, events, circumstances—and es-pecially they don't want culture in any deep sense—in the sense of "no demarca-tion" criteria and so on. They want to avoid anthropology of science by paying a price: they accept the collective character of science. *"C'est la part du feu."* So it's naturalistic in the classic sense, in that case, of going from one metaphor, from universal . . .

Callebaut: . . . from physics to evolutionary biology as a new paradigm of science . . .

Latour: . . . yeah. So science is not eternal; it's collective; and their models have a bit of evolution. But they are not yet historical and they are not anthropological.[20] So it's a way of avoiding the consequences of carefully reading science studies. That's my extreme point of view.

---

20. Historian of science Garland Allen (1991:702–3) even talks of—horresco referens—"Hull's 'es-sentialism'": "It becomes clear from *Science as a Process* that Hull thinks of the process of science as somehow eternal or universal, neither fundamentally different from culture to culture at the present time nor varying from one historical time period to another. In other words, the products of science—ideas or theories—evolve, but not the process. . . . In his mind the practice of doing science does not change; only the outcome, the theories of science, changes. Perhaps what I see as Hull's essentialism results more from omission than commission, but it seems a glaring contradiction in an otherwise consistently evolutionary view of science."

Callebaut: In Hull's defense, I could say: "Look where he's come from. He has given up a lot already! He has given up rationality . . ."

Latour: . . . No! . . .

Callebaut: ". . . in the sense of a rational method of science" (cf. Giere 1988).

Latour: No, it's not true, because in his book you have the equivalent of rationality, which is *the best allocation of resources*—a typically American myth. It's a completely American myth of economists: the optimization of resources, and now of disputes.

Callebaut: Wasn't that actually a French model, originally?

Latour: Yes (*laughter*), but the French don't believe in it! The whole idea is to have eventually replaced all the classic things of philosophers of science by the economists' equilibrium, using the poor Darwin in the process. That's kidnapping!

Hull: I suspect that Latour's dislike for my book comes from the credit/competition theme. In his book with Woolgar, Latour dismisses all the talk by the scientists whom he interviewed about wanting credit and competing for it as resulting from their being studied by social scientists. Scientists went on endlessly about competition and the like because they thought that this was what Latour and Woolgar expected. Talk about ad hoc. I think we all have to finagle a bit, but there are limits. I guess Latour and Woolgar are exemplifying their own principle that evidence does not matter. For them, apparently, it does not. Sorry to be so catty.

Callebaut: To the extent that Hull's analysis remains Mertonian, it will be an understatement to say, then, that American and Continental sociologists of science are rather different animals.

Burian: There seems to me to be a contrast in biology (at least in the 1950s, perhaps less so now) between the American structuralists who want to determine the structure of a molecule or of a cell and so on, and a series of rather diverse European physiological traditions. It is no accident, I think, that it was in France that nucleocytoplasmic relations were worked out in a way that would have been much more difficult in America at that time, because of the strength of the kind of bacterial physiology, micro-organismal physiology, which did not come from a genetic, structural tradition. In something like the same style, Mertonianism is too structural (Burian and Gayon 1991).

Callebaut: A similar point was made by Giere (1989a) in a 4S talk.

Burian: It seems to me that in a certain sense, the conflict between the European and the American sociologists could in part be parsed this way. You have a reward system, which is a structure *from* which functions get derived in the Mertonian vision as I see it. Whereas the functional interrelations in which even truth is negotiated—in which everything is functional through and through—is more in the style of what we call the "European school" in sociology (e.g., Luhmann 1990a). I'm suspicious of a too narrowly structuralist approach. On the other hand, to the Europeans I would say that unless you can put function and structure together, you haven't done the whole job right either.

7.2.7. Le roi est mort—Vive le roi!

Callebaut: Hull's evolutionary account of science and scientific change can be un-coupled, to a large extent, from the realism issue (cf. Hull 1988b).

Latour: Well, I've never met, really, a realist philosopher, so I don't know . . . I know they exist somewhere, but I think that's a bit of a dead issue.

Hull: Latour has met me and I am both a philosopher and a realist.

Callebaut: The strong connection between rationalism and realism which some people make when criticizing philosophy of science is not obvious. Rationalism and realism can be disconnected.

Latour: As I said before (4.3.1) there is a long-standing antirealist tradition in French epistemology, which is of course also very rationalistic. There is no realist French philosopher.

Callebaut: I agree—or almost (see Granger 1992). What I was rather hinting at was that nowadays, sociologists—and I am now not talking about French people— lump together realism and rationalism; for them, all philosophers are *by definition* realists. Ron Giere, in *Explaining Science* (1988), argues that there is no rational-ity in science in the sense of following a rational method . . .

Latour: . . . but realism? (*Laughs.*) What would be interesting is to study the avoid-ance strategies of all these people. I mean, how can you give up a *bit* of philosophy of science? That is something I cannot understand. *What should disappear is phi-losophy of science.* What will be left is philosophy. These people are deeply dis-trustful of philosophy. They think philosophy of science is different. That's the only good thing with France, actually: we don't make the distinction. The French know all of the important questions are in philosophy, not in philosophy of sci-ence. So I'm very interested in all those guys' avoidance strategies, because they want to save a bit of a discipline that has to disappear! Philosophy has to exist, and of course the whole field—whatever it is called—of studies of science in action. But philosophy of science as a sort of mixture of *pronunciamento* and a bit of history . . . You can't have an evolutionary stable strategy[21] in that. You either have to give it up entirely, or you don't. You can't be halfway. It's very amusing to see people trying to be halfway . . . a bit of an empiricist and historicist and a bit of a philosopher of science in the traditional sense (cf. Pickering 1991). They are sitting in between two chairs.

Callebaut: A last attempt to defend such an enterprise: Science studies are in too confused a state—that's perfectly normal, since they are very young and all that; but it can't continue indefinitely. So at some point some unity has to emerge, and that's where you can put in the Darwinism, according to some people. Darwinian

---

21. This game-theoretic notion (Maynard Smith 1982; cf. Kitcher 1985a:89–95, 1987a) captures the idea that a form of behavior may be maintained in a population under selection provided that rival forms of behavior cannot gain advantages in interactions with it: *I* is an ESS with respect to a set of rival strategies *R* if and only if there is no strategy in *R* that would be able to invade a population playing *I*.

thinking has provided biology with much of its present unity. It's a unifying framework . . .

Latour: "Unified" is not the word. There are hundreds of different interpretations of Darwinism, which are themselves in dispute with one another, disputes that should be studied by social historians and philosophers. They can't be used to throw light on history!

Callebaut: I'm talking about the modern synthesis, which was, of course, also a way of eliminating all kinds of alternatives (Provine 1988).

Latour: If we talk about hegemony and domination and simplification, OK. But, I mean, we know enough in the history of science to know that a conquest or a hegemony is not the same as being a unifying framework.

Callebaut: No, no; but that's how you call it. That's how you call it while attempting it and when you've succeeded . . .

Latour: We can attempt, but we won't succeed. We won't succeed to unify. We *should* not succeed in unifying.

Kitcher: Once again, I'm mystified. What's the criticism of David and me? I suppose that we try for generality where none is to be had. But that's simply assertion, to which I respond, "Let's give it a try." Think of people as part of the natural order. Among their goals are attempts to represent the natural order. Those attempts have a history, and the history shows them as improving their representations and improving their strategies for constructing representations. Philosophy has the task of trying to expose, given what we now believe, what are the best ways, individually and collectively, of going on. Latour just misses this point.

Giere: I must plead guilty to having advocated a "unified theory of science" (Giere 1984b, 1988). More recently I have begun to think there are real limitations on how unified a theory of science could be (cf. also Rorty 1988). Science is a complex phenomenon. It has all kinds of aspects that one could not expect to capture in a single family of models—even a fairly extensive family. There are *cognitive* aspects involving the creation of models and judging how well a model fits the world. There are *social* aspects involving communication and other forms of interaction among scientists. And there are aspects *of change,* for which an evolutionary model may be quite appropriate. I think an adequate theory of science will require at least these three sorts of models. The problem will be to get them to mesh comfortably together. Despite his protestations, Latour is the real unifier, in the sense that he wants a *monotheoretical* account of science—which is to say, his account and nothing but his account.[22]

---

22. Cf. 3.6.9 and Knorr-Cetina (1985:581–82) in her review of Latour (1984): "My major criticism of Latour's scheme is that it appears to be boldly reductionist, and hence runs against his pronounced goal to develop a non-reductionist analysis of science . . . Latour appears curiously unreflective in this matter, and never seems to notice that *his* reduction of social and scientific life to a particular form of political power struggle which rests upon alignment and translations does not escape the criticism he advances against other versions of reductionism."

Latour: I think that here Giere misses the point. There are two types of unification. One is by way of developing a metalanguage for science, which I think is what he tries and what Philip tries. But there is another way, which is the approach that I take: *infralinguistic resources,* which are just a way to move from one frame of reference to the next. So like theirs, my account is completely general—I agree. But unlike theirs, it is also completely empty. It does not say in which way sciences are made. That is, I think, our difference.

Kitcher: The dialogue between Bruno and me is an evolving one. His last and most recent remarks get something completely right. He *is* providing a framework which can be used not only to fulfill what *he* sees as the legitimate purposes of unification, but can also be employed by those of us who want a little more as a basis for going on. For example, from his maps of the ways in which networks change over time, one can try to give the kinds of explanations and prescriptive advice that I think both Ron and I are after and indeed to formulate things more precisely within the context of what Bruno thus provides . . .

Callebaut: . . . Actually Ron doesn't see much of a role for methodology (cf. 5.6.2), but he certainly agrees with you insofar as explanation is involved, as the very title of his book *Explaining Science* (1988) suggests.

Kitcher: I now begin to see something extremely valuable in Bruno's construction of the network approach to mapping scientific change. And I see it as a prolegomenon to a bigger enterprise. I suspect he is going to reject the extra parts that I want to put in! But nonetheless I think we now have a basis for discussing various things with one another and a clearer conception of what we are doing.

Giere: Among the alternatives for a *general framework* for understanding science as a cultural phenomenon, I would pick the *evolutionary* framework. There is room in that framework for both *cognitive* and *social* mechanisms—as there is room in organic evolutionary theory for both genetic and ecological mechanisms. At the cognitive level I think we need models both of modes of representation and of scientific judgment. At the social level I am less sure what is needed, but Bruno's network model is certainly a powerful contender. It is less promising as a unifying framework, even one that serves only as a linguistic resource with no empirical content.

7.3 A Tool for the Historiographer

Callebaut: Contrary to Bruno, Bob, you seem to regard an evolutionary approach to the history of science as a kind of panacea.[23] (You are, of course, invited to qualify that statement.) You have distinguished several historiographic models of scientific change: the *static* model of the Renaissance—which in fact proposed that science

---

23. Richards (1977, 1987:intro. and appendix 1).

does not develop at all;[24] the gradual *growth* model (e.g., Sarton 1927–47, 1952) that to some extent still perdures today; the *revolutionary* view—not to be confused with Kuhn's revolutionary perspective—which thought of a discipline as reaching a certain point in its history at which a conceptual revolution occurs that puts the discipline on the road to modern science. That view too has been reconsidered.[25]

Richards: Kuhn dispatched the older growth and revolutionary models by maintaining that a discipline can undergo any number of conceptual alterations, no one of which is more authentically scientific; each one is simply different from the others. And these revolutions occur not for good scientific reasons, but for, as Kuhn suggests, social reasons, psychological reasons—they are like conversion experiences.

7.3.1. *Tu quoque, Edinburgh*

Callebaut: That takes us to the present.

Richards: In more recent times, sociologists of science have had a say about the development of science, and sociologists from Edinburgh have maintained that the real factors in the development of science are social factors. They suggest, much as Marxists have done in other areas, that the causal relationships are to be thought of "vertically," that is, from the substructure of a culture, rather than "horizontally," that is, between knowledge claims. They claim that social factors really play the important roles in the origin, development, and validation of particular scientific ideas. Of course, the proper response to those who argue that science is completely socially determined is to ask why we should believe sociological theories formed by a dour Scottish culture.

Callebaut: Not all sociologists of science accept this "vertical" picture. How do you, as a constructivist, react to Bob's *tu quoque* argument, Karin?

Knorr Cetina: Here I find quite useful the analogy of the human mind which comes out of the neurophysiology of cognition (e.g., Maturana and Varela 1980). It conceives of the human mind as an informationally closed system (not an energetically closed system: energy comes in and out): information does not really get into the mind. Perception, for example, is an internal process, an internal reconstruction. Nothing going on "out there" really gets in through perception; it's always an internal reconstruction. When you use this analogy and say that what goes on in science can be conceived in similar terms as internal reconstructions, the question

---

24. According to this conception, says Richards, "knowledge that we call 'special knowledge about the world' is 'infused' at a particular time, usually given to some wise man in the past, for example Moses. That individual, then, would pass the secret knowledge to his successors."

25. Kuhn's paradigm model or Gestalt model, Richards emphasizes, is a notion that other historians and philosophers of science had adapted in their own ways, or had originally arrived at. "Michel Foucault (1972, 1973), for example: his notion of an 'episteme' is very much like Kuhn's paradigm. Hanson (1958, 1963), for another, saw the development of science much as Thomas Kuhn did."

of course is no longer whether something is an internal reconstruction or not—because it's all internal reconstructions. The interesting question now becomes, "What differences are there in terms of the kind of reconstructions that are produced?" This is a question you can address not only to the result, but to the methodology as well: what procedures do they use? The construction apparatus of the mind is very complicated, very differentiated, and very fine grained. Presumably, what gets done by this apparatus is different from what would get done by a different sort of apparatus. Then you can address the question whether what comes out of these studies is useful or not in terms of what procedures of reconstruction they have used, including such simple things as "Have they just sat in their armchair and produced a model?" or "Did they make an attempt to go in the field?" And although it would be acceptable to say, "Everything they do there is always a sort of internal reconstruction of what goes on, is always based on presuppositions which come from our field and is never simply a reflection of reality," you could still say, "OK, by way of the kind of method they have used they have provided an account of what goes on in the production of knowledge which is useful for some purposes." You can ask yourself whether it is useful, you can go into the laboratory and see whether you can recognize some of the things, whether the account laboratory studies have produced opens up the world of the laboratory for you. I think that would be the criterion. So the *tu quoque* argument in that crude form just says nothing; you can address it to everybody. I certainly wouldn't think that it suggests that you should stop doing these kinds of studies. It becomes a question of what sort of reconstruction procedure yields more interesting results for certain purposes.

Callebaut: *Tu quoque,* Bruno?

Latour: The first thing is that it's not a purely social argument. To suggest otherwise is precisely the danger and the implausibility of the Edinburgh approach. The *tu quoque* argument is a much stronger argument when coming from Ashmore, Woolgar, and other reflexivists, for whom it means that it applies to our account. Now of course it is seen as destructive only by those who believe that rationality is the basis of science. But for those of us who believe that even the hard scientists never had such rationality to ground their knowledge, it's not self-destructive. It's *self-exemplifying* our position, and I think there is nothing wrong with that.

Callebaut: Still you think there *is* a problem with the *tu quoque* argument.

Latour: The reflexivists believe that there might be a way out of this difficulty. In a way, I think, *there are no levels of reflexivity.* That's the problem I have with them. They think that if you add a second level of reflexivity, you somehow become more of a constructivist, more aware, more conscious. And if you add a third level . . . But all this levels stuff is an illusion, it's completely irrelevant. It's like saying that a painting that has the painter in it is more reflexive than one that doesn't. It's an amusing trick, but it doesn't mean anything for the painting. You can have a reflexive painting without painter in it. Deep down in the reflexivist's mind, there is this idea that somehow you might ground your knowledge on a solid basis. That, I suspect, is why the reflexivists are so busy with this.

7.3.2. *Evolutionary Eclecticism*

Callebaut: The evolutionary model that you advocate yourself, Bob, is eclectic in that it takes into account all these particular views and tries to make the best of them.

Richards: An evolutionary perspective in science maintains, just as in biological evolution, that such evolution occurs willy-nilly. Therefore one cannot predict what are going to be the most important factors in giving origin to, in shaping and maintaining sets of ideas.

Callebaut: Could you give an example?

Richards: For instance, Darwin's own views were largely shaped, I believe, by antecedent theoretical scientific views in biology. It is easy enough to think of Darwin as simply giving expression in the biological sphere to certain economic and political relations that were extant at the time of his writing. Everyone points to the fact that he read Malthus and that this gave him a theory by which to work (cf. 6.2). Many scholars then immediately conclude that Darwin's theory was largely shaped by economics or economic theory. But that is just simply a misreading of the historical evidence. Darwin's own ideas were largely formed by antecedent scientific views, his own observations, certain peculiar problems that he had to grapple with and found interesting. The development of his theory comes out of that matrix more directly and more proximately than out of considerations of the culture or the social ideas within the culture. Now this is not to say that the cultural matrix didn't have a role. It is just not quite so proximate as the scientific matrix.

Callebaut: How does your evolutionary model make sense of this?

Richards: It allows one to understand how such theories as Darwin offers us came to be developed. Darwin's ideas grew in *several overlapping environments*. The most proximate and the most forceful in shaping Darwin's ideas might be the environment of his scientific predecessors; the types of conceptual problems that Darwin received from them and his own efforts in solving those problems led, I think, to the views that he later maintained.

Wimsatt: Would you extend that to include natural theology? Now I am thinking of your paper on why Darwin delayed (Richards 1983).[26]

Richards: I think yes indeed that certainly Darwin's own conception of the important problems to be solved in handling the species problem were inherited from the natural theologians; not only William Paley (whom Darwin read as an undergraduate at Cambridge) and the kinds of considerations that Paley offers in his book

---

26. Cf. Ruse (1982:pt. 1) and Hull (1989:chs. 2 and 4). Contrast the moderately externalist explanation of Darwin's delay offered by Gould (1980:21–27)—fear to expose the materialist implications of his evolutionism—and Desmond's (1989) social-constructionist explanation—dread of being lumped with the dissenters and atheists in London's medical schools and secular university (University College London).

Natural Theology (1835–39), but such natural theologians as William Kirby and William Spence, whom Darwin read both while on the Beagle voyage and shortly thereafter (cf. Darwin 1839). These natural theologians were also scientists—they were entomologists—and their work on insects was terribly important in directing Darwin to develop his evolutionary views in the way he did. But the kinds of problems they set for him were also significant in shaping his ideas and in validating them.

Callebaut: Your evolutionary-epistemological model requires you to come to grips with a variety of causal influences on the shaping of ideas; it does not hold one set of ideas to be privileged. You don't think that "eclecticism" is a dirty word?

Richards: Watchful empiricists—people like Imre Lakatos—believe that it is really only those considerations based in immediate *empirical* evidence that ultimately are the telling ones. Others, like the Edinburgh sociologists of science, think it is really only the *social* factors that are the telling ones. Both views, I think, are too exclusive. They are not sensitive to the multitude of factors that might influence the thought of a scientist. And if the thought process of a scientist is a series of historical entities, then, like biological species, they are subject to a variety of causal influences, no one set of which you can predict ahead of time might be the important ones. You simply have to do a historical analysis to see which ones are.

Callebaut: You think not only that one can model historical processes such as the development of scientific ideas on the formal structure of evolutionary theory, but also that the analogy can be drawn out rather finely. Can you spell out the "logic" of your model in some detail?

Richards: I think of scientific conceptual systems as the analogue of species; so that, let's say, Darwin's idea of evolutionary theory, which exists in his head and in the heads of his disciples—say Thomas Huxley and others—is the analogue of species. And there are a number of individuals, as it were, instantiating that species type. Just as with species we can talk about a species insofar as we can refer to the genome—the genetic structure—of an individual organism and to the phenotypic representations of that structure, so also, I think, it is profitable to consider, let's say, Darwin's ideas of evolutionary theory as having different representations. When Darwin explains his theory to his children, that is one representation. When he writes it down in the *Origin of Species,* that is another representation. When he thinks about it while he is walking along the path, that is yet another representation. That is, depending upon the particular "intellectual ecologies"—the other conceptual systems that are initially inhabiting the vicinity of his ideas—those ideas will be differently expressed. You still want to say, I think, that though Darwin's evolutionary theory has different expressions, different representations, in his books, in his talks, in his letters, in his casual conversations, nonetheless there is one theory which Darwin "has" or "knows," which is called evolutionary theory; so that the *genotype/phenotype distinction,* which is the common coin of the talk of evolutionary biologists, is also useful in EE.

Callebaut: One could go even further, I suppose, and talk about conceptual systems as being themselves composed of ideas, just as genotypes are composed of genes.

Richards: This may seem to be stretching it a bit, but I think there is also some utility that could be gained by thinking of conceptual systems on the analogue of genotypes. So for instance, one can think of evolutionary theory itself as a genotype, which, as it were, finds specific instances in the ideas that exist in Darwin's head, in Wallace's head, in Huxley's head. This is the same as the relationship that exists between the geno*type*—say, *Canis familiaris*, the common dog—and the specific individual dogs that have a set of genes . . .

Callebaut: . . . what Bill Wimsatt (1981a) has called "geno*tokens*" . . .

Richards: . . . Genotokens, that's a good way of putting it. These sorts of relationships become extremely useful when thinking about the development of scientific ideas.[27]

7.3.3. *Contaminated by Your Object*

Callebaut: One thing the most elaborate and convincing applications of the evolutionary model to historical or contemporary scientific developments to date have in common is their subject matter: biology. In terms of the economics of research, it would seem perfectly natural to have your intellectual investment—in biology in this case—pay off twice: first, by providing you with the materials for your case study, and second by providing you with the source of your methodology. There is nothing wrong with "opportunism" (Campbell) of this sort in science. On the other hand, it is a well-known phenomenon that when you have been involved in some kind of research for a while, there is a "danger" (?) that you will become *contaminated* by your subject matter (cf. Vandenbrande 1979). What I am hinting at is whether the evolutionary approach to scientific change is as general as its advocates suggest. You have your doubts, Karin.

Knorr Cetina: One thing EE ignores is the disunity of the sciences. I can make some sense of the mutation/selection picture of how innovations come about and how problems are solved in molecular biology, because I find molecular biology has a habit of going about things in terms of trial and error: it uses a certain kind of procedural step; if it doesn't work, it uses a variation and it restarts the whole thing, sort of selecting the step that works in the end.

Callebaut: You have given the variation/selection scheme of EE a balanced treatment elsewhere (Knorr Cetina 1987).

---

27. Richards provides an illustration: "This allows one to say, for example, that Wallace and Darwin held specifically the same theory without being compelled to find perfect identity between theories, just as one might say that two human beings are of the genotype, *Homo sapiens*. But nonetheless, their specific genes—their genotokens, to use that language—won't be identical; they will differ from one another. And so too that if you had two individuals who were identical twins, it might be the case that their genotokens were identical but their phenotypic representations would be somewhat different. So these sorts of distinctions are terribly useful when talking about the development and progress of ideas."

Knorr Cetina: Now in particle physics, for example, you find a totally different situation. There you find a very reflexive, self-monitoring, "self care" form of epistemology, where it is very important to constantly study and understand *yourself,* by which I mean, for example, the detector you have made. Particle physicists spend a lot of time trying to understand the instruments they're creating. Now if you have that kind of attitude of studying and understanding what you're doing all the time, it is not at all the kind of mutation and selection thing, but something completely different: a deliberate, intentional effort to bend back upon yourself, unfold what you've done, and understand in detail what you've done. I have great trouble to somehow fit that into the EE kind of language. So I think the selection/mutation language fits some sciences but doesn't fit other sciences.

7.3.4. *How to Get at the Real Stuff*

Callebaut: Will you agree that there are some problems with the natural selection view, Bob?

Richards: I'll agree that there are problems, but they may be *solvable* problems!

Callebaut: Let's sort these out a bit. My main problem is: how do you *get at* these genotypes? Of course, the problem exists in biology as well: you never get to see the genes' "real stuff," as David Hull once put it. As a historian of science, you have to reconstruct the intellectual genotype exclusively on the basis of its phenotypic expressions.

Richards: That's absolutely so. I don't think you can get directly at what might be, as it were, the genotoken of ideas; that is, if you think of the conceptual system as a genotoken, in the sense that if you know evolutionary theory and I know evolutionary theory, we can represent it in several different ways, and those will be different phenotypic representations. Yet there is a genotype so that we might both want to say, "We know the same theory." But if you said, "Let's talk about exactly what that theory is," of course, to get at what I have in mind—even for myself to realize what I have in mind, I have to represent it or express it *to find out* what I have in mind! So there is no direct access to it. Neither is there direct access to what might be the genotype; that is, when you ask, "What is it that Darwin and Wallace have in common?"

Callebaut: How can you get, then, at what Darwin's theory really is?

Richards: To get at that—and as a theory, of course, it changes over time—you can only look at his expressions of it in his notebooks, in his publications, in his letters, and so forth. And it is through a kind of reconstructive effort that you do so. I don't think it is any different in evolutionary biology; it certainly wasn't before sequencing of DNA became a more ready and direct kind of access to the discovery of the genes, and even there one can quibble. But before that, the only way in which one could make some kind of guesses about what the genotype of a particular organism happened to be would be through the expressions in the phenotype and making assessments in that way. Well, I think a historian does the same thing; so that while access is difficult—it is always provisional and hypothetical; that is, we make guesses on the basis of our representations—nevertheless, those can be more or less accurate, more or less reliable.

Callebaut: I would like to push this a little bit, because I think this conceptual genotype/phenotype distinction is a really important idea and many things may come out of developing it. I would like to raise the problem of the *unity of the genome* (compare, again, biology; e.g., Mayr 1963). Darwin certainly wasn't born with Darwinism in his head. So, contrary to biology, there doesn't seem to be stability of the conceptual genome, even through the individual's lifetime.

Richards: I think that you are right. The genome itself evolves, it changes during the lifetime of an individual. Now if you press me to really cash out the analogy between evolutionary biology and EE, what I would have to say is something like this: *I regard the head of a scientist as a kind of ecological niche in which there reside different conceptual systems.*

Callebaut: How would this look like in the case of Darwin?

Richards: In his head resides a conceptual system loosely called "evolutionary theory." There is another conceptual system—a very small one—that has to do with certain religious beliefs. There is yet another conceptual system that has to do with finance and views about the prospects of his railroad stocks; and so forth. So there are any number of conceptual systems that might exist within the head of a scientist. These conceptual systems undergo continuous selection and development as they change relationships among themselves. That is, it is a dynamic environment in which they exist, and consequently they are always changing; so that Darwin's notion about evolutionary theory, if you look at Darwin at the very beginning of his career—let's say, in 1837—and then at the end of his career, in 1882, may have changed dramatically. We still call it "Darwin's theory of evolution," but the relationship has changed considerably. You call a snake a "reptile," but a snake is surely a different kind of creature than a dinosaur. A dinosaur is also a reptile. A lot of change has occurred in between the age of the dinosaurs and the contemporary age of snakes. I think it is the same case with Darwin. These conceptual systems continually undergo modification, selection, perhaps speciation. Sometimes, promising evolutionary developments, ideas come to a dead end in a scientist's head, and they go extinct, they're no longer thought about. The way I look at it is to look at the scientist's head, as it were—I hate to put it precisely in that way—as a kind of dynamic ecological system.

7.3.5. *You Really Have to Know What They Think They Are Doing*

Callebaut: Coming back to my question about the unity of the genome, where and how do you locate coherence? Already in your paper, "The natural selection model of conceptual evolution" (Richards 1977), you mention "logical coherence" along with empirical evidence and that kind of thing. There are two things I have in mind here. First of all, there is the issue—for some, quite a hot issue in the history of science—of what makes for most of the coherence in conceptual systems: is it the conceptual level or is it the social level? Hull, say, would claim there is more coherence to be found on the social level. Second, in older social psychology, you have of course Festinger-type views . . .

Richards: . . . dissonance theory, right (Festinger 1964) . . .

Callebaut: . . . or coherence "in one's head." Now if you're talking of an ecology of conceptual systems within one's head, that is of course departing from, say, Popper's Three Worlds view, to which I am strongly opposed, and I hope you are as well.[28] But isn't there a danger here that this is going to depart from what we already know about what's going on in people's heads? You now sound to me a little bit like the old Toulmin: it sounds a bit like free-floating ideas again. How would you relate your view to what we already know about what goes on in social groups?

Richards: To start off answering your question, let me just preface it by distinguishing my view from David Hull's. I think David does think that there is consistency in the social situation, and individuation that is obvious in the social situation, which allows you to individuate conceptual systems; that is, you will define the parameters of the conceptual system by a relationship to those scientists or thinkers . . .

Callebaut: . . . That was Kuhn's idea as well . . .

Richards: . . . Yeah. And I think that is wrong for the following reason: You really only know what constitutes a social group if you know what beliefs they have in common. It seems to me that it is the beliefs that define the social group rather than the social group that defines the beliefs.

Callebaut: Can you make this claim solid by pointing to evidence?

Richards: Yeah: If you try to do a kind of sociological analysis of interactions in, say, a laboratory, you might discover that a good many of the interactions are between scientist *A,* scientist *B,* and the guy who washes out the test tubes. Now it does not seem to me that you want to include the guy who washes out the test tubes—particularly if it is not a graduate student but someone who is simply hired to wash out test tubes—as part of the community. Yet, if you simply did a catalogue of the actual interactions between individuals in this room during a particular time, you might find that this person would be one of the most important people in the group!

Callebaut: Some people are willing to take this step. Karin Knorr-Cetina (1981a:153; cf. 1982), for instance, maintains that it is "variable transepistemic fields, traversed and sustained by resource-relationships" rather than professional member-

---

28. Campbell (1988c:375), who can be very diplomatic, writes: "I do not understand all of Popper's agenda in his concept of World III, but I am ready to posit a 'world' wherever independent invention/discovery and priority quarrels are possible, and thus a 'world' for mathematics and logic." For once I must disagree. See also Changeux and Connes (1989), one of only a very few attempts to apply an evolutionary-epistemological framework to the genesis of mathematical knowledge.

ship groups such as "scientific communities" that constitute the webs of social relationships in which the scientists situate their laboratory action.

Richards: I think that means that if you use the social group to instantiate the conceptual system and use it to find what coherence there is in the conceptual system, then you will have to include perhaps, in my case, thermodynamical chemistry—let's say it is a lab in chemistry. You will also have to include the scores of the Cubs' games, what the White Sox did, and so forth. That is, those too will be the conceptual concerns, particularly if you are going to include the test tube washer in your analysis. So it seems to me that that just does not make any sense. And there is no possible way of doing it. You exclude the guy who washes the test tubes, generally, because you have evidence that his beliefs are quite different and his conceptual concerns are quite different than those, say, of scientist *A* and scientist *B*. In other words, you find out what the group is by knowing what the group believes rather than the other way around. Initially, I think, that is not the way to do it. Although what people hang out together—social interactions—I think can be a clue as to the nature of a conceptual system. And social relationships can undoubtedly also cement conceptual systems, or—I should rather put it this way—cause them to develop in certain fashions.

Callebaut: Richard and Don, you have a problem similar to the one we are discussing with *Science as a Process*.

Burian: The trouble with David's book, in this respect, is that while it does very interesting historical and sociological things and plays with the notion of "conceptual variance" in very useful ways, it doesn't tell us where to draw standards from. To put it a little too harshly: in looking at the work of individual scientists and groups of scientists, you can't tell the people who are important scientists and well-grounded epistemologically from those who are kooks, to put it extremely. The grounds for evaluating the conceptual products are at least still slippery.

Campbell: I think David's use of his evolutionary formalism of interactors, vehicles, etc. does not help explain how scientific beliefs become more valid, since it fits sectarian Bible-belt protestantism just as well as it does the history of science (1988b; Maynard Smith 1988; Oldroyd 1990). Toulmin's model, applying evolution to the history of science, Hull's model, and Bob Richards' model as expressed in his wonderful book all fail to make clear how their biological evolutionary analogues make it more plausible that the beliefs of physicists have become more competent in their reference to unobservable physical realities.

Hull: The reason that Burian and Campbell can't see how scientists differ from Bible-belt protestants on the evolutionary accounts given by Bob and me is that the difference is so old-fashioned and commonplace. Scientists have a narrow conception of evidence. A statement in a good book does not count as evidence; neither does depth of belief. They also take evidence seriously. They really do test their beliefs. Typically they test the views with which they disagree much more seriously than those with which they agree. It is also true that the automatic response to apparent disproof of one's favorite hypothesis is not capitulation, but sooner or later, if they live long enough, scientists are forced to change their minds in the face of persistent counterevidence, usually under a smoke screen of some sort. In science, a little bit of evidence can go a long way. I am a flat-footed, possibly simple-minded realist. I continue to be puzzled how I can be read as holding views that are so much more sophisticated than those I actually hold.

Callebaut: Ron, you agree that we need a model of what is going on, because without one we just can't even delineate what we're going to investigate.

Giere: Right. You really have to know what scientists think they are doing. And the more you know about what they think they are doing, the better you can interpret what is going on.

Callebaut: You are one of the very few philosophers of science to have "gone native" (cf. Collins 1983:91) when doing the lab study you report on in your book (Giere 1988).

Giere: I was not trying to do an ethnography of the laboratory. In that part of the book I had very specific objectives and was really focusing on the realism issue. So I didn't feel that I needed to have a big ethnographic training, because I was not trying to make a big ethnographic study to make a big ethnographic point. Basically I had one issue I was focusing on and the method was not sophisticated at all. It was basically just, "Overwhelm them with data!" Make the antirealist position just look as silly as possible with a whole bunch of data about what is really going on. That was the "argument." In that case, my physics background did really help. That was the first time in my whole career that I've really used that knowledge explicitly, in a direct way. But again, I wasn't doing the foundations of physics, I wasn't talking about the physical theory they were using in nuclear physics, which I think they can do better than philosophers.

Callebaut: Back to Darwin: what provided his theory with the required coherence, Bob?

Richards: If you look at Darwin's theory, it has unity by reason of its historical past: it has a kind of *historical unity.* If you look at Darwin in 1882, you can see that the ideas he had in 1882 were very much like the ideas he had in 1872, which were very much like the ideas he had in 1862 and so I'm back to 1842. There is a kind of historical connection, and this provides a good deal of the coherence.

Callebaut: Also across individuals?

Richards: Yes, also across individuals. As it were, *common descent* is another cause of unity, similarity of ideas: they have descended from the same source. I think one has to maintain that what distinguishes these conceptual systems that we call "scientific" from others, however, are certain marks. Just as you want to say that what distinguishes the vertebrates from the invertebrates is that the vertebrates carry certain obvious signs: they have a backbone, and they have the genetic structure that allows the organisms to develop backbones. These are characteristics of everything we want to call vertebrates. Now in like manner, everything that we want to call a scientific conceptual system has a certain logical character to it; namely, that the ideas are not intrinsically inconsistent, or at least the effort of scientists is to eliminate as much inconsistency as possible. If the ideas of conceptual systems are manifestly or, as it were, *intrinsically* inconsistent, then one of two things will happen: they will migrate out of the scientific environment—I suspect that you find such ideas, for example, inhabiting the realm, right now, of astrology and parapsychology—or they will simply go extinct. But generally speaking one of the characteristics of scientific conceptual systems is that they are consistent.

Callebaut: You also mentioned empirical relevance in your (1977) paper.

Richards: Unity is also provided by the relationship to—at least as we think about it—the *empirical world*. Such systems have a logical character of being relevant to other conceptual systems which more grossly, less theoretically, more observationally represent the empirical world. I think that is another characteristic of what we call scientific conceptual systems. And these provide as much unity, as much stability as exists among scientific conceptual systems: they are labeled, they change, they alter. They don't have actually as much unity, I think, as we sometimes suspect they do. And the evolutionary approach emphasizes, or at least allows one to appreciate what unity does exist; but at the same time it also recognizes that they alter and change over time.

Callebaut: Similarity not being a transitive relation, isn't that causing the historian problems? If you wait long enough, you end up with something that is totally unrelated to the thing you started with in terms of similarity (see fig. 7.1; for illustrations of this in the context of the discussion of reduction, see Wimsatt 1976a).

Richards: Absolutely. I mean there is no reason to doubt, for example, that we are descendants of *Australopithecus*. But a lot of time has elapsed between the advent of Lucy, *Australopithecus afarensis,* and ourselves. And we do not look very much like Lucy. Nonetheless there is a continuity and unity of descent. So that what is at the end of the process looks rather different from what is at the beginning.

Callebaut: David Hull proposes to look at causal relations—"Who's got what from whom?"—in the first place; in his view, similarities are of minor importance.[29]

Richards: David thinks that one can virtually ignore similarities when talking about ideas as one can when talking about biological organisms. I guess this is again a way in which I would differ from him. I don't think you can ignore them in either case, because similarity becomes a clue, a sign of common descent. That is, if as a historian I look at, say, Wallace's development of evolutionary theory and Darwin's development of evolutionary theory, I see—as Darwin himself saw it—that they are very similar. Now similarity of conceptual systems suggests that there may be common descent. Maybe there isn't; but it is a clue, and it is the clue to follow up.

Callebaut: In one of several papers in which he criticizes the hermeneutic rendering

29. Cf. Hull (1976b:656): "I tend to agree with Wimsatt's [1976a] analysis of reduction as far as it goes, but similarity in substantive content, even interpreted as a serial relation, is not sufficient for individuating scientific theories."

Figure 7.1. Serial reproduction of pictorial material by the members of a group. Each participant redraws from memory the picture produced by his predecessor. From F. C. Bartlett, *Remembering: A Study in Experimental and Social Psychology* (Cambridge: Cambridge University Press, 1967), 180–81.

of psychoanalysis, Adolf Grünbaum has a fine analysis of the fallacy of inferring a causal linkage between thematically kindred events from their thematic kinship. He emphasizes that the issue goes far beyond psychoanalysis, and that its resolution "not only spells a major general moral for the human sciences, but also has instructive counterparts in biology and physics, for example" (Grünbaum 1989:189). My last question concerning your genotype/phenotype distinction. In biology, genotypes specify a "reaction norm." Can you possibly describe your conceptual genotypes in such terms; I mean, can you describe the possible sorts of expression you could get in different ecologies?

Richards: Actually there is a great deal of similarity, and I think the notion of reaction norms is rather poignant, particularly in my discussion right here. I can think best about this model with pen and paper at hand, with considerable time for reflexion. And what I might draw up in an essay or in a book chapter—which I take to be the typical environment for the expression of those ideas—is rather different from the environment of our conversation here, where I think the idea is a little distorted and grotesque (*laughter*), because the conceptual ecology puts different kinds of pressure on their expression to make them look a little bizarre.

Callebaut: What I was, again, hinting at is how, as a practicing historian, you proceed to reconstruct your genotypes from phenotypes? Is it more than a kind of pattern matching exercise: say, you use Wallace's version and Darwin's, and look at where they overlap and where they don't?

Richards: Well, it depends, I guess, on which questions you are pursuing in your historical analysis.[30]

7.3.6. *The Generative Entrenchment of Scientific Ideas*

Callebaut: Bill, your "developmental lock model" (cf. 9.1.1) is applicable to scientific change as well as to biological phenomena of development and evolution in the stricter sense.

Wimsatt: If scientific theories also have a generative structure, we would expect similar phenomena in that area. An analogue to von Baer's law, which operates basically by saying that mutations at later developmental stages accumulate more rapidly in evolutionary time than ones at earlier stages, would be the observation that *most changes in scientific theories are made at a relatively low level of abstraction* where not too much else depends on it: the changes are relatively local and so on. And the developmental lock would predict something like an exponential decline of rates of modification of features as they get increasingly generatively entrenched, or abstract, or general, or whatever.

Callebaut: Abstraction and generality aren't synonyms . . .

Wimsatt: . . . but greater abstraction breeds greater potential general applicability because the number of conditions required for application are smaller. Generative

---

30. Richards adds: "Certainly initially, when you look at the expression of Darwin's theory and Wallace's theory—compare them, for example, as was done when the essays of Darwin and Wallace were presented to the Linnaean Society in 1858—on that basis the historian might suspect, by looking at just those two expressions, that they are very, very similar. They are different in some important ways, but nonetheless similar enough such that you suspect that if you dug deeply enough, you would find common sources for those ideas. And as a matter of fact it turns out to be the case that Thomas Malthus was terribly important in the development both of Darwin's ideas and of Wallace's; Lamarck played a role, a more positive role in both cases than either Darwin or Wallace would have cared to admit; their travels in tropical areas—observing the lush biological productions in those tropical areas, both of them had that kind of experience . . . So that one can begin to understand the development of these ideas within a common context; so that, at least initially, some pattern matching goes on."

entrenchment breeds increased polyfunctionality—usefulness in a broader range of contexts. Increased polyfunctionality forces increased abstraction, since the same general schema has to be extended to cover a broader range of diverse cases. And increased abstraction potentiates increased generality, greater resistance to falsification, and a variety of other features of quasi-analytic statements. In fact, I mentioned abstraction, generality, resistance to falsification, and the like; I think you can get all the properties of the analytic/synthetic distinction out of this model as applied to scientific change.

Callebaut: Dudley, couldn't we, without committing ourselves to the hierarchical, "two-levels view" of science that you criticize, still account for the fact—which seems important enough—that certain items in a domain change more rapidly than others? Without precluding the possibility that all the items may change eventually, of course.

Shapere: The distinction I talk about is one in which ideas at a "metalevel," whatever they are, are independent of the scientific process itself and immune to revision in the light of its results. Some typical examples are the claim that there is a meta-scientific level at which logic lays down requirements for science (positivism); that fundamental concepts from the philosophy of language—meaning and refer-ence—can establish constraints on what science does independent of what science learns; that epistemology is a subject which analyzes the "concept" of inquiry or of knowledge in ways that are independent of, and presupposed by, actual inquiry of knowledge claims; and so forth. If I understand your question, you would like to admit that there is no "level" at which ideas are immune from change, but that there is a rough correlation between degree of generality of an idea and the length of time it takes for it to change . . .

Callebaut: . . . and that guiding principles are of that sort.

Shapere: Yes, such ideas as the gauge-theoretic approach, the mechanistic philoso-phy of nature, or the central dogma in biology can have the status of principles which guide research; and in that sense they often last longer than the particular hypotheses that are suggested in the light of their guidance. But this distinction can't by any means be made absolute, the basis of a comprehensive philosophy of science. It isn't just very general guiding principles that can last a long time; it can also be very specific sorts of things, or any scientific idea or approach intermediate between the very specific and the very general. Guiding principles, too, need not last a long time; the central dogma, for instance . . .

Callebaut: . . . according to which DNA makes RNA, RNA makes protein (the phenotype), but there can be no transfer from protein to RNA . . .

Shapere: . . . might not last very long. It's a contingent matter how well our guiding principles perform. We mustn't forget that there have been guiding principles in science that haven't lasted very long, and so fail to be noticed in retrospect. And in any case, they too can be quite specific in the guidance they offer, as in the case of the gauge-theoretic approach. *So it isn't a matter of generality; it is not a matter,* *necessarily, even of fundamentality!* So it's hard to make any sharp distinction into mutually exclusive classes, or even a kind of rough hierarchy, that will necessarily hold up.

Callebaut: The introduction of a new idea in science, say, a guiding assumption, is in many ways contingent: it could have happened earlier or later, in a domain that might have been better or less well consolidated, etc. So it would seem that the very point in time at which a guiding assumption is introduced can make quite a difference in terms of the impact it will have on other ideas.

Shapere: Yes, it can. But on the other hand I think that the looseness of constraints on when a guiding principle can develop exists largely in the early stages of science but much less in later science. For instance, four basic ideals (among others) of what an explanation ought to be are found in early Greek philosophy: the idea of *stability* or *permanence* (Parmenides, that whatever is explanatory must be unchanging), the idea of *agency* or *activity* (Anaximenes, Plato, and the Stoics, that whatever is explanatory must be an active agent), the idea of *definiteness* or *limit* (Pythagoreans), and the idea of *purpose* (Plato and Aristotle). There is a sense in which these could have been introduced "any time," or at least we can imagine them having been introduced at some very different time. But that's because there weren't very many constraints around. Although those ideas continued to influence and guide much of later scientific thought, even (except for purpose) well into the modern era, they had to be drastically altered in the light of scientific findings. And although there is at least a rational descendant of the idea of permanence and stability in the idea of symmetry principles in quantum field theories, nevertheless the descendant looks very different from its ancestors. So guiding principles become more and more shaped and constrained by what we have found out, by what we have come to have reason to believe.[31]

Callebaut: Bill, you are *not* claiming that scientific change is strictly analogous to biological change. I emphasize this, because some critics of EE always talk as if advocates of EE were making such a strong identity claim.

Wimsatt: In the cognitive realm, we have a number of heuristics for making very deep modifications which make deep modifications more frequent in cultural evolution than they are in biological evolution. As a matter of fact, if that weren't

---

31. Shapere explains what he means by "shaped" and "constrained" here: "It would have been impossible, for example, for the gauge-theoretic idea to have been presented a hundred years before it was. This brings out the deep ways in which what we come to believe about the universe on the basis of reasons (our 'factual' beliefs) shape, modify, and even change the ideals or standards of what we expect an explanation to be like. The interaction works in the other direction, too: how to understand the 'facts' is shaped in turn by what we have come to believe about what an explanation should be like. There isn't a bifurcation here. The same sort of interaction holds for methodological ideals, conditions of acceptance of a theory or of an idea as possible, the conditions of acceptability of an idea, the definite delimitation of domains, the description of domains, the selection of features of our experience or observation of nature as being important, ways of getting access to such information once you've decided that there is information to be obtained, and so on."

true, we'd have real problems, because biological revolutions are very rare; and while scientific revolutions are rare also, they are orders of magnitude more frequent than biological revolutions.

Callebaut: You also think the developmental lock points the way to making a genotype/phenotype distinction for cultural evolution, including science.

Wimsatt: What people who have tried to construct models of cultural evolution have done is they have looked at the gene and looked for analogues to the gene's *autocatalytic* function, that is, the way in which genes make more genes. They have emphasized reliable replication, high replication rates, and so on. But that is ignoring the other function of the genes, its *heterocatalytic* function, which is to produce a phenotype. In producing a phenotype, we are immediately talking about developmental programs and about generative entrenchment. What I would suggest is that those things that are cultural gene analogues are things which are deeply generatively entrenched: when you have them, you use them to generate a lot of other things.

Callebaut: Some people, especially in general sociology, continue to confuse development in the sense of ontogeny with evolution in the sense of phylogeny. They will talk, for instance, about the "evolution of the Roman Catholic church," without ever thinking of the Church as a population.

Giere: That is evolution just in a sense of changing and development. If you take it seriously, developmental models and evolutionary models are very different. They continue to be mixed up and that's a disaster!

Callebaut: Yet developmental models could have a legitimate place in studies of science. I am thinking, for instance, of *life cycle* type of views that are also of some use in economics. Although Diana Crane's (1972) reinterpretation of Kuhn's paradigm theory in terms of a life cycle (cf. De Mey 1982:ch. 9) may have been a particularly unhappy instance of a stage view, there seems to me to be nothing wrong *in principle* with developmental models and even stage views, as long as they are properly delineated from evolutionary models.

Giere: I would not really argue the developmental metaphor itself is a disaster. The evolutionary metaphor is much better, and they should not be conflated.

7.4 IN THE REAL THIRD WORLD, PEOPLE DIE

Callebaut: Popper has written a book called *Objective Knowledge: An Evolutionary Approach* (1972). His EE departs in fundamental respects from the sorts of view that have been discussed in this chapter. It relies on mind/brain interactionism and on an objectivist ("World Three") interpretation of the status of scientific knowledge as nonsituated—to mention its two most obvious deviant features. Popper's EE remains quite influential (Wuketits 1984, 1990; Riedl and Wuketits 1987; Radnitzky and Bartley 1987; Hahlweg and Hooker 1989).

Bechtel: The kind of argument that Popper makes is that consciousness could not have been selected for if it was identical to a brain process. His treatment of the mind-brain identity claim is itself peculiar since he thinks it makes consciousness into an epiphenomenon of a brain process. In any case, his claim is that consciousness would have to make an independent contribution to fitness for it to be selected for. But that's just not true (Bechtel and Richardson 1983). We know plenty of cases that violate this principle. Nothing selected independently for grass to be green, as well as for it to have other properties. It just happens that chlorophyll, which is useful for the plant for making energy, gives it a green color. So even if consciousness were not making any contribution, it could have been the product of evolution—which is what Popper denies. One thing we do grant to Popper is that probably there is a much richer evolutionary story to be told than the simple identity theorist would lead you to tell. If you want to ask seriously questions like, "What is the function of consciousness?" that's all do-able within a functionalist framework but is largely undone at present. But there is this additional advantage to doing it from a materialist framework: at least when you're dealing with material processes, you have the potential of developing mechanisms of heritability. How you account for heritability with a nonmaterial mind seems utterly inexplicable.

Callebaut: The theory of evolution applies only to *concrete* systems (Bunge 1983: 8); the mind must, in other words, be *embodied;* it makes no sense to talk about disembodied ideas (etc.) at all (cf. Campbell 1987a).

Bechtel: Right. The need for its embodiment in this argument is simply as a mechanism for heredity, genetic transmission being the obvious candidate that we had in mind.[32]

Callebaut: Dudley, you insist on saying "science thinks" rather than "scientists think." Why is that?

Shapere: I mean that scientists think the way they do, when they do it well, because there are *reasons,* within the science, and not vice versa: there aren't reasons because the scientists, as a social group, say that there are reasons. Saying that "science thinks" is just a shorthand way of calling attention to that, and for countering the relativist tendencies that lurk behind talking about scien*tists* as the primary object of philosophical study. I talk about science and not scientists; in other words, *ideas have a life of their own.* I don't want to introduce anything mysterious by such expressions as "science says . . ."

Callebaut: Talking about a Third World, which *is* Platonic, despite Popper's hand waving . . .

Shapere: . . . *does* sound mystical—it puts the point in a way that is irrelevant,

---

32. Bechtel explains: "If you are going to give an evolutionary explanation, it would be desirable and pleasing if you can show that consciousness is making a contribution to fitness. We grant Popper that even though it's not needed, it could be that it's just an accident of evolution. But we assume that William James is right: it seems that consciousness is here because of what it does for us. But then one must realize that not only do you have to have it *do* something for us, but it has to be heritable; we need some mechanism for the heritability of consciousness. And the basic vehicle of heredity between organisms is genetic. There is also a cultural transmission, but that presupposes consciousness, I think, in Popper's sense; so you've got to have heritability of consciousness first."

misleading, and false. What I mean is simply that the proper way to approach the philosophical study of science is through an analysis of science and scientific reasoning and its development. For many purposes it's *necessary* to focus on the content of science. For example, one can't attend only to what people have said or written in published or unpublished work, because it has to make sense for the good historian to make assertions like this: *that there was at some particular time a problem which nobody saw.*

Callebaut: How does this depart from Popper, then?

Shapere: I wouldn't talk in terms of a Third World in which all the ideas of that period existed—some Platonic entity in which that particular problem existed that nobody saw—but rather would ask, "On the basis of what sorts of evidence would we decide that?" And I don't even think that you necessarily have to have some abstract criteria for doing this, but only that you be able to specify in particular cases why you say that "this was a problem that nobody saw." That is, there are relations in *science*—as opposed to what *scientists* say or do—and the proper object of philosophical investigation is *science,* in a sense in which the science of an epoch or of a person or a group can be specified by historical inquiry. That I take to be the way to demystify the Popperian Third World and get at what I am driving at.

Callebaut: I would agree in thinking that looking at the actual way historians of science make these inferences is the best way to go.

Shapere: Sure; and one can *see* that, in a case like this: *nobody saw the contradictions between electrodynamics and mechanics in the 1890s.* You know *exactly* how to go about establishing that, at least in principle.

Callebaut: Popper wouldn't have to deny that either; on his view, problems exist "objectively" as well. Yet I see a very serious danger of reification in the construction of any transpersonal unity (cf. Church 1984).

Shapere: I wouldn't put the issue in terms of "reification" but rather in terms of the epistemic or methodological question, "How can objectivity in historical inquiry be achieved?" Without going into the details, I think we have to admit that there *is* such a thing as objectivity in historical inquiry.[33]

Callebaut: I am trying to push this: *scientists don't have anything else, apart from this subjectivity of living within their presuppositions.*

---

33. Shapere adds: "We *learn* how to do good history, and previous approaches and interpretations *can* be criticized in the light of what we have learned about historical methodology and historical facts. Historical reconstructions, even of transindividual events or attitudes, *can* be criticized. Even the background, what you take as having been learned, your methods, etc., can be criticized. Look at the alteration of our view in the history of science because of more careful, sophisticated, and critical historiographical approaches. Older ways of doing history, and the interpretations given, are criticized. Take Kepler: He used to be portrayed as a schizophrenic thinker, three days a week being a mystic and three days a good scientist. Now, through criticism of those views, we see that there is a much more integrated picture." Shapere suggests to compare, for instance, Dreyer's ([1906] 1953) treatment of Kepler with Koyré's ([1961] 1973) and with the studies included in Beer and Beer (1975). "What the historians have done has been to get a better idea of *how to do historical reconstruction objectively.*"

Shapere: But objectivity is not incompatible with using presuppositions. On the contrary, objectivity cannot mean "looking at the facts without any presuppositions," but rather "looking at the facts in the light of what we've learned" (1985).

Callebaut: Are you implying that scientists disagree only about the things that are at stake—at the focus of attention—at a particular moment?

Shapere: That is where the disagreements come; it is not in what I call the "background" of science, but in the *working* of science, that is, in the construction of hypotheses to meet actual, specific problems. *There* is where you find disagreement, variations in approach; and *there* has to be variation, because there are open possibilities, and perhaps the outcome will be something entirely different from any of the initial variations. The alternative approaches may "feed back" to produce variants on the interpretation of the background information—for instance, on what a gauge theory should look like. But still, there is a great deal of background which is agreed upon, and in terms of which these variations are constructed.

8

COGNITIVE APPROACHES TO SCIENCE AND PHILOSOPHY

IN MODELING COGNITION at both the individual and social levels, evolutionary epistemology is tributary both to the individualistic approach of traditional epistemology and (to a much lesser extent) to the collective approach of the current sociology of scientific knowledge. While sociologists are most indebted to the historicist line emanating from Kuhn's exemplary work in science studies, psychologists and at least some naturalistic philosophers hope that cognitive psychology—and ultimately even neurophysiology—will also be able to deliver important goods.[1]

The chapter opens with an experimental cognitive psychologist's perspective on the things that are going on in science studies today. The "procedural turn" in science studies prompts us to devote attention to nonlinguistic, e.g., visual or practical, features of scientists' work that were widely neglected hitherto. What explains the predilection of some students of science for the work of "diary men" such as Darwin or Faraday? Next, some *varieties of cognitivism* are distinguished, including connectionism. What distinguishes the neuroscientist's, the psychologist's, and the artificial intelligencer's perspectives? Once again, reductive hierarchies turn out to be crucial. How does the age-old mind/body issue look like from the perspectives of contemporary functionalist psychology? Is the dismissal of "folk psychology" by some "eliminativists" among the philosophers of mind liberating, or does it constitute a threat to human freedom? What is the "neurophilosophical" perspective (Churchland). And last but not least, we will catch a glimpse of what is going on in Searle's "Chinese room."

8.1 STEPS TOWARD A COGNITIVE SCIENCE

Callebaut: Ryan, your first conversion experience was the linguist Noam Chomsky (e.g., 1957, 1965, 1966).

---

1. "If beliefs can be produced by a process of causation lying in the province of physiology or psychology rather than of sociology," writes philosopher Roger Trigg (1978:298), "then we must turn to those disciplines, rather than to sociology for an explanation of at least some of the origins of belief. The more it is admitted that the sociology of knowledge cannot tell us the whole story, the less important the subject appears and the less 'strong' its programme can be."

Tweney: He had a grand theory, but when you take Chomsky into the laboratory, you discover it's not so simple. In the psycholinguistics lab you learn that the strategy of finding scientific accounts of human behavior is very complex (cf. Bechtel 1987). There is not likely to be a single theory that accounts for everything. You must assemble bits and pieces of this and that, hoping that your account is going to get closer and closer to the truth. When I worked on sign language with Harry Hormann and Ursula Bellugi, it became clear to me that the principles that underlie sign language were the same as the principles that underlie spoken language, except insofar as the former were adapted to a visual domain, which meant many specific things had to change. At about the same time, I realized that my interest in language was really secondary, that I'd gotten interested in Chomsky originally because he promised a way to understand the mind through language. Perhaps so, but it also became clear that you could study mental things—the workings of mind—*directly.* George Miller, for example, and Herb Simon (1979a), raised the cognitive issues that I'm interested in directly. And I've had a lifelong interest in the nature of science, since I was making stink bombs in my basement as a child . . .

Callebaut: . . . So was I . . .

Tweney: . . . and at one point I wanted to be in physics. What really is common to all these interests is understanding how scientific thinking works. It is *extraordinary* that it works! And it works so spectacularly well that it's incumbent on us to understand it.

Callebaut: Pat, you are similarly excited about the prospect of understanding the human brain (cf. Star 1989).[2]

Churchland: I think it's a very exciting time for philosophy of mind, for epistemology, for philosophy of science, and for neuroscience. There is no doubt that, barring some sort of social or political catastrophe—that is, so long as the research can go on—major and astonishing breakthroughs are going to be made. *We are going to get the story of how the brain works.* It's like Columbus discovering America: a whole lot of things are going to change. I find it absolutely thrilling. The research is often tough sledding, because neuroscience is technically difficult. Of course, there is a lot we do not yet know, so we are certainly in for some conceptual surprises, especially concerning how the dendrites integrate information and the nature of the dynamics of neural networks.

Callebaut: Tom, Herbert Simon's *information processing approach* to AI (Simon 1979a, 1992a; Langley et al. 1987) has been important in your development as well.

Nickles: Yes, Simon's stuff is very stimulating, although my emphasis is a little different. In my opinion Herbert Simon is one of the great American pragmatist

---

2. Churchland and Churchland (1983); Churchland (1987); Churchland and Sejnowski (1988, 1992); Sejnowski, Koch, and Churchland (1988).

philosophers. His work on satisficing versus optimizing fits very nicely into the pragmatist tradition and ties in strongly with problem-solving versus directly truth-seeking accounts of scientific inquiry and the whole issue of economy of research (Nickles 1988).[3] It was Simon (1977a) who first clearly related the mysterious subject of discovery to work already underway on heuristic search and problem solving. He and the other artificial intelligentsia (as Clark Glymour dubs them) first called attention to today's truism that scientific inquiry is shot through with problem-solving tasks. In fact, Douglas Lenat later wrote a paper called "The ubiquity of discovery" (1978).

Callebaut: You implied that you do have some differences from Simon.

Nickles: I think Simon and his colleagues tend to be whiggish with historical cases (but see De Mey 1992).[4] By failing to recognize the depth of historical novelty or conceptual change, they make it too easy on themselves. Roughly, it is a difference between Kuhn's "problems," which are typical of revolutionary science, and the well-defined "puzzles" of normal science. So far the AI people have dealt mostly with highly reconstructed puzzles. They would probably deny this, but I am *not* saying that they are only looking at "toy" problems. Better, let me use Simon's (1973b) own valuable distinction between well-structured and ill-structured problems. Now a historicist will try to avoid looking at innovative scientific work retrospectively and will find that the problems are often very ill-structured initially, much too unstructured to apply anything like the BACON computer programs, or even the KEKADA program, for example. The AI accounts that I have seen so far are more applicable in the civilized and refined atmosphere of mature, *Handbuch* science. But a well-formed question is already half answered, as they say, before AI is even brought to bear. In my jargon AI so far deals with "discoverability" (cf. 5.3.1) more than initial discovery. I am also less optimistic than Simon & Co. that AI programs will completely replace expert human judgment.

Callebaut: Simon himself would actually disagree on this, I guess (e.g., Kulkarni and Simon 1988, 1990).

Nickles: I readily grant that computers, programs, and automated procedures can provide tremendous assistance to scientists—they can do things that even the best scientists could not do ten or twenty years ago—but that is not to say that today's computers could have done all that work back then. I don't think human creativity in general will be easier to simulate than locomotion, as Simon apparently maintains. Now there is something to be said for parsimony—beginning with simple

3. Another pragmatist filiation that is relevant in this context is E. A. Singer, Jr.—C. West Churchman—Ian Mitroff. I am thinking in particular of their antifoundationist, "design" orientation toward philosophy of science (Churchman 1971; Mitroff 1974, 1976).

4. Cf. the symposium on "Computer Discovery and the Sociology of Scientific Knowledge" in *Social Studies of Science* 19 (1989):563–695, and the responses (including Simon 1991b) and replies in vol. 21 (1991):143–56; also Slezak (1991) and Collins (1991); and the comments on Simon (1992b) in the journal *International Studies in the Philosophy of Science* 6:1 (1992).

models—and I certainly want to avoid mysticism about the mind, but I don't think present approaches are anywhere close to replacing human judgment in its normal, open-textured form, in which it is sensitive to all manner of commonsensical as well as technical considerations. I will say in Simon's favor that I think the opponents of AI seriously underestimate the extent to which mature scientific problem solving is routinized and even automated. Look at all the amazing automation in a well-equipped biology lab, for example (Nickles 1990a). But while it depends heavily on these streamlined procedures, work at the frontier is almost by definition less well defined. So again, my gripe about Simon and his colleagues is that they tend to treat work at the cutting edge in too cut-and-dried a way, as merely "technical" problem solving. KEKADA is an improvement, though.

Wimsatt: Herb Simon was at Carnegie-Mellon when I went to Pitt, about a half mile away. I heard him give a talk very early on after I came there, was very fascinated by what he had to say, went over and collected a bunch of reprints either in my first or second year there. I read him and admired his style. I thought it was true, but I really hadn't incorporated it much into my image and bag of tricks. That happened much more strongly after I came here in Chicago. Probably there is no person whose work I cite and use more widely than Simon's. I think he is a true Renaissance man in the social sciences, though in the only sense in which there are any Renaissance men—that is, adopting Don Campbell's dictum, if you look for a successful Renaissance man, what you'll find in fact is a specialist in an unrecognized specialty which just happens to run across existing disciplines! Simon, like Campbell, had one major set of problems that he was interested in; and he tracked them wherever they cropped up, across a number of traditional disciplines.

8.1.1. *The Procedural Turn*

Callebaut: It's interesting to see how the linguistic turn that philosophers took around the turn of the century is now taken most seriously by historians and sociologists of science, whereas a growing number of philosophers are moving away from it. Van Fraassen, in *The Scientific Image* (1980:56)—I've quoted him before, but let me repeat this—says explicitly that the "main lesson of twentieth-century philosophy of science may well be this: no concept which is essentially language-dependent has any philosophical importance at all." He may be overreacting to some extent. But would you agree, Ryan?

Tweney: To some extent.

Callebaut: I'm not trying to enroll you!

Tweney: The philosophy that impresses me today is the work that is trying to go just beyond that. The linguistic turn in philosophy became excessively single-minded, as the effect was, "This is *all* we need to worry about!" Well, that's nonsense. My colleague at the University of Bath, David Gooding, is talking about the "procedural turn of the philosophy of science" (Gooding 1992). Like me, he has also been looking at Faraday for a number of years. He has replicated some of Faraday's experiments, and has found a great deal that supplements Faraday's diary accounts. In the nitty gritty of getting one's hands on that equipment and doing those experiments, a great deal of what is nonlinguistic or prelinguistic emerges. There are visual aspects, there are representational systems that are not like language; and in particular mathematics is a representational system that is used in science in ways that are completely unlike the use of language in science. Philosophy has not paid sufficient attention to this. And for that matter, many many philosophers of science strike me as knowing very little about science. They may know an enormous amount about philosophy—I don't mean to impugn their scholarship. But they do not know very much, in a *hands-on* sense, about the thing they claim to be trying to understand. That attitude baffles me.

8.1.2. *Let's Do the Two-Four-Six*

Callebaut: Could you briefly describe the problem situation in cognitive psychology: Why did some people at some time decide to do the kind of experimental work that you are interested in?

Tweney: The relevant work is by Peter Wason. Wason was, I think, the first one to see that logic as philosophers understand it—and in particular Popper's response to logical positivism—creates problems that lend themselves to psychological research (Wason 1966, 1983; Wason and Johnson-Laird 1972). And Wason with his two tasks—the four card task and the 2–4–6 task—at least posed the right question, which is, "How do people approach problems where the evidence potentially conflicts with their hypotheses?"

Callebaut: Could you describe one of the tasks?

Tweney: Let's do the 2–4–6 task. Suppose I tell you that I'm thinking of a rule which generates triples of numbers. You can find out what the rule is by guessing triples; I will tell you truthfully whether they do or do not fit my rule. To get you started, I'll give you an example that fits my rule: "2–4–6." Your task is to generate other triples to see if they fit. So you may guess "6–8–10." And I'll say, "Yes, that fits." Or "10–12–14." And I will say, "Yes, that fits." You then say, "Oh, your rule is 'three even numbers'!" And I say, "No, that's wrong." Then you go back and guess some more. But the rule I've used is, "any three numbers that get successively larger," and I start you off with a misleading example. In the process, you reveal something of your problem-solving strategy.

Callebaut: What did Wason find?

Tweney: He found that many subjects fall into a premature confirmation bias: They assume that their starting hypothesis, because it fits the data they have, must be correct. And they do not generate triples that are designed to *disconfirm* the hypothesis they started with. This is an interesting psychological phenomenon. Not everyone does this; but most people do. And the people who solve the task most rapidly tend to be those who've used a disconfirmation approach to their starting hypothesis.

Callebaut: What did you do in your own research?

Tweney: Well, my earliest research with Mynatt and Doherty at Bowling Green was using a much more complex version of a task to approach these same issues. We created a "complex universe" on a computer screen, with particles that flew around among a series of shapes (Mynatt, Doherty, and Tweney 1977, 1978). And subjects had to find out what were the laws that governed particle motion in this universe. We expected to find similar kinds of things in that context; that subjects who prematurely sought to confirm a hypothesis would do very badly. Well, in fact we found some surprising things. Indeed, subjects who only tried to confirm their hypothesis did badly, as we expected. But so did subjects who used a disconfirming strategy! Those subjects would always be rejecting any model they came up with and never seemed to make any real progress. The subjects who did well were the ones who *initially* used only a confirmatory strategy, who started with a hypothesis, using *only* the evidence that fit the hypothesis, and ignored the rest; and then, *after* they'd acquired a hypothesis with a lot of supporting evidence, switched and began using a disconfirmative strategy as a final *test* of their now well-confirmed hypothesis. Those subjects did quite well.

Callebaut: And you used these results to throw new light on the old philosophical dispute between the modern inductivists (the logical empiricists) and the modern deductivists (the Popperians)?

Tweney: We realized that what this meant was that the whole question of confirmation and disconfirmation had been approached in a simplistic way, that it was not a question of whether confirmation was "better" than disconfirmation, or vice versa, but that the best strategies were complex mixtures of both; that you could not understand either process in human problem solving if you did not look at the larger context (Tweney, Doherty, and Mynatt 1981; Tweney and Doherty 1983). That explains the limits on Wason's original work. It is not as simple as saying that people have an inherent bias to confirm whereas logically they should seek disconfirmation. It also greatly qualifies the kind of thing that comes out of Popper. His rejection of any kind of psychologism is too facile. You can't understand how science works unless you account for the complexity of these strategies, and you can only do so with research of the sort Wason did and we followed up.

Callebaut: Your work also has ramifications with respect to how one should interpret Tversky and Kahneman's work on heuristics.

Tweney: Their approach was very different in terms of the kinds of heuristics they were looking at. In fact, in many circles the term "heuristics" now means only those things that Tversky and Kahneman looked at, namely statistical biases in processing large amounts of data which has to be "averaged" in some fashion. In point of fact, the term should be much wider. I much prefer the way it was used by Newell and Simon in their *Human Problem Solving* (1972), to stand for a whole variety of strategies. Furthermore, Tversky and Kahneman produced the view that heuristics and biases are somehow similar things—that if we are looking at a heuristic, we are looking at a bias, in a negative sense. I'm not sure *they* said that, but they were read as having said that. That's wrong. What looked to us in the laboratory to be biases usually were biases in a specific situation. And when you look at the larger ecological context of human thinking, these so-called biases turn out to be *adaptive heuristics.*

Callebaut: We'll return to the whole issue of "ecological validity" below (8.3.2).

Tweney: The confirmation bias which you see in Wason's 2–4–6 task reflects a strategy that in many, many kinds of problem-solving situations works very nicely. If you have a hypothesis that explains some of the data, then, depending on the ecology of possible hypotheses, it may make great sense to *assume* that that's the correct one. At the very least, it gives you an explanation for the initial data; perhaps not the best *possible* one, or the most *inclusive* one; but it will get you a long way. Joshua Klayman of the University of Chicago has demonstrated this very clearly in the case of Wason's 2–4–6 task as such (Klayman and Ha 1987). Our work with the artificial universe shows something similar in a more complex task. And I think work like mine on Faraday shows the same thing.

Callebaut: How did psychologists react to your simulated universe studies? I imagine they must have looked closely at your methodology.

Tweney: One reaction was that it was not very well controlled, because the situation was so complex. Another reaction was that it was merely descriptive. A third reaction was that this was fine and good, but "do scientists really work this way?" Partly that's why I turned to case study methods. When I talk about Faraday, one reaction I sometimes get is whether or not he is typical. That is a misunderstanding of how we should use case study data, based on modern psychology's preoccupation with group studies and significant differences between groups. An account of how one individual works is so extraordinarily difficult that if we can get a good account of that person, that will represent progress. And the question of typicality or generality comes only after we know what it is we're going to generalize. I have a very different attitude about the group study and the idea of comparing group means than most psychologists do. I think this comes partly out of my recent reading of Gigerenzer and Murray (1986).

Callebaut: What, would you say, most singles out your approach from that of others in the psychology of science (see, e.g., Gholson et al. 1989; Gorman 1991)?

Tweney: I differ from many psychologists in what I see as a need for multiple methodologies in understanding something as complex as science. To say that we have to have a psychology of science, or better, a cognitive science of science, does not imply that we are studying a unitary thing.

Callebaut: Marc De Mey, who was one of my teachers, wouldn't disagree, I think (De Mey 1982, 1987, 1989).

"Ultimately a cultural mechanism produces good models of reality."

Ryan D. Tweney (b. Detroit, 1943) is professor of psychology at Bowling Green State University, Ohio, where he joined the staff in 1970. He started studying psychology at the University of Chicago and then moved on to Wayne State University to specialize in experimental psychology (M.A. 1969, Ph.D. 1970).

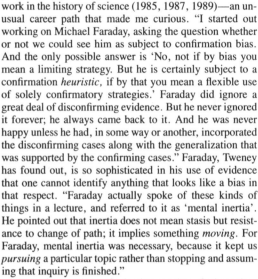

Tweney shifted from laboratory research in experimental psychology to work in the history of science (1985, 1987, 1989)—an unusual career path that made me curious. "I started out working on Michael Faraday, asking the question whether or not we could see him as subject to confirmation bias. And the only possible answer is 'No, not if by bias you mean a limiting strategy. But he is certainly subject to a confirmation *heuristic,* if by that you mean a flexible use of solely confirmatory strategies.' Faraday did ignore a great deal of disconfirming evidence. But he never ignored it forever; he always came back to it. And he was never happy unless he had, in some way or another, incorporated the disconfirming cases along with the generalization that was supported by the confirming cases." Faraday, Tweney has found out, is so sophisticated in his use of evidence that one cannot identify anything that looks like a bias in that respect. "Faraday actually spoke of these kinds of things in a lecture, and referred to it as 'mental inertia'. He pointed out that inertia does not mean stasis but resistance to change of path; it implies something *moving*. For Faraday, mental inertia was necessary, because it kept us *pursuing* a particular topic rather than stopping and assuming that inquiry is finished."

Part of the ecology of scientists consists of other scientists. How does a historian of science who uses the resources of cognitive psychology respond to people who insist that the student of science must always take its institutional context into account, or that science is essentially a collective, not individual, enterprise? He thinks that is correct: "Even if every individual scientist were completely irrational, the way science works culturally suggests that it could still move forward. The cultural context will do the falsification even if every individual in it is interested only in confirming his or her hypothesis." There are experiments that actually show this (Gorman and Carlson 1989). However, Tweney doesn't think scientists *are* irrational individuals. "What we do is *anticipate* that reaction, if nothing else, and do our own disconfirming, prior to putting it into the cultural or social context and having someone else do it for us." Any psychologist will tell you that such an attitude is adaptive as well: "Our egos are tied up with our hypotheses. We don't like seeing them

attacked. We would most like to make them resistant to attack. The way to do that is to do all the attacking on your own before you go public." Great scientists like Faraday do that. Tweney goes so far as to claim that if there is no social context in someone's thinking, they're not doing science but "off in some private fantasy world" of their own. (Some philosophers could learn from this.)

His analysis of Faraday's diaries represents a convergence between what historians like to do and what psychologists like to do. His aim is relevant to the subject of this book in the most direct way one can envisage: "In some ways I'd like to translate the folk epistemology of historians into the supportable terms of cognitive science, to get a richer, naturalized epistemology. That kind of an account, I think, will be much better able to be interfaced with the sociological account of science, which is very, very necessary. And it will bring us to the same kind of concern with the microstructure of science that we see on the part of some sociologists."

Tweney thinks that sociology of science has "very admirably" looked at the rich detail of the real work of science, and deserves to be emulated by cognitive psychologists at least in this respect. But his compliments are not always reciprocated. Thus he was attacked by a very eminent sociologist of science for his work on Faraday, "on the grounds that I was naively assuming that he worked as if there were no social context, merely because he was alone in his laboratory." Tweney's counterclaim is that nothing in what he said precludes any kind of social account whatsoever. "Clearly, in his laboratory, Faraday is affected by the work and ideas of others. Just as clearly there are mental events going on which he is keeping a record of, which can be understood as being a semi-autonomous set of processes. When I say 'semi-autonomous' or 'semi-independent,' I am acknowledging the effect of these social influences; which is not to say that we must talk about everything *only* in terms of social influences. That claim, it seems to me, is reductionistic in the worst sense of the word. And the real challenge is to take the dynamical account of the cognition of the scientist and map it onto the dynamical account that comes out of sociology." The most successful example of this that he has seen so far is Martin Rudwick's (1985) book on the Devonian controversy, which nevertheless he thinks still does not go deeply enough into the cognitive dynamics of the actors. But all in all, Rudwick, it seems to Tweney, "is practicing a naturalized epistemology very close to what we ultimately want at all levels."

Tweney is one of the authors, with C. R. Mynatt and M. E. Doherty, of *On Scientific Thinking* (1981) and has published extensively on the psychology of thinking and, since 1985, on the history of science.

I interviewed Ryan at the Club de la Fondation Universitaire in Brussels on April 22, 1989, at the occasion of a talk he gave to the Belgian Society for Logic and Philosophy of Science.

Tweney: Marc's book, *The Cognitive Paradigm* (1982), is, I think, the first call for a specifically cognitive approach, and so it's very, very important. I think that in our slightly earlier book *On Scientific Thinking* (Tweney, Doherty, and Mynatt 1981) we called for a psychology of science. And it seems to me that Marc was closer to the mark. It should be a cognitive psychology of science or a *cognitive science of science,* because it's the cognitive part that really has the contribution to make here. We are studying a series of very complex processes—the highest product, in some sense, of human culture. Again, we should not think one methodology is sufficient.

Callebaut: You distinguish "theories" from "frameworks."

Tweney: We have theories of perception, memory, attention, social process . . . We have to put all these theories together, and the laws that we derive from them, into a framework for understanding the complex, minute-by-minute reality of scientific thinking. By a "framework" I mean it is going to require an *interpretive* process rather than a theory-testing process. If all we do with the nitty gritty, or the molecular, or microstructure of science is to look for confirmation of our theories, then we are missing the most interesting part. The real world of science is so complex you can confirm or disconfirm any theory. All you have to do is select the right information. And that is most of what philosophers of science have done with historical episodes in science: They have simply selected. It's much more challenging and much more interesting to take a coherent body of stuff—a diary in particular—and see if we can't interpret it in as much of its complexity and richness as is possible, using whatever laws and theories we think we understand. If we can do that, then we gain faith that our approach is getting knowledge, and the interpretation is valuable in its own right. If we cannot do that, it means that our theories and our laws are still too restrictive, that we still need to work further on those. If they're *any* good at all, they ought to be good for that interpretative process; the theory should drive the selection of data, then we can see how well it fits and what doesn't fit.

8.1.3. *Micro Macro*

Callebaut: That takes us to the fundamental and much more general problem of interrelating an actor-perspective and a sociological approach focused on macrostructure (Knorr-Cetina and Cicourel 1982; Alexander et al. 1987; cf. Giere 1989a). You've already hinted at an answer by suggesting that the good sociological work also takes place on the microlevel.[5] Is that part of the answer?

---

5. Apart from the micro/macro debate, there is considerable disagreement in science studies as to the nature of the social. Slezak (1989a:580) is but one of many methodological individualists who maintain that "historical and sociological facts are supervenient upon the psychological ones" (cf. Elster, 4.1.6). Goldman's (1987, 1991, 1992) distinction between the domains of primary epistemics (the "basic," "native," or "natural" processes of perception, memory, attention, and reasoning), secondary epistemics (nonsocially acquired methods), and tertiary epistemics (which examines the social cognition involved in com-

Tweney: That's part of the answer. The good sociological work pays closer and closer attention to some of the details of the process; e.g., Knorr-Cetina's (1981a) book, which has focused very carefully on manuscript preparation, or the work of Harry Collins or David Edge. That's good; we should focus on that. What's missing is that the folk epistemology that *sociologists* frame too frequently assumes that the actor is something that is not problematic to understand, that the actor can be understood completely from a sociological perspective.

Callebaut: That's a criticism one can also make of Latour's work, I think (cf. 10.6.6).

Tweney: I challenge that stance. There are internal dynamics in the minds of the actors which do make a difference. Until we understand these internal dynamics, we don't know how to build that bridge between the social structures of science and the long-term processes that sociology can deal with and the short-term micro-processes—the moment-by-moment structures and changes in those structures that psychology is dealing with.

8.1.4. *Diary Men*

Callebaut: Did you pick Faraday (cf. also Nersessian 1984; Gooding and James 1985; Gooding 1990) purely for methodological reasons? And if so, doesn't your kind of approach restrict severely the sorts of material you can work on? I mean, people like Darwin (see, e.g., Gruber 1981; Kohn 1985; Richards 1987; Desmond and Moore 1991) and Faraday were prolific writers of diaries and notebooks; but how can one extrapolate from their work to the work of people—the overwhelming majority—who wrote much less?

Tweney: I got involved with Faraday purely for methodological reasons, as you say. There was a lot of it, it looked as close to think-aloud protocols as I was going to find with historically significant scientists. There was no mathematics, which made me feel better; and so on. All of the diaries are in one city—in London—so it was easier to travel. Subsequently I found that there is much more mathematics in there than I thought there was. There are no *equations*, but as I learn more physics and more mathematics, it becomes clear that it's necessary to talk mathematically about what he was doing (cf. Gooding 1985). David Gooding and I are writing a paper about that. I should add that Howard Gruber was very instrumental in persuading me to get really deeply into Faraday. It was a big leap for me, and Howard talked me into it.

---

municative acts and cognizers' beliefs about these) is equally controversial (Maffie 1991; Schmitt 1991; cf. Giere 1989a). The *social interests model* some sociologists have turned to as a causal-explanatory device (e.g., Woolgar 1981; cf. Hull 1988a) might be looked upon as a development that could bring the sociology and psychology of science closer to one another (after all, aren't interests psychological?) Yet on the whole, sociologists tend to dismiss psychological approaches (cf. Slezak 1989a, b; Simon 1991b).

Callebaut: Couldn't one claim that both science and its ecology have changed tremendously since the days of Faraday and his contemporaries?

Tweney: Sure. Of course I can't—I would never deny that. But our understanding of science must ultimately also include its history as a cultural institution. Perhaps I am writing only one chapter. Perhaps it's an early chapter. But it's a necessary one. That's a period when things were changing, and Faraday's epistemology is interesting in its historical context, that is, what *he* saw as the nature of epistemology, the epistemological assumptions *he* made. They are not entirely dead—much science still uses the kinds of strategies that he did, including science that has advanced beyond Faraday. Yes, if we stick only to diaries, there is a very limited sphere—there is Darwin, there is Faraday, and there are undoubtedly more contemporary cases. What we ultimately need to do, though, is think-aloud studies of ongoing science. The major constraint *that* will produce is the difficulty of analyzing such huge bodies of data, and the fact that most of what we collect will not be historically significant, at least not in the same scale of a Faraday. How do we identify a Faraday today? And even if we can do that, how do we get that person to collaborate? These are serious practical problems. But I have no doubt that, in time, people will begin doing this kind of thing. Perhaps I'll do something like that after I've finished with Faraday. People have sometimes asked me when I'm going to do another case, or another ten cases, or what have you. And the answer is "Probably never"—in the sense that I spent ten years doing one!

Callebaut: It struck me as quite surprising that you said that your work on Faraday was coming to a close: it would seem that the more you go into it, the more you discover . . .

Tweney: Oh, that's certainly true! My emphasis there should have been on the *my*. I have barely scratched the surface and I could easily spend the rest of three life times on Faraday alone and not exhaust all the problems! My involvement is coming to an end. I am writing up what I know, and I want to turn to a different problem. That leaves plenty still to do with Faraday. There are at least two other scholars active on Faraday right now, David Gooding and Frank James. There are others with interests in Faraday. And I'm sure that historians are going to continue to analyze this material for some time to come. Even so, it surprises me that so few people have worked on Faraday, especially when you see how big the Darwin industry is these days.

8.1.5. *Controversy Studies*

Callebaut: One obvious difference between Martin Rudwick's (1985) book and your own work is that he—as do many others in science studies today—focuses on a controversy (cf. Giere 1986).

Tweney: Yes, there is a lot of focus on controversy in science studies. Talking about Faraday is actually to involve yourself in controversies that *he* was involved in, in particular the life-long controversy over the nature of matter, force, and field. That concern really is simply complementary. By sticking with one individual across a longer span of time, I'm doing something that is more traditional, as far as historical study is concerned, than what Rudwick is doing. But again, if it will be part of the ultimate whole picture, we must have all these kinds of accounts before we can claim to understand what is going on. Rudwick, for example, spends relatively little time on the internal dynamics of his principal actors. He spends some, but not much. Some of those actors left diaries and notebooks which could be subject to the same kind of analysis.

Knorr Cetina: Controversy studies are important for obvious reasons. During a controversy you can observe, for example, the process of decision making about an experimental result or a theory or hypothesis. You can observe all the negotiations that go on and observe the interpretative flexibility that is part of it. You can see that the whole thing cannot be explained by a rationalistic formula. Whereas when the issue is closed—when you have scientific facts—you can't see *anything* any more. You have only two options, really. One is to look at the process during which these facts are produced, to insert yourself in this process of knowledge production—as laboratory studies have done—and participate in it. At that stage you can still see the uncertainty, the problems, the social structure, the negotiation—whatever it is. But once the fact is established, you can't see the thing anymore. The other alternative is to study controversies. That's another of those times when things are brought into focus: decision-making habits and mechanisms and argumentation patterns and the different rationalities of different participants.

8.2 VARIETIES OF COGNITIVISM

Callebaut: Ron, may I ask you to attempt a nutshell characterization of cognitivism?
Giere: That is difficult.
Callebaut: What do people associate with it? There are, for instance, East Pole and West Pole varieties (Dennett 1984).
Giere: I want to keep as catholic as possible about that. I think that the cognitive sciences, taken very broadly—all the way from neurobiology to cognitive anthropology, that whole spectrum—now provide the most promising set of resources for attacking the question of what kind of activity science is and how it works. That is in a nutshell the program; that is not any content. Within the cognitive sciences, when you look more closely, I guess I'm more comfortable at the lower end, that is, cognitive psychology and neuroscience. When you get to AI I start to get nervous. By the way, the cognitive sciences are not a continuum; they go off in several directions. One direction is from neurobiology, at the bottom, to psychology and then to AI. Another direction is to go toward sociology and anthropology. That kind of tree structure. I find myself uncomfortable working in AI.
Callebaut: What is it that makes you feel uncomfortable in AI (see, e.g., Haugeland 1981; Boden 1990)?

Giere: I am ambiguous about that. Many of the AI models, especially the very computational ones, I find too much like logical empiricism—a sort of logical empiricism with a computer. In the general camp, these people are taking a cognitive approach to the study of science. That is good, I like that. But then within that, I find differences. Paul Thagard and I have had a number of discussions recently, and I find his work ultimately uncomfortable. I have written a commentary for *BBS*[6] on some of his recent work, which he summarized in a target article (Thagard 1988, 1989; Giere 1989c). I don't think it is doing what it is supposed to do. I think part of the reason is that he has gone back to assuming that you can do it all on the level of theory, which is to say, what is programmable. That is very much against the evolutionary view. You've got to take real science, you've got to take the context. Let me put it another way: *AI decontextualizes an awful lot.* It gets back to trying to eliminate the content of science. In fact, of course, machines only have syntactic content. Humans are more complex than that; you have to add interpretation.

8.2.1. *From Crusoe to Connectionism*

Callebaut: At the East Pole, Paul Thagard (1989) and others are making a lot of fuss about the alleged fruitfulness of connectionism, also to approach a philosophical issue like explanation. Have you looked at connectionism?[7]

Giere: Yes, but you have to be careful here. Thagard uses propositional networks in which the representations—propositions—are localized and then linked together. That is a kind of connectionism. But the connectionism that has people excited is more radical. It utilizes *distributed* representations (see, e.g., Smolensky 1988; Clark 1989). There are no localized propositions. Like Paul Churchland (1989), I am enthusiastic about this kind of connectionism because it looks like something that might actually be instantiated in a human brain. It is more like the real thing than rule-based AI or even propositional networks.

---

6. The journal *The Behavioral and Brain Sciences.*

7. Connectionist models consist of networks built from large numbers of extremely simple interacting units. Ramsey and Stich (1990:189) nicely summarize what connectionism is all about: "Inspired by neuronal architecture [cf. note 11 below], connectionist units are typically linked in such a way that they can excite or inhibit one another by sending activation signals down interconnecting pathways. Networks commonly involve a layer of input units, a layer of output units, and one or more intermediate (or 'hidden') layers, linked by weighted connections through which a wave of activation travels. When the processing proceeds in only one direction, as in the case with 'feed-forward' networks, units modify and transfer the activation signal only to subsequent units and layers. In other, more complicated networks, activation may involve feedback loops and bi-directional communication between nodes, comprising what are often referred to as 'recurrent' networks. The units themselves may have threshold values, which their total input must exceed for activation. Alternatively, they may act in analog fashion, taking an activation value anywhere between 0 and 100%. Connecting links have varying weights or strengths, and the exact nature of the activation signal transferred from one unit to another (that is, its strength and excitatory or inhibitory value) is typically a function of the connection weight and the activation level of the sending unit."

Churchland: There is a tremendous amount of growth and development in connectionist models right now, some of which are addressing psychological phenomena and some of which are much more low-level. One of the most beautiful projects involves modeling very faithfully the neuronal organization in the leech that underlies the bending reflex (Lockery et al. 1989). It is crucial to get dynamics in the models, and Lockery has begun to do this. Other models capture aspects of language, stereovision, pattern recognition, and sensorimotor integration. There are many new ideas, recurrent nets being the most important, probably. The net result is that you can do all kinds of things that Jerry Fodor used to stamp his foot and say, "You can't do." For example, "You can't handle structure and compositionality in a net." Well, structure *is* handleable in nets, and so is compositionality; you just have to know how! Fodor and Pylyshyn's skepticism may have been applicable to the very early nets circa 1985. But those nets are the model T's of the field, and little can be concluded about the potential of all possible nets from these pioneering, makeshift nets.

Callebaut: But there are still lots of problems.

Churchland: One of them is that we don't have many models that are biologically realistic. We also need to find out how to scale up models, how to have nets interact with other nets, and how to get time—real time—into the models.

Giere: If you think of agents as processors that are parallel processors, connectionist machines of some kind, then—Paul Churchland is explicit about this—pattern recognition becomes important, and linguistics drops out. So you don't think of the mind or brain as processing language. That has always seemed to me crazy; that's not how the brain works. I'm enthusiastic about connectionism, because it pushes in the right direction (cf. Tweney 1992). And the older sort of straight computational AI is not a very good model for how science works. Simon says, "Oh, that's OK. For purposes of exploring these things, we can run connectionist programs on old hardware." Sure, why not? It will be slow, but maybe fast enough to see what is going on. That is just a methodological strategy. No problem there. But I'm uncomfortable with the models that make the human agent a standard von Neumann computer. I think pattern recognition is much more important in science, which is why now, what I am working on, are visual models.

Callebaut: Let's return to the social aspects of science. Could Robinson Crusoe be a scientist?

Giere: Try a more radical question? Could a *thorough* Robinson Crusoe be a scientist? That is, you know, dropped there as a baby and never socialized, and then raised by wolves or who knows what, and become a scientist. Well, in the end the answer looks like: no, because he never learnt any language. I think the social comes in because humans are social animals. They are social animals doing science. The attempt to abstract science from that context just won't work. Push it to the base level; you can't do it without language, and language is essentially social.

Callebaut: Cognitive ethologists distinguish between individual learning and social learning (cf. 9.2.1). Why should scientific language be of the second kind; or does it have *both* kinds?

Giere: Once you've got a person that already knows language, then you can set him off learning. Then it becomes so difficult that one person cannot do it. Actually I think it is a true generalization—or rather, one of those squishy generalizations— about the development of sciences historically, that it has become more and more social, in the sense that the solitary scientist is pretty much a thing of the past. Actually, I don't think the solitary scientist was ever as solitary as he has been pictured in reconstructions. Newton was a lot more social than he is often made out to be, and Einstein too . . .

Callebaut: . . . The popular Einstein picture of a solitary thinker and a humanist would seem to be a perfect example of what Marxists used to call "false ideology." *That is certainly not someone representative of twentieth-century "big science."*

Giere: Yeah. It is almost necessary, in talking about science, to talk about it as a social phenomenon. Then the other question—how it works as a social phenomenon—is an empirical one.

Callebaut: On your view, Kuhn is much less of a cognitivist than on De Mey's. There is his Gestalt view, of course . . .

Giere: Kuhn's official theory is the stage theory. But in fact, I think what he is groping toward is a cognitive theory. He picked Gestalt psychology because that was the best thing around in the late 1950s. He did not pick behaviorism, and that was smart because it was lousy! So in that sense he was striving for a cognitive theory; and there are all kinds of cognitive things in the theory. But he was writing before cognitive science had been invented. I think Marc is right to put Kuhn at the cognitive level. That is where he does belong.

8.2.2. *Ye Goode Olde Epistemology Again?*

Callebaut: Alvin Goldman's *Epistemology and Cognition* (1986) is probably the book coming as close to a naturalistic position as many philosophers are willing to go today.

Bechtel: The problem I have with Al is that it's still written too much from within a philosophical framework, and the issues are still the philosophical issues of justifying knowledge. My orientation would be to try to figure out first of all what cognition is. I would prefer to see the order being *Cognition and Epistemology*. Al has gone out and read a lot of the psychological literature, but I always have the worry that he isn't *shaped* by it. His philosophical interests shape what he's drawing from it. I guess that one of the advantages—or disadvantages—of having come into philosophy late and having left it early, in terms of where my own interests lie, is that I'm not gripped by traditional philosophical worries, by distinguishing "knowledge" from mere "belief." So I don't see the move to reliabilism as a major move. I'm interested in how cognitive systems work to produce—*in* *this sense*—reliable information, that is, information that makes it possible for organisms to get around their world fairly well. But then I must be willing in *some* respects to think of folk representations as knowledge, for they still enable the organism to get around in its environment.

8.2.3. *The Implausibility of Sentences in the Head*

Callebaut: How did *you* find connectionism, Bill (Bechtel and Abrahamsen 1991)?

Bechtel: One of the issues I had been harping on without knowing how to pursue it was the folly of a cognitive psychology, an information processing story, not motivated in any way by neuroscience and consciously cutting itself off from any information from neuroscience (1983a, 1984c). I soon discovered that most acting psychologists were unlike the philosopher Jerry Fodor. They didn't argue for autonomy of psychology in a principled manner; many of them simply said, "Well, we don't know where to turn; we don't see how to connect our work up with brain work; and we realize that's a problem, we would *like* to do it." You actually find a few psychologists like Michael Posner, who actively try to figure out how to connect psychological models and neural models.

Callebaut: What about "sentences in the head" (cf. Bechtel and Abrahamsen 1990)?

Bechtel: I never found sentences in the head terribly plausible either. In the terms that Bob Richardson and I use in our book (1991), I found that to be a *direct localization:* you took a story at one level and tried to use those categories to describe the system at the next lower level. Thus you use the vocabulary we use to describe people to describe what happens inside people. It has been done many times in the history of science. We took fermentation as a process of cells and then looked for the "fermenter" *in* the cell.

Callebaut: What do you think is wrong with that, precisely (cf. J. G. Miller 1978)?

Bechtel: It's not a silly move, in fact, because there are many times when you want to go several levels down before you decompose the process and explain it. But *direct localization never really generates an explanation,* because you're equally mystified when the homunculus does it as when the whole system does it. So I wanted something other than a sentential kind of account of cognition. But it wasn't until I was attending the Cognitive Science Society meetings in Boulder, Colorado in 1984 that I learned that there was a new game in town. And I learned it in a very dramatic way. I hadn't been to the Cognitive Science meetings for a couple of years at that point, but I don't think I missed any major transitions. But this meeting was different; it was *the coming-out party of the connectionists.* They were *en masse,* and they were shouting their position out in any context that was relevant. One invited symposium made an indelible impression on my mind, in which Dave Rumelhart, Geoff Hinton, Kirk van Lehn, and Zenon Pylyshyn were arguing the virtues of connectionism versus rule processing. All the reasons I had found rule processing to be uncomfortable—"too rigid," "arbitrary," "wasn't explaining what we wanted to do," "couldn't capture the variance well enough"—

were the kinds of issues that Dave was pressing. It also was very stunning to see Zenon characterize himself as once the young Turk, now the old guard trying to hold onto the camp, and give the view that Zenon has since maintained very well, that he is the defender of the territory and very frustrated with these young Turks on the territory.

Callebaut: Had you seen connectionism coming?

Bechtel: It was very clear that something new and very exciting had burst onto the horizon. In retrospect, one could have seen it coming, although I didn't see it coming. And interestingly, my spouse, Adele Abrahamsen, who was Dave's student and for whom connection became a natural approach, didn't see it coming. Looking back and seeing what Dave had been doing for the last few years, though, it represents a natural development. Adele was working with Dave when they were doing symbol processing models, but were frustrated with them. They were doing semantic networks, which are propositional, because even though you have nodes in networks that are similar to nodes in connectionist systems, all the links are assigned semantic content as opposed to just being transducers of energy. But they were well aware that they were trying to decompose linguistic structures and that they were missing a lot as they tried to fit them into the symbolic model.

Callebaut: People usually associate connectionism with the duo Rumelhart and McClelland.[8]

Bechtel: Certainly Rumelhart and McClelland brought popular attention to connectionism, but others had been developing network models for over a decade, including James Anderson, Stephen Grossberg, and Teuvo Kohonen. As I noted, Rumelhart had originally been developing symbolic models. During the 1970s he got interested in schema theory but was still trying to fit that into the traditional model, because he didn't have anything else. But there was a series of papers in *Psychological Review* in the late 1980s in which Rumelhart and more McClelland were starting to talk about cascade models—where processes would affect processes later on, which would come back and affect earlier processes—that all were foretelling. If you go back at the papers, you will find that "recognition model" was in *Psych Review* about 1981 or 1982. In light of this, the emergence of connectionism shouldn't have been much of a surprise. Connectionists are working at a different level of organization than the neuroscientists are, but it's clear that it's the kind of system of which you could ask, "How could you actually design a *real* nervous system as opposed to these model nervous systems, and could we use additional information about how *real* neurons work to inform ourselves as we put these units into our connectionist systems?" It has that potential. It has never been fully realized, although you have people more on the biological side of connectionism who are more interested than those on the AI side in modeling actual neuronal systems.

8. Rumelhart, McClelland, and the PDP Research Group (1986); McClelland, Rumelhart, and the PDP Research Group (1986); McClelland and Rumelhart (1988).

Callebaut: Pat (Churchland 1986a) has replaced sentences by vectors.

Churchland: The idea that representations are vectors is a common feature of most network modeling that's being done now, regardless of how the net is trained up—by back prop,[9] or Boltzmann,[10] or by Hebbian[11] mechanisms. In general, it looks unlikely that an individual cell represents grandmother. Rather, grandmother is the pattern of activity across a set, possibly large, of neurons. The activity of each cell in the set is an element in the vector; representations in perception, therefore, can be understood as activation vectors. Stored representations are understood in terms of weight vectors. That turns out to be an enormously important new way of thinking about what representations are. They aren't sentences, they aren't little pictures—they're vectors.

Callebaut: My impression is that in Europe, on the whole, neuroscientists just resent connectionists.

Bechtel: There is some of that, but more it is an interesting convergence of people coming from very different traditions who *do* have some resentments to each other—people coming from traditional AI, cognitive psychology, as well as people coming out of the neurosciences, *plus out of physics.* That's a very impor-

9. Churchland describes the *backward propagation of error algorithm* thus: "The input comes in, goes through the hidden units, and there is an output. In the initial stages of the training of the network, the output will not be correct. The network is trained by taking the error measure of the distance between what the output should have been and what it actually was. On the basis of that error measure, a signal is sent back down the connecting lines to adjust the weights of the units, so that they are a little bit more close to what they should be to give the right answer." See, e.g., Lockery et al. (1989).

10. Boltzmann, the brainchild of G. E. Hinton and T. Sejnowski, is an interactive network where instead of (as in back prop) putting an error message through every time, you hold the desired output stable and let the system relax and settle into a configuration that will later give that output for the input. It turns out that Boltzmann takes longer to relax into the configuration that produces the right answer than back prop. But the question regarding back propagation is, as Churchland puts it, "Who or what does the measuring of error? There has got to be something that knows what is the correct answer and can measure the error between what the actual output was and what it should have been. Of course, in the brain, there can't be any such thing. So we have to have a different way of handling, training up, a network, or at least that may use feedback but arrive at the error measure in a rather different way." (Modifications are being made to the original Boltzmann, mainly by Hinton; Josh Alspector has made a Boltzmann chip.)

11. Like behaviorist associationists such as Watson or Pavlov, Donald Hebb tried to develop a reductionistic system of psychology closely related to the data of neurology, but his emphasis on organizational units also shows an influence of Gestalt theory. His starting point is the "assembly," a functional unit of sensory and motor neurons formed by perceptual processes, in which the discharge of energy ("firing") is coordinated and synchronized. The assembly of cortical neurons can undergo changes through either "fractionation" (some cells "fall out of the line" and lose their synchronization with the assembly) or "recruitment." More assemblies are formed and become interconnected as the eye shifts from one item of an object to another—perception is learned not innate, but it proceeds from parts to the whole. (Hebb's experimental findings contradict those of Köhler.) When the separate assemblies become so well coordinated as to enable the individual to combine the respective elements he perceives into a distinctive whole, a "superordinate system" (or "*t* factor") has been formed. "The stability of a perception is not in a single pattern of cerebral activity but in the tendency of the phases of an irregular cycle to recur at short intervals" (Hebb 1949:62). In learning, "facilitations" are established in such a way that a certain neural unit (an assembly or a higher-order unit) makes the oncoming of some other units easier. Motivation is related to the persistence with which phase sequences take place; emotions are a sort of neurological disturbance. (Cf. Wolman 1981:92–95.)

tant link: a physicist such as John Hopfield enters the story and brings models of spin glasses and so on out of physics.

8.3 GIBSONIAN ATTRACTIONS

Bechtel: Even although I was writing about philosophy of psychology prior to 1982, there are two fundamental changes, that year and the next, that affect how I understand psychology. The most important in many ways is meeting my wife, Adele Abrahamsen, who is a first-generation cognitive psychologist trained by Rumelhart, Norman, and the San Diego school. We first met at a Jerry Fodor talk. We both had had Jerry as targets, but for quite independent reasons. I would say, "Psychologists believe *X*"; and Adele would ask, "*Who* believes *X*?" I would say, "But Jerry tells us that psychologists believe that X!" And she would say, "But Jerry is a philosopher!" I had assumed that since Jerry had an appointment in psychology, he must be a full-fledged cognitive psychologist.

Callebaut: The importance of informants . . .

Bechtel: . . . became very clear to me at that point. Here were a whole group of philosophers running around in the world, talking about psychology; and Jerry had really done a number on us. He had presented himself as the exemplar of how psychologists thought. And it helped that in 1981 he ran an institute for forty fairly late assistant professors or early associate professors at the University of Washington. This institute in a way was going to shape a generation of American philosophers of psychology, and he had had a chance to put his image on the whole discipline and shape the dialogue. Meeting Adele gave me an inside informant to a radically different view of psychology, and even more gave me entrée to whole numbers of cognitive psychologists who gave me a much different impression of what was the work of psychology (Bechtel and Abrahamsen 1990, 1991).

Callebaut: And the second move?

Bechtel: I also met Ulric Neisser, who at the time was at Cornell. In fact it was at Marjorie Grene's summer seminar in philosophy of biology (cf. ch. 6: introduction) that Adele took me over to Dick's office. Only during the next year did it become apparent that both of us were moving to Atlanta (he was due to take a Woodrow chair at Emory University). Dick is a maverick cognitive psychologist. While he is in many respects the grandfather of the field—his book *Cognitive Psychology* (1967) was the first textbook of the discipline—he also was one of the rebels against the way the discipline developed. *Cognition and Reality* (1975) reflects the influence of James Gibson[12] on Dick. Spending a lot of time over the last

12. According to J. J. Gibson's (1950, 1966, 1979) *psychophysical theory* of "direct" visual perception, the succession of retinal images, which are discrete when separated by saccadic eye movements and continuous when produced by bodily motion, contains all the information necessary for a proper construction of a three-dimensional representation. Perception is a function of the environment, as the information present in the environment is available to be "picked up" by cognitive agents, whose mental states it

"I'm not gripped by traditional philosophical worries."

William Bechtel (b. Detroit, 1951) has been associate professor of philosophy at Georgia State University in Atlanta since 1987. Having graduated in religion from Kenyon College in 1973, he couldn't make up his mind at first as to what to do; he opted for the University of Chicago. He recalls: "In the course of my interview at Chicago, Manley Thompson mentioned a student who was doing a thesis on Quine. And I, naively knowing nothing about twentieth-century philosophy, asked, 'Is that philosophers' shorthand for Aquinas?'" Much of his subsequent life at Chicago was spent learning who Quine was. His interest culminated in his dissertation, Intentionality and Quine's Epistemological Enterprise (1977), in which he argued that intentionality has a legitimate role in a naturalized epistemology (Quine tries to rule this out in arguing for a purely behavioristic kind of epistemology). Bechtel's dissertation was still straight epistemology, his only contact with philosophy of science at the time being courses taught by Wimsatt and Nickles.

In 1977, Bechtel took his first teaching job at Northern Kentucky University. There he shifted his research to philosophy of science. Having done little course work in science at Chicago, he realized that he had to master some science well enough to make use of it in addressing philosophical questions. He decided to focus on nineteenth-century physiological disciplines in biology, "partly because I realized just too much had been done with Darwin for someone who was in a high teaching environment, without enormous amounts of time to master the Darwin scholarship. Even then—that was before the last mushrooming of Darwinism—it would have taken more time than I had available to get anywhere." The same thing was true of genetics: "Too many people were already working in genetics; given the resources that I had, I couldn't get really into it." But an area like cytology was much more open to investigation. Kuhn may have been an influence on the shifting of Bechtel's interest toward science: "I was clearly doing philosophy of science by looking at historical material. One of the first things I came across that really attracted my attention were Schwann's drawings of cells and his account of the origin of cells. I was interested in why he would adopt the explanation he did. So it may be that already I was interested in Kuhnian-type questions without knowing it!"

One thing that facilitated Bechtel's career change was Northern Kentucky University's proximity to Cincinnati, which was just across the river: "Cincinnati has a very active group, called MOLE, which stands for "More Or Less

Exact Philosophy." "More Exact Philosophy occurs during two hours on Friday afternoon, when you actually discuss a serious philosophical paper. Less Exact Philosophy occurs the rest of the night while you consume incredible amounts of beer." Bechtel became a regular participant in that group, which included Bob Richardson and other members of the Cincinnati philosophy department. He and Richardson had not interacted much as fellow graduate students at the University of Chicago, as the latter was tending a farm in Ann Arbor while writing his dissertation. MOLE brought them closer together: "We read Dobzhansky, Ayala, Stebbins and Valentine (1977) to learn more about evolutionary biology. It was in the course of reading that book that we began our collaboration. Both of us had read Popper and Eccles (1977) and decided that there was a hideous evolutionary argument in that book. We wrote a paper which we first gave at an APA meeting and then revised many times until we finally got it accepted (Bechtel and Richardson 1983)." They realized that their ideas were quite complementary on a whole number of topics, although they were working in somewhat different areas (Richardson more in genetics and evolutionary biology, Bechtel more in the physiological disciplines). From 1980 to 1983, Bechtel taught at the University of Illinois' Medical Center, from which he moved on to Georgia State.

Bechtel's self-description of his work confirms Don Campbell's picture of science as an opportunistic endeavor: "It's not the case that I usually have a conscious plan in place. An opportunity strikes." Thus, working together with Richardson for the sustained project of their (1991) book, *A Model of Theory Development: Localization as a Strategy in Scientific Research*, came about partly because in the fall of 1983, Bechtel saw a grant announcement for collaborative projects. "Schemer that I am, I thought quickly, 'Who can I collaborate with? Since this is the first round of these grants, they are going to be far easier to get because they haven't announced them as widely as they will be in the future. So let's get a grant application in during the next month.' We put together the application for the localization book, and the funding we received has helped us make the effort involved in carrying out a long-distance collaboration. What I'm doing in much of my career is taking advantage of whatever opportunities put themselves in my place. The regrets come when I realize that there are more opportunities in front of me than I am ever to realize!"

In 1988, Erlbaum published Bechtel's *Philosophy of Science* and *Philosophy of Mind*, two books written with a specific audience—cognitive scientists—in mind. In 1991, he and Adele Abrahamsen published *Connectionism and The Mind*. Bechtel is joint editor of the journal *Philosophical Psychology*.

I first met Bill Bechtel at Bill Wimsatt's house in the spring of 1985. The original interview was taped at my home in Liège (near the Rue Théodore Schwann) on December 3, 1989.

six years in Dick's research group meetings at Emory gave me a much different impression of cognitive psychology than I would have gotten anywhere else.

8.3.1. *Concepts and Categories*

Callebaut: What specifically distinguishes Neisser's approach from Gibson's?

Bechtel: Unlike Gibson, Dick is really interested in cognition; but unlike most cognitive psychologists, he takes Gibson's enterprise of determining what's relevant in the environment to be important before going inside the system. To me what that means is that you're looking much more at the interface between the cognitive system and its environment in order to gain constraints on the internal system. Now Dick would disown that immediately—he thinks it's way too premature to do internal cognitive modeling. The first year I was in Atlanta, Dick ran an important seminar on concepts and categorizations. Doug Medin, Larry Barsalou, and Dick co-taught. A linguist, Naomi Baron, was visiting from Brown at the time. Adele was sitting in on the seminar; Bob McCauley was in on the seminar. There weren't many students in the seminar, you understand; it was just all of us! (*Laughter.*)

Callebaut: And you learnt a lot?

Bechtel: It was my first thorough exposure to psychological work on concepts and categorization and gave me a much different perception of what's important in psychology than I've had. Much of my intellectual contact since 1983 has been with psychologists and not with philosophers. Subsequently Barsalou left Emory to move to Georgia Tech. The three institutions—Georgia Tech, Georgia State, and Emory—are all within a few miles of each other; so now my intellectual contacts are split half and half between Emory and Georgia Tech and their psych departments, almost to the exclusion of spending time talking to philosophers. I'm deeply embedded in those two—cognitive science, psychology—communities that give me two different treatments of psychology neither of which is familiar to philosophers.

Callebaut: *Some* philosophers know this material, know the Gibsonian tradition.

Bechtel: That is certainly true. John Heil (1983) and Ed Reed (1988), for instance.

Callebaut: My teacher Leo Apostel (1989:106) is convinced that "for the realist I am, Gibson's work is a necessary source." Rom Harré (1986), who is also a realist, is another Gibson devotee; and so is Marjorie Grene (cf. Miller 1992).

Bechtel: My contact with Gibsonianism is within the framework of a distinctive psych department. It's not Connecticut Gibsonianism, which is more extreme. It's

---

causally generates (cf. also Dretske 1981, 1983, 1988). This theory is at odds with the prevailing *cue theory* of space perception, which has its roots in the empiricist philosophy of Locke and Berkeley. According to cue theorists, the three-dimensional visual world which is first reduced to a flat, two-dimensional, retinal picture (say, of a trapezoidal door), must be inferred through the use of picture clues that are part of the retinal picture (the door is seen as rectangular because one knows that this is factually so).

Gibsonianism, but treated within a framework of people doing (fairly) mainstream types of work in cognitive psychology. Eleanor Gibson has been a visitor at Emory for a couple of years, and together with Dick has helped shape an environment with a distinctive flavor.

Callebaut: What kind of people object most to Gibsonianism (Michaels and Carello 1981)? Selectionists for sure (e.g., Richards 1976) . . .

Bechtel: . . . I know Don Campbell is radically opposed to Gibson; I don't quite see why.

Campbell: In print (e.g., Campbell 1989; Campbell and Paller 1989), when I disagree with Gibson, it is with a great deal more respect than I do in private conversation. But his texture gradient affordances require neurological embodiments of assumptions about the world, and it is easy to generate optical illusions with artificial "ecologies" that violate those assumptions. Gibson's own visual cliff work is but one example. So his dismissal of the concepts of "cues" and of the Helmholtz tradition of "unconscious inference," "constructionism," and the importance of illusions in understanding how valid visual inference works is wrong, even though his emphasis on how very well vision works, and upon the cues added when the perceiver is in motion, are correct. For philosophy, his emphasis on the affordances being "out there in nature" is just plain wrong. His "direct" perception supports misguided foundationalist longings in philosophy.[13]

Richards: Who today reads Gibson? Mostly philosophers and historians of science, I think. James Gibson's work on perception in the immediate post-war years was exciting and forced psychologists of perception to examine the complex features of the stimulus array. His insistence that information was contained in the array and that no neural or cognitive processing was necessary perceptually to detect visual objects certainly harmonized with the temper of behaviorism, with its decrying of hypothetical entities. But the whole movement of contemporary perceptual theory rests, not on any Gibsonian foundations, but on the doctrine that he sought to bury—Johannes Müller's doctrine of specific nerve energies.

Callebaut: Please explain.

Richards: Müller argued that the properties ascribed to external objects were a function of the properties of our nerves and imaginative reconstructions in cognition. That is the view fundamental to virtually all of cognitive and perceptual psychology today. After all, the energy array at the retina consists in discrete photons, not in the "pencil rays of light" that Gibson liked to posit in the stimulus environment. Well-confirmed experiments demonstrate that the mammalian eye does not simply funnel absolute intensity gradients of light to perceptual consciousness; rather

---

13. Campbell confesses to "a partisan aspect" of his rejection of Gibson: "In conversation, he depreciated and disparaged my beloved Professor Egon Brunswik, if he acknowledged his existence at all. Tolman and Brunswik's (1935) 'The organism and the causal texture of the environment' is still exciting to read, and Brunswik would have been the senior author except that he still resided in Vienna and had written only in German. (See also Brunswik and Kamiya 1953.)"

those gradients are distorted by the retina—so as to produce sharp edges of objects. Brightness and color contrasts are rather common phenomena that indicate the neural system is not simply the passive receptor of information that Gibson believed. Gibson, in his own funny way, was a realist more naive than any Medieval Aristotelian. And I think most philosophers who try to make something of his views are indulging in a kind of necrophilia of defunct science. Müller lives, but alas poor Gibson!

Bechtel: I don't have the strong Gibsonian view. It seems to me that what Gibson is providing us is a good description of the environment in which the cognitive system can develop. Now Gibson does not want to talk about the cognitive system. For him all is perception; there is no memory. And he calls it "direct perception." I don't take that to mean "no processing," as many philosophers have. But it *is* to say that the information is specified in the environment. What the organism needs to do is figure out ways to, in Gibson's terms, "pick up" that information. That may require processing; but it tells us what the system must adapt *to*. I don't see it as incompatible with an adaptationist story and hence even with a selectionist story of the internal cognitive system. *It specifies the niche in which we're working.* But philosophers have reacted very negatively to Gibson. I am thinking of Fodor in particular. Because they hear "direct perception" as a denial—as it was *intended* by Gibson—of the relevance of any information processing story inside the organism. I don't see that as a necessary consequence. Probably that's what Don Campbell is reacting to, for he sees it as incompatible with a selectionist story. He thinks it involves some kind of . . .

Callebaut: . . . "clairvoyance" (in his idiolect).

Bechtel: I don't see it as implying that at all. I don't think the controversy between Gibsonians and information processing psychologists as it is going on within psychology has been particularly influential on the philosophical community. To me it was something to see within the psychological community, and there are many psychologists who just don't, or didn't, see much relevance in Gibson ten years ago. The expression—which sounds like a Gibsonian expression but hardly originates only with Gibson—"ecological validity" . . .

8.3.2. *Ecological Validity*

Callebaut: . . . "Only do experiments with ecological validity" . . .

Bechtel: . . . has now cropped up in all kinds of variants of psychology. When it is said by people like Don Norman, I'm assuming it's not coming from Gibson. It's again a self-recognition that we can't just do laboratory work on whatever task we want to put to the organism. We've got to direct our modeling work at trying to model the performance of *real-world* tasks. So I don't think it's really that strongly a Gibsonian point; and many psychologists still don't see the relevance of Gibson to their work. Probably those in perception will be the ones who will, because what Gibson has given is a richer description of the environment than anybody else has, and that is just food for perceptual psychologists to work with and say,

"OK, now what are the mechanisms by which we pick up this kind of information?" And clearly we are responsive to it. We are responsive to how fast we are approaching an object and it seems to be specified by the spread in the visual field Now how in the world is the internal system picking up that information?

Tweney: Bill's point about Gibson is really interesting to me, because it's only very recently that I've come to see Gibson as relevant, though Neisser has always appealed to me. Both make a lot more sense now that connectionism has given us models that let us move beyond rule-based flow charts with all the little boxes. *However,* even Neisser to some extent, and Gibson for certain, are missing the point that no model of cognition will succeed unless it includes an interactive, hands-on, loop. What we *see* in a "cup," say, has a lot to do with what we *do* with it. This point is really important for science studies. John Clement has a lovely paper (1989) in which he describes the hand waving of physicists as they solve a problem, and he concludes that this kinesthetic stuff is a critical part of their thinking. I've talked about Faraday's "push-pull" view of forces in similar ways (1987), and so has Gooding (1985). Nancy Nersessian's (1988) argument about the role of analogy in physics makes a very similar point.

Bechtel: For me one of the interesting contributions of Gibson comes in by the back door, because it's an old philosophical interest: intentionality. What is it about mental representations as they exist in people that makes them have content? For the information processing tradition that seems to be a real serious problem. What Searle (1980, 1983, 1985, 1992) nailed people on is that intentionality seems to be much more intrinsic with mental symbols than with linguistic symbols. It's not just something *added on*. It's not something instrumental such that, as Dennett would say, "you can *interpret* symbols in various ways." *Symbols, when they're mental symbols, seem to really embody their content.* The idea I'm playing with right now is that what we want at least for low-level categories[14] is that in some way, the objects themselves shape the representations; that there is some kind of constraint put on the nature of the representation so that it's not an arbitrary symbol for the object, as it is in language (you can take any word and make it the word for "cup"). The idea is that cognition is hooked up to the visual process, which is one source of constraint on what the internal representation will be.

Callebaut: What about action?

Bechtel: The other source of constraint is what we have to do with cups, which connects us with what Gibson called the "affordance"—objects afford certain kinds of actions to us, and our mental representation has to be sensitive to what we are going to be able to do with that object as well. That may tightly constrain whatever internal process could be the representation that we then use in other cognitive processing; so that at least for basic ordinary categories, concepts, they are tied to their environment.

---

14. Those categories that (can) directly describe things in the world: not names of fictitious entities but things like "cup," "bird," and so on.

Callebaut: That's the interface thing again (cf. 5.3.3).

Bechtel: The interface between the internal and external systems may be absolutely critical, because each will constrain the other in critical ways. In particular, the external may constrain the internal representation. Obviously that is going to break down with "Santa Claus" and terms like that, where there is no external referent. But what philosophers again ought to do is to go natural here and go study how *kids* learn "Santa Claus," study what factors in the environment control use of such terms.

Callebaut: You are expecting a lot of connectionism, aren't you?

Bechtel: I see it as an important tool in that it allows us to ask different questions and to focus on different things than five or ten years ago, when the only game in town—as Jerry Fodor was fond of saying—was symbol processing. From within the new game, everything looks different. We think of cognition as involving pattern recognition more than symbolic processing. Now we can start to do comparative evaluations. But we're still going to need a long time for connectionism to take shape and to know what the commitments are. It may turn out that there are radically different commitments coming from different connectionist schools. In fact, it's a great resource for sociologists of science to get into right now. My problem is that I'm looking at it for one purpose and noting at the same time that it's a tremendously fruitful area for a quite different inquiry into how a science develops. So I am trying to record material so as to go back in twenty years and look at the development of connectionism with the eyes of a philosopher cum sociologist and historian. Right now I appreciate that what I learned about the sociology of science applies in an area where I am a semiparticipant. But for now my chief interest is in participating in the development of connectionism, not in the sociological and historical questions.

8.4 RECONSIDERING THE MIND/BRAIN ISSUE

8.4.1. *So Many Isms . . .*

Callebaut: Is mind/body dualism or, more specifically, mind/brain dualism as espoused by Cartesians (who opposed mind as an immaterial "res cogitans" to the material "res extensa") still considered a respectable position today? (I know many of my undergraduate students in the sciences are stubborn dualists!)

Bechtel: Not by philosophers active in the philosophy of mind per se.

Callebaut: I take it that, say, philosophers of religion would take it seriously.

Bechtel: (*Hesitates.*) Right. One area in which people seem to point to a dualism is from the *qualia* issue (see also 8.4.2). People like Thomas Nagel (1974) and Frank Jackson (1982) want to say there is something more to mental experience than the causal account given by functionalists is able to get at. I haven't read all of their works well enough to know whether any of them comes down as espousing dualism per se. But clearly they're espousing the view that there is more to the story than the materialists have been able to tell to date. When I did the *Philosophy of*

Mind book (1988b), I was doing the same thing as Paul Churchland was doing in *Matter and Consciousness* (1984): trying to lay out the various positions. In one chapter I deal with dualism, and I don't think there's any contemporary dualism in philosophy of mind in the States of a serious sort. Popper and Eccles's (1977) interactionism got some discussion a few years back, but no longer.

Callebaut: Yet if one looks at the larger scene, things look rather different.

Churchland: *Substance dualism*—and property dualism, which I'll explain in a minute, as well—does have enduring appeal. It appeals to many people's intuitions. There are many psychological considerations here. People worry about life after death, so they want to believe that there is a soul that will survive the bodily death. When you explain "It does not make sense to believe that something is true because you *want* it to be true," they'll say "Well, yes, but you know, it's just something that's got a hold on me; I kind of can't let it go!" So it turns out that there are lots of people who still are substance dualists, but they often do not square that with scientific beliefs. Substance dualism is not just going to go away just because from a scientific viewpoint we realize that it's deeply improbable.

Callebaut: How about *property dualism?*

Churchland: Here it's not that there is a separate substance, but rather that physical material can be configured so that it has various special properties, where these properties are *never explainable* in terms of the organization of the underlying structures. The trouble with property dualism is that you can't really justify the claim that a certain high-level property is not explainable in terms of micro properties. How do you know that conscious awareness is not explain*able*—I mean, maybe not explain*ed* now, but never? The prediction is far too strong. Moreover, we are beginning to get clues regarding neural mechanisms of macro phenomena, including attention, awareness, and short-term memory.

Callebaut: In the first chapter ("Materialism Transcends Itself") of *The Self and Its Brain,* Popper rejects materialism for . . . ethical reasons—to secure free will in particular. He writes, among other things, that "human beings are irreplaceable; and in being irreplaceable they are clearly very different from machines. They are capable of enjoying life, and they are capable of suffering, and of facing death consciously. They are selves; they are ends in themselves, as Kant said" (Popper and Eccles 1977:3).

Churchland: That argument is unconvincing, to say the least. Hume, needless to say, had it straight: Even if you do have a nonphysical soul, either the events in that "spooky" thing cause the behavior or they don't. If they cause the behavior, then you have exactly the same problem as you do if the physical stuff causes behavior. And if the spooky thing does not cause the behavior, then the behavior comes about by mere chance rather than as a causal consequence of beliefs, hopes, desires, and so forth. If the latter, how can you be responsible for something that comes about not because you wanted it but because of mere random reactions? So it doesn't *help* the free will worry to postulate a nonphysical soul. Hume essentially said, "What we need to do is distinguish between the kinds of causes that underlie behavior for which we hold people responsible and the kinds that don't." Indeed, I think that's what we try to do. The task is very difficult, of course; and the more we know about the brain, probably the more difficult it will get! In any case, the normative point that we ought to treat humans as "ends in themselves" is independent of the factual matter that brain states *cause* behavior.

Callebaut: In the same chapter where you dismiss dualism, Bill, you also deal with Gilbert Ryle's (1949) "philosophical behaviorism," which tried to reduce mind talk to talk about actual or potential patterns of *behavior* of people.

Bechtel: Again, nothing going on there contemporarily. Dennett, as a student of Ryle, comes closest to the philosophical behaviorists' tradition. In the next chapter I deal with three views that still get some discussion. The identity theory—even that gets almost no discussion.

Callebaut: Yet at least some people still take type-type identity theory seriously: they somehow believe that it is possible to define a one-to-one correspondence between psychological and, say, neurophysiological predicate terms (viz. natural kind terms as they occur in law statements).

Bechtel: There certainly are; for instance, Enç (1983) or Hill (1984). But the two positions that get the most attention are eliminative materialism and the token identity theory. The *token identity theory* permits a different categorization at the psychological level than at the neuroscience level of events, but still holds that whatever is described in that vocabulary is an instance of some physical entity— that is, one physically described . . .

Callebaut: . . . whereas *eliminative materialists* like the Churchlands claim that ultimately, the only relevant categorization will be that given at the neurophysiological level. Before turning to that controversial position, I want to point out that Richardson (1980) has argued that contrary to what, say, Fodor (1974, 1975) has claimed, classical, Nagel-type reduction *can* accommodate some of the newer positions in the mind/brain debate.

Bechtel: What Bob tries to do there is say that Ernest Nagel was never committed to the claim that there would be a one-to-one mapping; that it is quite compatible to have many-many mappings between physical properties and mental properties as long as you could specify under what conditions a particular physical property would be a mental property, or under what conditions a particular mental property would be instantiated in a particular physical way. The extreme token identity theorists would be in opposition to that; that would be someone like Davidson (e.g., 1970), for whom there isn't any law-like connection at all; that is, you can't specify the conditions under which a physical system is going to have mental properties, and vice versa.

Callebaut: But as you said, you don't operate in mainstream philosophy of mind.

Bechtel: There may be a lot more credibility given to Davidson's position by pure philosophers than by those who take their lead from psychology. As I see it, the issue breaks down on whether in fact you are going to preserve some level of psychological discourse other than one based directly on neuroscience, and worry about how the two connect . . .

Callebaut: . . . that is, for instance, Fodor's position . . .

Bechtel: . . . or whether you are only going to deal with one—the neuroscience level.

Callebaut: That takes us to San Diego eliminative materialism, which is usually opposed to functionalism, to be understood here as Fodor's thesis of the "disunity" of psychology and the more fundamental neurophysiology. Couldn't one say, somewhat provokingly, that within current philosophy of mind, dualism survives in a new guise: functionalism?

Churchland: *Theory* dualism often does. But notice that we must distinguish between Fodor's version of functionalism and what Paul and I think of as cognitive-neural functionalism. I take it as obvious that we need macrolevel categories, if only for the reason that networks of neurons have properties that result from both the intrinsic properties of neurons *plus* their interactions. My eliminativism is really my *revisionism*. That is, I make the empirical prediction that current macro concepts will be revised as neuroscience and experimental psychology progress. The revisions, I predict, will in many cases be extensive. This does *not* predict that we shall end up with no macrolevel concepts. Rather, it says that in all empirical probability, more adequate macrolevel concepts with a neurobiologically informed basis will emerge.

Bechtel: Functionalism really is a way of characterizing mental states in terms of their causation that is compatible with different stances on the identity thesis. You could be a classical identity theorist, thinking that each of the units in your functional story are going to map one-to-one onto some neurophysiological units. You could be an eliminative materialist. Or you could be a token materialist, token identity theorist about it. Those two issues can be handled separately; so the Churchlands call themselves functionalists, as will Fodor, as will someone like Armstrong who is closer to the traditional identity theory. Lycan (1987) makes this point very clear. The kind of move Richardson makes, Armstrong is likely to make too: that the same mental state may have different physical instantiations, so OK, well, I'll tell a different story in each case. And in fact the Churchlands are making that same move; that is, if there is a psychological category they're worried to preserve, it doesn't have to map onto just one neural state.

Churchland: As an eliminativist (revisionist), I really do have much more in common with most functionalists than with dualists. There is a sense in which the digital computer metaphor has led us astray here. Of course, Paul and I are also functionalists in the sense that at a given level—for example, the level of the network—you can specify the properties of the network so that you're giving a functional description. Such a description is instantiated in terms of neurons. For neurons you can give a functional characterization too: they are instantiated in terms of molecules: lipids, proteins, amino acids, and so forth. So you cannot help being a functionalist, but there is not one single functional level, and the autonomy of psychology does not follow.

Bechtel: As the Churchlands have pointed out, temperature doesn't map similarly when you're dealing with fluids, solids, and gases; it has a different instantiation in each case. So everybody is comfortable with the mapping of mental descriptions being very complex at best. No one is insisting on a narrow interpretation of how physical events are going to relate to mental events, as one might have thought naively—as at least *I* thought naively when I first got into issues of the identity theory back in graduate school.

Churchland: The real issue between functionalists like Fodor and us has to do with two things: whether research in neuroscience will be relevant to the questions we're interested in, and whether there will be revision of the high-level concepts that we use to explain behavior. We say "yes" to both. We certainly expect that there will be revision—how much remains to be seen. Concepts like perception and memory are just the most vague kind of outline of what is going on, and there will have to be much more fine-grained, high-level, functional, neural-based descriptions of those things. We don't think—and many people have a hard time seeing this—that you will explain psychological capacities such as "remembering to put out the garbage" in terms of the behavior of an individual neuron. To assume that you will is to look at the problem in the wrong way. Explanation will be in terms of the behavior of a whole set of networks, because representation is certainly a widely distributed phenomenon. So we must not imagine that there will be a reduction or an explanation that goes from psychological capacities, in one gigantic leap, to neurons or molecules. The reduction will go stepwise through all these other intervening levels—systems, subsystems, macronets, micronets, neurons, and molecules.

Callebaut: . . . as classical reduction (Oppenheim and Putnam 1958) was already stepwise . . .

Churchland: . . . Exactly so. I don't ever imagine I'm going to want to explain my dreaming about catching a plane in terms of an individual neuron spiking at a frequency of 200 Hz, for example. But anti-reductionists sometimes parade this straw man as reductionism, and then knock down the straw man with a fast scoff.

Wimsatt: I've been railing for years against eliminativists like the Churchlands and Stich's confident pronouncements that with the progress of neurophysiology, we would ultimately junk folk psychology and its categories, and consult the future descendants of neural atlases to figure out whether the preceding speaker's remark was intended as an insult. Then recently I thought: What if we discovered/decided that most thought was basically visual, gestaltish, massively parallel, and the like? That would be a deep revolution (at least in Academicians' beliefs about folk psychology), might change some beliefs about thinking, propositions, intentions, beliefs, etc., but is something I could live with. I think most people could. I also

"Science overrides our intuitions."

Patricia Smith Churchland (b. Oliver, British Columbia, 1943) is professor of philosophy at the University of California, San Diego, and adjunct professor at the nearby Salk Institute. She studied philosophy, first at the University of British Columbia (1961–65, B.A.), then at the University of Pittsburgh, where she stayed only one year. The critical thing that happened at Pitt, she recalls, was being in a seminar that worked through Quine's *Word and Object* (1960): "Quine made tremendous sense to me; everything in it was something I could endorse. By and large, his perspective that philosophy and science are continuous, his idea that there is no first philosophy, that everything is in principle revisable, seemed to me to be right. *It seemed to me to make sense of philosophy for the first time.*" Feyerabend was to become a second major influence: "I was enormously impressed by a talk he gave. Paul and I read his papers and were taken with his views about the theory-laden character of observations and the conservatism in much science and philosophy. Of course the later Feyerabend had some rather odd things to say, but it isn't like buying a car—you don't have to take the whole package."

Photo courtesy of UCSD

Wanting to see Europe, Churchland left for Oxford, where she was one of only a few students who thought that Quine's (1951) views on the breakdown of the analytic/synthetic distinction were "absolutely revolutionary." After all, "if there are no truths that are somehow independent of science, what the hell do the philosophers think they're doing when they say they are doing 'conceptual analysis?'" From her first tutor in Oxford, Elisabeth Anscombe, she learned Frege and the Fregean tradition, which "allowed her to understand the ideas behind the conventional wisdom." But nobody was very sympathetic with her infatuation with Quine and Sellars: "Strawson was thought to be the great figure of the time, the one with the really deep insights, and he still maintained an analytic/synthetic dichotomy." Yet *Word and Object* continued to seem to her more plausible than Strawson and Austin. "I was really a fish out of water the entire time I was at Oxford." But that is only one part of her story, the other being that Paul came to Oxford to visit her. "At first, Paul still thought that maybe the analytic/synthetic distinction could be saved. We had these wonderful rows day after day after day, walking through Oxford market, in the meadow, in the pubs. Finally he gave in and said, 'You've got to be right!'" Patricia ended up doing a dissertation on theory of action and the nature of behavior, inspired in part by Donald Davidson's idea that mental states are causes of behavior. She was allowed to do pretty much what she liked: "In certain respects, Oxford was also very tolerant. They didn't make it impossible for me to follow my instincts, and Arthur Prior, on my examining committee, actually liked it."

Paul and Patricia then began to explore the implications of the Quinean perspective they now shared, which eventually led them to rebut so-called *folk psychology*. By the time they were in Canada and had their first jobs, "Paul was already thinking the unthinkable: if, as Sellars, Feyerabend, and Quine point out, the framework we use to observe and organize the world is like a theory, then maybe it's a *false* theory, and maybe our ideas about the way we think about knowledge and mind are either false or at least need revision, and we need to think about a new science of the mind." Initially Patricia thought he was wrong; she couldn't really understand what a replacement might be like. "But eventually we began to agree on the skepticism about sentential epistemologies. So that was a case where I shifted to his ideas. He played around and played around with anti-sentential ideas. It was *very* difficult to get published. Other philosophers thought he was wacko! Respectable journals would not publish stuff like that, especially since we were from the Canadian outback." Kuhn, she acknowledges, also had a really big effect as she and Paul tried to understand the relation between psychology and neuroscience. "It's odd, in a way, that I don't refer more to him in the book. Maybe the reason is that he had a greater influence on Paul, and Paul sort of taught me Kuhn as he saw him."

Churchland finds working at the Salk Institute very exciting: "I love it. It also makes it easier if I want to place students there to do a rotation or two working in that lab." With Francis Crick, she does a seminar on Consciousness and the Brain: "The focal point is the convergence between neuroscience, cognitive science, and neural nets." The lab has tea every afternoon, which is really a seminar very often, and lab visitors come as well. "The tea time is by now a legendary institution, and I do more science *and* philosophy there than anywhere else." Others at the Salk she interacts with are Ursula Bellugi, Helen Neville, Charlie Gray, and Jonas Salk. "Francis Crick is very important to us all. He has a great breadth of vision, and at the same time he is cautious; he is both imaginative and skeptical. He is interested in consciousness but he is canny about narrowing the problem down to what can be neurobiologically approached."

In addition to many papers in philosophical and scientific journals, Churchland has published *Neurophilosophy: Toward a Unified Science of the Mind-Brain* (1986a). She is collaborating with Terry Sejnowski (UCSD) on a primer on computational neuroscience.

The interview was recorded in Pat's office at the University of California, San Diego on June 4, 1990. It was our very first conversation.

think we wouldn't junk folk psychology because I don't think it was ever as propositional (intentional, yes) as most (linguistically skewed) philosophers supposed. So now I am waiting for the Churchlands' revolution to see what will happen. But I hope that they will continue to let me write in English after it does!

8.4.2. *Consciousness Is Not a Unitary Thing*

Callebaut: In addition to rejecting Popper's arguments against physicalism, you and Bob Richardson (1983) also advocate the idea that consciousness is probably not a simple phenomenon.

Bechtel: There is a real risk of reifying consciousness as a single thing. Looking at consciousness simply from a functional perspective, it performs different tasks. The notion of a *selective mechanism* is highly important: we are able to screen out information. Already in that paper we pointed to some work of Neisser's on selective attention. There is good evidence that there are mechanisms in the brain that are capable of focusing attention. But there are probably other aspects of consciousness with different functions. Once you stop thinking of consciousness as this single entity but start asking what would be the various components of consciousness, you open up the possibility of getting a serious mechanistic explanation of consciousness, not one that seems to be dead in the water because it's got all consciousness in at once or you fail to deal with consciousness at all.

Callebaut: It's the same attitude you take toward the "qualia" problem some people have talked about.

Bechtel: Qualitative experience is obviously important, but to give it a name, to reify it that way, may block our inquiry into it. We want to know *aspects* of qualitative experience. I think Dennett once again has the lead, and that is: If we can show physiologically that we can alter our qualitative experience piecemeal, we can find its parts. You know, pain experience is not a single thing, because various medications for pain make people say such weird things as: "Well, I still feel the pain, but it doesn't bother me now!" Obviously, then, we split the consciousness of pain into components. And that seems to be exactly what you have to do as an opening move. Now you're on the road to a mechanistic explanation and not on the road to positing autonomous consciousness, a self separate from any interacting with the brain, and so on.

8.4.3. *Real Intentions*

Callebaut: Dennett (e.g., 1987) wouldn't object to that, I guess.

Bechtel: I don't think so. Obviously my views in many respects are close to Dennett's, although I'm a realist about intentionality where he is not.

Churchland: Paul and I have disagreed with Dan Dennett for a long time on the instrumentalism, though we agree on just about everything else. I think what's right about his approach is the idea that when we do try to ascribe intentionality to someone, we have to do it from our own perspective, and we do it on the basis of what we think would be a reasonable hypothesis as to what that person means.

Bechtel: The issue I take with Dennett is I don't think intentional ascriptions are merely something we choose to make at various times to various systems (Bechtel 1985). I take it that Searle is on target insofar as he has pointed to the flaw with at least what has gone under the name of "functionalism" in contemporary philosophy of mind: The story purely of the internal causal manipulations of representations doesn't capture their intentionality. And it seems that clearly, mental systems *are* representing the world. I can't go as far with the Quinean game as to reinterpret everything. I can reinterpret what you say by forming an alternative translation manual, according to Quine (1960). I can try doing that for my own statements. But for any level of interpretation I want to give, I've got to start somewhere and take the home language at face value. Quine never tells you what it means to do that; but it seems that *right there something critical is happening*. To take it at face value means that you have established a semantic interpretation. That you can always reinterpret that, by taking another frame of reference at face value, doesn't answer the question what it is to take it at face value. That seems to me where there is something wrong with the instrumentalism of someone like Dennett: We've got to say what it means to have an interpretive framework.

Callebaut: In the paper you wrote in *Cognitive Science* (Bechtel 1985) . . .

Bechtel: . . . I only took one step in this direction when I argued that in fact the realism of intention ascription came from the fact that the system was in an environment. That's still part of my current view, but I hadn't realized that the next move was to say how that external environment would affect the internal processing system. Now what I don't want to do is go to the point of saying, "Mental representations are discrete things in the head." That may be true, but I see no reason to believe it's true. But I do want to go so far as to say that the structure of what's in the head, even if it has got to be very holistically analyzed—which connectionism might give one avenue to—is still affected by what's out there in the world. And so I want to maintain what I said in my paper, "Realism, instrumentalism, and the intentional stance," but then say, "We've got a real shaping process; my ability to talk about cups is based in part upon how I've experienced cups by interacting with them." And here the Gibsonian point of view has to be taken even more seriously. As I suggested before, it's not just how they occur to me while I'm sitting stationary but how I interact with them as a moving entity in a world in which they are. So I experience them through different senses, from different points of view; I see them as I move and all that. I think that's important to the kind of representation that would be in me. That's what I mean by realism: it's a real feature of my cognitive system that I have thoughts, beliefs, etc., *about* cups. It's not just a matter of arbitrary interpretation.

Callebaut: Pat, how would you go about tackling the intentionality issue?

Churchland: My own research strategy would not be to focus in on the debate that currently goes on in philosophy, because I don't think enough is known at this point to make significant progress. I think *mostly it's a mess.* Were I working on intentionality, I would focus on questions about the relation between *the brain* and the world. I would look very closely at research on child development, not just linguistic development but in general; I would look very intensely at language acquisition data and approach it with a combination of cognitive linguistics plus connectionism. But until we know much more about nonlinguistic representation in actual nervous systems, I suspect progress will be slow.

Callebaut: Bill would agree with that.

Churchland: Yes, I think Bill would. Now it's hard to know how successful the empirical approach is going to be, but it is clear that the developmental literature is understudied by philosophers who seek a theory of meaning. In my humble and jaundiced view, most philosophers haven't got anything very interesting to say about intentionality right now. Searle basically just reiterates our commonsense ideas; Dennett tells us the story on instrumentalism (and the intentional stance), Stich tells us something about it involving syntax, but this does not take us anywhere because he doesn't have any idea about what his "syntax" really means in terms of cognitive neuroscience. I suspect we might approach the problem more fruitfully by understanding the nature of representation in simple organisms and in mammalian subsystems.

8.4.4. *The Dismissal of Folk Psychology: Liberation or Oppression?*

Callebaut: You and Paul like to compare the demise of "folk psychology," which you expect to be the obverse of the progress of the neurosciences, with the disappearance of witchcraft in Western culture (P. M. Churchland 1979, 1984; P. S. Churchland 1982, 1986a; Stich 1983). You are also convinced that neuroscience can be liberating. Please explain what you mean by that.

Churchland: Detailed knowledge of the brain may well be liberating. It may help us understand things about ourselves and our behavior that we often find perplexing. For example, it's very liberating for people with obsessive-compulsive disorder to discover that it's not that they have a character defect or are screwballs, but that there is a neurochemical (serotonin) imbalance. And it's certainly liberating for them to be able to be administered serotonin enhancers and become normal humans again. We can imagine how it would be liberating, on a more fine-grained scale, for people to discover why they have certain fears or habits or a long-standing depression or whatever. Or if somebody is a psychopath, it may be very important to know exactly what that means in terms of brain structure and function. A psychopath can be a very, very dangerous individual. They're dangerous if they're white-collar individuals working on Wall Street and they're dangerous if they're wandering around the streets or in Congress. It might be very useful to know the neurobiology of those things, but we also have to be careful about human rights.

Callebaut: The other side of it is that people say, "I don't want too much to be known about me. I want my privacy. I am an individual."

Churchland: Well, those are things that we would need to work out very carefully. I of course agree that we don't want to have our minds read or micro-manipulated. We do want to be free to think what we want. But it's the most arrant science fiction to suppose any such detailed access is on the cards in the foreseeable future. It's simply not. There is not even a remote possibility that we shall be able to read other people's minds by means of some extraordinary machine that is outside the brain. If you know any neuroscience at all, you must realize how deeply ludicrous that idea is. Consider: what kind of a machine could it be? An EEG machine? No, that technique cannot show anything specific about thoughts. It may show that you're having an epileptic seizure or that you're in REM sleep, but it's not going to reveal a thing about the *contents* of REM sleep. What other technique would you use? Implant a batch of electrodes? Intracellular recording destroys the cell after a short period, so you cannot record from enough cells, even were the technical means available. Which they are not. But any kind of thinking or desiring or decision making is going to be widely distributed over hundreds of thousands of neurons in the brain. So privacy invasion on the imagined scale is just a deeply implausible fantasy.

Callebaut: Would you be prepared to defend folk psychology on political grounds, Helen?

Longino: To the extent that folk-psychological concepts are going to be replaced by purely mechanistic concepts, yes, I would defend it on political grounds. I would say: "Look, when we start out making an inquiry into this, we already have a conception of the sort of thing that it is, under which we're inquiring about it. Let's start, then, with a certain notion of what it is to be a human being. Then we can operationalize or make more scientific that concept" (see Longino 1990 on the "object of inquiry"). But why suppose that a correct description of objective inquiry is going to emerge from a study that begins at the lowest level of reduction that we can get and then builds up from there? I am sympathetic to focusing on neurobiology as opposed to thinking in terms of cognitive systems and then doing the neurobiology from there. I'm quite happy to start from analyzing the material base. But let's be clear what kinds of activities this brain is capable of. If that's what we want to understand, we want to understand how a human brain can produce symphonies or mathematical theorems which are meaningful and whose meaning can be grasped by the brain. Furthermore, folk-psychological categories are those we use to talk about ourselves as well as about others. To the extent that their dismissal as comparable to witchcraft or other discredited superstitions involves a covert plea that they be replaced by categories whose meaning is understood only by experts, it's another way of taking power—in this case the power of self-definition—from ordinary people. I find this very troubling, politically.

8.5 Neurophilosophy

Callebaut: Can machines think? In the 1930s, Alan Turing proposed that we judge a machine's "intelligence" by seeing how it copes with certain questions we ask it ("Turing's test"). A loud minority of people[15] on the fringe of AI are pessimistic as to the prospects of machine intelligence. Some of their objections have roots in phenomenology. Dreyfus's emphasis on the role of "inarticulate background knowledge" may also be likened to Polanyi's (1962) "tacit knowledge" dimension. From there it's a small step to the second Wittgenstein and historicism generally (Collins 1991, 1992; Dreyfus 1992).

Churchland: That's right, Dreyfus's insistence on inarticulate background knowledge came out of his European perspective. But we had an interpretation of it that fit with psychological data, which indicated that humans organized their concepts along the lines of *prototypes* rather than in terms of necessary and sufficient conditions. If that was true, and if, as Philip Johnson-Laird (1983) suggested, even human reasoning involves extrapolation from a paradigm case rather than the use of formal logical rules, then, we thought, there might really be a lot to Bert's hunches and ideas. Now he wasn't crystal clear and precise at the beginning; but then people *aren't* at the beginning of a surprising new idea.

8.5.1. *There Isn't Any Meaning Juice in the Chinese Room*

Callebaut: According to John Searle, a human mind has meaningful thoughts, feelings, and what have you, whereas a digital computer is just a symbol-manipulating device, making no reference whatsoever to meanings or interpretations of these symbols by itself (cf. also the exchange between Giere and Simon in *Social Studies of Science* [1991]).[16] You and Paul are not impressed. Your own "luminous room" (Churchland and Churchland 1990) perfectly parallels Searle's Chinese room argument.

Churchland: Paul has a way of working up an analogy so that you can see the relevant parts of the original problem and see it anew. It is very difficult to deal

---

15. Dreyfus (1979); Dreyfus and Dreyfus (1985); Weizenbaum (1976); Searle (1980, 1983, 1985, 1991); Winograd and Flores (1986); Collins (1991); cf. Hofstadter and Dennett 1981).

16. Searle's classic "Chinese room" argument against AI stages a monolingual English speaker who is locked in a room with a rule book for manipulating Chinese symbols according to computer rules. "In principle he can pass the Turing test for understanding Chinese, because he can produce correct Chinese symbols in response to Chinese questions. But he does not understand a word of Chinese, because he does not know what any of the symbols mean. But if he does not understand Chinese solely by virtue of running the computer program for 'understanding' Chinese, then neither does any other digital computer because no computer just by running the program has anything the man does not have" (Searle 1989:45).

with this powerful intuition that Searle draws on: no meaning can come out of just the assembly of bits of paper and a person passing paper in a room, and so on. The reason we think something must be wrong with that intuition is that when all is said and done, *that's all our brains are:* little bits of this, little bits of that, passing signals to one another. There isn't any meaning juice in there, any soul in there to have meaning—just calcium, sodium, etc., going back and forth across cell membranes. A number of people in the commentaries to Searle's original *BBS* article (1980) had made essentially that point. Paul suggested that you could address the intuitive appeal by considering a problem entirely different from the problem of how brains have meaning. He said: "Suppose that someone was given an explanation of Maxwell's theory on the nature of electromagnetism and the claim that light *is* electromagnetic radiation. They find it hard to believe. Imagine them scoffing: 'Well, just look, here's a magnet. Wave it back and forth over your head: it does not cause light! It is obvious; my intuition reveals that it's impossible that light *is* electromagnetic radiation! It's unimaginable that it is so.' *And yet it is so.* Moreover, there is an explanation why, when you wave a magnet about, you do not see any light."[17] Pauls's analogy shows how, although our intuitions can seem to be tremendously compelling, *they can be wrong.* It's a Quinean point: the science overrides our intuitions.[18]

Callebaut: Your *Neurophilosophy* (1986a, b) is basically a plea for the integration of top-down approaches—philosophy of mind, cognitive psychology, AI—and bottom-up neuroscience.

Churchland: There appear to be a variety of levels of organization in the nervous system: molecules (where there are important peptides and amino acids, for example, that interact with neurons and bring about changes in the neurons)—structures of neurons (the synapses or channels within the synapse)—whole neurons (they have receptors onto which the molecules attach)—small networks of neurons—larger networks of neurons—systems[19]—the whole brain. We want to be able to explain mental capacities such as being able to remember and store information, or to have stereoptic vision, in terms of, first of all (presumably), systems and large networks. Then we want to be able to explain how that works in terms

---

17. The wavelength of the electromagnetic waves produced is far too long and their intensities far too weak for human retinas to respond to them (see Churchland and Churchland 1990).

18. Dennett, the Churchlands, and others also think that the Turing test might not be entirely adequate for the purposes at hand: on the one hand it may be too tough; on the other, it may not be enough to mimic the in/out behavior in order to simulate or create something intelligent. Says Pat Churchland: "To do that within the relevant time constraints and in a way that permits the relevant kinds of behavior, you need to mimic the computations of the brain as well. Consequently, you need to mimic certain kinds of organization; not down to the last detail (you might, for example, not need to use neurotransmitters), but you might need to mimic it in considerable detail. We take that to be an empirical question: *maybe* you can make something that is conscious and intelligent, that doesn't mimic the organization of the brain; *maybe* the most efficient way will be to do it the way the brain does it. At the moment it looks to us that it's the best research bet to try to find out how the brain does these things and apply it to a machine, rather than pursue the isolated strategy that has been typical of conventional AI."

19. It's not quite clear how many levels of organization there are.

of the ever smaller networks. And of course, molecules can be studied biochemically, and those results can in their turn be explained in terms of the physics of matter. The process of discovery—the research—can proceed at many levels at the same time.

Callebaut: You consider it a unity because for each level of organization, the next level down should provide the explanation for the properties and the capacities.

Churchland: That, I think, is a very exciting possibility when one thinks of the unity of theories of the mind/brain: one hopes to get an explanation of psychological capacities—of what it is to have a character of a certain kind, or to see motion, or to be able to remember certain things and not others—in terms of the structure and organization of the brain. Now *it could turn out that no such explanation will ever be found.*[20] But fortunately, the prospects do not now look like that. On the other hand, we don't want to rule *anything* out a priori, of course. One has to adopt an empirical attitude towards all of these questions; we have to wait and see.

Callebaut: You acted as a kind of go-between, linking the various groups of people that are involved.

Churchland: When I wrote the book I said something to the effect that neuroscientists are often very isolated from philosophical considerations and computational techniques. But that has changed enormously since my book was written. Many neuroscientists are now very keen to develop computational models of one kind or another—of stereopsis, or the vestibula-ocular reflex, or echolocation in bats. There is a very different attitude towards computer models now that we see how we can make them brain-like (at least to a first approximation) than there was when they seemed totally remote from anything biological, as they were in the heyday of AI. Guzman's SEE program or Allen Newell's SOAR, for example. Neuroscientists are saying, for instance, "I can see how I could use a network to model learning in the hippocampus or walking in the spinal cord." You couldn't use good old-fashioned AI to do anything remotely like that.

Callebaut: There is also interest in seeing how these kinds of developments are fitting with a larger philosophical perspective.

Churchland: Scientists want to understand the arguments about qualia and what they really need to worry about. I had a neurosurgeon a couple of days ago ask me that: "What exactly are the arguments about the inexplicability of qualia? I want to be able to answer my colleagues." That's relatively sophisticated. To take another example: Francis Crick and I jointly run a seminar on "Consciousness and the Brain." We wanted to address philosophically sophisticated questions in a scientifically informed context. The students who are in that seminar, and the faculty too, are from philosophy, psychology, computer science, neuroscience—all over the

---

20. There could be a huge explanatory gap between, e.g., networks and neurons, such that one could never explain the behavior of the network in terms of the constituent neurons.

place. We all know we have to talk to each other in a jargon-free (well, almost) way, and we teach each other. So, for example, we look at what is known about attention by psychologists, and neurobiologists.

Callebaut: You found Crick's "searchlight theory of attention"[21] quite important.

Churchland: Francis's searchlight theory was important partly because it was a wonderful example of co-evolution of psychological and neurobiological hypotheses. It was also important because it took something that many people doubt you could get—a neural explanation for attention—and said not "I'm too scared to tackle this!" but "What the hell! I'm going to see if we can get a plausible explanation, in neural terms, for a psychological phenomenon like attention." Many aspects of his idea have survived; some have not. The oscillations that he referred to in the thalamus turn out to obtain only when someone is in deep sleep—unfortunately, they aren't the oscillations that one wants to use as an attentional mechanism. Other oscillations, however, have turned up. Since then there have been incredible developments both at the physiological level—that is, finding individual neurons in the visual cortex oscillating in a coherent way to a stimulus—and changing properties of neurons as a result of shifts in attention. There is new work on the way bearing on a set of issues—attention, visual awareness, pattern recognition, and so on. I think it is breathtaking stuff. There are many questions and a host of experimental issues that have to be settled. On the other hand, the research does constitute an attempt to discover something about the neurobiological mechanisms relevant to visual awareness. Many people would have thought that was too tough to handle. But it looks empirically addressable.

8.5.2. *The Devil Must Have Taken You!*

Callebaut: Do you see the major breakthrough in our understanding of the mind/ brain happening in your own lifetime? . . .

Churchland: . . . Some, undoubtedly . . .

Callebaut: . . . Or do you think it is happening right here and now? It might also be that there will be many small steps being taken that more or less imperceptibly turn into the big change.

Churchland: It's hard to know. It may be that there will be a lot of small steps and that in fifty years, people will look back and say "Hot damn: this is what this added up to!" It's really hard to tell. But the pace of research is accelerating. Many people feel that major breakthroughs, if not colossal ones, are on the cards, and that it's just a matter of time. It *may* not be within one's lifetime; as long as we can keep doing the research, we are going to get the answers.

---

21. Cells in different areas of the cortex appear to be specialized to respond to distinct dimensions of the physical stimulus, say movement or color; but our perception does not show such disunity. Crick (1984b) has suggested that in order to express the particular conjunction of properties in the stimulus, there must be temporary associations of cells, where each represents a property from a distinct dimension.

Callebaut: Do you see happening in neuroscience something similar to what happened in the 1930s and 1940s: the massive migration of physicists to biochemistry and later molecular biologists?

Churchland: Oh yes, I do. Post-docs and doctoral students in physics here, for example, are turning to neural modeling, working in Terry Sejnowki's lab, for example. I see students shifting back and forth between physics and neurobiology, or between psychology and neurobiology. Whereas five years ago, when I first came to San Diego, most psychology students would have thought it was unimportant for them to know about the brain. Now it's just taken for granted that you've got to know the basic neuroanatomy and neurophysiology. Similarly, amongst neuroscientists it's now taken for granted that it is essential to know the central literature on the psychology of memory and attention.

Callebaut: Also about philosophy?

Churchland: I think it is important to know about philosophy. Neuroscientists are not patient with substance dualism by and large, or with a view that says "Neuroscience is a waste of time, because it's all happening at the program level." They know very well that the distinction between software and hardware does not apply at all to the nervous system. But they do want to know about what reductionism is, or what qualia are, or what really are the problems about consciousness, what representations are.

Callebaut: Aren't you afraid of obstruction of some kind coming from groups similar to the creationists?

Churchland: Well, I get some mail from people saying, "How can you deny the existence of the soul? The devil must have taken you! God will damn you for this!" and so forth. Most people, however, are intrigued and fascinated; they want to see how the arguments all go. After a television interview I did with Bill Moyers, many nonacademics wrote or phoned to say they were relieved and fascinated by the idea that the mind is the brain.

9

DEVELOPMENT, LEARNING, AND CULTURE

JOHN MAYNARD SMITH (1986:v) lists "development" and "cognition" among the major unsolved problems in biology. Cognition, the naturalistic study of which was the main theme of chapters 7 and 8, is biologically relevant for many reasons, but maybe primarily because "the ability of like to beget like is the most fundamental characteristic of life" (Maynard Smith 1986:9; cf. Lewontin 1989, 1990; Tauber 1991). Development, in the sense of ontogeny, is not only important per se (E. S. Russell 1982; Thompson 1917); there is an increasing awareness that evolution cannot be understood accurately without taking into account the developmental constraints on it.

Development is the first major topic to be discussed in this chapter. The new techniques of molecular biology and of computer modeling are beginning to open new vistas here. They should eventually allow us to answer the question, "What would happen to minimally complex organisms in the absence of selection, given the constraints on their development?"—thus pointing the way to a better insight in what the null hypothesis for biological evolution is. The next issue to be addressed also has to do with "morphotype" or "Bauplan" considerations. If a complex phenotype is viewed as a structure generated from a small number of simple components, interacting with one another and with the environment, all kinds of biologically meaningful questions can be usefully rephrased in terms of the dimension "upstream–downstream." Thus the chance that a random mutation at a node that has a lot downstream depending on it will be adaptive is small, because too many things can go wrong. This idea of *generative entrenchment* may be equally applied to the perennial debate about what is "innate" and "acquired" and the more specific issue concerning a cultural equivalent of the genotype/phenotype distinction.

Many creatures do not just develop; they *learn* as well. Why is an entirely preprogrammed behavioral capacity not sufficient for all animals? The second section is devoted to the biology of behavior and learning, a topic which rejoins the bioepistemology discussed in chapter 7. Plotkin posits an identity of process across the four nested levels of evolution he distinguishes: genetic evolution, variable epigenesis, individual learning, and social learning. His view that these levels can best be characterized in terms of the pertinent information storage sites is a controversial one. Development and learning are, of course, also "gender-sensitive" issues. More generally, the participants in the discussion in this chapter suggest that a number of methodological (rather than political) lessons can be learnt from the recent sociobiology debate.

The third section is devoted to the emerging field of *evolutionary psychology*, which is often expected to provide the "missing link" that was the blind spot of sociobiology: a model of the evolutionary origin of cognitive capacities in humans, which can itself be used to try to explain certain human behavioral regularities. Is each and every one of us equipped with a "cheat-detecting social contract algorithm," as Cosmides and Tooby want to have it?

The discussion of multilevel models of evolution leads us naturally to reconsider, in the fourth section, the meaning of the notion of *fitness*—the "bugbear" (Plotkin) of all theoretical biologists. Extrapolating from biology, we will also ask what *cultural fitness* could be like. If fitness is defined in terms of *efficacy of match*, we can further ask how this biological notion of fitness is to be related to the philosophical notion of *truth*.

Everybody in their serious moments agrees that culture makes the difference in the human species, but what does this difference specifically amount to in terms of our extrapolations from noncultural animals to a cultural species like us? In the final section, Brandon argues that conditions that favor *phenotypic plasticity* (i.e., nongenetic determination) also tend to favor cultural transmission. Also, in a rapidly and capriciously changing environment, the strategy of "staying put" genetically is better than trying to track this environment. A promising approach to the study of cultural transmission is through studying human language, which approaches the properties of an ideal cultural transmission system. What, then, are the differences between biological and cultural evolution? How important is cross-lineage borrowing in culture? The chapter is rounded off with a discussion of the question whether cultural evolution may be said to be "Lamarckian" in any meaningful sense.

9.1 WHY GENETICS IS NOT ENOUGH, I: THE RETURN OF ONTOGENY

Callebaut: Bill, why are an increasing number of biologists agreed that development is so important to understand evolution (Maynard Smith et al. 1985)?

Wimsatt: We know an awful lot about development already, and there is a lot more we need to know, of course. But evolutionary biologists have become increasingly convinced in the last decade that the course of evolution is strongly influenced by developmental constraints, that you can't make any phenotype you like out of any given phenotype, that some variations are more likely than others.

Callebaut: Is this consistent with the randomness of mutations postulated by neo-Darwinism?

Wimsatt: It is not inconsistent with the fundamental assumption of the randomness of mutations with respect to whether or not they are going to be adaptive. But it does mean that mutations are directed; some mutations are more likely to happen than others; and of these mutations, some are more likely to be adaptive than others (Sarkar 1991a; Foster 1992; Keller 1992). There is *no orthogenesis.*[1] But denying orthogenesis does not force you to say that anything whatsoever can happen.

Callebaut: At least some people see more of a threat coming from some of the new studies of development than Bill is suggesting here, Dick.

Burian: Stu Kauffman (1992) has a book in press which is probably going to be of ground-breaking importance. It is too strong to claim that it will gain as much notoriety as Darwin's *Origin,* but it is likely to stir up a great deal of controversy. What is at stake is an issue posed by Darwin. Consider the end of chapter 6 of the *Origin:*

All organic beings have been formed on two great laws—Unity of Type, and the Conditions of Existence . . . On my theory, unity of type is explained by unity of descent. The expression conditions of existence, so often insisted on by the illustrious Cuvier, is fully embraced by the principle of natural selection . . . Hence, in fact, the law of the Conditions of Existence is the higher law; as it includes, through the inheritance of former adaptations, that of Unity of Type. (Darwin 1859:206)

The current issue is roughly this: Given both lineage-specific and quite general constraints on development as a potential cause of unity of type, the question that Darwin seemed to have resolved in this passage is being reopened by some biologists, including Kauffman. There are many open questions to be examined here, but we are beginning to have the tools to deal with them, thanks in good part to the techniques of molecular biology (cf. Burian and Richardson 1991). I don't want to make a strong claim about what will result from the current reexamination of the interrelations between conditions of life, natural selection, and unity of type, but the new work on these interrelations will need to be looked at very closely indeed.

Callebaut: What is it that Kauffman purports to show, specifically?

Burian: Kauffman has at least shown that you can pose the questions in the following form: *What would happen in the absence of selection, given just the character of metazoan developmental systems?* Until you answer that you don't know what the null hypothesis is, you don't know how to ask what base line selection departs from. If that can be asked clearly, we have a wonderful new question to cope with. That's what he's pushing for.

9.1.1. *The Developmental Lock Model*

Callebaut: Bill, you've been working for a decade and a half or so on a model of developmental constraints which you call the "developmental lock" (1986a, n.d.a;

---

1. It is not the case that only adaptive mutations happen or that there is a strong bias towards adaptive mutations happening.

Glassman and Wimsatt 1984; Schank and Wimsatt 1986; Griesemer and Wimsatt 1989; cf. Burian 1986a). What do you mean by that?

Wimsatt: The basic idea behind the developmental lock is this: Imagine the phenotype as a generated structure, which is generated from a relatively small number of elements—call them "genes" if you like—which interact with one another, and with input from the environment in the developing organism, to produce a much more complicated structure. Now suppose we represent this structure by some kind of network in which nodes are traits, or characteristics, of the developing phenotype, and arrows are causal relations between the production of those traits and the production of other traits. Now ask the question, "For a given node in this picture of the phenotype, how many other nodes are downstream of them?" What, as it were, is its causal scope through the flow of arrows to other nodes, and so on? The intuition of the developmental lock—and it is an intuition that other people have had before—is that *if a lot is downstream of a given node, the chance that a change in that node will be adaptive is very small, because an awful lot depends on it,* and to assume that it will be adaptive you're assuming that you can change that node without screwing up things all the way downstream. What the developmental lock does is just to quantify this in effect (see fig. 9.1). In the developmental lock, it turns out, very roughly, and with some qualifications, that for mutations acting earlier in development there is an exponential decline in the proportion of those that will be adaptive.

Callebaut: Could you say something about some of the features of the developmental lock model? What does is *do?*

Wimsatt: It does several things. Thus, if a much smaller proportion of mutations that act early in development are adaptive than mutations that act later in development, you would expect evolution to be much more conservative early in development. But if evolution is more conservative for earlier developmental stages, you get von Baer's law—that *differentiation proceeds from the general to the particular,* and that traits that appear earlier on are much more likely taxonomically distributed than traits that appear later on.[2] You get a lot else out of it. Part of what you get is due to the fact that the developmental lock model is not limited necessarily to talking only about biological phenotypes. It turns out to be a truth for any generative structure whatsoever.

9.1.2. Beyond the Nature/Nurture Dichotomy

Callebaut: One of the things you have applied your model to is the innate/acquired (or inherited/learned, or nature/nurture) distinction (Wimsatt 1986a).

---

2. Nineteenth-century evolutionists tended to view evolution as a kind of embryological, i.e., as a developmental, process. Haeckel's (now defunct) *biogenetic law*—that ontogeny recapitulates phylogeny—is best known in this respect, but he was not exceptional. In fact, Darwin himself was "guilty"—to put it whiggishly—of endorsing the recapitulation thesis, if we are to believe Richards (1992). On the decline of recapitulationism in twentieth-century biology, see Rasmussen (1991).

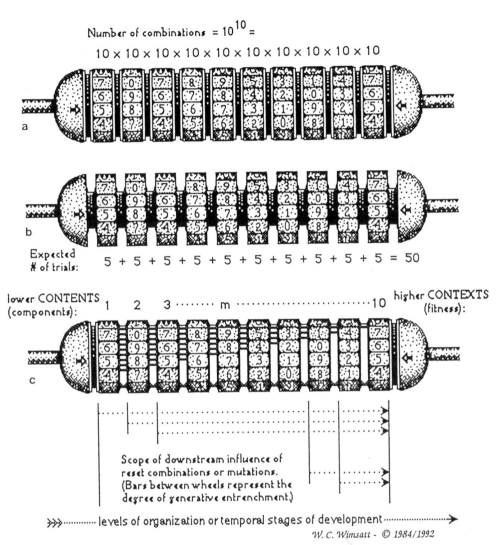

Number of combinations $= 10^{10} =$

$10 \times 10 \times 10 \times 10 \times 10 \times 10 \times 10 \times 10 \times 10 \times 10$

a

b

Expected # of trials: $5 + 5 + 5 + 5 + 5 + 5 + 5 + 5 + 5 + 5 = 50$

lower CONTENTS (components): 1   2   3 ······· m ···················· 10   higher CONTEXTS (fitness):

c

Scope of downstream influence of reset combinations or mutations. (Bars between wheels represent the degree of generative entrenchment.)

⋙ ·········· levels of organization or temporal stages of development ··················⟶

W. C. Wimsatt - © 1984/1992

Figure 9.1. The developmental lock

(a), Simon's "complex lock": Ten wheels with 10 positions per wheel. In the "complex" lock, the correct combination is only discoverable as a complete solution. (No clues are given for partial solutions.) Expected number of trials $= 10^{10}/2 = 5 \times 10^9$.

(b), Simon's "simple lock": Just as above, but a faint "click" is heard when each wheel is turned to its correct position, allowing independent solutions to parts of the combination. (The advantage of near-decomposability in problem solutions is the ratio of expected number of trials for the two locks: $5 \times 10^9/50 = 10^8$.)

(c), The developmental lock: This lock is a hybrid of Simon's two locks. Suppose a "click" is heard when each wheel is set to its (conditionally) correct position, but what position is correct is a function of the actual positions (whether correct or not) of any wheels to the left of it, so that a change in position of any wheel randomly resets the combination of all wheels to the right of it. (Simple if worked from left to right, since the partial solutions to the left are not disturbed by work on wheels to the right, but complex if worked from right to left, in the sense that partial solutions are not preserved.)

Wimsatt: Suppose you replace the notion "innate" by the idea of generative entrenchment. A couple of things happen. One is that since there are different degrees of generative entrenchment, the innate/acquired distinction is no longer an either/or opposition; there are *degrees of innateness.* A second thing that happens is that a node is just represented as a trait. Now influences can come in from the environment and have substantial later effects. What this means is that the reception of certain environmental information can be deeply generatively entrenched, in the sense that whether you get it or not strongly affects what you are doing—so the reception and use of environmental information can be innate.

Callebaut: Please give an example.

Wimsatt: Whether a greylag goose just coming out of the egg imprints on its mother or on Konrad Lorenz is going to affect its behavior in all sorts of ways. If it imprints on Lorenz it is going to hang around his household, probably going to be well fed, but it is going to have trouble when it comes to mating, because its conception of a proper sex object is Lorenz or people rather than greylag geese. In effect, by the way, biologically, that's a lethal behavioral mutation, although everybody has said, "Oh, you can do that, it doesn't hurt the goose!" That is not looking at it evolutionarily! If the environment were full of too many Lorenzes, geese would probably go extinct or the imprinting mechanism would be changed by selection.

Callebaut: What problems does interpreting the innate/acquired distinction in terms of generative entrenchment solve?

Wimsatt: I think it solves virtually all of the traditional problems of that distinction. It has interesting consequences. One is that *it separates being innate from being genetic,* because environmental things which are design parts of the phenotypic program can also be innate. So it can be innate that you see mother when you come out of the egg if you're a greylag goose.

Callebaut: And you are prepared to draw "political" conclusions from this.

Wimsatt: This strongly undercuts the arguments of many nativists, at least in the U.S., who hold that what is innate is genetic and what is genetic is unmodifiable, and therefore say, "If intelligence, or a certain degree of intelligence, is innate, you shouldn't have Head Start programs or programs that enrich the environment, because Blacks who have a lower IQ are somehow doomed to that and there is nothing you can do about it." What you say in fact, by contrast, is that they have been deprived of a natural part of their environment and that you need to enrich their environment so that their phenotypes can develop normally, and then you will expect that they will have more IQ—I shouldn't say intelligence, since I don't think IQ is intelligence.

Callebaut: Henry, you claim that one of the consequences of your multiple-level model of evolution, to which we will turn below, is that it radically solves the nature/nurture problem—it simply eliminates it.

Plotkin: That certain things are innate and other things are acquired has been argued for fifty or sixty years to be false. A common reason given goes by the name of *interactionism*. You cannot argue that a particular phenotypic trait is acquired *or* innate, the interactionist argues, because it is both; these two sources of influence interact to produce the outcome. That seems to me to be a very strange kind of answer to what is a very fundamental problem in psychology, because it brings to mind the image of two entirely different spheres of influence—*the* innate and *the* environment—which don't meet until somehow they mix in the production of a particular phenotypic attribute.

Callebaut: What is your alternative, then?

Plotkin: Living systems are *highly integrated* knowledge devices in which it is *conceptually* wrong to consider an attribute as having been formed by information coming from entirely separate sources. The reason for that lies in the notion of why living systems have evolved multiple evolutionary loci in the first instance, why they have a number of different kinds of information-gaining devices. That is because of the *sampling limitations* of the genetic system (see 9.2.1), which form the selection pressures for the evolution of a subsidiary knowledge-gaining process, the *developmental* system, which limitations in turn form the selection pressures (together with appropriate rates of environmental change) for the evolution of the third level in our model—*individual animal learning*. And the limitations, presumably, of individual cognition itself—that is the strong argument—are the selection pressure for the evolution of the fourth level of *cultural* knowledge gain, exchange, storage, and so on.

Callebaut: With Lorenz, you insist that learning is an innate capacity . . .

Plotkin: . . . and then it becomes manifestly absurd to argue about the possibility that some things are learned and some are innate, because *everything that is learned is innate*. No attribute of any creature is ever to be seen as being "dichotomizable" (if we could use that word). We are talking about something which is so complicated, I think, that in the end will we come to see that the notion that we originally had—that the genetical system is in some way the primary information-gaining system—may be historically correct; but if we take a living system after a very long period of evolution, it may be that the genetical system does not have this primacy, this kind of absolute sense of "being the boss" (cf. 6.4.3) of all the knowledge-gaining systems.

Callebaut: Your favorite example is the work that is being done on *bird song learning* and the way in which certain species of song bird learn dialects (regional dialects, variations) on their song patterns.

Plotkin: These regional dialects, it is thought, may be involved in mate choice in a causal fashion. Which bird is mated with which other bird is, it is thought, directly caused in part by attributes of dialect. If that is so, and if the dialect is learned, then what it means is here we have the third level of knowledge gain, individual animal learning, directly affecting the constitution of the gene pool of the breeding population to which these birds belong! That is to say, *evolution as it was tradi-*

"Living systems are highly integrated knowledge devices."

Henry C. Plotkin (b. Johannesburg, South Africa, 1940) is Reader in Psychology at University College, London. Having started out in science as "a perfectly ordinary brain behavior scientist" who produced lesions to study the effects on learning in animals, he became increasingly unhappy at the naivety of the experimental situations in which animals were placed. In the early 1970s, Plotkin worked with Karl Pribram's group at Stanford—one of the most interesting and innovative brain behavior laboratories in the world—only to become even more disenchanted: "The technical facility they had for the brain side of the thing was stupendous, but the behavior was just as limited and simple-minded as anything that I had been doing before." Konrad Lorenz's essay on the innate bases of learning (1969) would be an eye-opener. Back in England in 1973, Plotkin and a then graduate student, John Odling-Smee, began to ask what it is which allows certain animals to survive adequately *without having to learn.* "You must understand that such an interest is *very* unusual for a psychologist," Plotkin insists; "evolutionary thinking has almost no place in contemporary psychology." Plotkin and Odling-Smee set out to analyze more formally than Lorenz and others had done the "sampling limitations" of the information-gaining capacity of an animal who is operating only on the basis of genetically acquired information. The result was their multiple-level model of evolution, which looked attractive to many as a nonreductionist alternative to sociobiology (Plotkin and Odling-Smee 1979, 1981, 1982; Odling-Smee and Plotkin 1984).

The reason why sociobiology has not worked, Plotkin diagnoses, is that it attempted to reduce social science to what is known about genetical *mechanisms.* Now social science—for him, roughly, the study of culture—has no mechanisms, "at least not in the way in which biologists talk about mechanisms—as something palpable, point-at-able." Plotkin thinks that social science works on a different metaphysics—a "metaphysics of process," as he argued at the 1984 Ghent Evolutionary Epistemology Conference, where we first met (Plotkin 1987b). He suggests the problem be turned on its head in order to try to achieve "some kind of synthesis, some kind of meaningful relationship between biology and the social sciences by looking for appropriate *processes* in biology." In one of his rare philosophical moods, Plotkin may come close to "a rather Bohmite view of things" (Bohm 1969a, b), but without endorsing the mysticism one associates with the philosopher of physics David Bohm (see, e.g., Weber 1986). He assures us he is in reliable company there: "Marjorie Grene

(1969) comments on that particular piece of writing, and she says (I paraphrase): 'I have seen the light and the way; this makes current thinking in biology to me banal; this is magnificent!'."

Usually Plotkin thinks more soberly, though. He describes himself as an evolutionary epistemologist, but like Hull (1982), he is not interested in evolutionary epistemology *as epistemology,* insisting that it should be more scientific and less philosophical (Plotkin 1987a). And it took Plotkin a while before he realized which enterprise he was indulging in: "Donald Campbell pointed out to me that what had happened to me has in fact happened to almost everybody who comes to evolutionary epistemology: It is not the kind of thing that you're likely to be exposed to in any kind of undergraduate or graduate program, at least not at very many universities. We enter science not knowing about it; some of us 'discover' this set of notions. We get terribly excited about it, and then we also discover that a number of other people have had similar ideas as well."

I was surprised to learn from him (Plotkin 1979, 1981) that even creatures as primitive as beetles (whose study used to be his hobby in the summer months) display a "very, very limited" form of learning. He chose a predatory ground beetle because predatory animals are almost by definition very active. One of their behavioral characteristics is that they are "positively thigmotactic"—they like to move with some part of their receptor apparatus, mainly in their antennae, in contact with a surface. "They tend to hug surfaces; they move along cracks in the earth, or rocks, or bushes. I found that these animals will lose their thigmotaxis for a relatively short period of time if they are exposed to water, away from the surface that they are in contact with. But the thigmotactic response will not be weakened by any kind of food that I can administer to them." This is a kind of a beetle equivalent of nausea conditioning in the rat. "It happens to one thing—water—and not to another—a whole variety of foods." But does it make sense to try to account for these beetles' behavior within the same kind of theoretical framework as what, say, a chimpanzee will do? The stock answer of comparative psychologists is that they have got nothing to do with each other at all. Plotkin, who would prefer to see a truly general theory of learning developed, insists that the evolutionary analogy, because of its flexibility, "gives one the possibility of at least thinking about a theory of learning that has very wide applicability rather than being very restricted in scope."

Plotkin is the author of some sixty book chapters and journal papers (a number of which are listed in the bibliography), and the editor of *Brain, Behaviour and Evolution* (with David Oakley, 1979), *Learning, Development and Culture: Essays in Evolutionary Epistemology* (1982), and *The Role of Behaviour in Evolution* (1988a).

The material published here consists of updated excerpts from a series of Belgian Public Radio broadcasts that were recorded in Henry's house in Islington, London, on September 2, 1985.

tionally defined by so many biologists—changes in the frequency of genes in the gene pool—*becomes in part caused by learning.*

Callebaut: I would like to go back to the concept of "innateness" for a moment. That concept is much older than genetics, Bill.

Wimsatt: The introduction of criteria for saying that *something is innate if it is genetic* is in fact a twentieth-century phenomenon. Roughly: genetics worked very well; "innate" had always been read to be "inborn"; "inborn" elements are determinants of other things and are hereditary; and lo and behold, we had genes which were determinants of other things and hereditary, so you just had to identify them! But this is a mistake.

Callebaut: Why?

Wimsatt: On my analysis something can be genetic without being innate, namely if it doesn't do anything, or if the consequences of that gene are not very big, if it does not have much generative entrenchment. And it can be innate without being genetic, namely if it is a reliable feature of the environment which is utilized by the developmental program in a way designed by selection in producing the phenotype.

Callebaut: Someone like David Hull would go a long way with you but would object to associating things being relatively deeper generatively entrenched than others with them being automatically more general.

Wimsatt: He would?

Callebaut: Oh yes! He told me, for instance, that the genotype/phenotype distinction should not be associated with the idea of an opposition between "more general" and "more particular," respectively.

Wimsatt: Well, that's not exactly the same thing, but if he thinks that generative entrenchment has nothing to do with taxonomic generality, he's wrong! My student Nick Rasmussen (1987) has applied the developmental lock model to *Drosophila* development. What he has found is very strong confirmation of the model, and in fact predictions—not only predictions but confirmations—that things that are deeply generatively entrenched in the developmental program are more conservative evolutionarily and therefore more widely found taxonomically. So they are more general in that sense. There are two other senses of generality appropriate here. One would be morphological generality; that is, a general *Bauplan* is more general than any of its specific realizations and differentiations. And thirdly, functional generality: more deeply generatively entrenched evolved things are more general purpose than less deeply generatively entrenched things. Rasmussen has found very definite and suggestive confirmation of both of those ideas as well! So I just have to disagree with Hull on that.

Callebaut: Some of the ideas behind generative entrenchment are very old.

Wimsatt: The main thing is that people haven't yet exploited them. I would say that there are intimations of the idea of generative entrenchment in Gould's book *Ontogeny and Phylogeny* (1977), and going back earlier, to Schmalhausen and de

Beer on the basic form. I haven't been able to check it, but there are similar ideas. I assume the only difference is—I may be wrong in that—I think I am the first person trying to model this. (*Bursts out enthusiastically.*) Sorry! There is one important exception to that which I should mention. There is a beautiful book by Wallace Arthur, called *Mechanisms of Morphological Evolution* (1984), which argues for a very similar model. I think his views and mine are complementary in a variety of ways. And Leo Buss (1987) has since written some very interesting stuff on this, and is in the process of constructing a very powerful and general model of development which should be capable of further testing many of these ideas.

Callebaut: You have also exploited the idea of generative entrenchment in the context of gene control networks.

Wimsatt: We've been looking at variants of Kauffman's (1969, 1974, 1986) model of the evolution of gene control networks. We find that his models, when modified to give you information about generative entrenchment, strongly confirm the idea that more deeply generatively entrenched features stay around longer. I find that delightful, because at least at one early stage Kauffman saw his ideas as in opposition to this. The fact that his models and his structures actually turn out to be a confirmation of it is particularly delightful. That is work I'm doing with Jeff Schank, who does all of the programming work; we collaborate on the design of the simulations. It's very, very exciting stuff! (Schank and Wimsatt 1986; Wimsatt and Schank n.d.)

Callebaut: How—if at all—can all this be linked to the European tradition in comparative anatomy and evolutionary biology, which ever since Goethe has argued for the importance of form or *Baupläne* constraints on evolution (Gould 1977; Riedl 1978)?

Burian: Obviously, some people think it can; Brian Goodwin is an extremist in that regard (Goodwin, Holder, and Wylie 1983). I think there is less of a link than meets the eye, though Kauffman talks out of both sides of his mouth on this issue, and some of his allies, including Goodwin, talk very strongly as if there were a close link between Kauffman's results and the *Bauplan* tradition. The connection, however, is not obvious, and I think it is incorrect, for Kauffman's results drop out of the formal treatment of the statistics of complex systems. They suggest that the organisms of a lineage will lock onto some stable configuration or other, but also that there does not need to be any intricate virtue of the particular *Bauplan* which is involved. That is, lineages lock into some configuration, but not in virtue of the architectonic guarantees or benefits that come with it, not necessarily because of the influence of selection—but simply in virtue of what is to be expected on statistical grounds. At least, this is the sort of thing that Kauffman's models suggest.

Callebaut: Is it possible to illustrate this in a more hands-on way?

Burian: A very simple model which blew my mind was developed by Joel Cohen (1976); it can be used to raise a question about the *Drosophila* population cage experiments. It's a simple experiment that has been done both with simulations and analytically, of an urn-drawing sort: you draw balls with replacement out of an urn. Suppose a very large box initially contains one green ball and one blue ball. Choose one ball at random and look at its color. Place that ball back in the box and add another ball of the same color to the box. For any number of runs, you have a certain distribution of green and blue. What will the distribution of green and blue balls look like after, say, 1,000 runs, or 10,000 runs? It turns out to be a complete surprise. Obviously it is symmetrical: you have to have green and blue come out equal *over the long haul*.[3] Is it that most of the time you get something in the middle or rather something mostly green or mostly blue? It turns out that almost always the series of drawings comes to an equilibrium fairly quickly; *but the equilibria are absolutely uniformly distributed*. That is, a single run is as likely to be highly biased as not, but, on average, *any* equilibrium frequency is as likely as any other. So it's symmetrical and *the outcome of a single run is not informative about the intrinsic structure of the system*. Well, if that's the statistical base line for population cage experiments, you need to ask some sophisticated questions to know whether this population in a cage experiment arriving at some particular equilibrium is a significant result or not. That's the analogy on which you need to know what developmental processes work like: to ask whether this outcome reflects selection or not. That's not Stu's analogy but my own; but I think the point stands.

9.2 WHY GENETICS IS NOT ENOUGH, II: THE BIOLOGY OF BEHAVIOR AND LEARNING

Callebaut: You already intimated before, Henry, that many organisms do not only develop but are capable of *learning* as well. You insist that learning, as defined by psychologists, is akin to biological evolution. And you talk about animals and even plants as "localized observers." What is meant by that?

Plotkin: Any living system has a finite extension in space and in time. The information it acquires is going to be limited to its localized position in space and time; it will not be able to transcend that. It's a perfectly simple point . . .

Callebaut: . . . but one that is often neglected or even consciously disregarded.[4]

9.2.1. *Genetic Cost/Benefit Analysis*

Plotkin: What happens if that creature moves out of its original spatial boundaries or extends its life span, that is, if the sample of information it has been working on

3. Compare: In an actually performed coin-tossing experiment reported by Cohen in the same paper, the proportions of trials which came up heads after 10, 100, 1,000, and 10,000 runs were .400, .440, .502, and .507, respectively.

4. Cf., in a somewhat different context, the discussion of *satisficing* in 4.1.6, and Herbert A. Simon, "Rational choice and the structure of the environment" (1956), in Simon (1979a).

becomes applicable only to its previous history? It clearly becomes inadequate. An animal or plant that is working on the basis of information acquired entirely through a genetic system is picking up information at only a certain point in time—at conception. But during ontogenesis, the world the creature is interacting with is changing. If the condition in the world is different from the nature of the world which was responsible for the genetic instructions picked up at conception, then that creature has got a problem. The resolution of the genetic system is equivalent at the very least to one period of generational turnover for that particular kind of animal or plant. Any change which is occurring on a faster time base will simply not be "seen" by the genetic system.

Hull: There is an interplay between how long your phenome exists and how rapidly you can reproduce. For instance, sequoia trees last five thousand years but can reproduce very, very rapidly. In other species, like salmon, the reproductive period is the life period. There is an interplay between the two. Now, as far as adaptation is concerned, Darwinian evolution is a function of how quickly you can reproduce. You cannot evolve faster than you reproduce. That the modifications in an organism do not get transmitted directly to the progeny means that if you have got a good phenotype, you are able to reproduce more frequently, but there is an upper limit on how fast you can change: reproductive time. That means, say, that if in the morning, the sun comes out and it gets hotter during the day, and at night it gets cold again, you have got to adapt to that change phenotypically. There are lots of short-term phenomena—relative to your reproductive span—that leave your genes untouched. Now in Lamarckian inheritance, in the daytime your genes will be changed, at night they will be changed back again, the day after they will be changed again. That would be incredible: constantly changing your genotype. If you can know in advance which are going to be the long-term changes, which the medium-term changes, and which the short-term changes, OK; what you do is react to the long-term changes and ignore the short-term. But there is no way for organisms to do that. In Darwinian evolution you develop abilities to react to short-term changes. The price for that is that long-term changes require at least a generation.

Plotkin: And if the changes in the world are important for the survival of that creature, then unless it can evolve additional information-acquiring processing and storing capacity, that creature is going to be disadvantaged relative to other creatures that are not subject to these kinds of changes or have acquired the appropriate additional knowledge-gaining capacity. You can actually produce a formal set of statements about the sampling limitations which will give rise to these additional requirements and devices.

Callebaut: It has been objected by some people that it is not obvious that in this kind of hierarchy, the control function that you obviously need can actually be displayed (e.g., Corning 1983:121). What is the "logic of command"? In your 1979 paper with Odling-Smee, you pointed to parsimony as allowing one to proceed in a nonarbitrary way. What kind of an animal—I mean, mechanism—does the "choosing" of the appropriate level?

Plotkin: Right. That, in a nutshell, is the heart of the criticism that is made against us: We do not even begin, we are told, to answer the question of what it is that is making that decision. In fact we have formulated at least the beginnings of an answer both in the 1979 paper—which Corning seems not to have read—and more recently (Plotkin 1988b, c). The "decision" is made by the whole system operating a set of rules which constitute a tradeoff between the ability of a level of the hierarchy to detect change of a certain frequency on the one hand, and the costs of storage and retrieval on the other.

Callebaut: And the rules involve the assumption that information which is stored genetically is cheaper than information that is stored, say, in an animal's brain, in its central nervous system (cf. the Esposito quote in note 33)?

Plotkin: Timothy Johnston (1981) actually delineates quite clearly the cost of information stored in genetical terms as opposed to information which is stored within the CNS of a learner. Williams (1966) had already said—I'll paraphrase it—that if information can be stored at the genetical level it will be, because its costs are very much less. The argument is, roughly, that if an animal has to have a particular pattern of behavior in its repertoire, and that behavior can be hard-wired into its CNS as a result of particular kinds of genetic instructions, then the amount of genetical space that will have to be used up in making those instructions, in setting those instructions, will be less than if that behavioral pattern has to be learned. Remember: *all learning is innately given.* It therefore makes more sense, in terms of amount of actual information that has to be stored genetically, for it to be left at that level and not to include a component of learning whereby the CNS becomes involved as well.[5]

Callebaut: With respect to learning processes, there is the additional complication of linguistic competence.

Brandon: Individual learning sometimes does involve linguistic competence, but not necessarily so. In *Homo sapiens* it oftentimes does; in other animals it usually does not.

---

5. Other considerations are also important. Given that the amount of information is less, there is likely to be less of a problem in terms of rates of mutation or various kinds of disruptance which might occur at the genetical level, which would then have deleterious repercussions at the behavioral level. And there is also the matter of the amount of actual CNS space that is needed for pre-wired behavior as opposed to learned behavior. It should be noted that Tierney (1986), a Canadian scientist, argues quite the reverse. Says Plotkin: "She found it difficult to get her stuff published, which is I think very distressing and tells us a lot about the sociology of science. Anyway, she eventually succeeded. She really takes us by surprise. She argues that the most primitive form of information storage in the CNS is one of plasticity, and that it is only with subsequent evolution that the nervous system becomes concretized, settled into a much less flexible form of behavior. I think it is an important kind of counterargument to the sort of assumptions we were making about the multiple-level model of evolution. There are many other reasons why it may be wrong as well."

Callebaut: In his paper, "The naked meme," David Hull (1982) distinguishes between individual and social learning in terms of cost-effectiveness.

Brandon: Hornstein and I make a different distinction. We distinguish observational learning from instructional learning. *Observational learning* may often involve certain other learning rules, but it *will* involve some interaction between the organism and its environment. It involves the observation of—usually—conspecifics doing something, and learning from that. For instance, you might watch your neighbor eating some food source and observe its effects. If he does not get ill, then you eat that food. This does not necessarily involve any linguistic communication. We distinguish observational learning from what we call *instructional learning*, which involves some sort of linguistic competence—a "communication system," I should say, such as the human language. Instructional learning explicitly involves a communication between teacher and learner. If you want to find something like this somewhere, it might be in *Programmed to Learn* (Pulliam and Dunford 1980). We were much influenced by that book, and something like that distinction may be in their book; I can't remember right now. In any case, we try to look—and this is very much influenced by Pulliam's work—at the cost of various methods of learning, and argue that under certain conditions instructional learning is more cost-effective than observational learning or than simple trial-and-error learning. You know, you can imagine cases like learning about the dangers of rapidly spreading fires. You can observe your conspecific being burnt, but if you are close enough to observe that then you are in danger too! Much better to be told of that by some language system.

9.2.2. *Is It a Boy or a Girl?*

Callebaut: We will look in more detail at how phenotypic plasticity impinges on cultural evolution below (9.5). But first I would like to consider the same issue at another level: the final wiring of the human brain which occurs after birth and is governed by early experience (e.g., Aoki and Siekevitz 1988)—a sensitive issue, as one would expect. Helen, you have taken a good look at the value aspects of work in the biology of behavior.

Longino: One of the most interesting features was discovering that the biological-determinist[6] account of human sex differences was structured by a variety of dif-

---

6. Both Lewontin, who shuns gene-reductionism, and Dawkins, a prominent gene-reductionist, have embraced the view that causal determination and human freedom are reconcilable. Acting as a trouble-shooter, I like to juxtapose the following quotes. Lewontin (1983b:183): "A process is free from . . . some set of causes if it is extremely weakly correlated with any one cause or small subset of these causes, although its movement may be perfectly determined by the conjunction of all of them . . . As a result of the social organization produced by our material brains and hands, our individual lives are the consequences of a bewildering variety of intersecting causal pathways. In this way, our biology has freed us from our biology." Now compare Dawkins (1982a:11): "Human nervous systems are so complex that in practice we can forget about determinism and behave as if we had free will . . . If you are a full-blooded

ferent background assumptions laden with different kinds of cultural values.[7] Gender ideology, I thought, was implicated primarily in the description of behaviors, the classification of hormones, and so on. The inference to causal explanations, to causal hypotheses, I thought was mediated by what my colleague Ruth Doell and I came to call the "linear-hormonal model" (Doell and Longino 1988; cf. Longino and Doell 1983). According to this model, human gendered behaviors are an outcome of prenatal brain exposure to certain (levels of) hormones. The idea is that these hormones, which are influencing the development of male or female reproductive anatomy or physiology, are simultaneously sexualizing the brain, so that the brain is programmed to respond in a set of distinct ways to later stimuli, depending on whether it has been exposed to higher or lower levels of a given hormone.

Callebaut: So the linear-hormonal model has a particular picture of the relation of the brain to behavior embedded in it, which becomes clear when contrasted with the selectionist picture that Gerald Edelman has worked out of brain development (cf. Changeux 1985; Plotkin 1991).

Longino: In Edelman's (1987, 1988b, 1992) selectionist view functional connectivity in the brain is not an outcome of processes *imposed on* the brain, like hormone exposure. In his picture, as brain and neural tissue is first developing, it consists of masses of neurons and neuronal groups that are very densely interconnected. There is a kind of massive connectivity that is genetically determined in the brain. That connectivity is not itself functional. The functional connectivity—the connectivity that is going to underlie various neural functions, whether of the autonomous or the nonautonomous nervous system—is selected from this genetically determined massive interconnection. In Edelman's view, it's in a sense a haphazard matter which initial connectivity gets utilized for a particular function. But its utilization reinforces a particular synaptic pathway, and other pathways that do not get utilized are eliminated from the system. So *functional connectivity develops as a consequence of function itself.* The connectivity does not preexist the functioning; rather, the connectivity develops as the functioning itself develops.

---

determinist you will believe that all your actions are predetermined by physical causes in the past, and you may or may not also believe that you therefore cannot be held responsible for any sexual infidelities. But, be that as it may, what difference can it possibly make whether some of those physical causes are genetic? Why are genetic determinants thought to be any more ineluctable, or blame-absolving, than 'environmental' ones?" (See also Dawkins 1982b.)

7. Longino agrees that biological determinism is misread and misused in the popular context (cf. Segerstråle 1992). "That's why it is important to talk about it and why it is a political issue. It has nothing to do with the political motivations of the individual. This is another point on which I disagree with Hull's (1988a) treatment of the claim that social ideologies play a role as a claim that the social allegiances and identities of individuals have a causal bearing on the scientific views they promote. But *the political allegiances and motivations of the individuals are irrelevant to our understanding of the political importance or implications of biological determinism.* It has to do with the logical-conceptual connections between the biological-determinist approach and political philosophies."

Callebaut: What are the major implications of such a view?

Longino: This has some really interesting consequences when we think about some of the issues that the linear-hormonal model is designed to explain. The initial selective process is going on for at least two years after the birth of a human infant. Edelman distinguishes "primary" and "secondary" repertoires. The secondary repertoire, which is the functional repertoire, is in constant reformation throughout an organism's lifetime. This is in quite direct contrast with the linear-hormonal model, which holds that the brain is programmed, prior to any experience, by hormones to respond in certain ways to stimuli in its environment. So we can see, for example, how, in the selectionist view, sex differences in behavior could be understood as the product of very early interactions which the human organism has. I mean, the first question that's asked when a human is born is "Is it a boy or a girl?" And the answer to that question determines the whole history of how that infant is going to be treated from that moment through its adult life. There are all kinds of interesting studies about differences in the ways people will hold male and female infants (Seavey, Katz, and Rosenberg Zalk 1975). So *everything* about that infant is going to be dependent on how its genitalia have been identified at its first emergence. In the selectionist view, we can understand, then, that whatever connectivity underlies sex differences in behavior could be forming in these very early social experiences. It could be different if those experiences were different.

9.2.3. *From Politics to Reaction Norms*

Callebaut: Any discussion of the biology of behavior and learning in animals (Dewsbury 1985) and man must come to grips with the sociobiology controversy that mobilized and polarized scientists, philosophers, and humanists for a decade or so after Harvard entomologist E. O. Wilson (1975:4) proposed to "codify sociobiology"—the systematic study of the biological basis of all social behavior—into "a branch of evolutionary biology and particularly of modern population biology." For the present, sociobiology focuses on animal societies; but it "is also concerned with the social behavior of early man and the adaptive features of organization in the more primitive contemporary societies." Extant sociology, Wilson suggested (he did not use the phrase), was *human-chauvinist*.[8] "It still stands apart from sociobiology because of its largely structuralist and nongenetic approach. It attempts to explain human behavior primarily by empirical description of the outermost phenotypes and by unaided intuition, without reference to evolutionary explanations in the true genetic sense" (E. O. Wilson 1975:4). Thus a new biological imperialism was born (cf. Rosenberg 1980). The collision with established social science seemed programmed. Each side in the debate has accused the other(s) of

---

8. As far as I am aware, this expression was coined by the environmental ethicists Routley and Routley (Mannison 1980).

using arguments contaminated by ideology—I think rightly so.[9] Maybe the controversy generated more heat than light. Some of the issues, especially the political ones, have been talked to death and will not be rehearsed here.[10] But a number of pressing questions concerning research strategy and methodology remain with us (cf. Fetzer 1985). For instance, it has become clear that sociobiologists tended to neglect development.

Plotkin: John Odling-Smee and I got to thinking about the famous chapter 5, where E. O. Wilson (1975) quite explicitly downgrades development amongst a whole lot of other, fairly basic aspects of living systems. Yet the selection pressures on living creatures can be so fearsome as to drive some of them to evolve additional information-gaining capacity. The most obvious form of this is in terms of a plastic developmental system, which we see in plants and which produces quite astonishing results (cf. 6.3.4). Nothing is more amazing than the kind of form that a plant will adapt if you put it under certain circumstances. And it does this without a nervous system at all. Its flexible response to environmental demand is very powerful indeed!

Callebaut: Philip, you are convinced that both sides in the sociobiology debate mischaracterized one another's political purposes (Kitcher 1985a, 1987b).

Kitcher: The Sociobiology Study Group attributed to Wilson just an incredibly simplistic position, and Wilson attributed to his opponents an incredibly simplistic position. The issue, I think, is an issue about the shapes of *norms of reaction*—it is as simple as that!

Callebaut: What do you mean?

Kitcher: A norm of reaction can be thought of, for the present purposes, as a graph which shows for a given genotype the way in which a phenotypic property varies with the environment. So for example, suppose that we had a bunch of plants (tea plants or corn plants or whatever) that all had the same genotypes. Then we could grow them up in different environments and look at the way in which some particular phenotypic properties such as the height vary with the environment. Of course, what we would expect is that in an environment with very poor soil and little water, the plants would be stunted; and in the kinds of environment that we try to provide in our gardens, the plants would grow beautifully. In a similar fashion, one can take any characteristic of any organism in which one is interested and look at the way in which that characteristic varies for given genotype with the environmental variables. The way the Sociobiology Study Group interpreted Wil-

9. Needless to say, I have my own *parti pris* on this matter. I was a member of the Dialectics of Biology and Society Group which, in the heat of the debate, edited books with the telling titles *Against Biological Determinism* (Rose 1982a) and *Toward a Liberatory Biology* (Rose 1982b). And I do not recant what I wrote at that occasion (Callebaut 1982), though I hope to have learned a couple of new things about reductionism since.

10. The following should be a representative sample of the literature: Burian (1978, 1981–82); Caplan (1978b, 1984); Gregory, Silvers, and Sutch (1979); Jacquard (1978, 1989, 1991); Kitcher (1985a, 1987b, c); Lewontin (1976, 1977, 1983b); Lumsden and Wilson (1981, 1982, 1983); Montagu (1980); Ruse (1979b, 1986c); Sanmartín (1987); Segerstråle (1985); Stent (1980); Thuillier (1981); Wilson (1978).

son was in effect saying: "Norms of reaction are absolutely *flat!*" And Wilson's
reaction was: "Norms of reaction go all over the place; you can get just about any
point on the phenotypic trait axis by adjusting the environment." What the Sociobiology Study Group *ought* to have said was the sort of thing that Lewontin said
so clearly and elegantly and effectively in the IQ controversy: "You can't get information about the shapes of norms of reactions from the kind of data you're
using!" (cf. Lewontin 1982a; Rose, Kamin, and Lewontin 1984). That, in a sense,
is at the heart of many of my objections to pieces of Wilson's program.

Callebaut: Your own view, I take it, is that it's not that simple at all?

Kitcher: Very subtle issues can come up in this area. There is a comment that I use
in my book, which comes from a paper called "Monogamy in mammals," in which
Devra Kleiman (1977) says that some behaviors that feminists want to achieve are
biologically inconsistent. Don't worry about whether or not feminists want to
achieve these behaviors, or whether climate is right, or anything like that. What
does this mean? It is the sort of provocative conclusion that made the Sociobiology
Study Group see red, of course!

Callebaut: What do *you* think it means?

Kitcher: Something like this: You've got norms of reaction for human genotypes,
loosely lumped together as human male and female genotypes. They are shaped
in such a way that as you try to create an environment in which you get one human
behavior in which you are interested, you automatically move to a region of the
environmental variables in which you go away from the state that you are trying
to achieve with respect to the other behavior in which you are interested.

Callebaut: Sociobiologists have also been chastised for blurring the distinction between "social" and "cultural."

Plotkin: A *society* is an aggregation of conspecifics. A *culture* is defined at least in
part by the extragenetic transmission of information between individuals.[11] If you
don't make the distinction you are in danger of making the error of thinking there
is little difference between, say, a flock of birds and a scientific community. A
society, as a mere grouping of animals, only becomes a (proto-)culture when information acquired by learning is transmitted to others in the group by an act of
learning. This probably happens just in a handful of species.

9.2.4. *Needed: More Sophisticated Evolutionary Theory*

Callebaut: Another lesson to be drawn from the sociobiology episode is the inappro-
priateness of the gene-selectionist version of evolutionary theory that sociobiologists used—and continue to use—as a general tool for studying behavior.

Hull: Gene selectionism is really not good enough to apply to unproblematic traits
like eye color. The sociobiologists extrapolated it to very difficult properties like

---

11. The biologist's definitions of "society" and "culture" thus depart from those of most traditional
sociologists, which consider sociality an exclusively human affair.

miscegenation, xenophobia, and what have you (Hull, "Sociobiology: Scientific bandwagon or traveling medicine show?" [1979], reprinted in Hull 1989:263–84). Although many people reacted negatively to the whole enterprise, I think that it is worth doing. It is just that you are going to need a very sophisticated version of evolutionary theory if you want to have any legitimate sorts of inference. One of the things that I have been trying to do is help improve evolutionary theory, so that when we do make these extrapolations, they have some foundation.

Burian: There is both wonderful and horrible work under the label "sociobiology." One would be ill-advised simply to take the label and run away with a diatribe about what goes under the label.

Callebaut: Let's talk about the serious stuff. Philip.

Kitcher: You've got an evolutionary theory supplemented through the theoretical insights of Hamilton and Maynard Smith primarily, Trivers, George Williams, and a bunch of other people. Certainly as far as the main theoretical machinery goes, it seems to me that the stuff ought to be uncontroversial. The issue is: *how to put this to work in concrete evolutionary situations?* It is at this stage that you start seeing lots of splits in various types of sociobiology, because there are people who practice the enterprise rather carefully, focusing on very small groups. The two examples I use in the book are Parker's work on dung flies and the work on helping at a nest in Florida scrub begun by Woolfenden and continued by a bunch of other people. There are lots of others—people working on social insects, for example—who are doing that same kind of rigorous work.

Callebaut: You distinguish yet another kind of quite rigorous sociobiological theorizing, which you don't give so much prominence in your book.

Kitcher: That is the kind of thing that goes on in, say, Charnov's (1983) monograph on sex ratios, where you take a general phenomenon and you look very carefully and in great detail at a large number of diverse animal groups in which it occurs— actually in his case to plant groups as well. Again, you apply the techniques of your evolutionary theory, and you hope in this way to build up generalizations. I was concerned to contrast this with what I called the "hasty non-human sociobiological studies," in which one looks from the start for very grand and simple models that cover defensive behavior, or all organisms that have social groups, or sexual strategies across all organisms. If you compare the work of people who are doing piecemeal looking at sexual strategies to the kind of thing that goes on in, say, E. O. Wilson's big *Sociobiology* book (1975) and even more in his *On Human Nature* (1978), or to the kinds of things that Barash (1977, 1979) has written, there is just a world of difference.

Callebaut: Dick wants to amplify that.

Burian: By and large the attempts I have seen to reach humans looked shoddy. What I have not seen is the kind of natural history, conducted on the necessary scale and with the appropriate sophistication regarding the interaction of genetic and cultural evolution, to deal with the transition from a prelinguistic primate community to the sort of bi-evolutionary structure typical of human societies. The conceptual footing for such an enterprise is still very insecure. In addition, there has been an unfortunate and avoidable contamination of our language in sociobiological discourse, as Lewontin, Gould, I (Burian 1978, 1981–82), and many others have pointed out. It's not that one should not study these things, but that one should clean up one's act in doing so.

Callebaut: In your view, Philip, there is a fundamental confusion in what someone like Alexander (1979, 1987) has been trying to do.

Kitcher: One cannot begin to understand the evolutionary explanation until one has formulated the evolutionary *explanandum*. The explanandum isn't why people in Northern India murder their female offspring or why men give money to their sisters and to the children of their sisters. The explanandum is the basic *dispositions* that individuals have, which, in combination with the current environmental factors, lead them to act in these ways (cf. 9.3). There is a sense in which what is being explained evolutionarily in the work of Alexander and the work that he has inspired among anthropologists seems to be quite the wrong thing!

Callebaut: Another methodological caveat, several of my interlocutors insist, concerns the unconscious use of *reductionistic problem-solving strategies.* [12]

Lewontin: How do we decide how to cut up the description of "human nature" (Wilson 1978) into individual bits and pieces that are separately evolving? How do we decide to take human social behavior and divide it into individual units (of "aggressiveness," "dominance," "territoriality," "xenophobia," etc.)? This is something which in evolutionary biology can't be done even for anatomy. We don't know whether the hand is the unit of evolution, or the finger, or the fingernail, or the whole arm, or the whole body. The problem of dividing up an organism into separate atomic bits and pieces each of which has its own evolutionary history is a problem that sociobiologists have never thought about when it comes to behavior. Yet for behavior it must be at least as difficult, if not more difficult, to make a reasonable division than for anatomy. There is no evidence whatsoever that any of the specific behaviors talked about by sociobiologists are "coded in the genes." We don't know that there are genes for aggressiveness, or xenophobia, or entrepreneurship; yet sociobiologists postulate those genes constantly. They *have* to postulate those genes, because if they don't say there are genes for those characters, then they can't talk about evolution of those characters (since evolution requires some kind of inheritance). Those traits which are known to have the greatest similarity between parents and offspring in social traits are clearly not inherited; for example, the language spoken. North America is a laboratory which has proved that the ability to speak Dutch, French, Russian, Japanese, and so on is not coded in the genes.

12. Cf. the general discussion of reductionism in 4.2 as well as the exchanges on the units of selection controversy in 6.5.

Richard Levins (b. New York, 1930) is John Rock Professor of Population Sciences at the Harvard School of Public Health. He is best known for his work in ecology (several of his papers and his book, *Evolution in Changing Environments* [1968] have become classics, but his concerns extend to such issues as public health and the political economy of world agriculture as well as

other complex systems (Puccia and Levins 1985). Levins is an original who sparks many ideas, some of which transpire in publications by others.[1] He has also invented all sorts of things which he has later discovered to have been invented elsewhere. "For example, Levins has a number of beliefs about hierarchies and levels of organization that can be mapped one to one onto Simon's, but he didn't get them from Simon!" (William Wimsatt).

Richard C. Lewontin (b. New York, 1929) is Alexander Agassiz Professor of Zoology at the Museum of Comparative Zoology, professor of biology and professor of population science at Harvard University. The main focus of his scientific research has been on genetic variation in natural populations of insects and human beings, which he is studying both experimentally and theoreti-

cally. He has also published extensively on philosophical and historical issues in biology. In addition to more than one hundred papers in scholarly journals, he has written *The Genetic Basis of Evolutionary Change* (1974a), *Human Diversity* (1982a), and, with Steven Rose and Leon J. Kamin, *Not in Our Genes* (1984). Wimsatt thinks that "quite apart from his support of a long list of philosophers who wanted to learn some evolutionary theory and popu-

lation genetics, Dick Lewontin's best philosophical work has been some of his more reflective methodological work on evolutionary theory—though he has done elegant work on almost everything he has touched."[2] *The Genetic Basis of Evolutionary Change,* Wimsatt emphasizes, is not just a classic: "it falsifies Popper[1] through about Popper[13] on the elaboration—not falsification—of the selectionist and neutralist theories."

As Wimsatt sees it, "Levins and Lewontin are both natural philosophers, in that they naturally do conceptually and methodologically interesting things, or perhaps Natural Philosophers, in the classical historical sense in that they do science in a way that is not unnaturally separated from doing philosophy." L&L's scientific collaboration took shape in the early 1960s when both were members of a small group of evolutionary biologists around Robert MacArthur. It was tightened when they became congenial to a mysterious "Isidore/Isadore Nabi" of Bourbaki fame but more controversial (Levins and Lewontin 1985:127–31; Hull 1988a:223–30), and got to work together institutionally, first at the University of Chicago, then, from 1972 on, at Harvard. (Lewontin [1979] has described their attempts to organize a workplace within Harvard University that would be "democratic and collective.") Considering their longstanding record of scientific and political cooperation—both profess to be Marxists—it seemed only appropriate that I interview L&L jointly and that their intellectual biographies, which are to some extent interwoven with the history of the Science for the People group, be described in the same note.

Science for the People was formed in the early 1960s to oppose the involvement of U.S. science in the imperialist Vietnam War. Later the group or subgroups campaigned against the view that there are inherited differences in intellectual ability between individuals and races, and that this was responsible for their different social positions. The Science for the People group argued that the equation of 'inherited' with 'fixed' and 'unchangeable' was a biological error, and that whatever the causes of differences might be between individuals *within* groups had nothing to do with differences *between* groups; that in fact there was no evidence of any biological difference in intrinsic ability for intellectual tasks, either between blacks and whites, or between men and women, or between people in the working class and people in the middle classes. L&L consider the involvement of the Science for the People movement in the IQ controversy a success in the United States and Great Britain. "Unfortunately, the theory is still very much alive, for example, in France with

1. In a world in which "fish gotta swim, birds gotta fly, people gotta talk, and academics gotta write" (Lewontin in Pinker and Bloom 1990:741), it is good to remember that intellectual stature is captured but very imperfectly by the length of one's publication list.

2. At some point in the editorial process of this book, Lewontin (personal communication, February 21, 1990) wrote me that "I note with some amusement your continued valiant efforts to introduce the

claim that I have influenced various philosophers here and there, and their singular resistance to your attempts to suggest this to them. I don't really know what the origin of people's ideas is, and I am not sure that I care. It is a fact that many of the things that Bill Wimsatt and Elliott Sober talk about were developed in our period of work together. So, in proper dialectic style, I would claim that one should not talk about who influenced whom, but rather [about] the consequences of a unique contextual situation."

the New Right, where it is being combatted by Albert Jacquard (1978, 1989, 1991) and others (e.g., Thuillier 1981, Frankel et al. 1986). In this case it may take a little longer on the Continent" (cf. Sanmartín 1987) (Lewontin).

A second campaign of Science for the People dealt with the claims, especially made in Britain, that human criminality was somehow "in the chromosomes"—that males with an extra Y chromosome were biologically predisposed to be criminals. "It is interesting," Lewontin remarks ironically, "that nobody ever went to look at the chromosomes of presidents of universities, or major corporations, or members of the Senate, or cabinet officers, all of whom are said to have the same qualities which, when exaggerated, were said to lead to criminal behavior (aggressiveness, dominance, self-confidence, and so on), in view of the fact that it is not always easy to tell the difference between the normal entrepreneurial behavior of a corporate president and criminal behavior." (At the time Lewontin said this, of the hundred largest corporations supplying goods and services to the military in the U.S., forty were under criminal investigation for charges against government accounts.) In this case, Science for the People was instrumental in preventing plans for screening newborn children for the extra chromosome.

In the mid-seventies, the Sociobiology Study Group of Science for the People engaged in dealing with the problem of human nature and human sociobiology[3] as one recent manifestation of "biological determinism" (Lewontin 1976, 1977, 1981, 1983b). They argued that the description of "human nature" as explained by sociobiologists was really a description of what sociobiologists saw around themselves in society rather than being drawn from any general principles. When sociobiologists say that people are afraid of strangers, are territorial, or selfish, that "entrepreneurship is a natural part of human nature," that "the desire to increase one's capital comes naturally in the genes," and so on, they do in fact describe modern middle-class capitalist society, "with little or no looking back in history, and with an attempt to make various primitive societies appear to be really identical with the society in which we live" (Lewontin). This is a kind of reification: "People talk about 'entrepreneurship' as something that can evolve; but entrepreneurship as we now understand it—indeed the notion of property as a whole—is only, say, seven hundred years old in European society. Seven hundred years ago, the notion of 'owning the land' on which you were working, and being able to sell it to some neighbor when you felt like making some money, simply didn't exist." Likewise, the notion of *individuality,* which we all consider to be primary, is a relatively recent notion. A fundamental philosophical error of sociobiology, L&L maintain, is that the properties betrayed by nations and by collectivities in general are nothing but the summing up of individual properties of individual members of the group. Anyone who has ever gone to war knows perfectly well that you don't go into the army because you want to shoot a national of the other country; you go into the army because you are forced to, because you are drafted, because if you don't do that you yourself will be put in prison or shot. Once you're in the army, there may be a great propaganda effort in order to make you, as an individual, hate individual people of the other country. But this follows and does not precede aggression" (Lewontin).

Looking back on the sociobiology debate, Lewontin summarizes: "Our work there has been largely to develop a complete critique of biological human nature ideologies, and to attempt to reveal to the public at large how they are being fooled by this kind of pseudoscience." That struggle, although it has calmed down from the earlier days in the U.S., is very much alive in Europe today—especially with the various New Right movements. "The recent upheavals in Eastern Europe—the violent manifestations of nationalism and the reactions against state socialism—are models said to show that xenophobia and entrepreneurial activity are really part of biological human nature and cannot be long suppressed."

Needless to say, Lewontin vigorously rejects Alex Rosenberg's claim (see biographical note at 3.3) that he and his Harvard colleague Stephen Jay Gould commit the genetic fallacy: "I have never nor do I believe Gould has ever made the silly argument that one can judge the truth of a claim about nature by looking into the ideology or idiosyncratic properties of those who make the claim. Our method is precisely the reverse. We begin by finding that many claims about nature are either counterfactual, or without any observational foundation, or illogical on their own terms, or based on methodologies that are insufficient by the canons of evidence already established, or in some cases actually fraudulent. Second, we note that these claims are made by trained scientists who are supposed to know better. Only *then* do we ask how it comes about that trained, successful, and often prestigious scientists come to make statements about nature that are unacceptable on ordinary epistemological criteria. We then offer ideology, social class, personal history, etc., as an explanation of what would seem to be irrational behavior on the part of rational people." Lewontin is tempted to carry out this analysis on Rosenberg's statements about the genetic fallacy: "He cannot give a single example of my having committed this fallacy, because it is crystal clear from my writings that I do the reverse. But Rosenberg is a trained philosopher. How can we explain his counterfactual statements in this matter? Perhaps he is suffering from professional territoriality, resisting the intrusion of a nonprofessional onto his turf—note his remark that my best work is owed to my association with Sober—or perhaps he is a right-wing ideologue whose politics are offended. I do not know, or care to know, the truth of these possibilities, but if it were important enough, I might work on them and other possibilities. I only offer them as illustrative of the logic."

The original joint interview with Levins and Lewontin was recorded at the Modern Language Center studio at Harvard University on May 22, 1985.

3. In the United States, sociobiology is identified rather more narrowly than in Europe with the views of E. O. Wilson, Barash, Alexander, and some others. Many people in the U.S. who think that knowledge of biology has definite implications for analyzing human social behavior would call themselves sociobiologists if the term hadn't that kind of connotation; they don't accept the implications that the Wilson school draw out of it.

Wimsatt: Here again, reductionistic research strategies are very often very effective tools. But when applied to the wrong kinds of problems, the problems with the wrong kind of structure, they will lead you down the garden path in the wrong direction (cf. Howe and Lyne 1992; Lewontin 1992). I will mention one example: *Selfish* and *altruistic* interactions are only two of four possible interactions that you can get. In addition there are also *spiteful* interactions and *cooperative* inter-actions. My own bet is that social organization in humans and in any other social beings involves primarily initially cooperative interactions, that is, interactions in which you benefit yourself while benefiting others (cf. Caporael et al. 1989). After you have a cohesive social organization tied together by cooperative interactions, altruist and selfish genes can then invade and live off of the diffused general co-operative benefits. One of the delightful paradoxes of this picture, however, is that many genes which try to be selfish will, in doing so, lose some of the cooperative benefits and end up being spiteful, and thus the population may be protected from many selfish invasions.

Callebaut: How do you explain the fascination of many people—Richard Dawkins (1976, 1978, 1982a, 1986) being the most obvious example—with selfishness and altruism?[13]

Wimsatt: The fascination stems in part from their loaded character to us, and to the cult of the individual in Western society—*selfish genes interpreted as ethical ego-ists*. One of the reasons the altruism/selfishness interactions are interesting to evo-lutionary biologists is the issue of their testability. Truly altruistic behavior—in the biological sense—of an individual cannot be selected for at that level of orga-nization, and so it is a useful detector of when selection at another level (lower: kin selection;[14] higher: group selection[15]) must be invoked. One unfortunate con-sequence of this *testability bias,* however, is that the study of cooperative interac-tions in the game-theoretic sense, in which free riding is impossible, has been somewhat neglected. People who have studied cooperation usually try to sneak it in through the back door, like reciprocal altruism. This is not taking cooperative behavior seriously, and reciprocal altruism (and other such ploys) are subject to the free-rider problem, which true cooperative behavior is not.

---

13. See, e.g.—the choice is quite arbitrary—Hamilton (1963, 1964); Trivers (1971); Campbell (1972); Phelps (1975); Doolittle and Sapienza (1980); Stent (1980); Axelrod and Hamilton (1981); Derlega and Grzelak (1982); Margolis (1982); Axelrod (1984); Kitcher (1985a); D. S. Wilson (1987); Sober (1988b, 1989a, 1992); Caporael et al. (1989); Elster (1989a:ch. 5).

14. The advocates of sociobiology I have talked to (both in-and outside the context of preparing this book) all referred to kin selection theory as a potentially fruitful resource for explaining social phenomena. E. O. Wilson (1975:578) calls altruism (the biological definition of which is "self-destructive behavior performed for the benefit of others") the "central theoretical problem" of sociobiology: "How can altruism, which by definition reduces personal fitness, possibly evolve by natural selection?" His answer is kinship: "If the genes causing the altruism are shared by two organisms because of common descent, and if the altruistic act by one organism increases the joint contribution of these genes to the next generation, the propensity to altruism will spread through the gene pool. This occurs even though the altruist makes less of a solitary contribution to the gene pool as the price of its altruistic act" (E. O. Wilson 1975:3–4).

15. Cf. 6.5 and Vrba (1984), Schull (1990), and the work of M. Wade and D. S. Wilson in general.

Callebaut: There is an attempt, I think, but it remains quite speculative in spite of being extremely well documented: Peter Corning's *The Synergism Hypothesis* (1983), which stresses the nonlinear and "punctuational" character (threshold effects) of many combinatorial effects in nature and society. But many biologists won't like it because of its progressivism.

Wimsatt: In any case, evolution of sociality and cooperative interactions deserves at least as much effort as do selfish or altruistic interactions. I think the sociobiology program can be attacked internally in terms of the uses and biases of reductionistic problem-solving heuristics without ever raising flags of politics or ideology. And then, after you've done this, of course you can say, "Well, why should all these people have been so ready to accept these dumb conclusions of the model?" And that's where the cultural context, including the politics and the ideology, comes in.

9.2.5. *Why Children Hate Spinach*

Callebaut: Dick, you and your colleague Steve Gould (1978) have been particularly severe with the "adaptive stories" sociobiologists invoke to argue that presumed genetically determined features are themselves the product of natural selection. Now what is wrong with that? I mean, aren't those sort of stories the common coin of all evolutionary biologists?

Lewontin: The problem is that anyone can sit back in his or her chair and tell such stories. *The problem is not to tell the story, the problem is to give it some testable substance.* The theory of sociobiology depends critically on the ability to tell a kind of "just-so" story (cf. Rudyard Kipling) about the origin of particular human traits.[16] Almost all sociobiological stories are of the same nature: a generalization is made of how people behave; that generalization is then said to describe something that is in the genes; and then a suitable story is concocted.

Kitcher: The idea is that once you've got the evolutionary analysis, you've got the claim that under certain conditions organisms would maximize, usually, their *reproductive success,* which is supposed to be a measure of inclusive fitness, by behaving in a certain way. So the first stage you have got is supposed to tell you

---

16. Lewontin delighted in giving me what may seem an absurd example, but one that actually has been published in a textbook for high school students (DeVore, Goethals, and Trivers 1973): "Why do children hate spinach, but why do adults like spinach? The sociobiologist can explain this. 'Everybody knows that it is a true characteristic of people—you only have to look around and you see that all children hate spinach and all adults like spinach. Secondly, suppose there were a gene such that, if children hated spinach but adults liked it, the gene would result in that characteristic.' Why would that be a good thing in evolution? Why would people who carried that gene leave more offspring? 'Well,' the sociobiologist says, 'because spinach has in it a substance that prevents the absorption of calcium. If children eat spinach, they will not succeed in absorbing calcium, their bones will be soft, they will get rickets or bow legs, and it will not be good for them. On the other hand, once they reach adulthood and their bones are not growing anymore, they don't need the calcium so much, and so they are at liberty to eat the spinach and get the various other things that spinach has, and that will be good for them.' So the argument is that any gene which makes children hate spinach and adults like spinach will be selected by natural selection." The reader is invited to detect for herself or himself "the absurdity of this argument at all levels."

how the animals *ought* to behave. Then you claim that they *do* behave in this way. Then you infer from this that there has been a history of selection. Then you infer that there must have been some kind of genetic basis that has become fixed in the current population. And then you infer that because there is a genetic basis, it would be difficult to modify behavior.

Lewontin: In the end, all such adaptive stories have one thing in common. They assume that there is *some fixed kind of environment* in which human beings find themselves, and that the problem for human beings is to solve problems set by nature—fixed problems—by developing genetic traits that will enable them to leave more children.

Callebaut: David, your judgment is milder.

Hull: "Just-so" stories can be true! The trouble is that there are so many different entities that you can tell "just-so" stories—scenarios about particular adaptations—forever. Testing such scenarios is difficult because they concern events in the distant past and the evidence is hard to get, etc. But that is not the big problem. The big problem is that there are *millions* of species. How many "just-so" stories are you going to tell? What you really are after are *underlying mechanisms* which regulate these processes. But I have got nothing in principle against historical narratives, which "just-so" stories are examples of (Hull, "Central subjects and historical narratives" [1975], reprinted in Hull 1989:181–204).[17] Someone like Steve Gould, who comes out hard against "just-so" stories, has spent a large part of his professional career producing "just-so" stories.[18] Now he thinks that his narratives are more carefully and critically developed than some others, and he is arguing against the kind of prejudices one is prone to when making these "just-so" stories. But there is nothing *in principle* wrong with "just-so" stories.

9.2.6. *Man the Hunter, Woman the Gatherer*

Callebaut: "Just-so" stories have been attacked by feminists as well. Donna Haraway (1985) has labeled primatology "politics by other means" (cf. Latour and Strum 1986; Strum and Latour 1987).

Longino: The example of human evolution is really quite interesting, because what we have here as data is just bits of fossilized bone, stones collected in urn beds, a

17. Generally speaking, all sorts of stories have been of paramount importance throughout human history. Herodotus's *Histories* are but one obvious, Western, example. Somehow stories fell in disgrace as history came to be thought of as "too sublime a subject to be presented in the same way as myths and fairytales" (Feyerabend 1987:115). The modern believers in grand deductive-nomological explanation schemes had no use for narratives either. The "distrustful and condescending attitude" of most philosophers towards anything nonliteral, especially the visual (Ruse 1989c:503; cf. Wimsatt at 8.4.1), should also be mentioned here. But stories are becoming fashionable again. A historian of science like Martin Rudwick (1985:xxii) avowedly aims to "interest *and entertain*" his scientific readers (italics mine): he argues that even the development of (scientific) ideas is better served by narrative than by a "conceptual account." Feyerabend (1987:115) refers to Rudwick when claiming that "narratives are the only form adapted to the complexities of human thought and action." Cf. also Lewontin (1967, 1990) and Latour (1988c).

18. Gould (1980, 1983, 1984, 1985, 1988, 1990, 1991); also Eldredge (1987).

few skeletons. One of the issues is how it is that the capacities that we take to be distinctively human, that characterize *Homo sapiens,* evolved. That story has to be told using this very fragmentary data. One of the standard accounts has been the "man-the-hunter" account of human evolution, which holds that human anatomy and distinctively human forms of intelligence, certain kinds of cooperative abilities and so on developed as a consequence of male hunting behavior. The evidence for this is the finding of certain kinds of objects that are identified as hunting implements that can be dated to certain time periods and are found in certain locations where various hominid species are also known to have been active. You look at these stones and ask, "What could they have been used for?" The answer is given in the context of taking into account what the male members of the species might have been doing, and the assumption is that the male members were hunters. So a wonderful story is told about these males going off, hunting in packs, using these stones, developing cooperative abilities. We can also then tell a story about human male aggression developing in this context. There is lots that can be attributed to this hunting behavior, including anatomical changes (dentition, bipedalism). All that is distinctively human gets attributed to this male hunting behavior.

Callebaut: How about the feminist story?

Longino: In the seventies a number of women entered physical anthropology, many of them trained by an eminent "man-the-hunter" theorist, Sherwood Washburn: Adrienne Zihlman (1978), Nancy Tanner (Tanner and Zihlman 1976), who took a quite different approach to the very same body of data. But they started out thinking about what the female members of these species were doing as the forests receded and these hominids found themselves in the savannah. What challenge faced these females in their tasks of bearing and nurturing young? They came up with a very different story about the stones, which in the other perspective are seen as hunting implements. For the "woman-the-gatherer" theorists, these same stones are taken as implements used to beat tough vegetable matter into edible material, or used for digging roots.

Callebaut: I think it is also important to mention, as you do in your book (Longino 1990:108), the tools that left no traces, like wooden sticks.

Longino: That's a good point about this "man-the-hunter, woman-the-gatherer" story, because the "man-the-hunter" theorists take the date at which these stone implements would have been used as a date for the discovery of tools. *Homo faber* dates from the time of the stone tools, whereas in the "woman-the-gatherer" story the stone tools are just part of a long history of tool use with organic tools possibly being in use long before the stone tools. But the organic tools, of course, leave no trace. The "woman-the-gatherer" theorists point to the fact that our contemporary chimpanzees use organic tools; they use digging sticks to go after termites and so on. So there is a little empirical support for this notion as well. So it gives a very different picture about the timing of the development of tool use. Tool use in the "woman-the-gatherer" picture is attributed to females, at least within this hominid line. Again, the kind of distinctively human forms of intelligence or cooperative behavior or tool use—tool invention or inventiveness—are seen as outcomes of female behaviors rather than male behaviors. So a complete story can be told that accounts for the data in either perspective. When you put them together, the "woman-the-gatherer" story is somewhat more compelling than the "man-the-hunter" story: it's more recent, it takes account of more evidence, and it has this nice property of scientific theories that it kind of suggests new things to look for. But it strikes me as really just as much of a "just-so" story as "man-the-hunter." Probably, when we have more gender equality, we'll find another story to tell that appeals to traits or behavioral patterns that are characteristic of both males and females.

Callebaut: What would be required precisely, you think, to go beyond the limitations of the current stories—supposing that substantially better data won't be forthcoming soon?

Longino: Well, I don't know; the way that we would decide, if we were just being physical anthropologists about this, is we would look to see which picture could somehow take care of the most data. But since theories of some of the grosser features of the "man-the-hunter" hypothesis have been abandoned, certainly physical anthropologists have been developing new accounts of evolution that are still androcentric, that still try to place these distinctively human traits as somehow a consequence of male behavior patterns.

Callebaut: One could also bring in here your point that sexual dimorphism is a form of sexism (Longino 1990). Both of the "just-so" stories you mentioned seem to me to be sharing a number of assumptions about, for instance, the way males and females would have spent their time in these societies.

Longino: Yes, both of these are strongly shaped by our gender-dimorphic ideology: males do one sort of thing, females do another sort of thing, they are radically different from one another.

Callebaut: They spend most of the day—maybe not of the night—apart . . .

Longino: . . . Maybe of the night too: if we look at other species, males and females don't—in some they do—spend a lot of time together. Which contemporary primary species are we going to use as our model? Well, we have absolutely no guidance as to which contemporary species is the best model. So we use our own contemporary experience as a guide to what might have been the case in the past. But there is no more reason to pick that than to pick any other. Going beyond the "just-so" stories would require developing a different, non-gender-dimorphic, view of human personhood.

9.3 THE MISSING LINK: EVOLUTIONARY PSYCHOLOGY?

Callebaut: Jon, how do you, as someone who views social science as an enterprise which is rather different from natural science, feel about sociobiology?

Elster: I think sociobiological explanations have some validity, but less validity than has been claimed for them by their most extreme defenders. I think they are valid or useful in two respects. First, there are certain forms of behavior which are closely linked to reproduction, which—also in humans—can be explained in terms of biological fitness. Secondly, biological evolution has led to many kinds of *general capacities*—capacity for language, for problem solving, for cooperation and so on—that certainly are important in specific social situations. What I object to is the attempt of some sociobiologists to explain social behavior directly by fitness arguments. I think in most cases you need a two-step argument. First, you need to argue that general human capacities have in fact been used by natural selection; and then a second argument about how these general capacities are brought to bear on specific social situations.

Callebaut: Could you give an example?

Elster: When somebody argues that the behavior of presidential candidates in American elections, such as handshaking, etc., can be explained directly by sociobiological arguments, then I am very skeptical. But if someone says that the kind of problem-solving behavior that humans deploy have in fact emerged by natural selection, then I would certainly tend to agree. Let me give an example which I think has some important consequences. *Natural selection in itself is myopic or opportunistic.* It tends to accept only small, stepwise improvements; it resists mechanisms such as "One step backwards, two steps forwards" essentially because the step backwards—a temporary reduction in fitness for the organism—would eliminate its descendants from the population, and therefore eliminate the possibility of taking two steps forward at a later time. So the kind of problem-solving behavior that is produced by natural selection—the adaptation of organisms to their environment—will tend to occur through local, incremental improvements leading to a local, not a global maximum of adaptation. There may be solutions to the problems of organisms that are simply out of reach for natural selection, because to reach them there would have to be an intermediate stage with a temporary loss of fitness.

Callebaut: In *Ulysses and the Sirens* (1979:4–9), you talk about a "locally maximizing machine" with respect to natural selection, and you oppose it to human society as a "globally maximizing machine."

Elster: One of the most important capacities of human beings is their ability to use such indirect strategies, and indeed to achieve global and not simply local maxima. The most important example, perhaps, is the ability to *postpone gratification,* to save and invest instead of consuming immediately. It is a capacity that has been produced by natural selection, and when deployed, it leads to certain specific forms of behavior. But these specific forms of behavior cannot themselves be explained, I think, as directly produced by natural selection.

Callebaut: In "A Critique of Optimization Theory in Evolutionary Biology," a chapter of their book *Caste and Ecology in the Social Insects* which has become famous, John Oster and E. O. Wilson (1978) wrote things like the following (this is about Fisher's "fundamental theorem" of natural selection): "The solution of the equations maximized a quantity that Fisher sagaciously called 'Fitness.' Never mind the delicacy of such a result—constant environment, limited types of density or frequency dependence, no linkage or epistasis, and so forth—teleology had been given mathematical respectability!" I've always found this chapter, which seems so self-critical, quite puzzling in light of many of the other things Wilson has written and done. (On optimality see also the papers by Kitcher and Lewontin in Dupré 1987.)

Kitcher: People have told me it was written by Oster. I can't believe that actually Wilson would just have let that go in without having *some* contribution to it. People who know Oster have told me that he wrote that chapter, and that the reason he gave up and went on doing development after that was because he became convinced that optimality modeling is too difficult! I don't know whether this is true. But what is amazing is that in the same year Wilson, jointly with Oster, produced this book—a very sensitive discussion of the limitations and difficulties of using optimality models—and another book, in which he just went most casually from optimality arguments to claims about selection and thence to claims about genetic determination.

9.3.1. *Gene-Culture Coevolution?*

Callebaut: Advocates of sociobiology sometimes accept many of the sorts of criticisms we have been reviewing, but then go on to say that the *coevolution* model of genes and culture of Lumsden and Wilson (1981) is what is really at stake—implying that the more recent models somehow overrule most if not all of this criticism.[19]

Wimsatt: That's nonsense!

---

19. Janzen (1980) defines coevolution in general as an evolutionary change "in a trait of the individuals in one population in response to a trait of the individuals of a second population, followed by an evolutionary response by the second population to the change in the first." Like others before them (including Campbell 1972, 1975, 1979a; Durham 1976, 1979; Feldman and Cavalli-Sforza 1976, 1979; Boyd and Richerson 1976; Pulliam and Dunford 1980), Lumsden and Wilson (1981, cf. 1982, 1983) have investigated the reciprocal effects of genetic and cultural evolution. Their theory focuses on (1) *gene-culture translation,* i.e., the effect of "epigenetic rules" (regularities that channel development) of individual development on social patterns, and (2) *gene-culture transmission,* i.e., the transmission of "basic units of culture" ("culturgens"), in which the choices are not all equiprobable. The epigenetic rules are regarded as "ultimately genetic in basis, in the sense that their particular nature depends on the DNA developmental blueprint" (Lumsden and Wilson 1981:370). "Culturgens" are defined as "relatively homogeneous sets of artifacts, behaviors, or mentifacts (mental constructs having little or no direct correspondence to reality) that either [share] without exception one or more attribute states selected for their functional importance or at least [share] a consistently recurrent range of such attribute states within a given polythetic set" (p. 368).

Callebaut: There aren't so many people who have tried to rebut the Lumsden and Wilson models in print (see the open peer commentary in Lumsden and Wilson 1982).

Wimsatt: Well, we have Lewontin's beautiful review, "Sleight of hand" (1981). The Lumsden and Wilson book, as Lewontin points out correctly, involves mathematical overkill for conclusions based on a model whose basic presuppositions have not been examined closely enough. There are far more productive goings towards increasingly more productive accounts. I would mention Pulliam and Dunford (1980) and Cavalli-Sforza and Feldman (1981). And far more better and advanced than any of those is the work of Robert Boyd and Peter Richerson, *Culture and the Evolutionary Process* (1985). That is an exceedingly important book, an order of magnitude more sophisticated than any of the ones before. (*Resumes after a brief pause.*) I suppose that if you are a genetic reductionist, you could read their book somewhat reductionistically; but that's not the way to read it.

Plotkin: Boyd and Richerson (1985) is, in my view, the most important work published in this area since we first spoke. But their entire analysis assumes, takes merely as given, what the multilevel model tries to understand, viz., why and how individual intelligence evolved at all.

Kitcher: I think Boyd and Richerson have made a nice step forward. But I am beginning to think that is also in a sense profoundly wrong, because it is another way of avoiding proximate analysis.[20]

Callebaut: What do you mean?

Kitcher: What you do à la Boyd and Richerson is you *simulate* certain features of the proximate mechanisms and the developmental situations by seeing these as isolated "taggings on" from previous generations' culture which are melded to the phenotype in certain ways. But that is not the right way to look at things. The right way to respond to the problems is: you've got to bring in the evolutionary considerations *where they belong,* at some very fundamental abstract level of human propensities, dispositions, capacities, or whatever, which we probably don't yet know how to describe. So I am beginning to think that any modeling of the evolution of human social behavior has to be a *real synthesis,* not just a "new synthesis." And that has just got to involve an enormous amount of psychology and developmental studies, which may be both developmental biology and developmental psychology.

Callebaut: Why are you looking severe now, Lisa?

Lloyd: Boyd and Richerson really made an advancement in the field. I think a lot of their mathematics, their models actually are in many cases identical to the Cavalli-Sforza and Feldman models . . .

---

20. Mayr's (1961) distinction between "proximate" and "ultimate" causes and effects in biology has become commonplace. Cf. Tinbergen's (1968) catalog of questions ethologists can meaningfully ask.

Callebaut: . . . That is one reason why I am asking you this! . . .

Lloyd: . . . But they do a couple of things which are an advancement on Cavalli-Sforza and Feldman. I would be happy if everybody would read Cavalli-Sforza and Feldman more carefully instead of looking at the math and then have their skin crawl. That was my first reaction to the book as well! But this is making a virtue out of ignorance. You know that it's worth a try, because Boyd and Richerson make something interesting out of these models that wasn't in the original Cavalli-Sforza and Feldman. So why not make the extra effort and actually learn the math?

Callebaut: I propose that from now on we call this *Lloyd's canon:* "Work harder!" (*Laughter.*) Back to Lumsden and Wilson.

Kitcher: Trivers wrote in the *Los Angeles Times* that the Lumsden and Wilson stuff was—I think he called it "an arcane system of logic that nobody believes." Trivers is presumably quite angry with Wilson. I guess there is a big difference of opinion among sociobiologists about the exact merits of the Lumsden and Wilson material. It does seem to me to be an utter fiasco. It seems, in a sense, that they've got the right problem. They see that it's important to relate evolutionary forces (what they call "the assembly of the mind"—say, cognitive science, neurophysiology, and cultural transmission); there are three components of the picture.

Callebaut: How do they compare to Boyd and Richerson?

Kitcher: I think Richerson and Boyd are less good at putting mind into the picture in their programmatic notes. But Lumsden and Wilson just do it wrong. There is no really serious use of cognitive science. Maybe there isn't the right kind of cognitive science yet done to be useful. That is a possibility. But the models, when they are probed . . . I think Maynard Smith and Warren (1982) did the world an enormous service by starting to probe them. I read the Lumsden and Wilson book quite thoroughly when it first came out. When I saw the Maynard Smith and Warren review, I thought, "I'm going on the right lines." That was an amazingly penetrating review and it made just the right kinds of points about the book, which are that the models are in part confused, in part unnecessarily complicated, and in part designed to generate particular results through cunning artifices with parameter values.

Callebaut: Are you suggesting that there was any intent there?

Kitcher: I don't think so. I think this is one of those areas in which you can get out of your depth; and I think that's exactly what happened to Lumsden and Wilson: that they didn't see how they were getting the results they were getting; they just *got* them! They played around and played around and played around and they got the kinds of results they wanted to. I don't think either of them had a sufficient grasp of what was going on in the system to see how it was all generated. What I try to do is show how all the tricks get done. Once you see it, it seems to me the enterprise is ludicrous. On my own view, this program is just *dead*—you just have to go back to the position from which they *start:* the insight that we are dealing with a creature that has a mind, a complicated psychology, and has culture. Now how are you going to model this? I think that's as far as they got with the problem!

9.3.2. A Cheat-detecting Social Contract Algorithm

Callebaut: Leda Cosmides and John Tooby are attempting the formation of a new field of inquiry they call "evolutionary psychology," which is meant to be a response to a number of problems in sociobiology. Basically, evolutionary psychology relates evolutionary explanations in terms of adaptive function to psychological explanations in terms of proximate mechanisms (Cosmides and Tooby 1992b:17).

Lloyd: In many senses it is a really good and reasonable response, because they say, "Sociobiology has had a lot of trouble because it tries to directly connect evolution with behavior. This is not the right idea: you have to look at the connection between evolution and the psychological mechanisms producing behavior, not at the behavior itself." This strikes me as a very interesting and good thing to say. Many people have said it before.

Callebaut: Philip Kitcher, for instance. In other words, you are willing to accept the "missing link" argument?

Lloyd: Yes. I mean, *if* you're going to do sociobiology . . . I don't personally have a lot of confidence in the reasonableness of sociobiological ideas; not because I think we didn't evolve, or aren't animals, or aren't somehow biological, or have some traits that are not genetic, but rather because the kinds of social traits that the sociobiologists are interested in don't correspond to anything that is actually biologically on! So it's a very particular kind of complaint.

Callebaut: Are you suggesting somehow that we have to climb up the ladder one more step—the step of evolutionary psychology—to find out then that we must throw away the ladder?

Lloyd: Yeah, and what worries me is that the psychology may not be very interesting. There isn't any guarantee that the kind of psychological mechanisms that the researchers find interesting and provocative and explanatory will be biologically the mechanisms that play a role in evolution! I don't have much confidence about that, partly because of the extremely poor track record on this. But that's not really fair; let's give them a chance to clean up their act! I think that's fine. I think it's pretty dicey, and I think a lot of the stuff that is going on has been irretrievably horrible science in the sense of empirically inadequate science. At any rate, I'm interested in this project of trying to clean up the act and getting to do some serious work. I was interested in investigating this more closely. I ended up spending some time on the cognitive psychology literature out of which the Wason selection test (cf. 8.1.2) came that Cosmides is using (Symons 1979; Cosmides 1985, 1989; cf. Tooby and Cosmides 1988, 1990 as well as Sober 1981a on the evolution of rationality).

Callebaut: Her hypothesis is that there is a "cheat-detecting social contract algorithm" in the brain: a domain-specific "Darwinian algorithm," that is, a specialized mechanism that organizes experience into adaptively meaningful schemas or frames (Cosmides 1985; Cosmides and Tooby 1987), equipped with a cheat-detecting procedure.[21]

Lloyd: That hypothesis strikes me as the sort of wrong-level description that one often sees in sociobiology. But it's worth a try. Cosmides lays it out and then does Wason selection test experiments. She hypothesizes the existence of a special type of cost/benefit analyzing module in the brain. She then uses this module to explain the origins of certain psychological responses, such as behavior in the Wason logical reasoning tests. The theory is that social contracts are enacted by engaging this cost/benefit analyzing module, and that the selection pressures on this particular ability were quite strong.

Callebaut: What do you think is wrong with this argument?

Lloyd: Her use of a lot of assumptions about human interactions and psychological interpretations of human interactions, I think, is empirically very risky. She and Tooby make some very strong statements that run into many problems. Still, basically you can understand why it was set up that way. What bothers me about this paper is that although it is supposed to be an application of evolutionary theory to cognitive science, the evolutionary theory that is included is just amazingly naive. And the level of exposition of the evolutionary theory, the claims that are made about the evolutionary theory, the outdated material that is used . . . ! Trivers' (1971) reciprocal altruism has been superseded, as Trivers recognizes, by various game-theoretic approaches, whose empirical underpinnings are uncertain. The fact that this was made quite public a few years ago seems to have passed them by completely. They make a very classic and old mistake about kin selection theory (Cosmides and Tooby 1987), in which they assume that kin selection theory predicts that you must be able to recognize your kin; and therefore we can infer a cognitive module in the brain that can recognize your kin. This is one of the most basic errors in kin selection theory that was pointed out most visibly by Dawkins in his "Twelve misunderstandings of kin selection" (1979). The fact that they seem to have completely passed by even the relatively accessible work in evolutionary theory is really quite shocking to me. So the level of evolutionary biology in that research is to me inexcusably low, inexcusably ignorant, it is astonishing to me that they actually apply this. Also, there is their very puffed up way of speaking—"Fortunately, evolutionary theory has shown"—and then they claim that evolutionary theory makes some very strong predictions about what human life must have been like.

---

21. Cf. Chomsky's (1966, 1975) "cognitive competences" or "mental organs," or Fodor's (1983a) "modules." Cosmides and Tooby (1992b:25) think that these "psychological or cognitive adaptations" are (usually) most usefully described on the cognitive level of *proximate* causation, and that they are evolved adaptations.

"You advance knowledge by pushing a line, even if it is later shown false."

Michael Ruse (b. Manchester, 1940) is professor at the University of Guelph, Ontario. He got interested in philosophy soon after he went to the University of Bristol in 1959. In 1962 his professor, Stephan Körner, got a query for a graduate student from McMaster University in Canada. Ruse soon found himself on the *Empress of England,* heading for Montreal, to do an M.A. in philosophy. In 1965 he started teaching at Guelph, an agricultural university, without a Ph.D. He went back to Bristol in 1967, for one year, to write a thesis. He was supporting himself, so he "had to get cracking." He wanted to do something with a "hands-on" aspect to it. Peter Alexander, his supervisor, suggested biology. Although Ruse had never done a biology course, that sounded like a good idea ("When you choose a Ph.D. topic, you want some literature on it, but not too much and not too good!"). Now he found his niche. He started reading evolutionary biology seriously: Maynard Smith's [1958] 1975 "wonderful little book on evolution," Mayr (1942, 1963) on systematics, Julian Huxley (1942) on the modern synthesis, Dobzhansky (1951) on genetics, Stebbins (1950) on plant evolution. There was little written in the philosophy of biology at the time. There was Grene's (1959) paper, which he is perhaps more sympathetic to now than then. Flew was good but lightweight. Hull had just started publishing on species (1965). There were "one or two quite bad arguments" by people like Manser (1965), who were mainly using evolutionary theory to attack the logical-empiricist paradigm. Ruse was more attracted by Braithwaite's (1953), Nagel's (1961), and Hempel's (1965) views on explanation. He basically defended the neopositivism of those people in the biological realm, arguing that evolutionary theory was at least in sketch their philosophy in action, against critics like Scriven.

Both the sociobiology and the creation/evolution controversies were important for Ruse. When sociobiology became the big thing after Wilson's book was launched in 1975, Ruse became wary of the critics' attacks: "I thought and I still think that Wilson, but even more, that *Darwinism* was getting a rough deal. I've always been a socialist at some level, but I'm not a Marxist, and I've certainly never been a radical in that way. I never felt that biological ideas and left-wing ideas are incompatible. Neither did Haldane, although he was a Marxist" (see Werskey 1978, Sarkar 1992a, d; Ruse finds himself "very empathetic with a lot of the positions Haldane took"). Ruse thought that he was sitting on the fence. "I've never felt that I was defending Wilson as 'right'. I'd always felt that the critics' case

wasn't well taken." Even today, Ruse feels that "the proof of the pudding in sociobiology—particularly human sociobiology—is whether it works or not. No amount of philosophical argument is going to answer that. *You can go on criticizing stuff. What I say is, leave it, and get on with the job now.* Ruse views himself as a Popperian only in the very specific sense that he does favor bold conjectures: "The way to move science forward is by pushing out and seeing if you can grab something, and if you can, then 'Jolly good, now let's go from there!'" That is why he likes John Oster's hypothesis about *Archeopteryx*—that the feathers were used to catch flies, and the beasts then lifted off: "It's *wrong,* but it was a good hypothesis to carry your thinking forward." Wilson's way of doing science is also to say, "Let's jump in," knowing that "we may be wrong." But can human nature theory be seen on a par with speculations about *Archeopteryx*? Ruse will not disarm so easily: "Before some intellectual wet says 'Yes, but you don't see the social danger of even toying with an idea like human sociobiology,' I will counter that this century has seen just as much evil done by those who favor the cultural end of the gene/culture spectrum" (see 5.1.2 for Ruse's involvement in the creationism debate).

In addition to spending many a semester at Cambridge University, Ruse has visited at Indiana University (1976) and has spent a year (1983–84) at Harvard's Museum of Comparative Zoology, working with Wilson. He is the author of nine books and has edited several more volumes on evolutionary biology and related matters, some of which have become quite popular; he has also written innumerable articles and reviews. *Taking Darwin Seriously* (1986a) is his own favorite: "I really tried to work out what is philosophy, and what it means to me. For the first time in my life I faced the real questions, and didn't just try to do something which is intellectually respectable. Of course, the fact that it is in its fifth printing and going into half a dozen languages also helps. I'm told that you can buy it at the tobacco kiosks in Argentina." Ruse—who is a Fellow of the AAAS and of the Royal Society of Canada—is also the founder and editor of *Biology and Philosophy.* "I don't have many graduate students at Guelph; I've got to make my mark in my own way." He thought there was a place for him to start a journal "which would go beyond the narrowly epistemological, like *Philosophy of Science,* into the ethical, social, and historical." The "Booknotes" at the end—the first thing many people look at when they see the journal—are his own baby. Ruse promises he will do the journal for ten years only: "Editors shouldn't stay in the same rut. However good an editor is, he or she has got their own ideas, and there comes a time when, even if it's not as good in the future, you should change and get a different perspective."

I interviewed Michael on a bench under the trees of the University of Western Ontario campus (London, Ontario) on June 24, 1989, at the occasion of the HPSSB meeting there.

Callebaut: Where do you see this thing going (cf. also Fetzer 1990)?

Lloyd: In the long run, it can go two ways. People can read this stuff, see the short-comings and say, "We've got to clean up our evolutionary theory, we've got to make it have some resemblance to modern evolutionary theory (*laughs*), we've got to have it be more sophisticated; then, maybe, we can do this program of looking at the evolutionary mechanisms!" That would be an OK way to go. It can go another way. The specter of naive adaptationism and very primitive, genetically unsubstantiated and implausible evolutionary theory that has been popularly prac-ticed among sociobiologists can simply be passed on to these more highfalutin sociobiologists, who are then going to make the same mistakes everybody has been complaining about since 1975. It's a theoretical worry, it's a methodological worry. The fact that many of these worries are dismissed as being motivated polit-ically strikes me as scientifically irresponsible. It's a dirty move, and it's a false move. That's not what is going on. The complaint is quite internally scientific. And the kind of Galileo defense . . .

Callebaut: . . . meaning? . . .

Lloyd: . . . "Everybody believed that Galileo was wrong when he first went out, but he bravely continued to have an unsupported hypothesis"—that is a very popular defense among sociobiologists. OK, fine! Well, it's fifteen years later now. The fact that these people are publishing something in 1989 (Cosmides and Tooby 1989), which refers to a paper published in 1971 (Trivers 1971), which has been rejected by the establishment that she claims are her allies is not a good sign! Now, as far as how this reflects on the overall project of sociobiology . . . Look, my problem, and people's problem in general, with this—Feldman's worry for sure, also sometimes Lewontin's worry—is that the questions are important, they are interesting. Nobody I know, even on the extreme other side of the fence of socio-biology, thinks that we didn't evolve. Well, I've been accused of this; they've been accused of this. Nobody thinks that. But the fact is that these are still the terms of the debate. Just last week in Paris, somebody came up and said, "Oh! Gould is really a sociobiologist!" And I said, "Oh really? Why do you say that?" And he said, "I caught him making an inference that something was actually genetic" (*laughs*). I said, "I'm sorry, but I really don't think that's what the debate is about!"

Callebaut: That sort of thing happens often in Europe.

Lloyd: Well, he was an American actually!

9.3.3. *Overhearing a Coffee Room Conversation*[22]

Callebaut: Lisa has qualms.

Lloyd: I've seen plenty of psychologists with an interest in evolutionary biology. But

---

22. CRC, *entre amis*. The realization that CRCs are of paramount importance for the advancement of Science, *and* a constant burden for the test tube washer who is lowest in the pecking order of the lab (but

I usually see people getting it wrong. Evolutionary biology has a very special status. It's like writing a novel in that everybody thinks that because they can read, they can write. Everyone thinks that because they took Intro Biology, they know all about how evolution works. They reinvent the wheel. It's a risk you run into every time you go into an interdisciplinary context: you are going to reinvent what people in the other discipline already have done. You have to reinvent tools. For instance, why doesn't Plotkin use the multilevel genetics that are out there? Why doesn't he use the multilevel selection theory?

Callebaut: For tactical or strategic reasons, I guess: he thought that in order for his basic point not to be misunderstood—the need for a multilevel approach to evolution—he better stick to a fairly orthodox, conventional view of genetics, being the lowest level in his hierarchy.

Lloyd: I think that his fairly orthodox view of selection has been basically completely discredited in the last fifteen years. I understand what you're saying: rhetorically, strategically, he had a particular idea about what he wanted to do.

Callebaut: He also lives in a different culture.

Lloyd: But many British biologists don't understand American genetics; they don't know American genetics. I'll tell you something else that I've become convinced of in the last year (this is relevant to this point): As far as I can tell, very few British evolutionary theorists actually know the relevant math.

Callebaut: I once heard Maynard Smith, one of the pioneers of the game-theoretic approach to evolution (1982; cf. Lewontin 1961, Sober 1985b), claim himself that his mastery of the "sums" involved there was of the most rudimentary sort.[23] But that was certainly an English understatement.

Lloyd: I've come to believe that this is a thing in their educational system, that they don't read the mathematical models because they never learnt the relevant mathematics, because it is not part of their education. This allows the Americans a certain edge, because Americans can do both the theory and the mathematics. Whereas the British end up in this funny situation where they misinterpret the mathematics; they don't read the mathematics, and they're sort of driving themselves into a funny place in the theory where they remain ignorant of some of the main, important explanatory structures in the actual theory, because of some glitch

---

someone has to clean up the mess!) is a major achievement of the sociology of scientific knowledge (SSK). CRCs were at first incorrectly classified as part and parcel of Knorr-Cetina's (1982) "variable transepistemic fields" (VTFs); cf. my exchange with Richards on the Cubs and the White Sox (7.3.5). But a growing number of ethnographers of science now consider the CRC a truly endemic institution of their objects. Those of a Marxist bend—especially the more vulgar ones—deduce this from the invasion of coffee room space by strange machinery (cf. the sections 4.3.7, "What is the Purpose of all this Equipment?" and 4.3.10, "Stepping Back from the Tohubohu"). It seems safe to hypothesize that the more the philosophy of science will be naturalized—and nobody is going to stop that process now—the more significant CRCs will become in that dismal field as well—mutatis mutandis and ceteris paribus, that is.

23. Conference on Biological and Economic Competition, Facultés Universitaires Saint-Louis, Brussels, 18 December 1984.

in their educational system. This isn't a claim about capability; it's a claim about what they actually learn.

Callebaut: The French would know the mathematics.

Lloyd: I don't know much about them. But then, they don't believe in Darwin![24]

Burian: In fact, the Catholic countries generally resisted Darwin. It's an interesting sociological question why this is so. I had an interview with Philippe l'Héritier, the inventor of the population cage and a French mathematical population geneticist, who connected the difficulty of Darwinism and the favoring of so-called Lamarckism (and even Lysenkoism) in Catholic countries to some extent with moral training—"the sins of the fathers shall be visited on their sons." But the neo-Darwinian mechanism is inconsistent with that stance! It illustrates a conundrum to which I don't know the answer. I think there is in fact sufficient difference between the Catholic and non-Catholic receptions of Darwinism that one wants to ask that question.

Callebaut: (*Desperate.*) Now please, don't also bring in Piaget (1978, 1980) here with his allegedly post-Darwinian "phenocopy." Don't make matters more complicated than they already are!

Campbell: (*Malicious.*) Piaget was a Christian socialist with liberal Christianity, but it had a definitely religious streak, and he went through a phase of trying to find Henri Bergson's emergent evolution coherent. It left him with a residual bias against this utterly nontranscendent materialistic Darwinian and neo-Darwinian theory (Vidal 1987).

Lloyd: (*Was never listening.*) That reminds me of what happened with Jonathan Schull (1990). He has done a lot to learn evolutionary biology, but he still likes to set up this straw man of evolutionary theory that he got out of Fisher. (*In an outburst of élan vital.*) Fisher is not the only geneticist! This is just madness! The fact that the only genetics some people have ever learnt . . . (*Somewhat hesitant.*) Not Schull. He knows more. That's what surprises me about the kinds of generalizations he makes: he has read and he understands Wright's interdemic selection models—which is not an easy thing; and he also has a good feeling for species selection. (*Resuming her previous pose.*) They all have this kind of rhetorical ploy of using the simplest sort of evolutionary theory which is accepted by virtually none of the evolutionists I know personally. They make it up, and then they use it as a straw man to say, "Now we need to do something new!" That stuff is all over the place in the genetics literature that they're not looking at. That's why I don't like it; because of the sort of chest-beating that goes along with the presentation of the "new" stuff! Everybody wants to be more new than they actually are. Obviously they're not talking to geneticists who will tell them about the pertinent models. That's what I don't like about their work. I object to spreading ignorance. If you

---

24. Thus Gérard Jorland, the editor of a series in the history and philosophy of the sciences published by Hachette (Paris), recently talked about René Thom's combat against "la débilité théorique du darwinisme," which has received no answer hitherto ("Avertissement," in Thom 1990:8).

claim that something is not going on when it is, and you claim that it has to be reinvented, you spread ignorance. I will put the same criticism against Maynard Smith; I will put the same criticism even more strongly against Dawkins. To dismiss something about what you don't understand and then claim that it's not important is to make virtue out of your own ignorance. I don't like it intellectually. That's the problem I have with Odling-Smee.

Callebaut: In their defense, I could point out that when you go into hierarchical views of culture . . .

Lloyd: . . . you have to know everything. So basically I would respect everybody who tackles the problem with a certain amount of subtlety, because it is a very difficult problem. I was just reading a paper by Dan Sperber (1990). He follows a general intuition of Feldman and Cavalli-Sforza—trying to do the *epidemiology* of culture. I think that in his anthropological interpretive work he's got something to offer that should be listened to by the formalists. But again, it's such a complex picture that everybody has a contribution to make. Everybody's got to try. But that makes it especially worrying if somebody tries and then says, "Anthropologists ought to look at this!" or "Geneticists ought to have thought of that!" or "Let's have this sort of model!"

Callebaut: A reaction, Henry?

Plotkin: Lloyd appears to suffer from a severe case of genetics worship when it comes to evolutionary theory. She should not think that others ought to be similarly afflicted. I am interested in the evolution of learning and culture and how such phenomena relate to other, fundamental, features of creatures that can learn, such as their genetics. But what I do is not genetics and never has been, and so I would never claim to have made a contribution to that subject. I don't know why Lloyd thinks that I would have made such claims.

Callebaut: I recommend that readers who are distressed, disappointed, or disenchanted by this episode check out Dr. Campbell's paper on "Ethnocentrism of disciplines and the fish-scale model of omniscience" (1969), of which our little incident here is a nice illustration (or let us pretend so). (*Clapping hands.*) And now coffee klatch is over, my friends; let's get back to work!

9.4 FITNESS, THE BUGBEAR OF EVOLUTIONARY BIOLOGY

Callebaut: Jon Elster (cf. 4.1.5) thinks that a basic asymmetry between biology and social science is that the latter lacks the equivalent of what is often presented as the common coin of evolutionary biologists: the notion of fitness.

Plotkin: Fitness is the bugbear of all theoretical biologists. It is arguably one of the most difficult problems in biology and in the social sciences: what does one exactly mean by fitness? Let me try and answer it in a rather oblique way. One of the interesting, fruitful things about the evolutionary analogy is that if you depart from the notion of evolution as a consequence and think of it more in terms of a cause—

that is, a process or set of processes that causes things like changes in gene frequency in a breeding population—then one is able to be consistent in departing from what I think has become a very simple-minded notion of evolution: that evolution occurs only at a certain level in living systems. I'd see that as a fundamental tenet of evolutionary epistemology: the idea that *evolution occurs at multiple loci within a living system.* What that means, then, is that whatever one's description or whatever one's definition of fitness is, it has to be something that you can utilize at more than one level; so that in *that* sense, something like literal reproductive competence—producing more offspring—is no reasonable measure of fitness at all.

Callebaut: How would you define fitness in general, then?

Plotkin: I would much rather define fitness, if it is going to have a multilevel application, more universally. I would be inclined to do it in terms of something like, perhaps, *efficacy of match* between information stored—somehow embodied within a living system—and those aspects of environmental order that have been the precipitating factors producing that knowledge gain in the first instance. Sommerhoff began to formalize such an approach in 1950, but his work has never been followed up. Williams (1966) mentioned that he considered using Sommerhoff's (1950) work as the basis for his treatment of adaptations in that book but decided not to because it was then so unfamiliar to others. Now, forty years on from its original publication, Sommerhoff's work is even less familiar.

9.4.1. *Digressing on Truth*

Callebaut: Doesn't this bring in again, in a biological guise, most of the problems philosophers have been struggling with when trying to define truth in terms of correspondence?

Plotkin: Truth is a philosophical matter. Biology is not concerned with truth. In this particular case the problem is one of accounting for adaptation. That in itself is a very complicated process, as you know; it is one which other people have approached from a very different kind of view—Lewontin, for instance: the notion that the animal defines its environment, and therefore is in some way an active agent in its own evolution (cf. 6.3). I think he is correct at the unilevel explanation of the evolution of the organism. But if we go to the evolutionary-epistemological position where evolution is occurring at multiple levels, then we simply multiply the complexity. When Piaget (1971) talked about the dialectic of accommodation and assimilation, he was talking about the same thing, but at the cognitive level. I don't doubt that it works at the developmental level, at the cultural level, at any other level which becomes defined by the presence of the process of evolution.

Callebaut: What can one say about the dialectic of accommodation and assimilation in general?

Plotkin: It is very complicated. Living systems evolve in very active ways—Lewontin's point; my point being that this is applicable at all levels. And it seems to anyone who knows anything about psychology that that is what cognitive psychologists from Bartlett (1932) onwards, and perhaps even earlier, have been saying, but in an entirely different kind of language, using a different kind of set of ideas as their references: it is that we take information into our heads, and because we are active, the information which is gained then changes in accordance with what is already inside our heads, and as a consequence our view of the world is altered.

Callebaut: Dick, you are a *philosopher* who has written about adaptation (e.g., Burian 1983); you don't have the psychologist's easy way out!

Burian: Fair enough. I am tempted by Alex Rosenberg's (1983b, 1985) response: Concerning fitness in contemporary settings, we are typically asking for systematic biases in the distribution of phenotypes at later times in the history of a population. Those occur—if they are systematic—for causal reasons. But the causal reasons can't be pinned down by that description. To that extent, the application of the concept of "fitness" is a matter of good natural history, good experimental biology. But you are going to come up with immensely diverse answers, because you are matching a problem setting, or matching an environment, or matching a set of competitors. And those matches can't be predicted a priori; they depend on who is doing what to whom.

Callebaut: It's the supervenience point again (cf. 4.2.1).

Burian: Right. And in that sense I'm tempted by a supervenience account of fitness. What I'm heading toward is the claim that evolutionary causes are whole bunches of little causes; how they add up to big causes is very unclear. In that sense, fitness is an extremely important and useful concept in evolutionary theory, which always has some fill-in or other in biological realities. But it is not going to come across with a correspondence account of truth where you have fitness as such entering as a state into a type-type account of what makes some type fitter than others in a given environment.

Callebaut: Don, which part of your intellectual personality wins here: the philosopher or the psychologist?

Campbell: Since the first footnote in my paper on creative thought (1960), I have called for a redefinition of the problem of knowledge as a subtype of cases of fit between x and y. In my comment on Meehl (Campbell 1990b), I argue that we philosophers (neither Meehl nor I are technically philosophers) should give up epitomizing knowledge or truth in the form of statements, and that the concept of *map* is a much better one (cf. 5.5.2). Maps can be right or wrong no matter how oversimplified, no matter how single-purposed. We can have a highway map and a railway map of the area between here and New York; one can leave out the other; each can put in routes that don't exist, or erroneous connections—the degree of simplification or abstraction is separated from the concept of validity. When we get away from propositions and get into the map metaphor instead—and an *n*-dimensional map is probably a model—we find that a "map theory of knowledge" hasn't burdened itself with the notion of *complete exhaustion of the referent*; unlike many versions of the statement theory of knowledge, particularly when they get into indirect reference.

Callebaut: "If I know something about the Morning Star, do I know it about the Evening Star if I haven't yet identified these as the same?"

Campbell: Those kinds of issues come from a nostalgia for a concept of "captured" truth as *complete*. The visual metaphor also leads in that direction: vision *seems* so complete; we forget it is *highly selective, superficial reflection*. So we've got to get rid of the notion that anything that is *usable* knowledge is complete. Mark Brown, the philosopher at Syracuse University, told the ERISS conference of 1981 the Lewis Caroll/Charles Dodgson anecdote about the two cartographers in a competitive conversation. The Englishman says: "We've mapped all of England 1 mile to the inch!" The German says: "That's nothing! We've mapped all of Germany 1 inch to the inch, but the farmers won't let us unroll it!" (Carroll 1898.) It's a marvelous *reductio at absurdum* of the concept of "complete knowledge."

Callebaut: Let us please go back to Henry's point.

Campbell: If you have "fit of belief to referent" as your epitome—and that's so pervasive in the adaptationist jargon—you have what Linnda Caporael comes close to calling a "foundational principle"—if a belief has competent reference, it is because the referent has participated in selecting the belief.[25] *This cup has participated in selecting my belief that the cup is present, due to the broadcast diffusion of light and selective reflection.* What I want to keep of the concept of truth is the goal of getting it right. That goal, it seems to me, is necessary to motivate curiosity, science and the like, so I don't want to give that up (cf. 7.2.5). But I certainly don't want to keep the notion that any single way of packaging or capturing truth will be complete or exhaustive. (It will often be multi-purposed, or even omni-purposed: rats, once they've learned the maze for food, turn out—to the surprise of some early learning theorists—to already know the maze for water, even though they always run for food perfectly satiated on water.) In this perspectival selectivity, since no knowledge can ever be exhaustive, we can have congruent knowledge that fits, but not identical abstractions.

Callebaut: How would all this apply to the EET program that you feel is based too strongly on a biological evolutionary analogy?

Campbell: In EET we have slipped into thinking that all the selective forces increase the one kind of fit that I call "competence of reference." But we know from sociology of science and from our own experience with rejected manuscripts that many selective forces winnow our beliefs. Only a few selectors represent the referent of the belief. What is needed in the EET program is to do what my friends and fellow

---

25. Linnda Caporael took part in this conversation (cf. the biographical note on Campbell at 7.1.1).

admirers of science Toulmin, Richards, and Hull have failed to do: specify how the social selection process makes it plausible that the physicists' beliefs are being selected by the physical reality to which they are trying to refer.

9.4.2. *Cultural Fitness*

Callebaut: From philosophy to culture shouldn't be too big a leap. What could *cultural* fitness be like?

Brandon: One does need to address that question, and I think this comes up in particular when you are thinking about asymmetric transmission systems (see 9.5). What Boyd and Richerson (1985) and Cavalli-Sforza and Feldman (1981) show is that cultural fitness need not correspond to genetic fitness. A cultural parent who has high cultural fitness—meaning that he is transmitting his traits more than other cultural parents—may have no genetic offspring. They don't correspond. I think that basically it is just going to be a pretty direct analogue with genetic fitness: you are going to define it in terms of the number of cultural offspring. Now the problem is, how you determine cultural offspring? It is more difficult than in sexual organisms, where you've got two parents. In cultural transmission you may have one parent, you may have three parents, you may have twenty-five parents.

Callebaut: People, even within the small evolutionary epistemology community, notoriously disagree about the importance of cross-lineage borrowing in cultural evolution (cf. 9.5.2). Another relevant idea here is the difference between local and global maximization (Elster).

Kitcher: It is an interesting idea. Obviously, local maxima are the kind of things that evolutionary modelists look for. For human beings there have to be some kinds of mechanisms that get rid of local optima. I mean, suppose one were an Alexanderian and tried to respond to Elster's point. Then, I think, one would respond something like this: there are inclusive-fitness maximizing mechanisms in humans which allow them to prevent themselves being blocked at what would otherwise be local maxima. I don't believe there are such things, but that is part and parcel of my suspicion about Alexander's program. The root objection that I would have to the response that I just put in Alexander's mouth in response to the Elster point is that *there is no reason to believe in the existence of these fine-tuned capacities of maximizing inclusive fitness.* I don't think we have these. I think the whole system is jerry-built, and it is like the internal fertilization systems of orchids! What we should expect to find if we ever get a complete proximate analysis of human behavior is that a lot of the behavior we produce isn't inclusive-fitness maximizing, and that we can envisage alternative strategies for doing better in terms of reproductive success; but those just would involve dispositions that, if they'd actually arisen in us in our ancestral hominid environments, wouldn't have turned out to be fitness-maximizing across the board. I have this image of this very complicated internal network, and you've got a whole spectrum of outputs. The evolutionary task is somehow to maximize the reproductive returns on all of the

outputs; that is quite compatible with many of the channels being much less sufficient than they might be on some alternative scheme. What matters is the net
output across the network.

Callebaut: Jon would say, I guess: "On this general level, you have fitness versus
many other sorts of things (outputs)."

Kitcher: The sociobiologist wants to respond to that by saying: "Sure, there are these
other things, but these are also on the leash of fitness maximization. These are
ultimately controlled by fitness maximization." That is what ultimately makes the
Boyd and Richerson book so powerful: they not only talk about systems of cultural
transmission that would lead one to *diverge from biological optima;* but they show
how such systems might in fact be maintained. That is a very important point, I
think.

9.5 GETTING A GRIP ON CULTURAL TRANSMISSION

Callebaut: In your article on phenotypic plasticity (Brandon 1985a), you show how
considerations about questions of genetic determinism are connected with the idea
that culture makes a difference in the human species. The upshot is that extrapolations from noncultural animals to cultural species is problematic. But what is the
argument, Bob?

Brandon: The basic argument is really fairly technical, yet easy to summarize. It is
that a certain range of the conditions that favor *phenotypic plasticity*—nongenetic
determinism—also favor cultural transmission. The argument involves some theoretical population genetics. Most people who have thought about the conditions
that would favor phenotypic plasticity agree that it is favored under environments
changing rapidly and capriciously. So the environment is going all over the place;
and it is better to produce multiple phenotypes—some of which have a chance of
being well adapted in these environments—than one set phenotype which may be
well or poorly adapted, putting all your eggs in one basket. What I have argued,
based on considerations of heritability, is that *if phenotypic plasticity increases,
then the genetic heritability decreases.* So the more plastic are genotypes, the more
their phenotypes overlap; which means that selection at the phenotypic level is not
translated down to the genetic level; which means that genetic heritability goes
down.

Callebaut: Theoretically, this is clear enough; but do you have any empirical evidence?

Brandon: No, it is a purely theoretical argument. Then there is a suggestion due to
John Roughgarden (1979), a population geneticist at Stanford, that under rapidly
and capriciously changing environments, phenotypic plasticity is good for another
reason. *The more plastic are the genotypes, the lower the heritability,* and so that
results in: no genetic change. So the idea is that the strategy of staying put, genetically, is better than trying to unsuccessfully track this capriciously and rapidly

"Chomsky has changed his mind on the evolution of language!"

Robert N. Brandon (b. Concord, North Carolina, 1952) is professor of philosophy at Duke University, where he began to teach in 1979. He studied at the University of North Carolina at Chapel Hill (B.A., 1974) and at Harvard University, where Hilary Putnam and Robert Nozick supervised his Ph.D. thesis, Philosophical Investigations in Evolutionary Theory (1979).

Although the center of gravity of his published work has been evolution and adaptation, our conversation centered on the differences between the mechanisms of biological and cultural evolution, and in particular on the role of human language as our primary cultural transmission system. How and why is our species significantly different from noncultural species? Brandon (1985a) has explored the reasons why cultural evolution would be selectively advantageous and points out that cultural transmission raises the possibility of asymmetric transmission systems. Once that happens, genetically maladaptive traits can evolve. Brandon is able to show how questions about genetic determinism (which a lot of people have worried about in thinking about human sociobiology—cf. 9.2.2) and questions about culture are intimately related: "Under conditions that favor phenotypic plasticity, if environments are changing rapidly and capriciously but not *too* rapidly, it would be useful to have this communication system—language—that allows for cultural transmission." Having looked into the environmental conditions they describe, Brandon and Hornstein (1986) conclude that such a communication system must have three basic features. (1) It obviously must be able to *transmit information;* (2) it must *not be stimulus-bound,* that is, it must be possible to transmit information about things that are neither close in time nor space; and (3) the transmission system must be *semantically recursive.* Brandon illustrates the second feature: "If you are going to make use of the advantages of instructional learning over observational learning in trying to teach your offspring about the dangers of rapidly spreading fires, then you need to be able to talk about something that is neither close in time nor space. Many investigators both from human linguistics and from animal behavior have noted that difference. Brandon and Hornstein (1986) try to explain it in an abstract way. They describe a continuum of animal communication systems ranging from "perceptually iconic" communication systems which through *ritualization* can become "phylogenetically iconic"—a genuinely symbolic system where the signs lose their iconic relation to their referents. The vast majority of animal communication systems only allow reference to the here and now. Some people argue that the waggle-dance system in bees is an exception, but the evidence seems inconclusive to Brandon. In any event, "a species like *Homo sapiens* is not going to be accurately characterized by the central tenet of Wilsonian sociobiology—that the characters that evolve will be those that maximize fitness." In a species like our own some things are very plastic, others less so. "It may be that phobias about snakes and spiders are something that are not transmitted culturally at all. So Wilson's explanation of those sorts of things may be perfectly correct." Brandon and Hornstein are only arguing against "the apriori argument" on Wilson's part that suggests that cultural evolution *must* obey the constraints of genetic evolution, i.e., must always go in pathways that maximize genetic fitness.

They also want to offer a plausible evolutionary argument for the origins of culture and of symbolic languages. "What a biologist coming to the planet Earth would find most striking about *Homo sapiens* is their language." And there are some important differences between human languages and other animal languages, like the *semantic recursiveness* of the former—their capability to transmit messages about things that "one generation ago we did not know we needed to talk about." Remember that things are changing rapidly and capriciously, so "you may find something that is important now that was not important before, and so your communication system must be semantically open-ended." (Brandon's estimate of the number of basic messages that can be conveyed by the semantically closed communication systems of animals is about 600.)

"Chomsky has said in print more than once that no light is to be shed on the evolutionary origins of language. He has read our paper (Brandon and Hornstein 1986) and he thinks that it is very good and that we have set out on the evolutionary path! At least temporarily he has changed his mind—maybe he will change his mind again!"

Brandon is the author of *Adaptation and Environment* (1990) and of a limited but quite remarkable series of articles on adaptation, levels of selection, holism/reductionism, cultural evolution and some other topics. He is the editor, with Richard Burian, of the reader *Genes, Organisms, Populations: Controversies over the Units of Selection* (1984).

The original interview was taped *impromptu* on April 16, 1986 in an office in the Sociology Department of the University of Chapel Hill, North Carolina. Bob, whom I only knew at the time from some of his papers, had been so kind to come over from Duke University to make my acquaintance at the initiative of sociologist Rachel Rosenfeld, a friend of Bill Wimsatt's, who had invited me to Chapel Hill for a talk in her department. Only there and then did I realize that Duke is close to the Raleigh-Durham-Greensboro area to which Chapel Hill also belongs. As the conversation got going, I suggested I could as well get my tape recorder running (I was carrying it along to tape the discussion after my talk), and Bob didn't object. Talk of serendipity!

changing environment. Putting these two things together, that's an argument—two separate reasons why phenotypic plasticity is favored. There are also conditions under which it is not favored. And the final part of the argument is that if environments are changing in a rapid and capricious way, but not *too* rapid,[26] then, if you had a transmission mechanism that was capable of transmitting all and only those traits which proved useful in the past environment to the present generation, that would be selectively advantageous.

Callebaut: And genetic transmission couldn't do that?

Brandon: No, it would always lag behind. But cultural transmission could do that. This conclusion was arrived at independently, and the arguments were slightly different, but it really is the same conclusion that Boyd and Richerson (1985) come to. Then the idea is that what we have shown is that under a range of conditions that favor phenotypic plasticity, a *cultural inheritance system* would be favored as well. Once you get a cultural transmission system—this has been argued extensively elsewhere—you raise the possibility, almost the inevitability, of asymmetries in the transmission systems.

Callebaut: Asymmetries?

Brandon: Genetic transmission works from genetic parent to genetic offspring. Cultural transmission could work from the same parent to the same offspring but might also work with other patterns, so that a nongenetic parent might be your cultural parent. Transmission systems are asymmetric if the patterns are different. What Boyd and Richerson point out, and also Cavalli-Sforza and Feldman, is that once you've got asymmetric transmission systems you get an entirely different evolutionary dynamic—one that allows for the evolution of traits that genetically are quite maladaptive.

Callebaut: So the basic point of your (1985a) paper really is that in a species living in a (not too) rapidly changing environment one should expect high phenotypic plasticity and cultural transmission.

Brandon: If you think about the human species, this seems to fit pretty well. If that is the sort of species we are dealing with, then the major tenet of sociobiology—that is, that the social patterns that evolve will be ones that maximize fitness—will not be true.

9.5.1. *Human Language as an Ideal Cultural Transmission System*

Callebaut: Have you looked at the problem of what inheritance really is? We know a lot about it in the genetic case—the classic biological case. But what is cultural transmission about?

Brandon: Well, in fact, this leads to this other paper that I wrote with a linguist, Norbert Hornstein (Brandon and Hornstein 1986). What we argue in that paper is

---

26. That is, there still is *some* correlation between environmental states, so that knowledge of what was good in the past environment would be useful in the present environment.

that human languages, symbolic languages, are in fact cultural inheritance systems; and that moreover, they have exactly the properties that an *ideal cultural transmission system* would have.[27]

Callebaut: So you characterize cultural transmission abstractly.

Brandon: I like to think about it in terms of the two ways that a parent can influence an offspring's phenotype. One way of course is by genetic transmission: information in the genome is directly transmitted from parent to offspring, and that influences the offspring's phenotype. The other way, described very generally, is just by modifying the offspring's environment.[28] The offspring's phenotype is a product of its genome and its environment! The second way is cultural transmission, very abstractly characterized. Then the question is, "How is this realized in actual animals, in particular in humans?"

Callebaut: Conspecifics would obviously be the most important aspect of that environment. Boyd and Richerson very early in their book made the decision not to define social learning in terms of *effects* of the environment, but in terms of teaching and learning by conspecifics.

9.5.2. *Super Sex*

Callebaut: David, how much importance do you confer to cross-lineage borrowing in cultural evolution (the economist Kenneth Boulding has referred to this as "super sex")?[29]

Hull: In biological evolution as well as in cultural evolution, the two manageable situations are: very little cross-lineage borrowing or a whole lot. Very little means you can treat the lineages as separate entities; a whole lot, you treat them as one. What about the intermediary cases? What if there are lineages where there is intermediate borrowing? They are going to be unmanageable. In most biological evolution, things are clustered at one end or the other, except in plants. There are some plants that have fairly consistent, long term interlineage borrowing, but not enough to call them one lineage. That makes studying these species really difficult. That is the way that things are and there is nothing we can do about it; but it makes it a more difficult phenomenon. In conceptual evolution . . .

Callebaut: . . . the instance of cultural evolution you've looked at in detail . . .

Hull: . . . I find that there is much less interlineage borrowing than people claim.

Callebaut: Why *would* people think there is a lot of interlineage borrowing?

Hull: They see similar ideas, and they think that if they are similar, they have been borrowed. But when you look, you find that they have not been borrowed at all! *Source* matters in borrowing. Just because you have two people who have very

27. Some other recent discussions of language in an evolutionary framework are Lieberman (1984), Millikan (1984), and Pinker and Bloom (1990).

28. Cf. Levins and Lewontin on the "elusive environment," 6.3.

29. "With the advent of the human race something that might almost be called 'super sex' came into evolution, whereas biological evolution never got beyond two sexes" (Boulding 1981:16).

similar ideas, it does not mean that they got the ideas from the same source. Hence, history of science is much, much more important. You've got to find out who got what from whom. This is not giving credit where credit is due. This is to understand the sort of warring factions that occur in science; because origins make a difference in those disputes. From my experience, there is much less cross-lineage borrowing than it looks on the surface, and that is good because it makes the study of science more manageable! The synthetic theory of evolution is not very synthetic. There was not as much borrowing as you would be led to believe!

Callebaut: As David has noted, from the modeler's perspective, having twenty-five parents may be viewed as having just one parent—the asexual case again (although this is not necessary). Such a reduction of multi- to uniparentiality would certainly simplify matters; but do you agree with David here?

Brandon: That is the difficulty I see in trying to precisely define cultural fitness, but I would want to approach it basically in terms of the number of cultural offspring.

Callebaut: As an empirical matter, just how important would horizontal transmission or cross-lineage borrowing be?

Brandon: I don't know! You have lots of anecdotal evidence in cases of big scientific revolutions: the thought that the reason it happens is that someone is bringing ideas from other areas and so on (cf. 4.2.5). I don't know if anyone has carefully studied that in a controlled way—*anything* like that. But one of the things I would suspect is true—and this may be the sort of question that is just hidden from us *forever*— is that in the early period of human evolution, transmission was primarily vertical, parent to offspring. But once you did that, it allowed uncle as well as father to teach child. I am sure that happened in small groups of people. You would learn from different models. But still it would be primarily vertical transmission, I think, in small groups. And that is probably where the knowledge that was *important* would come from. Now what is more important in the nineteenth and twentieth centuries may be entirely different! Maybe now, horizontal transmission is extremely important.

Callebaut: Parents can also learn from their children to some extent.

Brandon: There is an *indirect* transmission: just as a grandparent can indirectly transmit his genes to grandoffspring by means of the parent, one can indirectly transmit culturally to future generations by writing books or whatever. But the other way around it is limited . . .

Callebaut: . . . so that the complete picture has a directionality to it.

9.6 Is Cultural Evolution Lamarckian?

Callebaut: David, you definitely don't like the word "Lamarckism" (Hull 1982, 1988a)—how come?

Hull: The name Lamarck is a very useful name and has been used from Lamarck's day to the present in wars between scientists. Everything that could have been called Lamarckian *has* been called Lamarckian; so when you put "Lamarckian" in front of another term, it really means nothing. It certainly does not mean any of these people have read Lamarck and have the faintest idea of what Lamarck said (Lamarck 1984; Corsi 1988).

Callebaut: But there is one fairly prevalent use of the word . . .

Hull: . . . which is that *acquired characteristics are inherited.* An organism has a phenotypic character, interacts with the environment in such a way that that character is changed, and somehow that changed phenotypic character changes its genetic material and gets transmitted. Let us say that a plant that normally grows in the shade is grown in the sun, and the leaves are changed accordingly. That change is transmitted to its genes, so its offspring are already predisposed to larger leaves. It is necessary that the phenotype changes in ways that make the plant better adapted, that those changes get transmitted to the genes, and that the genes then get transmitted to offspring, which makes them already preadapted to their environment.

Callebaut: In biology, it is generally held, it does not work that way.

Hull: In biological evolution you would think that would be a great way to evolve. Whenever anybody first thinks of evolution, they think of it that way. But it does not work that way. So far we have found no mechanisms for Lamarckian evolution.

Callebaut: A decade ago, Ted Steele's vindication of Lamarckianism in the light of molecular virology, *Somatic Selection and Adaptive Evolution* (1981), caused some commotion (cf. Sarkar 1991a).

Hull: Does it mean it is impossible or unscientific? No. It just means in point of fact that this is not the way biological evolution occurs. It is interesting why not. I think the answer is: because Lamarckian inheritance would be *maladaptive.* It looks superficially to be a great way to evolve; but in point of fact it would be a disastrous way to evolve. Probably, it has developed numerous times and rapidly gone extinct.

Callebaut: Please be more specific.

Hull: It has to do with the time lag argument mentioned before (9.2.1). Lamarckian evolution reacts too rapidly to changes in the environment, and reacts to all of them indiscriminately. Whereas the Darwinian way gives you two types of reaction time: short term—phenotype; long term—genotype. It is the interplay between the two that produces evolution. Now that is guaranteed to be very wasteful: lots of death, lots of extinction. But at least it is not instant modification.

Callebaut: How do things look like in the case of conceptual or, more generally, cultural evolution?

Hull: To produce a comparable claim for conceptual evolution, you're going to have to distinguish between conceptual phenotypes and conceptual genotypes, because without that distinction you cannot even begin to draw the contrast. Let us say I invent something and then teach it to you; and then you teach it to someone else. That is an acquired idea which is being passed on. But these ideas are analogues to genes, not analogues to phenotypic traits. So what you've got, if anything, is the inheritance of acquired memes or genes; so that it does not make any of the appropriate cuts. It does not help to call it "Lamarckian." It is important; social learning is very different from individual learning, where you learn it on your own. But, calling one Lamarckian and the other not does not draw the appropriate cuts on any definition of the word. All it does is introduce needless confusion.

Callebaut: Why are so many people, but especially the French, keen on calling cultural evolution Lamarckian?[30]

Hull: Because it is a great propaganda device. I just would like the word "Lamarckian" to be removed because it induces confusion. There is nothing especially Lamarckian about cultural evolution. Anyone who thinks that conceptual evolution is Lamarckian will have to come and tell me what they mean. So far, nobody has. Nobody has ever said what they do mean by "Lamarckian" other than "I can learn from you." But after all, you can get fleas from me. Does that mean that you inherited some of my acquired characteristics? No! You know, it has got to go through the genetic material. And as far as I know, nobody thinks that social learning goes through the genetic material. So I do not know in what sense it is Lamarckian. Though a lot of people in print have called it "Lamarckian," they are going to have to do a lot of terminological finagling to make it sound like they thought they knew what they were talking about. You have got to save face. You have to throw up a smoke screen and scurry out from under. You cannot say "Whoops, I just was repeating something that I heard but did not really understand. It just tripped off my tongue." You cannot say that. It would be nice if you could, but you cannot.

Callebaut: Bill, you are willing to disagree with David here.

Wimsatt: I don't agree that these ideas can be simply regarded as gene analogues without saying more about the case. The other side of this is Weismannian inheritance, from which our bowdlerized image of Lamarck derives. Weismann's views involve a denial of the inheritance of acquired characters, but in a particular way: he distinguished between germ plasm and soma, and said that there is no inheritance of acquired characters *through the germ plasm*. (He didn't actually deny the inheritance of acquired characters *tout court,* for instance, through culture.) For Weismann, the germ plasm generated both the germ line (that is, it continued itself) and the body (which, with anything it acquired, failed to make it into the next generation). These two roles later became the *autocatalytic* and *heterocatalytic* functions of the gene.[31]

Callebaut: You agree with David that making a genotype/phenotype distinction is crucial for cultural evolution, but unlike him, you think you can do it . . .

---

30. E.g., Barloy (1980:16): "Tout le monde reconnaîtra que l'évolution culturelle se fait, comme le voulait Lamarck, par transmission d'une génération à l'autre, des 'informations' acquises."

31. E. B. Wilson's (1896:xi, fig. 5) classic (partial) representation of Weismann's views probably did much to influence later views of Weismannism, as well as suggest a way of making a genotype/phenotype distinction (Griesemer and Wimsatt 1989).

Wimsatt: . . . and this bears on the Lamarckian/Weismannian distinction. The biological notion of the gene is anchored in terms of its generative roles. Mendelism (and Morganism, and the later development of genetics) elaborated the combinatorial processes and rules through which genotypes, as combinations of genes, made other genotypes—*autocatalysis*. If you look at classical genetics, genes seem to be defined in terms of these combinatorial properties, and, for the Morgan school, in terms of their linkage relations, which also just served to modulate ratios of offspring. Cultural evolution has no such neat combinatorial rules for transmission, and so it might arguably not seem to have any genes. (Dawkins et al. talk about "memes" as gene analogues without having established any basis for doing so.) In biology, for many years, we elaborated the workings of the autocatalytic functions of the genes, without making much progress at all on their heterocatalytic functions. This is why developmental biology is now getting so much attention—because we are beginning to get somewhere on that. *Suppose we choose to emphasize the heterocatalytic role, and define a gene as "any element which plays a significant role in generating a large structure whose degree of adaptation determines the number of replicates or close variants of that element that are produced in the next generation."* This is defining a gene in terms of its role in producing a phenotype, and this is the way we need to do it for culture. This gives the phenotype David seeks for his memes, but points out that Dawkins et al. haven't succeeded in defining either gene analogues or phenotype analogues for culture adequately.

Callebaut: So your proposal is once again to consider the generative entrenchment of an entity as a measure of how gene-like it is (cf. 9.1.1)?

Wimsatt: Right! Many cultural entities play important roles in generating other behavior—norms, behavioral practices, and the like. For scientists, important generators include theories, laws, experimental paradigms, standards, and procedures, models and model organisms, etc. The important thing to ask in determining whether a cultural entity is gene-like is, "What else hangs on it?"

Callebaut: I am tempted to invoke here Latour's "first principle" in *Science in Action* (1987:259): "The fate of facts and machines is in later users' hands; their qualities are thus a consequence, not a cause, of a collective action."

Wimsatt: Right! An idea that doesn't (much) affect one's behavior is not a cultural gene, no matter how often and reliably it is transmitted, because it has no heterocatalytic function. And this same distinction gives a cultural phenotype, namely the adaptive structure of behavior resulting from or affected by having the set of generative elements in question. Thus, for example, a major adaptive structure of theories is their set of applications—for these are major, though perhaps not the only, determinants of their fitness.[32]

---

32. Some important residual questions about how we individuate and distinguish between different cultural genes and phenotypes are ignored here.

Callebaut: You also think the multiparentality of cultural evolution makes for a major difference between cultural and biological inheritance.

Wimsatt: Cultural inheritance is both developmental and multiparental. We receive our cultural inheritance from different sources in chunks and pieces throughout our lifetime—there is no well defined point in our lifetime when we can say, "Now that's it—we have our completed cultural zygote." We get cultural chunks which affect our receptivity to other cultural chunks, and some cultural chunks will not take root unless preceded by a long series of other cultural chunks which prepare the way. (Thus college physics uses differential equations, which requires as prerequisites math, algebra, and calculus, not to mention vectors, a little matrix algebra, and the like.) If we take the first major bolus of culture (and what would that be—language perhaps, and the cognitive skills that go along with it?) that we receive in ontogeny as the cultural germ line, then everything else that comes later appears to be Lamarckian. But if what we inherit is a bunch of generative complexes which are elaborated and prepare the way for the reception of other such complexes, and if some of these can be inherited or not independently of others, I prefer to regard this as a kind of multilevel Weismannism, with multiple-germ tracks. If we imagine that Lamarckian inheritance is just directly inheriting a trait, transferring it whole from phenotype to phenotype, and Weismannian inheritance involves inheriting generators of some sort which produce an elaborated set of traits, then inheritance looks more or less Lamarckian or Weismannian depending on the generative fruitfulness of what we inherit, and some less fruitful things may be Lamarckian while other more fruitful things may be Weismannian (see fig. 9.2). I don't much care whether cultural evolution is seen as Lamarckian or not. The points I regard as most important are about how to make a genotype/phenotype distinction for cultural evolution, the existence of multiple cultural germ tracks for each individual, and the role of the generative concept of the gene in making these distinctions.

9.6.1. *Propaganda and Contagion*

Callebaut: Is cultural evolution Lamarckian, or can it be, on your analysis, Bob?

Brandon: Well, I *called* it Lamarckian. In fact I have been talked out of this by a number of people. Because the primary advantage that we appeal to in arguing for the selective advantage of cultural transmission is that it does allow for the inheritance of acquired variation. Boyd and Richerson also used "Lamarckian inheritance" originally, and also because it allowed for the inheritance of acquired variation. The term "Lamarckian" seemed to be a good descriptor of a transmission system that allows the inheritance of acquired characteristics. Boyd and Richerson used the term "Lamarckian" in their original manuscripts and we were using the term "Lamarckian" quite independently; and we were dealing with some of the same people, who were all telling us: "Don't do it!" I was finally convinced not to do it by sending the paper to a number of people and finding all sorts of misunder-

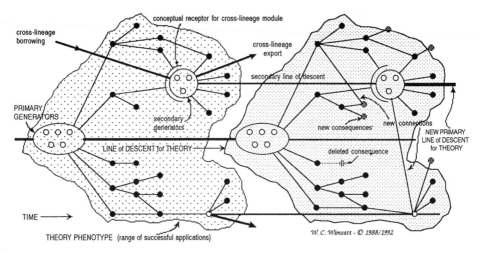

Labels within the figure:

conceptual receptor for cross-lineage module

cross-lineage borrowing

cross-lineage export

secondary line of descent

PRIMARY GENERATORS

secondary generators

new consequences

new connections

NEW PRIMARY LINE of DESCENT for THEORY

LINE of DESCENT for THEORY

deleted consequence

TIME

THEORY PHENOTYPE (range of successful applications)

W. C. Wimsatt - © 1988/1992

Figure 9.2. Inheritance and evolution of scientific theories.

Inheritance in scientific theories is via a kind of hierarchical Weismannism. The primary generators (axioms, heuristics, and exemplary problems, solutions, and procedures, represented by 5 unfilled circles) are transmitted to a scientist and provide the initial basis for the elaboration of a theory (represented by other [black] nodes) which is generated from them. Some of these consequences provide a natural context or receptor site (the semicircle) for the incorporation of other generators (the smaller circle with 3 unfilled circles) which may be transmitted later from other sources, but are subsequently reinterpreted, elaborated, and transmitted and taught as part of the theory. (They may, as here, remain separate from the primary generators of the theory, either because they are thought of as part of a separate subject matter, or because they cannot be taught until after other things are taught first.) Some consequences (the unfilled circle at the bottom) may be sufficiently noteworthy that they may themselves be transmitted independently (with or without modification) either to later generations of that theory, or exported to other lineages. Thus, think of Darwin's principles, as applied to organisms, as the primary generators, Mendelian genetics as the imported secondary generative complex at the top, which becomes population genetics, and the idea that selection maximizes fitness as the unfilled generative node at the bottom (revised in the second generation with input from population genetics) to become maximization of inclusive fitness. The

phenotype of a theory is the whole of its support apparatus (including all of its primary and secondary generators, and local models) together with the full development of its consequences, both internal theorems which have no direct application, the explicit attempts to apply to problems, the range of technologies that depend on it, and their immediate consequences, and the range of data and phenomena for which it provides a useful classificatory, explanatory, predictive, or exploratory tool. The global character or status of the theory can change as a result of the development of these new linkages and consequences, and the developing theory as a whole will tend to become richer in its consequences and more finely tuned to its applications (cf. the many new consequences and the smoothed edges and finer fill pattern, respectively, for the theoretical phenotype).

This diagram is grossly oversimplified, so it cannot, at this level of magnification, capture the fine structure of historical detail. Three kinds of inheritance are represented here: (1) derivational linkages within the theory, by the thinnest lines; (2) the inheritance of the primary generators by the thickest lines; and (3) transmission of secondary generators, both within and across theoretical lineages by medium lines. All kinds of inheritance can be multiparental, since a consequence may be derived from the conjunction of several others and generative complexes may be derived from a variety of sources. Also represented in the diagram is the production of new consequences (denoted by gray nodes) in the second generation, and loss of problematic consequences. Not represented in the diagram is the role of the conceptual environment (both intra- and intertheoretical) in selection, mutational changes in the meanings of key elements, and perhaps most importantly, major changes in the structure of the theory. Thus, e.g., it is arguable that population genetics has in many contexts replaced or become co-equal with Darwin's principles as the primary generators of the modern theory (represented by the wider band issuing from it at the right, signifying its greater generative entrench-ment—what is *not* represented is the consequent change in the inferential structure of the theory). The complexity of this diagram should be compared with the simplicity of Wilson's classic diagram of Weismannism (E. B. Wilson, 1896:11, fig. 5), but the point of both is the same: what is transmitted is generators, not the whole phenotypic structure of the theory. The latter is reliably generated (and elaborated) in the right context from these primary generators.

standings being produced by using the word "Lamarckian." People read too much into it. Everyone has their own view of what was central to Lamarck's theory of evolution. What we were thinking of was the primary differences between Lamarck and other modern theories of evolution, not Darwin's, but modern theories . . . We don't allow for the inheritance of acquired variation. Weismann said that this is impossible, and that is one of the central tenets of modern evolutionary biology. Cultural transmission is different in that it does allow one to acquire a trait during one's life time and then pass that trait on to offspring.

Callebaut: According to David, neither on a literal nor on a metaphorical reading can cultural evolution be Lamarckian *by definition*.

Brandon: I disagree with David there, and in both of these papers you'll see that I follow the practice of certain quantitative geneticists who define heritability in a purely statistical fashion that makes no assumption about the mechanism of inheritance. Heritability is defined simply as a regression between parent and offspring, deviation from the mean. Defining heritability in that way—and that is the central idea of heritability in quantitative genetics—the mechanism may be genetic or it may be something else. So I disagree with saying that talking about cultural inheritance is metaphorical: it is literal inheritance.

9.6.2. *Learning Rules*

Callebaut: The genotype has been interpreted cybernetically as a kind of (evolutionarily functional) self-simplification of the phenotype in a part of itself.[33] If this idea is extrapolated to cultural evolution, phenomena such as Arrow's (1974) "code" of an organization can be approached in its light, and—I take it—Nelson and Winter's (1982) claim to have identified Lamarckian evolution in the case of the behavior of business firms can be confirmed: their "routines" (genotype), when consistently resulting in deficient behavior (phenotype), *may* be (of course, they

---

33. "The function of the genotype is supposed to be one of storing information for the phenotype so that the negentropic commitment (in the form of complexity) represented by the phenotype can be somewhat minimized by allowing lower negentropic factors (chemical bonding, for example) to be at the service of replication. As a result of the economy brought about by self-simplification of the phenotype in the genotype, higher phenotypic levels of complexity can then become possible" (Esposito 1975:136). Cf. also Conrad (1983:ch. 10).

mustn't be!) intentionally changed—for better or worse—and this change *may* (although it mustn't) then be inherited.

Brandon: I would agree with Hull's premises; I just would not draw the same conclusion. It is clear that when we transmit a cultural phenotype, it is not the phenotype itself, it is rather the information that allows for the rebuilding of that phenotype. If it is a simple behavioral phenotype like eating the same food source, you don't transmit that, you're transmitting information.

Callebaut: Lamarckian evolution as I just described it ought to be distinguished from contagion as envisaged by Shrader (1980).

Brandon: I think Boyd and Richerson and Hornstein and I were talking about Lamarckian evolution as you see it. It is definitely not a contagion model! It really was the inheritance of acquired traits that we wanted to label as Lamarckian. Let me talk about the sort of example that I think we all had in mind: the inheritance of certain *learning rules*—rules that tell you what phenotype to adopt. By trial and error, a cultural parent might adopt a certain learning rule that is an appropriate one, and then transmit that learning rule to offspring. That is inheritance of an acquired trait. Let me just make a little side remark which may or may not interest you, about why we all decided to drop it, or at least why I decided to drop the reference to Lamarck. First, Robert Boyd and I talked about this, because independently we had decided to use the term "Lamarckian." We were really talking about very similar things, and this was done independently. He was at Duke for a while, and we got to know each other, and we talked about this; and finally he was convinced not to use it. I think David Hull was one of the people arguing that he should not use it. So here was one person saying: "Well, even though it is quite appropriate I am going to drop the term." And then the other thing that influenced me was that a lot of my friends from biology, when they read this paper, agreed the term "Lamarckian" would sort of raise a red flag. It had all sorts of connotations that we didn't want to carry along. So we decided just to call it phenotypic transmission, a much more neutral term.

Callebaut: When I talked to David, he developed an argument to the effect that biological evolution in the strict sense cannot be expected to be Lamarckian: environmental changes would be tracked too quickly, and you might end up with something maladapted again.

Brandon: Again, I think one cannot make that statement in an unqualified way. There are certain conditions under which it would be advantageous to track the environment very rapidly and other conditions where it is not advantageous to do that. So the way Boyd and Richerson do it, it really depends on the degree of correlation between subsequent environments.

Callebaut: It would seem that because of the very point David was making, Lamarckian evolution should be a useful thing to have on the cultural level, where the time scale is different . . .

Brandon: That's what I thought too. I think the term *is* quite appropriate, but all I can say is that a lot of people will read into that things that you don't want to read into it.

Callebaut: But that happens in other cases as well: When one says "Darwinian," that usually doesn't refer to what Darwin said or thought either!

Brandon: I don't know. I found that whole episode somehow amusing, just finding out how people would misunderstand you. I was ultimately convinced to drop the term.

10

PHILOSOPHY MOVES ALONG

IN THE FINAL CHAPTER of this book we will briefly look at several discussions and issues which, for various reasons, command special attention. The first discussion concerns the resurrection of *evolutionary ethics,* i.e., the application of evolutionary theory to the normative realm of morality. Long relegated to the status of a pastime for (retired) biologists and considered dubious—to say the least—by professional philosophers, evolutionary ethics is at present imposing itself with a vengeance. Our next concern is metaphilosophical in character. We will inquire into the foreseeable consequences for the practice of philosophy of the various *centrifugal movements* (many of which are related to the naturalization of philosophy) that have been documented throughout this book. How real is the danger that we will all become isolated on our own turf? But also the flipside of this: What are the chances for general philosophy (which is also being revived currently) to succeed; what are the risks involved? These and related questions bear directly on the way people are *educated* in philosophy— our third subject. Is the "technical imperative" creating a new barrier? How much science should one have mastered before being in a position to address philosophy of science issues competently? Is there room for both specialists and generalists? What will be the importance of the massive use of computers in schools and philosophy classes at universities?

Naturalism is a "weasel word" (cf. Hull 1988a:6f.), meaning different things to different people. In particular, Quine certainly has a different philosophical agenda than the historicists or someone like Nickles or Shapere. A historical perspective on the *Methodenstreit* between "hard" naturalists and "softer" historicists may be instructive here.

Gender, we should all realize by now, is a basic way things are socially constructed. Yet despite the fact that there is both a booming demand for and supply of *feminist science studies,* the *fact* "gender" is still perspicuously absent from the mainstream literature in philosophy of science (where the situation is worse than in sociology or history of science). Lloyd and Longino point out several gender biases in recent biological research, in particular in the context of evolutionary explanations of human nature and human female sexuality.

432

It is only appropriate to end our tour on some speculations on the future of science studies. Will the current, conflict-ridden division of labor between philosophy of science and the more empirical approaches of sociologists and historians settle into a more harmonious relationship? Have we really learned to take into account the full importance of technology when we are thinking about science, and, related to this, does it make sense for sociology of science to consider the "delegation of actions to nonhumans" (Latour)? And if we *do* take the idea of "de-anthropomorphisation" seriously, what are the prospects of an alien science?

10.1 Evolutionary Ethics

Callebaut: Bob, you have done a great deal to resurrect interest in something which many people thought was dead: the application of evolutionary theory to the realm of morality or ethics.[1] Could you briefly describe, first, the origins of evolutionary ethics, and then, the reasons for the bad press it has had until very recently?

Richards: In the nineteenth century quite a number of biologists, psychologists, and philosophers attempted to understand the development of morality in evolutionary terms—Darwin being one of the first, although really not the first, but certainly one of the most influential. Darwin, in *The Descent of Man* (1871), thought he had to account for all the distinctive features of human nature, of the human organism.

Callebaut: What distinguished man from other animals, according to Darwin?

Richards: What distinguished man most perspicuously, Darwin thought—and the trait that most of his contemporaries would recognize as that which was distinctively human—was moral judgment and moral choice. Human beings can make moral decisions; animals, we suppose, cannot. These biologists and psychologists did not believe that reason, for instance, was distinctive of man—at least most of Darwin's contemporaries did not. And those philosophical contemporaries who were brought up in British empiricism would not have thought it extraordinary to say that animals can think or reason. Many of the natural theologians whom Darwin read were not at all embarrassed to talk about the rational decisions and actions of bees, wasps, and ants. But what they could not do was to make moral judgments. Only man can do that.

Callebaut: So if one is to give an evolutionary account of human beings . . .

Richards: . . . then one has to give an account of man's moral sense; at least that is what Darwin believed. I won't go into the details of his explanation except to say that he believed that at the basis of conscience, of the moral sense, was social instinct. Man, being a social creature, underwent a kind of social evolution in which certain sentiments and attitudes about working for the welfare of the group (most particularly the family, but also the larger kin group) would have been se-

---

1. Richards (1986, 1987:appendix 2, 1988, 1989); cf. Flew (1967).

lected for by natural selection, and would have produced instincts of a social kind in protohuman beings. As evolution continued to mold these beings into the kind of creatures we're now familiar with, it produced a moral attitude that led men to act for the common good of the group they found themselves in. Their intelligence and language ability would help codify these urges, these instincts, and they would be codified into those rules of morality that you find existing in even the most primitive of human groups now—the ones we think whose behaviors and whose societies are at least vaguely resembling those of ancient man. They act according to certain rules, and these rules at least have both as their validating character and their causal motive certain kinds of social instincts.

Callebaut: Darwin—whose broad view on these matters you share—was attacked . . .

Richards: . . . by those who dismissed evolutionary theory altogether, but even by those evolutionists who were of a theological bent. By the time he came to write *The Descent of Man,* his view about evolution had persuaded a good many of his contemporaries, particularly those who were active in biological research and natural history, but also, more generally, the intellectuals of his generation. From 1870 to the end of the nineteenth century and even at the beginning of our century, there were a number of evolutionists, however, who dissented from Darwinism. They were believers in evolution as a developmental process, but gave a somewhat different account of the causal factors that led to it. So, for example, St. George Jackson Mivart: he was a lawyer, an anatomist, an evolutionist, but was quite convinced that ultimately God had established the laws of evolution, and that evolution was regulated according to divine law. This allowed him to maintain that the moral sense in man was not simply a natural product but a product of divine consideration.

10.1.1. *From Spencer to Haeckel to Hitler*

Callebaut: Later on, Herbert Spencer's views became more prominent.

Richards: Spencer was really the first to indicate how evolution—in his own version—might be applied to ethics. When his ethical views were widely broadcast, they seemed to prescribe a rather individualistic ideal for human development, in which men were competing with one another. Ultimately, Spencer thought, this would lead to a kind of utopian society, not unlike the Marxist society that was forecast at the end of the nineteenth century. Both Spencer and the Marxists thought that as a result of certain natural processes, a kind of utopian social structure would be finally achieved; Spencer believed this would occur through evolutionary forces. Part of that evolution, however, would lead to the development of conscience that was molded in certain ways. Now given Spencer's individualistic proclivities and his seeming hard-heartedness about the less fortunate within society, many objected to an evolutionary point of view insofar as it seemed to be inhuman and cruel.

"Either evolutionary ethics is justifiable or no ethics is."

Robert John Richards
(b. St. Louis, Missouri,
1942) is professor of history, philosophy, and be-
havioral science and
chairman of the Commit-
tee on the Conceptual
Foundations of Science at
the University of Chi-
cago. Those reading the
cartoon on his office door
in which the headwaiter in
a distinguished restaurant
tries to find out whether
the "Dr." who is calling to
make a reservation is a

real one—i.e., holds a medical degree—ought to know
that Richards studied philosophy at St. Louis University
(Ph.D., 1971), biological psychology at the University of
Nebraska (M.A., 1974), and history of science at the University of Chicago (Ph.D., 1978). He joined the University
of Chicago staff in 1978 and has been a visiting teacher in
the Department of History of Science at Harvard in the
Spring of 1983.

As a student of Wilfrid Sellars, Richards's roots in the
American naturalistic tradition are indisputable. By their
very nature, his historical works on nineteenth-century
evolutionary theories (*Darwin and the Emergence of Evolutionary Theories of Mind and Behavior,* 1987) and on
recapitulationism (*The Meaning of Evolution: The Morphological Construction and Ideological Reconstruction
of Darwin's Theory,* 1992) are impossible to summarize in
a few lines. Of late, his version of a Darwinian ethics has
been a main concern. If one grants that humankind has
evolved to have a moral sense which is directed toward the
benefit of the group or the community so that we all feel
urges to promote the community welfare, then that has
moral consequences. From such facts we can draw moral
imperatives, Richards thinks, without committing the naturalistic fallacy. He is eager to defend this daring claim at
any place and time. Sometimes his example is the Ayatol-
lah Khomeini's death sentence against Salman Rushdie;
today he invites us to consider a hypothetical community
of conservative Christians in which two members are discussing extramarital affairs. "One says he thinks they are
OK; the other says they are prohibited. If they both accept
the rule, 'Whatever our preacher says is morally prohibited, is morally prohibited,' then when one of them says,
'Didn't you know that the preacher gave a sermon yester-
day, saying that extramarital affairs were wrong?' that is a
fact. From that fact one can draw, following the rule they
both accept, the conclusion that extramarital affairs are
wrong. That violates no principles of logic. What you have
drawn from that fact, via that rule, is a normative conclusion."

In the same way, if evolution has instilled in us certain
rules like "Promote the community good!", then, "like mo-

dus ponens, they will become the justificatory rules for
drawing conclusions from certain facts." Richards thinks
that Hume's problem was really a blind alley. "What *does*
give ground to any ethical views? One usually appeals to a
kind of *empirical intuition.* You say to someone whom you
think is violating common ethical principles: 'What would
happen if everyone would do that?' or 'Can't you see that
this is incorrect?' Let us say that the key ethical principle
that informs all your moral maxims is 'the greatest happi-
ness for the greatest number,' and that you believe that is
the principle that any rational and ethical person ought to
adopt. One typical way to justify that principle is to do it
as Moore did, that is, to consult your intuitions about reality. What is that, other than an empirical request? How is
it that Kant justifies the categorical imperative? Well, by
showing that our moral experience can be explained by the
categorical imperative. Again, this is simply to appeal to
the moral experience, in Kant's case, of early nineteenth-
century German burgers and pietists." In any justification
of a general principle, Richards concludes, one is forced
to recognize that that principle is either derivable from another principle—but then one has to inquire about the justification of that more ultimate principle—or substantiated
in the way philosophers are apt to substantiate most general principles. They give you certain uncontestable
cases—the murder of an innocent child, for example, as
something everyone recognizes as unequivocally immoral—and point out that their principle can demonstrate *that*
is immoral. They will also show you that their principle
will imply or suggest that certain acts that everyone commonly regards as being perfectly moral and advisable *are*
moral. If this implicit appeal to experience is incorrect,
then no ethics can be justified." Richards' position has recently been challenged by Williams (1990).

In addition to the aforementioned books, Richards has
written numerous articles on historical and philosophical
subjects. In 1988, he received the Pfizer Prize of the History of Science Society for the best book in the previous
three years in history of science.

A first interview, in the spring of 1985, had to be cancelled due to Bob's many duties at the time. I gave him a
series of written questions to ponder when returning home
to Europe. My (admittedly weird) idea was for Bob to go
to the student radio station and record his answers all by
himself. Then Bob and Bill Wimsatt got this brilliant idea:
Bill was so kind to sit in for me on June 24th, adding some
questions of himself to my list. So far so good; but they
used an office with a noisy computer on, taped the interview on one of those unreliable mini tape recorders, and
had streaming colds that day. The disaster was complete
(technically speaking). A Belgian radio technician later
became convinced that this interview had been recorded
"in hell." Well, isn't the Univeristy of Chicago supposed
to be a "Tierra del Fuego of the mind" (Richards)? (Thanks
anyway, Bill and Bob.) Fortunately I was able to talk extensively with Bob in his office on March 24, 1986.

Callebaut: What happened after Spencer and the other "social Darwinists," as they continue to be (misleadingly) called?[2]

Richards: Toward the end of the nineteenth and at the beginning of the twentieth century, when German biologists like Ernst Haeckel came to describe man's evolutionary development and particularly his moral sense, they were influenced by the racism of German culture.[3] It is easy enough to see, I suppose, in miniature, the kind of excesses that led to Nazism in the 1930s and 1940s already implicit in Haeckel's evolutionary views. That is one reason why evolutionary ethics has gotten a bad reputation—its association with historical movements that we regard as repulsive. On the other hand, philosophers have decried attempts at constructing evolutionary ethics, because they have maintained, as G. E. Moore (1903) is popularly thought to have, that the application of evolutionary theory to ethics commits a simple fallacy, namely the naturalistic fallacy: the view that one can get moral imperatives or *ought*-statements from factual descriptions of nature.[4] For these two reasons evolutionary ethics has been thought to have died some time ago, and most philosophers still believe that it would be quite foolish to attempt to resurrect it in the current period (e.g., Cottingham 1983).

10.1.2. *A Collective Illusion of the Genes*

Callebaut: Michael, your version of evolutionary ethics is inspired by sociobiology.

Ruse: In 1983–84 I spent a year in Wilson's lab at Harvard. Although my relationship with Wilson was a lot happier than my relationship with Bob Young ten years previously, in a way it was also the same letdown at one level. Young was no longer interested in Darwinism by the time I got there; when at Harvard, I really felt that Wilson's interest in human sociobiology was gone by then. So there you were. At this stage I'd started to get interested in the possibility of evolutionary ethics. Wilson of course talks about it, but I just didn't think that was the way to do it. I really couldn't see the way until the beginning of the 1980s, when a number of people—especially the late John Mackie (1978) and Jeff Murphy (1982)—started to make moves in a way that I felt could be done. Cutting a long story short, I wrote a book on evolutionary ethics. But I felt that if you're doing that, you have to do evolutionary epistemology to balance it.

2. On the question of the "social Darwinism" of Darwin himself, see Greene (1981), Richards (1987), and Ruse (1991b).

3. Kitcher (1992a:53) suggests that Haeckel-the-naturalist (in the philosophical usage of the term), the intellectual star of late-nineteenth-century Jena who "continued a philosophical tradition by drawing on science to address the great questions of epistemology and ethics . . . would have been surprised to learn that one of his relatively obscure colleagues"—Gottlob Frege, the founding father of linguistic philosophy (cf. 3.5.1)—"would help to overthrow that tradition."

4. According to Moore (1903:39), the naturalistic fallacy is based on the mistaken assumption that "ethics is an empirical or positive science"; it is the attempt to deduce ethical conclusions from purely empirical premises or to define ethical characteristics in nonethical terms.

Callebaut: That's not obvious.

Ruse: No, it's not obvious, but put it this way: I felt that if you're going to write on evolutionary ethics, you have to explain why you are not going to write on evolutionary epistemology as well. I felt that, in fact, the same approach could be applied to both. This eventuated in *Taking Darwin Seriously* (1986a), where I apply Darwinism to both. In a way this is my midlife crisis book. Although it is on evolutionary ethics and evolutionary epistemology, I feel it's very much a book on straight ethics or epistemology—on Philosophy, if you like—where what I am trying to do is to make sense of my life: I am not a Christian any more; I'm not a Marxist; how can I understand what makes us tick?—and everything like that. In a way it's a self-indulgent book, because I wrote it for myself. It's also—and this goes back to what I was saying about my personality—a book that most of my fellow philosophers, including my fellow philosophers of biology, will find, if not deeply offensive, *radically wrong*. Even the evolutionary epistemologists you've been interviewing will think this, because I have about as much time for their thinking as does the regular philosopher.

Callebaut: Back to your evolutionary ethics (Ruse 1984, 1985, 1986a, b, 1991b), please: it purports not to violate Hume's rule and also to dispense with justification.

Ruse: My position on ethics is really quite simple. The traditional evolutionary ethicist—from Herbert Spencer through Julian Huxley to E. O. Wilson and Bob Richards—thinks that you can justify moral norms by reference to the facts and processes of evolution. Although what those norms might be tend to be as varied as the norms of Christianity! I think that all attempts at justification run afoul of Hume's law—the illicit transition from *is* to *ought*. What I argue rather is that modern sociobiology gives a fair indication that our moral understanding is an adaptation put in place by evolution to make us good cooperators. Of course, I don't deny that culture has a place too, especially in making for differences between societies. But at root morality is a function of natural selection working on the genes. Moreover, I argue that sometimes when you have given a causal analysis you can see that the call for reasoned justification is otiose.

Callebaut: Could you be more specific and maybe mention some examples?

Ruse: The claim of the spiritualist is one example. The claim of the moralist is another. Moral rules have no justification. They are in a sense a collective illusion of the genes, put in place to make us efficient social animals. Note that I don't cross Hume's law, because I'm not trying to give a basis for morality. Indeed, the distinction between *is* and *ought* is crucial for me, because I think that only by having moral urges will we break from our usual selfishness. Of course, I'm not saying that moral laws don't exist, nor am I preaching license. I'm just saying that they have no foundation. What makes morality truly illusory is that our biology makes us think morality is externally based. In Mackie's (1978) words, we "objectify." If this didn't happen, we'd all start to cheat and morality would collapse. Obviously I'm a genetic determinist inasmuch as I think morality is imposed on us. But who ever thought that we have a choice about morality, other than philosophers like Sartre, and he didn't really. (Mackie suggested this too sometimes before he took up biology.) The point is that we have a choice about whether we're going to be moral. In this sense, we have a dimension of freedom lost on the ants. I admit that my position is not that original. It's Hume brought up to date via Darwin. As such, it's open to me to take other Humean options. For instance, I'm a soft determinist.

Callebaut: Alex, E. O. Wilson seems well aware of the naturalistic fallacy. He has written, for instance, that "the fallacy of my critics is that to know where we have come from is not to prescribe where we are going," and also that "when any genetic basis is demonstrated, it cannot be used to justify a continuing practice in present and future societies" (quoted in N. Wade 1978:330). Nonetheless he has been accused of committing the naturalistic fallacy himself: he seems to reject only *certain* transitions from *is* to *ought* but not others (for instance, in his defense of human rights). In point of fact, wouldn't the whole enterprise of human sociobiology become pointless otherwise (Apostel in Raes and Vanlandschoot 1984:ch. 8)?

Rosenberg: No, because it's not a normative enterprise. Besides, Ed Wilson rejects the naturalistic fallacy argument anyway (Rosenberg 1988e, 1991).

10.1.3. *The Refusal of a Slice of Kidney Pie*

Callebaut: Bob, how does your evolutionary ethics apply to actual practices—say, to the controversial issue of the morality of clitorectomy? Performing a clitorectomy on women is a practice in many native tribes in Africa.

Richards: If you take a particular example like that, it seems to me that from our perspective, what *we* would want to say is: "In our society we would never do that." We think it is wrong, that sort of mutilation serves no purpose or perhaps it serves only political and subjugatory purposes. But what are we to say about, say, a priest or a medicine man in a primitive tribe who performs these exercises, and of the women who submit to them and of the tribal elders who sanction it? What we have to say is: if they sincerely believe that this is conducive to the community welfare, what they are doing is a moral act. It is a moral act which is performed in light of inadequate knowledge, of superstition, of erroneous belief. But nonetheless, though those other conditions obtain, they are yet motivated by altruism.

Callebaut: What would you say, then, to women who—perhaps under Western influence—object vehemently to such a practice and appeal to "universal" human rights in order to have it discontinued? The reason I picked the example was because of its cross-cultural dimension, which seems to me relevant for much of what is actually going on in current ethical debates.

Richards: Well, I don't believe we Westerners could hold the tribal elders as morally guilty. They are acting under the motive of altruism, but acting, certainly, under the burden of false beliefs about what really does advance community welfare. On the other hand, I think we would be morally obligated to instruct them in correct beliefs, and even to restrain them if their acts are causing physical harm without any real commensurate benefit.

Callebaut: In general, you wouldn't say that your evolutionary ethics as it stands allows one to approach this kind of problem of conflict between different moralities?

Richards: No, this is only at a very high level of abstraction.

Callebaut: A foundationalist enterprise?

Richards: Exactly so; it is a kind of meta-ethical perspective meant to deal with certain problems in the conceptions of ethics, namely problems related to the naturalistic fallacy.

Callebaut: Can one be an ethical foundationalist and an epistemological naturalist without becoming inconsistent?

Richards: Well, I don't think they are incompatible. Even as an epistemological naturalist one must be concerned with problems of justification, and ultimate justification as well. After all, as an evolutionary epistemologist one cannot be satisfied with merely explanatory descriptions of the beliefs people hold; one must also attempt to show that some beliefs are better than others, more effective, more likely to realize one's goals. Consider, by contrast, Michael Ruse's approach to ethics. He offers an evolutionary account of why people might hold the beliefs they do, but he doesn't think some beliefs and practices are morally better than others. According to Ruse—who thinks the "naturalistic fallacy" is a fallacy—no beliefs can be justified. Ultimately all he can do is say he *prefers* some beliefs over others. But to give up justification in morality for the reasons he offers is to give up *all* efforts at justification, even in epistemological matters. If Ruse urges us to do that, he is simultaneously asking us simply to accept his personal preferences. Now I happen to know he likes kidney pie; anyone with that sort of preference simply cannot be trusted. His own principles allow all of us to reject his nonjustificatory attitudes in theory of knowledge and morality—as well as refusing him when he offers that slice of kidney pie.

Callebaut: Don, you don't see much of a difference between ethical and epistemological justification either (cf. Chisholm 1957, 1982; Brandt 1967).

Campbell: For ethics, we have to make an *unproven choice* of values. I suggest *human survival under humane conditions:* we don't want humans under r-selected conditions (as many offspring as possible, most of them dying, earlier and earlier pregnancies, etc.).[5] We don't want human survival with fundamental species

---

5. In population genetics, r designates the intrinsic rate of increase of a population, whereas K is the symbol for the carrying capacity of the environment. The demographic parameters r and K are subject to evolution. An "r strategist" or "opportunistic species" relies on a high r to make use of fluctuating environment and ephemeral resources (MacArthur and Wilson 1967). E.g., fugitive species are constantly wiped

change. We probably don't want survival without cities and urbane life—or at
least, I don't, although I'm not sure how New Yorkers feel about that.

Caporael: That's why we're still in New York, I suppose!

Campbell: The goal of human survival under humane conditions we choose. Then
we can do most of our ethics as *mediational* ethics (1979a). It is only meaningful
to a community that shares that ultimate goal. But for those mediational ethics, we
then use scientific hypotheses about human nature and the nature of the environment—we are doing *hypothetical, contingent* mediational ethics. Similarly there
is no proof that one should want to know. But if one chooses the value of mapping
(unobserved) physical reality better and better, then a *hypothetical, mediational,
normative epistemology* that is contingent as to our guesses concerning the nature
of the world and the problem-solving capacities and tools available to man is available. It is contingent, as science is contingent. People still reject the cultural evolutionary ethics of the last century as thoroughly disproven. That's absolutely
wrong. They were contingent, hypothetical, *scientific* ethics. The fact that they
did not provide apodictic grounding for moral norms makes them no worse than
the critics, who haven't provided that either. It's a different ball game.

10.1.4. Consequence, Intention, Free Will

Callebaut: When I first read the second appendix in your book (Richards 1987), Bob,
where you interpret ethics as a victim of the dominant "passive" view of evolution,
I thought there was a conflict between being an ethical intentionalist (consider your
reference to what the tribal elders "believe") and an ethical consequentialist in
adopting the evolutionary stance. Could this be related back to what you say in
your introductory chapter—that the nineteenth-century evolutionists you discuss
all shared the view that *behavior* and *mind* ultimately drive the evolutionary process? Are you suggesting *that* is the way to be an intentionalist in matters ethical?

Richards: Yes, that's right. I actually had not thought about it in exactly those terms.
The reason I called it "a revised version of evolutionary ethics" is that from a
contemporary philosophical point of view, the system that I develop, which has its
roots in Darwin, is in Darwin himself undeveloped; it does not meet all the requirements for a sound ethical theory on contemporary philosophical standards. So that
this difference between intentionalism and consequentialism, I think, actually you
can find it in Darwin, but only in a rudimentary form. You're right about that: it is

---

out of the places they colonize; they survive only by their ability to disperse and fill new places at a high
rate. "The *r* strategy is to make full use of habitats that, because of their temporary nature, keep many of
the populations at any given time on the lower, ascending parts of the growth curve. Under such extreme
circumstances, genotypes in the population with high *r* will be consistently favored" (E. O. Wilson
1975:99–100). In contrast, a "*K* strategist" or "stable species" lives in a longer-lived habitat. Its population, and those of the species with which it interacts, are at or near their saturation level *K*; for this species,
having a high *r* is no longer advantageous. Here, those genotypes will be favored that are able to maintain
the densest populations at equilibrium (see, e.g., E. O. Wilson 1975:99–103 and passim).

a necessary consequence of what was the nineteenth-century perspective—which is to see mind at the root of the evolutionary process in ways that most of us no longer recognize. That mind and behavior were the things that really drove the evolutionary process was a deep and abiding perception of nineteenth-century evolutionary thinkers.

Callebaut: Michael, you and Edward Wilson did a paper on ethics together (Ruse and Wilson 1986).

Ruse: It's a bit of a hybrid tension paper, because we've got pretty different positions. I wish we'd written the paper as a dialogue, with our differences not papered over but made much more explicit. But he felt very strongly that it was important that we present a united front together; so we've got some ambiguous statements about foundations which he thinks mean "justified" and I think mean "explain away"! Let me say that for all I would criticize Wilson (or Popper for that matter), they saw intuitively that Darwinism matters for the foundations of philosophy. Like my fellow professionals, I used to sneer at such attempts. But they were right and I was wrong. Hence, for this reason if for no other, I was proud to put my name alongside Wilson's. Most professional epistemology and ethics produced today is compatible with creationism, and that has to be deeply wrong.

Burian: That's an inference I will not draw. I have a piece Michael hasn't seen—a lecture for undergraduates on Darwin and creationism—which makes the following point: in Darwin's day, there really were serious and good creationist arguments which he had to cope with as a serious scientist. I do not think there are such now, given the evidence and how the evidence is treated. But this is not because "if it looks like creationism, it is bad science." It's bad science because of how one goes on with it from one's starting point.

Callebaut: Michael, you disapprove of Philip Kitcher's treatment, in his *Vaulting Ambition* (1985a), of evolutionary ethics; for instance, of Wilson's views on free will . . .

Ruse: . . . which I would be inclined to criticize myself. Philip doesn't talk at all about the approach on ethics, for instance, of myself, or Mackie, or Murphy, or others who are taking a neo-Humean line. We may be wrong; but I think our way has a possibility of working in a way I don't think Wilson's has. Admittedly, others would disagree with me about this. In fact, Bob Richards, I think, is much more empathetic to Wilson's kind of approach than I am.

Callebaut: Philip, in talking about ethics, points out that Wilson has not much idea, for instance, of the position of John Rawls (Kitcher 1985a:426, 433).

Ruse: I think that's true; I think Wilson's treatment of Rawls is off the wall. But if you look at what Rawls has to say, you see, for instance, that Rawls is himself quite sympathetic to certain sociobiological ideas, although admittedly, I want to take up where Rawls leaves off. Perhaps Philip would say that wasn't his purpose in the book. I think his is a sin of omission rather than of commission: he should have looked at those things and said: "All right, Wilson's wrong; now what about

what others are trying to do?" I know he *knew* about these alternatives and exten-
sions, because he came to a seminar that I gave on the topic at Minnesota. That is
my criticism of his book. Much of it is very clever and very well taken, but I felt
that it was dated when it appeared!

10.2 THE EXPLOSION OF PHILOSOPHY OF SCIENCE

Callebaut: Philosophy of science seems to be exploding: there are philosophies of X
and of Y and of Z. Some day these things will have to be reunited again—or not?
Elliott.

Sober: One of the dangers that we face in philosophy of science now is that we will
all become isolated from each other; that we will become so involved in the details
of one scientific theory that no general philosophical lessons will emerge; that we
will recede into these separate compartments and never emerge again. There is no
general preventive recipe that I can state for avoiding that outcome. I just think
that we have to remember what philosophy has traditionally been: a discipline
which has tried to integrate different areas of knowledge and not become specialized in the details of any single one. This is, I think, something that is possible
within philosophy of biology, within philosophy of psychology; indeed it has been
possible and continues to be possible within philosophy of physics. My interest in
the details of certain problems in evolutionary theory leads me to think about
issues of much more general philosophical significance, like the nature of causality
in science, of explanation, of confirmation and testing. One has to look at the
problems at hand at two levels: understand the details of the scientific problems so
that what one talks about is really science and not some papier maché model of
science, and at the same time remember the tradition of philosophy which has—
for better or worse—posed questions at a very abstract and general level; and
somehow try to combine those two.

Callebaut: It's the problem of defining the "third generation" all over again. Dick?

Burian: I think we're fairly due for a crisis in philosophy of science. The history-
and-philosophy-of-science types have brought about something approximating
a schism in philosophy of science between those who held some variant of a
Carnapian or a Hempelian or a even a Kuhnian program for a unified view
of the dynamics of science or epistemological evaluation of science and those
who have been fighting battles for some kind of contextualism. Contextualism
has been carried to its extreme. We need a dialectical response: how do we get
anywhere from where we are? If we are still asking to do a normative enterprise, we have to get *somewhere*. One of my complaints about David Hull's book
(1988a) is the lack of adequate grounding for anything normative. Now maybe
I'm wrong about the book—one can argue about the specifics of it. But the demand suggested here is, I think, an important one which philosophers are going to

have to face. The problem of reestablishing a normative role for philosophy of science is going to require a great deal of attention from philosophers of science in the next decade.

Callebaut: Which direction do you see it going?

Burian: Lindley Darden has published a book on the history of genetics (1991) which, although it has a wonderful lot of history in it, is on *reasoning strategies* in genetics. On the one hand she identifies very helpfully and interestingly a number of strategies in a way that I think scientists will actually latch on to. On the other hand, worrying about just this particular issue, she does very little to provide guidance as to when and why one should approve the adoption of one strategy rather than an alternative. My own view is that the answer to that will in the end have to be contextual; but *there better be an answer!*

Callebaut: Alex, you are more of a stoic on this score.

Rosenberg: There are a lot of problems in philosophy that aren't problems in the philosophy of *X* or *Y* or *Z*. They are the traditional problems of philosophy that have been with us since Plato, and no matter what the vicissitudes or the popularity of the philosophy of biology or physics or economics, these problems will continue. And they are at least as important and at least as interesting, I think, to the best of philosophers of biology or economics or whatever as any other—problems in logic, in metaphysics, and in the philosophy of language, the problem of universals, free will versus determinism . . . Right now, I am teaching a course on this subject, and I am knocking my head up against it. The problems in this area are not going to be solved by any new disclosures in the philosophy of psychology or of biology or anything like that; and they are real, serious problems, they are not pseudoproblems. It is important, in reporting on these very lively areas of philosophy of science, to always keep in mind that there is still the problem of what *possibility* is, and *necessity,* and *causation,* and things like that.

10.2.1. *Supersmart Individualists?*

Callebaut: Granting this, do you think it is possible and useful for philosophers to have "personal" views on all these issues?

Rosenberg: Useful?! I think it is unavoidable! So the question whether it is useful is probably irrelevant.

Callebaut: What I am really hinting at is whether, from the perspective of the field as a whole, the pressure to be "original" in philosophy is not counterproductive sometimes. Isn't this demanding too much of ordinary human beings, considering their limited time, the level of technicality of discussions, and so on and so forth. You mention necessity, I could mention probability as well, etc. etc. etc. I know there are still people trying to do it all: metaphysics, ethics, epistemology . . . In fact that trend seems to become more and more pronounced with books like Nagel's *The View from Nowhere* (1986), Nozick's *Philosophical Explanations*

(1981) or Parfit's *Reasons and Persons* (1984), which are all considered very good philosophy. There was an article in *The Economist* (1986), showing these were the guys making it big in Anglo-American philosophy: Nozick, Nagel, Parfit, Dennett . . . These people shift subjects easily in one and the same book. Yet isn't there a danger that they are going to make silly statements one day about subjects they don't really master?

Rosenberg: There is a danger. But interestingly enough, they have made very few silly statements. By and large, the people who involve themselves in these really traditional problems, like David Lewis or Saul Kripke—Lewis is probably the best example: he does not deal with problems in the philosophy of biology or the philosophy of economics, but I'll wager he knows more about economics and biology than I do! These people don't make silly mistakes about the sciences, by and large; and either they don't because they have been lucky so far or because they're supersmart, and they've learnt as much as we've learnt about these subjects and seen that the problems in them are relatively inconsequential. And they are keeping their eye on the main problem. The people who aren't working in the philosophy of biology but working in the philosophy of mathematics, on the problem of what is a number and the problems Plato dealt with and that we are still dealing with— *those* are the really smart people in our business. Not me!

Callebaut: Aren't they lonely individuals?

Rosenberg: Lonely?

Callebaut: I mean, don't they work alone? Take an issue like group selection: if you come to think of it, hasn't it really become team work, in a sense? Of course, people involved in this do not (yet?) publish as *teams* . . .

Rosenberg: I don't think so. I think it's still pretty individual. People do sit around and argue about it a lot, more than they agree about possible worlds. But that may be because it is a subject that lends itself to half-baked thoughts more easily than, say, what it means to say that a possible world is actual. If you are going to get up and say something about that in front of a lot of smart people, you'll better have thought about it a long time by yourself.

Callebaut: To the extent that these philosophical discussions become more naturalized, that is, more difficult to distinguish from science—which, with the possible exception of some Slezak (1989a, b, 1991) or other, everybody knows is a collective enterprise—I would expect this to have consequences at the social level.

Rosenberg: Well, I'm just trying to think. Rawls has been accused of dragging more economic machinery into *The Theory of Justice* (1971) than he needed, in order to impress people. I myself don't think that that is true. (I don't think it is false either; I just don't have an opinion on it.) But there are certainly no callow economic mistakes in *The Theory of Justice*. Dennett has not spent a lot of time on biology either, but there aren't any callow mistakes about evolutionary theory in his work, though it depends very heavily on results in biology. So I think that these people who aren't working in these areas haven't stepped on minefields or anything like that.

10.2.2. *Philosophy of Science is Wandering Aimlessly*

Callebaut: What bothers me in a way, Dudley, is the striking contrast between the way science is done collectively—there it's really Newton's "Standing on the shoulders of giants"—and the way philosophy is done, by individuals building their own systems.

Shapere: Your criticism of philosophy seems to me correct: philosophers do build too much in isolation. But I don't think it's so much a matter of "individuals building their own systems," because most philosophers today aren't very systematic and certainly aren't building systems. In fact, most of them are writing for and in response to other philosophers, reacting to their claims and criticisms. It's the error the old positivists fell into all over again: of talking too much to *each other. (Editorial interpolation: I guess I should have blushed here.)* I don't mean that they are ignoring science. But their use of the science tends to be highly restricted, specialized, and fails to come to grips with the larger philosophical issues in a way that is both comprehensive and detailed. As to this, the situation is even worse than it was with the positivists: today's philosophers of science tend too much to throw out very general, hazy suggestions that aren't worked out in detail, and to stop with that.

Callebaut: It is important for philosophers, for professional reasons (cf. 2.3.3 and 5.1), to have ideas about next to everything. Yet I don't see any a priori reason why philosophy couldn't be done more the way science is done, in the sense of a more similar division of work.

Shapere: My interpretation of what you are saying is that philosophical ideas about specific issues must be put into relation to ideas about many others. We need a new *systematic viewpoint,* answering many questions in the light of answers to many others. It would be a viewpoint in which we have a *connected* understanding of the knowledge-seeking enterprise and its roots; the sources, nature, and development of its subject matter, problems, and expectations; its understanding of rationality, objectivity, and observational evidence, and the evolution of that understanding; the significance of its explanatory claims; and many other such issues. We mustn't accept slogans as solutions: all aspects of such a view must be worked out in detail, and with close attention to what goes on in science. The analytic tradition in philosophy has made such approaches unpopular, but philosophy of science is wandering aimlessly today. Only if we get such a viewpoint will it be possible to have any sort of cooperative direction in philosophy. But it would make that possible only if it were a radical revision of the current ways of doing philosophy: it would have to take the best conclusions of science as its framework ("naturalistic philosophy" of science in the sense that I would use that expression) and the broad scope of the knowledge-seeking enterprise as its object of study.[6] There would still be room for disagreement, variation—as there is in science—but there might be agreement as to the framework within which disagreement is to be formulated.

10.2.3. *New Imperialistic Temptations*

Callebaut: I remember seeing a paper in *Philosophy of Science* quite a long time ago in which someone argued that philosophy of science could be a model for philosophy in general (Bushkovitch 1970). This kind of "reductionism" would seem to be outdated now, Bob.

Brandon: At a philosophy talk at Duke, Thomas McCarthy, a guy from Northwestern, was talking about Richard Rorty and the end of philosophy. A lot of the things that Rorty and some other people talk about happen in philosophy of science, and looking at the change in philosophy of science in the last thirty or forty years provides a nice microcosm for some of the things that happen in philosophy in general. McCarthy was arguing—and I probably put words into his mouth—that there needed to be some new paradigm—I know he didn't use that term—some new way of doing philosophy. I remember saying to him that *we philosophers of science are doing very good work now, and we don't know what the hell we are doing!* It is not that we are working under one tradition or another, some paradigm or another; we identify some problems as real problems.[7] And then we try to solve them. Doing good work in philosophy—as in any other area of inquiry—depends primarily on the ability to recognize "good" problems. Someone asked me "What school of philosophy of science do you belong to?" I said "I don't know!"

Callebaut: Today the sociologists of science are certainly more imperialistically minded than the philosophers (cf. Schmaus, Segerstråle, and Jesseph 1992). Bruno Latour, as we have seen, proclaims philosophy of science "dead"; on the positive side, he is arguing that the development of science studies has an important message for political theory as well.[8] Karin, you also think that part of the future of science studies is outside of science studies.

Knorr Cetina: In sociology, for example. Many sociological concepts, like *structure, action, interaction, network, system, context,* and whatever, are not useless concepts. They catch up some of the patterning of social activity. Within sociology, all these concepts have a definite place: "interaction" is associated with the life world, with everyday life; "structure" is associated with the macro-sphere;

6. Agreement about background is what makes it possible for the science of one period to build on that of its predecessors.

7. Cf. the Darden quote in 2.1.1.

8. Still under the spell of Shapin and Schaffer (1985), Latour suggests: ". . . the science studies field is often considered as the extension of politics to science. In reality, case studies show that it is a redefinition of politics that we are witnessing in the laboratories. To the political representatives (elected by humans) should be added the scientific representatives (spokespersons of nonhumans)" (Latour 1991:3).

"action" is allocated to individuals, "context" is allocated to space and time—and so on and so forth. One of the things which science studies, for example, show is that *these things may pop up in all kinds of places.*

Callebaut: Please give an example.

Knorr Cetina: Take the concept of "interaction." Interaction (for example, conversational interaction) is something that science studies show is used as a *technical instrument* in the laboratory. There, it has nothing to do, necessarily, with the life world. Or the body, which is something sociology relegates to the family and maybe to sexual activities, is something that is used in the laboratory as a *device . . .*

Callebaut: . . . You have several people working on that issue in Bielefeld.

Knorr Cetina: You have to consider these reallocations of categories and actors. You might find that social action is actually better exemplified by the insides of a computer program and what goes on in the program—is *built into programs*—and less exemplified by what individuals sometimes do in science. That's something students of science can show and have shown, and that sociology doesn't realize or recognize. In addition, there is the constructivist point of view: the constructivist approach brought to bear upon sociology *itself,* which is something that is not done in sociology at all. Sociologists like Luhmann, who have taken over constructivism, don't take this reflexive stance either or accept a more flexible concept of social reality. This is again the point about ontological flexibility I made before (cf. 2.2.1). You recognize that patterns can be activated and deactivated, enhanced or not enhanced, assembled or disjoined; that there is an issue of penetration and relocation of things. This more flexible conception of social reality, I think, is brought into focus by science studies but is not part of sociology (yet).

10.3 PHILOSOPHY EDUCATION

Callebaut: Alex, do you think that philosophical education as it now exists prepares people well for their new tasks as philosophers of science?

Rosenberg: Good American philosophical education either does or should. It does because very many of our philosophy students started out as science students, unlike in Europe. I started out as a physics student. And those students who didn't start out in science but become interested in philosophy of science find themselves pretty well obliged to study either by themselves or formally one or another science. In addition, the emphasis on mathematical logic and the amount of mathematics that a good American undergraduate education provides should be enough for an able and intelligent philosophy student to allow him to come to grips with, say, Newtonian mechanics or the basic mathematics of population biology in a way that makes him semiliterate scientifically. So students are pretty well prepared to do that.

Callebaut: You wouldn't make taking a major in, say, biology a prerequisite for a philosopher of science?

Rosenberg: Certainly not. The more you get into the philosophy of science, the more you realize you need to learn more and more science. But you can certainly get your foot in the door without first having learnt up all the science. In fact, some philosophers of biology have spent too much time on biology and not enough time on philosophy, so they can't recognize the philosophical problems, whereas they know too much about the details of biology.

Callebaut: This is an interesting idea that I had not previously thought of—that one could know too much about something; it sounds almost biblical to me! Anyway: the people you weren't mentioning—couldn't they be considered as being on the one end of the "generality continuum"—the nitty gritty work end? Their work can be very useful . . .

Rosenberg: . . . Oh yes, their work can be useful, but it is not philosophy. They're not dealing with philosophical problems.

Callebaut: There have to be people doing this.

Rosenberg: But some of these people, of course, who don't have a nose for the philosophical problems, waste a lot of our time, and they duplicate the work of those who aren't interested in the philosophy at all and just interested in the sciences. But most often they throw up problems that are not of interest either to the biologists or to the philosophers.

Callebaut: Bill, do you think that more traditional epistemology and philosophy of science continues to be useful?

Wimsatt: I think I would rather punt on that (cf. 4.2.6). What I would rather say is: "If people want to do it, fine!" I don't want to argue that there is nothing there to be found. I don't have much interest in it. I found it to be generally profoundly irrelevant to the kinds of problems I work on. It's just that I want to say: *I wouldn't want a department full of only my type of person,* only people who are interested in close analyses of science, or law, or whatever. I can learn from people who approach things differently than I do.

Callebaut: It's clear that your work, Pat, is really being done in symbiosis with the scientists. Do you think that is the way for philosophy of science to go in general? Or do we need a division of labor within philosophy of science, with some people doing this sort of work and others doing—je ne sais quoi?

Churchland: I don't like to tell other people what to do. I have a certain approach that satisfies me, at least to a degree. I am curious about certain things. Doing research the way I do helps me to understand the things I want to understand.

Callebaut: But you're a teacher as well. Inevitably, by osmosis some of this is bound to transfer to at least some of your students.

Churchland: That's true. I encourage students to have hands-on experience. If what they want to do has anything to do with philosophy of neuroscience, they absolutely have to, otherwise they don't really understand what they're talking about.

It doesn't necessarily mean that they have to be doing physiology. They could be doing anatomy. For example, I have one outstanding student (Steve Quartz) in the Sejnowski lab, who is working with the confocal (3D) microscope and getting data that are quite stunningly wonderful, things that nobody has ever seen before. Another student, who is working on consciousness, has gone to neurology rounds (clinical meetings at the hospital) and seen autistic children and Williams syndrome children. I do encourage them to see things first hand, because I don't think there is any substitute for it. But of course I can conceive of somebody doing a philosophy project which doesn't involve actually having any hands-on experience.

Callebaut: They've also got to get the possibility to go into the lab, which is possible here but not obvious at all in many places. I am talking as a European now.

Churchland: Here it's possible, that's very true. The neuroscientists here have been exceptionally open and welcoming to our students. It's tough, because to walk into a lab and not know very much means that you're not only useless, but that you're probably a menace. But a number of the students take courses in mammalian neurobiology, neurochemistry, neuroanatomy, and so on. Hence they're not really green and they can go and do useful things in a lab. But other students are afraid. They're afraid that they will be ham-fisted and that people will laugh at them. You do have to have guts to be able to go into a lab and start making a fool of yourself! But so far all my students have really flourished with lab experience.

10.3.1. *The Technical Imperative*

Callebaut: If we look at the discussion of the units and levels of selection and evolution in the last two decades, there has been tremendous progress. The price paid for this is, of course, the quickly increasing technicality of the debate. This has coincided with a rapid expansion in terms of the number of people involved in philosophical and other types of studies of biology. Viewed in purely economic terms, this would seem to be the most obvious way to grow in an initial phase: "Grow extensively, and the ensuing increase in specialization will allow you to cope with the increasing intellectual demands that are due to the ever-growing technicality." *But you can't grow extensively forever,* as the Club of Rome reminded us in the 1970s. What is next? Will opportunities for *intensive* as opposed to extensive growth persist? That is, will it be possible to compensate increasing technicality by ever "deeper" philosophical (or other) accounts? What I am pushing is the question: What are the sociological consequences of the current explosion of philosophy of science going to be? At a PSA meeting, a lot of people— well, quite a few—sit in at a session on the units of selection. How many of them do have a real feeling for what is going on there, Lisa?

Lloyd: Well, not very many. Again, the comparable field is philosophy of physics (cf. 6.1). It has never been the case that some person in another field could walk into a session on measurement theory and understand what is going on. People shouldn't be surprised about that. My first experience with a PSA meeting was in 1982 in Philadelphia. I went down there as a baby graduate student and absorbed what was going on. I went to a philosophy of biology session. Later I heard Bill Wimsatt tell this story about a philosopher of physics who had attended the session on units of selection, walked out in total amazement, and said, "God! I hardly understood a word of that!"—in complete amazement, because "Everybody knows evolutionary biology; this is not a difficult thing!" I think that the answers to your questions are to be found in the philosophy of physics: how philosophy of physics gets done, how philosophers of physics are trained, things like that.

Callebaut: Philosophy of physics seems to be deficient in osmosis with the science, though (cf. 5.2). How should philosophical education be organized so as to cope with the problem of technical expertise as effectively as possible?

Lloyd: (*Getting impatient.*) Look, I take my own education as what I know best. It worked for me . . .

Callebaut: . . . Alex wouldn't require philosophy of science students to take a major in science, because there are all these big philosophical issues to be taken care of. How about you?

Lloyd: Well, I spent about a year of my graduate education in a science department. I learnt a lot from that. It was invaluable to me. And would I recommend that philosophy graduate students go out and take undergraduate and graduate courses in the science that they're meant to be doing? Absolutely, without a doubt!

Callebaut: This seems to me to have tremendous implications. Two-thirds of the more visible philosophers of biology today have spent time in the very same lab (Lewontin's at Harvard). How many scientists are willing to grant philosophers access to their labs?

Lloyd: There are a few more. You don't even have to be a member of a lab. As far as I am concerned, some of the people writing on population genetics ought to have taken at least one upper division population genetics course (*laughter*), I mean to have actually done the math. I am talking about a very basic level of scientific education, where you may begin into the beginnings. Most graduate seminar instructors, if you've taken the prerequisites, will allow you to sit on a seminar. You raise a sociological problem which is very serious with our field. But at the same time I don't grant that it is as impossible as people make out. You don't have to go and be spoon-fed by Lewontin, as many of us were, to actually make a dent in the science. Look, if you are going to be a scholar in classical philosophy, you have to be able to read Greek. It's the same thing. Now there is a certain reaction against that, which is that people tell me, "Well, I can't understand your stuff, because you have all this technical population genetics in it!"

Callebaut: We probably ought to apply Campbell's (1969) "fish scale model" to the organization of philosophy of science itself. Also, it seems to me, not every philosopher of science needs to be as specialized as his or her colleague next door.

Lloyd: I don't think everybody should be a specialist. That's what is so curious to me: I've got a reputation of being a hard core specialist, right? But obviously, with my background (*laughs*), this is a ridiculous thing to assume. I assume that's also true for other people that I talk to. Anyway, do I think graduate students need some fairly high-level scientific education? Without a doubt. Mostly I taught myself population genetics. I would go to the people who were nice to me and I would check whether I had made some horrible error. Many people have taught themselves in philosophy of science. It doesn't have to be formal, but the level has to be there, the expertise has to be there.

10.3.2. *The Use of the Computer in Class*

Callebaut: You stimulate your students to make quite heavy use of computers in the classroom, Bill. That's not obvious in philosophy teaching—certainly of undergraduates—and rather exceptional even in philosophy of science.[9]

Wimsatt: I'm interested in strategies and tactics of model building—how you can be led down the garden path as well as how you can use it effectively. There is no substitute for hands-on experience—doing a simulation with a model, finding out how the system or set of equations works. You can do that much faster with a simulation than with just giving the students the equations and letting them read all about it—and more effectively too, because *they* have done the manipulations. Once you get a hang of how the model works, you can also learn what makes it break down, and this is essential for knowing where to be careful with it, and how to improve it. Asking how it could be improved (whether it gives you the information you need, whether the information it generates is probably right, how you could find out if it is wrong, whether the model itself is easily modifiable) are variants of questions that people who design programming languages and think about problem solving are asking.

Callebaut: Does this also have implications at a more theoretical level?

Wimsatt: It certainly does. Jeff Schank is doing state-of-the-art simulations of menstrual synchrony (a tough problem, because it requires detail about the interacting systems at at least three levels of analysis) that are helping him to figure out the physiology and to think about the process of building and testing mechanistic explanations via simulation and experiment. He and another student, Stuart Glennan (who is designing a software experimental shell for a speech perception lab) have had to think very carefully about programming and experimental design, and have come up with very convincing arguments that new trends in object-oriented programming have things to teach us in general about mechanistic explanation. I think their stuff will have implications for what Alex Rosenberg might call mainstream philosophy of science, but I can't imagine it having happened in either case

---

9. Regarding the educational use of computers in U.S. schools in general, *MacUser* columnist John C. Dvorak (1992:372) is even more pessimistic: "educators . . . are apparently the only group in the country that isn't benefiting from computer technology."

without their having gotten hands-on experience both at programming and in their respective scientific laboratories. (Learning programming without a problem to pursue can be pretty sterile too!)

Callebaut: The simulations you use in class are in PASCAL.

Wimsatt: This is an introductory biology course, and I have undergraduate students doing models of two and three alleles at one locus with inbreeding, two alleles at two loci, n species Lotka-Volterra systems, population growth models and chaotic behavior, and so on—these are topics that are not normally discussed except at a graduate level. The students can understand them and can play with them. While they are for class use, they have actually been an essential part of my own research. I understand a mathematical model far better if I can write a program for it and simulate it. With a recent simulation package which Schank and I have finished for teaching model building,[10] you can investigate chaotic behavior more fully and more visibly than with anything now available. The Mac has beautiful graphics and is fast enough for the purposes, and we are able to teach (and more importantly, the students can learn for themselves) all sorts of things about strategies—and pitfalls—of model building.

Callebaut: How did all this come about?

Wimsatt: The use of the computer in the simulations pioneered by Alex Fraser and Dick Lewontin, about twenty-five years ago, made a major revolution in how evolutionary biologists were able to think about theory in development, and led to an enormous growth in mathematical modeling. We are about to see the next generation of that: the percolation of this down to the level of the PC, which research biologists are already doing (cf. Fontana n.d.). You can't talk about research in population biology today without talking about computer simulation. But the PCs are getting powerful enough; you don't need the mainframe any more except very rarely. And because the computers are getting cheap enough, undergraduates and even high school students can do this. I think we are on the edge of a major revolution of increasing sophistication in this area.

10.4 LET US NOT FORGET GERMAN HISTORY

Callebaut: Let us return to the naturalism issue for the last time in this book (well, almost). Dudley, would you agree that philosophers of science can learn not only from scientists themselves but also from sociologists or psychologists of science—yes or no? Can't we make use of whatever psychology and sociology is available, and hope that it will improve in quality as one goes along. Wouldn't this strategy be in line with one of your holy principles, the "rejection of anticipations of nature"?

Shapere: There is no question that we ultimately will have a lot to learn from soci-

10. For use on the Macintosh (Wimsatt and Schank n.d.).

ologists and psychologists. But when you look at the disagreements between schools of psychology and either the questionability or the utter triviality and irrelevance of so much of their experimental results, you can't expect much right now. Sociologists of science tend to overstress the social influences in science, to the extent that they often fall into—or rather presuppose by their very assumptions—relativism. There is something to learn there, but nevertheless it can only be learned if we disentangle what is involved in scientific reasoning from the social influences. Good sociology of science must recognize that distinction, not deny it from the outset (Shapere 1987b).

Callebaut: Tom, you told me that you are leery of the label "naturalistic turn," especially when, like Ron Giere and many other people nowadays, I speak of the historicists as "naturalists."[11] Why so?

Nickles: Today there are a lot of naturalistic types running around saying that they are historicists, and a lot of historicists who say that they are epistemological naturalists. I should know, because I was one of them. Unfortunately, these are loose ways of speaking, for a strong naturalism is incompatible with a principled historicism. What they have in common is the rejection of precognition or clairvoyance and other divine attributes; and that suits me fine. But the old *Methodenstreit* over the *Geisteswissenschaften* versus the *Naturwissenschaften* among German intellectuals around the turn of the century should warn us that beyond this common ground lies a minefield of mutual incompatibilities. At that time the Germans were the world leaders in both the historical sciences and the natural sciences. We don't live in the nineteenth century, of course, but we must expect the incompatibilities to pop up in new forms even if we refuse to identify naturalism with positivism or with scientism of some kind (cf. Pickering 1991).

Callebaut: Are they in fact popping up today (see also Fuller 1989, 1991b, 1992; Sorell 1991)?

Nickles: One of the forms in which the opposition is beginning to express itself is as a debate between biologically or psychologically oriented philosophers who take a fairly hard-line "naturalistic" approach to epistemology and the new-wave sociologists, who are in some ways the counterpart to the German historicists and cultural anthropologists. Roughly, it's sociology of science against psychology of science. So far, philosophers have found the naturalistic account far more attractive than the social, as did Quine, one of the founders of naturalistic epistemology. One reason is that many philosophers have admired the simplicity of reductivistic positions. Another is fear of relativism. But I am not sure these two reasons fully explain why sociology has received such a bad press in comparison with psychology and biology. Today the latter are far more popular than sociology even among American

11. Cf. Slezak (1989a:572): "Through the strong programme's proponents have espoused the virtues of a 'naturalism', they seem not to have taken seriously the possibility of other, equally naturalistic, yet quite different accounts of science. There is some irony in the fact that the epistemological background of the work in AI and cognitive science is Quine's well-known 'epistemology naturalized.'"

undergraduates. Those facts together with the fact that the naturalistic approach is much better financed than the sociological one leads me to worry that the latter will not receive an adequate development or a fair hearing. It seems to me sufficiently obvious that, whatever their biological nature may be, *human beings are such cultural creatures, so self-defining, or self-programming if you will, that taking too sharp a naturalistic turn may put us on the wrong road.* Biology alone does not distinguish between *essen* and *fressen.* As Marx observed long ago, the development of the five senses is the work of all of human history (Nickles 1989a).

Callebaut: Isn't Tom overreacting a bit?

Knorr Cetina: I don't like the kind of division between historicism and naturalism that's introduced here. I don't think I would want to be completely on the historicist side if that means hermeneutics and a certain kind of interpretive paradigm. Foucault (1973) has this argument against a very interpretative form of history that produces interpretations of interpretations all the time. That is also something I would reject. I'm too ethnomethodological for that. Anthropology is somewhere in between. In terms of the content of what I do, I think I am more indebted to Foucault and to ethnomethodology than to Kuhn, for example.

Callebaut: What would you say to Tom's point concerning the differential funding of naturalistic and historicist projects?

Knorr Cetina: I don't see why that should have anything to do with the naturalism. In cognitive science, it certainly has something to do with informatics and AI, the whole computer movement getting a lot of money. But there ought to be naturalistic studies that don't get money! It depends . . . When you mean by "naturalism" a source of model construction which searches only for output equivalents and not for process equivalents . . . That is something that AI, for example, does frequently: produce a model without caring whether the model actually represents anything going on in the mind, as long as the output can be used or as long as there is some output equivalence between an expert and the expert system. Not only you don't care; if you were to care, you actually wouldn't be able to produce the model. You have to refrain from trying to simulate what goes on in the mind. If that is naturalism—output-equivalent models that are not process-equivalent—and if historicism means "concern with the process," then, of course, I would have to be in the historicist camp. Because the mechanisms are interesting for us; we are not in the business of producing programs that replace activities in the social world and that only have to be output-equivalent. We're interested in the mechanisms, and to that extent we have to use a sort of "point-by-point" methodology. We cannot go out to the field a long time to produce our models and then only check them at the end, because it is the model which is interesting in itself and should include a reconstruction of the process observed. In that sense I would be a historicist. On the other hand things are associated with historicism which I don't very much agree with.

Callebaut: Tom's view about the divergence of naturalism and historicism can be read, at one level, as a critique of the sort of work "San Diego" advocates and that you yourself are engaged in. A reaction, Pat?

Churchland: It is useful to make the point that human creatures are cultural creatures, and that many aspects of our behavior are the results of the way we interact with other people. But having said that, of course, it doesn't really mean that the naturalistic part of the story relevant to the social domain should just drop out. The reason is this: the way an infant interacts with other people changes its brain. It acquires certain kinds of network configurations (cf. 9.2.2 and 10.5). So it seems to me that you want to have the naturalistic part of the story to account for "social representations," broadly speaking. But you also want to have the sociological component. It provides the wider perspective, and it sometimes tells you why certain people believe certain things: for example, their library burnt down, and so they had no choice but to be illiterate. You want to know about such things as the rise of fundamentalist religion in America in the 1980s, about the end of the Cold War in 1989–90. So there certainly is a place for sociology and for the sociology of science. I see them as complementary to psychology and neurobiology.

Nickles: Despite the provocative tone of your questions, Werner, I basically agree with Pat Churchland that both naturalistic and social components are needed. There are many varieties of historicism, and my own version, though relativistic, is not *radically* relativistic. Nor do I wish to drop out the naturalistic part of the story. I certainly believe that we can find natural mechanisms and regularities operative in the world. My point was that the relations of historicism and naturalism need working out rather carefully. The fact that these positions have common enemies does not make them automatically compatible with each other. Moreover, it does seem to me that in the present flourishing of cognitive psychological and neurological studies, there is some danger of overemphasizing "nature" and underplaying "nurture." As Pat would surely agree, even brain structure, at a detailed level of description, depends significantly on "social" conditioning. By this I mean that the ways in which our synaptic structures and such grow depends, among other things, on our "education" or enculturation into the various communities in which we participate. The old idea that the brain is simply hardware and that what we learn is purely software was too simple (cf. also Churchland 1983). Learning is not just programming a completely pre-existent structure. I see no disagreement between Pat and I on these points, and I certainly am not objecting to neurological studies.

10.5 GENDER STUDIES

Callebaut: Lisa, you criticize the new philosophers of science for not having properly incorporated the feminist perspective.

Lloyd: They like to talk about social things and everything except something as trivial and nonfundamental as gender!

Callebaut: Hasn't this been a persistent feature of philosophy in general? Reading just the classics in the history of Western philosophy as a window on our society and culture, one wouldn't be able, as a Martian, to figure out that there are two human sexes on Earth, except for the works of Ludwig Feuerbach and some others (cf. Grimshaw 1986 on the "maleness" of philosophy).

Lloyd: This is a fundamental weakness of current philosophy of science. If you are doing socially and historically sensitive philosophy of science, gender is a basic way that social things are constructed—there can't be any doubt about that. And the sort of protection of conceptual life and scientific life that's implicit in these assumptions that do not mention gender . . . It's bullshit! It cannot be sustained on intellectual grounds.

Callebaut: Michael Ruse wrote a book about sexism in science (1981).

Lloyd: Yeah, he first gives a survey of feminist complaints and then says, "But sociobiology has shown us the following facts. Feminists don't like those facts; and therefore feminists are being unrealistic about, sort of stamping their foot and being hysterical about the science." Now to me this is a non sequitur and also a non-argument (*laughs*). So I was very disappointed in that book. Well, Michael has funny views about feminism. He mischaracterizes feminism. So I don't think one should take that seriously.

Callebaut: He told me he felt it was "one book too many," actually.

Lloyd: It is more important to look at people who are more responsible about their uses of gender and their attempt to actually incorporate gender into their analyses, just to see where it might go in analyses of science. Even Ron Giere, who wrote this nice book on theory structure and explores some of the cognitive science aspects of it . . . Gender is all over the place in there . . .

Callebaut: . . . In the lab, for instance . . .

Lloyd: . . . yet it's not really incorporated into his story. I see that as a weakness. I don't see his book as, "I'm going to throw it out of the window because it doesn't have gender!" I don't feel that way about it. But I do think that, programmatically, as a field, it is a real mistake not to get into gender issues. And that's why I try to do what I consider to be a very hardcore gender analysis on this case study that I'm looking at (see 10.5.1). Because every time we are going to talk about human beings, gender is going to come into it.

Callebaut: How are gender differences being produced, Helen?

Longino: My own view is that they are not simply socialized into us by some form of conditioning. I can't give you any real reason for this, except that it is a view that seems compatible with the rest of my views about human nature. *We try to be good exemplars of what we think we are.* And if we are told that we are boys or told that we are girls, we try to be good exemplars of boys and girls. We have tons of messages to tell us what a good male or a good female is. It's not just a matter of being conditioned to respond in certain ways; it's also a matter of our own trying to be good sorts of persons, responding and acting in a social context which tells us that there are two kinds of persons, males and females. What is nice about

Edelman's selectionist view (cf. 9.2.2) is that it is a theory about the *material base*—about the brain and brain development—that is quite compatible with this kind of story about the development of gender differences. It's a theory of brain function that is designed to give an account of the material substrate of intentionality, higher cognitive capacities, and so forth. So it starts out with a certain conception of what it is that humans can do. It does not rest on a well elaborated psychology; but it starts with a certain conception of what it is that humans are capable of, and asks "What sort of account of neural functioning do we have to give in order that it can be an account of something that can engage in these sorts of activities?"

10.5.1. *Biased Explanations of Human Female Sexuality*

Callebaut: Lisa, you are working on evolutionary explanations of human female sexuality, exposing all the sexist biases that are hidden there. How did you get into this line of work?

Lloyd: The summer that I was finishing graduate school I was having a margarita—well, about three, maybe four—with a friend of mine, who just sort of said, "Listen, what is the function of a female orgasm?" I said, "Pfff . . . That's a good question; I don't know!" I looked it up and found the most amazing stuff, just bizarre, simply beyond science fiction. It was all very deeply male-centered in its sort of ethos; so it was an obvious case . . . Actually I have to thank Evelyn Fox Keller for this. I met her shortly after that and she said, "Are you working on any feminist case studies?" I said, "I suppose I could work on this one, it's the only one I've really thought about." So I started working on it. I went to a conference and presented it about a year later. I was poking around in it but didn't really know what to do with it—it wasn't the sort of thing that I usually work on; and I didn't have a theoretical framework to put it in or anything. I've taken it to the publisher because I became convinced, after trying to write up a paper that turned out to be a hundred pages long, that I couldn't present my case in less than a book-length manuscript. The evidence and argument are very complicated.

Callebaut: Steve Gould got involved as well.

Lloyd: I was talking with him about this project, and he said, "Why don't you write up the summary of what you found out, and I'll write one of the *Natural History* articles about it," which he did (Gould 1987). It was really nice, because he drew a certain amount of flak for what he said there, which was what I was saying in my stuff. So I got to find out how people would respond and became convinced that it's even more complicated than I had ever anticipated. I'm actually quite bogged down right now in the middle of the details of neurophysiology, hormones, primate sexuality, sexology . . . It turns out that you have to know everything! The only saving grace really has been Helen Longino, who has just published what is the best book on feminist philosophy of science so far written by a philosopher of science (Longino 1990). She presents an analytic framework that I think is going to be very good for me to use in writing this book. The various points of insertion of social values into science and basically the argument all the way down I think is right.

10.5.2. *Confusing Generations*

Callebaut: It is your feeling (cf. also Longino 1989) that feminist critics who talked about your work have sort of missed the ball. Why is that?

Lloyd: Sandra Harding, for instance, in her book on feminist philosophy of science (1986), and other people are arguing against a previous generation of philosophers of science (cf. Harding and Hintikka 1983; Harding and O'Barr 1987). They aren't arguing against the kind of work that is being done now, and certainly not against me and my assumptions or Helen and her assumptions. So it's a problem, because the responses in sociology and history of science have not caught up to the actual philosophy of science that is being done.

Callebaut: This book may help change that!

Lloyd: I bet it will too! You cannot assume, for instance, that your opponents are realists. Half of the arguments—Harding's as well as some others' objections—only work against realists. *Feminist empiricism*—I would categorize myself as a feminist constructive empiricist—does not have, as far as I can see, any of the traits that they *attribute* to feminist empiricism. It does not have the realistic aspects, it does not have the sort of imperialistic aspects—nothing! There is a lot of noncommunication going on in the feminist community about philosophy of science, exacerbated by Harding's book, and I hope that Longino's really outstanding book will correct this. I have high hopes for this, partly for totally selfish reasons: I don't want to redraw all the turf when I write my book on my case study!

10.5.3. *Interactionism*

Callebaut: Helen, you talk about such features of an alternative approach to science as interactionism (as an alternative or, minimally, a complement to reductionism). But what is specifically feminist about it? Cannot interactionism also be meaningfully related to other "alternatives" like, say, ecological thinking (cf. Levins and Lewontin in 6.3)?

Longino: It's hard to say what is specifically feminist about interactionism. Or at least, there are different views about that. There's Keller's (1983) view. Speaking specifically of Barbara McClintock, she says that McClintock faced a particular paradox: her scientific culture spoke of scientific inquiry as a masculine activity and spoke of the relationship between the scientist and nature in metaphors taken from sexual life—heterosexual sexual life, where the scientist is a male, courting or subduing a female. As a woman, McClintock couldn't adopt that view of herself. It didn't fit her relation with nature, and so she had to invent a different kind of relationship between herself and the natural world. Rather than seeing herself in a relation of dominance to the natural world, she saw herself in interaction with the natural world and used quite other kinds of metaphors to describe this. Keller quotes things like *feeling for the organism* (the title of the book), or *identifying with the material*—not seeing herself as separate but as somehow "one with" or "part of" the subject. So there, interactionism is a kind of solution to a problem that the woman in science faces. There is nothing specifically feminist about it, really, although many feminists have taken Keller's description of McClintock and of her solution as an exemplar of feminism, which McClintock rejects. McClintock says it had nothing to do with gender, that gender "just drops away." Keller also rejects the notion that she has described a feminist science in describing McClintock's work or that McClintock's interactionism is a direct expression of gender. So the account that I just gave is a reconstruction of what McClintock was doing, an attempt to explain why she engaged in the kind of research she did; it's not something McClintock would really say about herself. What McClintock has said about gender is that when you go into the laboratory, gender just drops away; though she certainly was quite clear about the role gender played in her not getting recognition, not getting university posts, and so on. So she certainly was not blind to the male domination in science.

Callebaut: If, as you suggest, Keller's account can't be used as an account of the relation between feminism and interactionism, are there others?

Longino: Other feminists have wanted to say that there is something about femininity that predisposes it to interactionist accounts, using mostly Nancy Chodorow's (1978) account of gender development based on the *object relations theory* of human individual development. Chodorow takes gender difference to be a kind of fact that requires explanation, talks about women's gender identity as characterized by connection with others, which is a function of their identification with their mothers, as opposed to men's gender identity depending on separation from others, and because their developmental task was to distinguish themselves from their mothers. So if femininity or being a female psychologically involves connection with others, then—the argument goes—it's connection with the natural world that will be emphasized by the female scientist. I myself am a little worried by the kind of essentialism involved there. I am worried about the empirical claims that are being made about women, and also that does not tell us why interaction should be particularly feminist! It says it is an outcome of "female nature," if you will, but that doesn't make it particularly feminist.

10.5.4. *Man the Unmoved Mover*

Callebaut: As the systems engineer Jay Forrester (1975) once put it, when thinking about action we tend to think in terms of first-order negative feedback loops, simply because that's the way we mostly act when doing simple things. (I realize this is moving far away from feminism now.) If I use a hammer in trying to hit a nail, I can hit my finger instead and I am aware of that. Even driving a car may basically involve skills at that sort of level. Whereas in a more complicated inter-

"The 'free' flow of information really is from center to periphery; it produces a global uniformity."

Helen E. Longino (b. St. Augustine, Florida, 1944) is professor of philosophy at Rice University, Houston, Texas. She studied philosophy at the University of Sussex in the United Kingdom (M.A.) and at Johns Hopkins University, where her Ph.D. thesis, "Inference and Scientific Discovery," was supervised by Peter Achinstein. She previously taught at Mills College, Stanford University, and at the San Diego and Berkeley campuses of the University of California.

To characterize Longino's stance vis-à-vis the general issues discussed in this book, one can conveniently start from a reading of her paper, "Feminist critiques of rationality: Critiques of science or philosophy of science?" (1989). Feminist scholars have undermined many women's and men's confidence in scientific knowledge and methodology. Longino distinguishes three strategies in feminist critiques of rationality: (1) a direct critique of science, (2) one which tackles scientific methods, usually through a critique of philosophy of science, and (3) a critique of the bureaucratization and industrialization of science. While providing an important initial insight into contemporary science, each of these moves also has "distinctive weaknesses," Longino thinks. The paths between critiques of science and a critique of rationality are "treacherous." But if one treats the cognitive processes of scientific inquiry as truly social processes, one can see "how social values and ideology can be expressed in so-called 'good science' as well as in methodologically deficient inquiry." *Underdetermination* thus becomes the keyword to thinking about feminism in science in a way that avoids these weaknesses.

As a radical feminist philosopher (she is a founder and associate editor of *Hypatia: A Journal of Feminist Philosophy*), Longino rejects the idea that there is some basic human nature. Her view that "we humans make ourselves, and we make ourselves in some way what we want to be" reminded me of Simone de Beauvoir's *The Second Sex* (1953). Longino admits "a trace of existentialist influence there," but does not think "we are as transparent to ourselves or as radically independent of others as some existentialists seem to have thought." (This, she adds, "is why Simone de Beauvoir is a good reference. She has a somewhat more social concept of human life and possibility.") That we are embedded in and formed in interaction with our social environments "does not negate our ability to think about our world and ourselves, to reflect on our values, and to make choices based on our beliefs and values.

The most exciting, interesting thing to be is a kind of creature that can make and act on decisions about itself and the world based on what it thinks." Longino thinks it's really not worth being a human being if we can't do that, and urges us to "try to be the most interesting thing that we can be" and "develop theories about ourselves as the most interesting kind of thing that there is to be. Let's not let a metaphysics developed to enable us to do physics determine how we think about ourselves scientifically. Let's let what we can hope for ourselves play a role in what we think about ourselves scientifically!"

Her anti-essentialism leads Longino to a different view of the connection between feminism and interactionism than the views arising from the McClintock story as told by Evelyn Fox Keller or from Nancy Chodorow's story of femininity. Hers is basically a political point: "One of the things that feminists have done is actually uncover dependence where previously we had seen autonomy. In particular, feminists have shown the dependence of the 'public' sphere on the private sphere. The activities of the public sphere are only possible given that certain things are going on in the domestic realm. The domestic realm is also dependent on the public realm." The feminist search for interdependence supports an interactionist view in the sciences. "That's what I think the connection is; it's an extension of an analytical orientation that has revealed previously overlooked features of the social world to a new domain—the natural world."

In addition to a series of papers in philosophical, biological, and feminist journals, Longino has written *Science as Social Knowledge: Values and Objectivity in Scientific Inquiry* (1990), ideas from which are discussed at some length in several chapters of this book. Longino has also co-edited, with the novelist and critic Valerie Miner, *Competition: A Feminist Taboo?* (1987), a volume in which twenty-six women who are critical of male power games reveal different experiential, conceptual, or analytic points of view of "the competitive struggle that is characteristic of the male and capitalist world" (Longino) and its flip side, cooperation. In her contribution to the volume ("The Ideology of Competition," 248–58), Longino notes that women engage in competition even while espousing ideologies that condemn it, ideologies that lead them to reconstruct their behavior "in false, misleading, and sometimes self-serving ways" (p. 249). The radical rejection of competition is *au fond* "a rejection of the postulate of scarcity" (p. 256). Scarcity, which in the eyes of radical feminists is an artifact of masculine psychology, is understood by socialists as an artifact of the requirement of individual (or private) accumulation in a capitalist economy. Alternative ways of coping with scarcity need to be spelled out.

The original interview was recorded in Helen's office in Mills College, Oakland, on May 28, 1990, sometime before she moved to Rice.

action like environmental pollution, the "boomerang effect" is not obviously and immediately perceived, and therefore far less get-at-able. There may be something to this idea that there is a relation between the structure of our action—which has us perceive ourselves as the "unmoved movers," pace Aristotle—and the "linearist" attitude.

Longino: This is another way to make the connection between an interactionist perspective and gender: the different social experiences of men and women in Western middle class societies. Men are able to have a much more instrumentalist or "unmoved mover" relation to the natural world . . .

Callebaut: . . . You caught me there! . . .

Longino: . . . whereas women whose place has been in the domestic world—even women who go out and are active in the public world—have grown up with a particular gendered experience. But the domestic world is not one in which one can see oneself as the "unmoved mover." There is too much going on that is really out of the control of a single individual, and you can see your experience as really the product of multiple interacting factors. You can see your experience as an outcome of the needs and actions of others using the domestic space, of those who provide services to it, all in a repetitive daily cycle. One woman the other day was saying that one of the things she learned as a female child was to kind of pay attention to lots of different processes happening on parallel tracks that would come together and then separate, as opposed to her brother who could pay attention to just one thing (*laughs*). Now I don't know what one can make of this. But I do think there is something about the character of the experience in the public and the domestic realm that facilitates—you're right—certain sorts of model building as opposed to others. I don't want to push that too much, because clearly, men can come up with complex models as well. But we need to know a lot more before we can make claims like this.

10.5.5. *Biases of Late-Twentieth-Century Feminism*

Callebaut: In addition to interactionism, a couple of other features are relevant as well.

Longino: What I was talking about before had to do with the intellectual orientation of feminism. From a political point of view, interactionist perspectives would be preferable because they provide an alternative to accounts which attempt to understand natural processes in terms of the domination of a process by another process or by a kind of entity, and, as many people have observed, that has the effect of naturalizing domination. *"That's the way in nature, so why shouldn't it be so in human societies?"* Well, if we can find other ways of understanding those very same processes which don't privilege relations of domination, but in which the participants are viewed as somehow equal to one another and as having mutual effects, in which the participants are understood as interdependent, then we have a different model of natural processes which does not privilege relations of control.

So the idea that domination is natural and ought to be expected is undermined. That's a political reason for any egalitarian, democratic view to prefer interactionist accounts when they can be given. Also it gives us some impetus for trying to find and construct interactionist accounts. We're always constrained by what there is in the world, but if all these others previous philosophical arguments are correct, that constraint has its limits and there is room for invention and construction as well. This is one area where it ought to be applied. So I think feminism has this in common with other views; I don't think it is unique to feminism necessarily.

Callebaut: What I'm really trying to get at are the contingencies of the current alliances between the women's movement, the ecological movement, and other things going on at the periphery and aspiring to become central. In other times or other circumstances, these alliances would have looked different. Or is that something we can only speculate about?

Longino: That's an interesting question; maybe one can only speculate about it. Surely the shape that feminism has taken now is a function of its emergence at this particular historical moment when these other tendencies and political movements are afoot. And what we can identify as feminism in earlier historical periods looks different from contemporary feminism—Christine de Pisan in the sixteenth century, Artemisia Gentileschi, Mary Wollstonecraft.[12]

Callebaut: Exactly.

Longino: Why should we say that today's feminism is the real feminism as opposed to the other? Well, I don't think there is any reason to say that. This just is late twentieth-century feminism, which makes alliances with other political movements. And there are really problematic kinds of power relations. A lot of feminism is, in spite of the superficial rhetoric, quite blind to racial and class differences. Feminism is spoken, in many minds, on behalf of white, middle-class women. Working class women, Black women, or Chicano women get dropped out of the equation; when we're thinking about "the working class" or "racial minorities," we tend to think in terms of the male members of those social groups. So feminism is both shaped by the particular movements it is in interaction with, but it is also in some tension with those—an unfortunate tension in some cases, but that is produced because we are still very much a middle-class-dominant and racist society, and feminism is marked by that still. It shouldn't be!

10.6 THE FUTURE OF SCIENCE STUDIES

10.6.1. *Evolution Is Not Everything*

Callebaut: At this juncture I propose a *tour d'horizon* of the main science studies fields we have been discussing in this book to try to see where they are going. We

---

12. Christine de Pisan (1364–1430) wrote ballads, historical works (*Livres des faits et bonnes moeurs du roi Charles V*), and poems (*Ditié de Jeanne D'Arc*). Artemisia Gentileschi (1593–about 1652), a follower of Caravaggio, was famous for her gruesome paintings of decapitations. Mary Wollstonecraft (1759–97) is the author of *A Vindication of the Rights of Woman* (1792).

begin with the field that has been our primary concern, the philosophy of biology. Alex, how representative would you think is philosophy of biology as currently practiced—with its heavy emphasis on evolutionary theory—of what goes on in biology at large?

Rosenberg: It is hard to say. I think the philosophy of biology goes where the smoke is in biology. That is, where the biologists are in disagreement with one another, the philosopher sticks his nose in to see what sense he can make out of it, and produces results that are sometimes valuable and sometimes not.

Callebaut: Why the obsession with evolution, Dick?

Burian: A minimum of three answers. One: historical accident. Dick Lewontin, Steve Gould, and Ernst Mayr were damn influential, and they have been very welcoming, as people in many other biological disciplines haven't been. Two: there are some obvious conceptual problems that don't require deep knowledge of detail. It takes a long time to learn enough molecular biology to understand the mechanisms! Three: there are some long-standing larger philosophical concerns about the status of man and so on which feed into the evolutionary stuff in an obvious way (more than into anything else). And maybe just fashion as well: evolution has been an easy target.

Callebaut: Coming back to an issue that I discussed with Bob Richards before (3.1.2), Alex, does philosophy of biology do much more than reflect on theses raised by a few highly visible people who are pursuing goals external to philosophy of biology as currently defined?

Rosenberg: Some of the people in biology are extremely brilliant, in particular Lewontin and Gould. Despite what I say about them elsewhere (cf. biographical note at 3.3), I have the greatest admiration for their intelligence. They are brilliant people, and so is Richard Levins. Anyone who comes within their orbit is soon likely to become a protégé. The philosopher of biology—someone like Kitcher or Sober—who has had the opportunity, finds himself within their ambit because of the smoke, because of the heat, because of the controversy that he sees these people involved in, and soon begins to take their side, or at any rate to apply his philosophical talents to teasing out what's really serious and consequential in the disputes that they find themselves in. Sober and Kitcher have done that very well; that is, despite their close association with Lewontin, they're not guilty of the kinds of genetic fallacy that I want to accuse Lewontin and Gould of. (Cf. Lewontin's reply in the biographical note at 9.2.4.) And indeed, I think, much of Lewontin's best work has been done in association with Sober, through Sober's influence (Sober and Lewontin 1982). Whether that is representative of what is going on in biology? Probably not. I think that at least an exciting area of biology is molecular biology, and I only know of one philosopher of biology who has made it his business to stick his nose into that area, and that philosopher of biology has found quite as exciting a set of problems in the philosophy of molecular biology as Sober may have found in the levels of selection.

Callebaut: I'm afraid I don't even know his name.

Rosenberg: It's me!

Callebaut: I see. There is no philosophy of chemistry for that matter either, except for one book (Primas 1981).[13]

Rosenberg: I know.

Burian: As far as molecular biology is concerned, that's not quite true, but it is true that Alex is one of the relatively few.

Callebaut: You have done some work in that area yourself (Burian 1990b; Burian, Gayon, and Zallen 1988, 1991; Burian and Gayon 1990) . . .

Burian: . . . and so have Olby (1974), Fantini (1987, 1988, 1990), Sarkar (1988, 1989, 1991b, 1992c, n.d.), Zallen (1989), Culp and Kitcher (1989), and maybe a couple others . . .

Callebaut: . . . like Fujimura (1988), and Abir-Am (1982a) in history of science.

Rosenberg: There are some interesting problems about the relationship between molecular biology and evolutionary biology, about functional explanation at the level of the macromolecule, where you think it would disappear and it hasn't (Rosenberg 1985; cf. Thompson 1989a); and about problems at the intersection of neurobiology and psychology—problems which are now being dealt with at the level of the primary sequence of the macromolecules and that few philosophers have thought seriously about.

Callebaut: Would you venture a prediction concerning the future of the philosophy of biology, Michael? There has been a concentration on evolutionary biology to the detriment of other subjects.

Ruse: My own feeling is it's going to continue that way, because evolution is an area which calls for philosophical discussion more than any other area. I mean, how can you have a big philosophical discussion about anatomy, for instance? Maybe you can, but it's not as obvious that cutting up dead fish gives an insight into the *Ding an sich*. Taxonomy has had its philosophical moments. But certain issues—and evolution clearly is one—invite philosophy more than others. So I think that will always be a major thing.

Callebaut: Other areas of biology might become more and more important philosophically by the roundabout of their importance for *society*.

Ruse: I'm dubious about philosophers making predictions. I see areas like ecology coming up, with the growing interest in the threats to the environment. I would like to think that evolutionary epistemology, certainly evolutionary ethics, will develop and flower. Moreover, I do see entirely new prospects for the philosophy of biology. At the 1989 and 1991 HPSSB conferences, we've heard lots of talks about illustrations. I've always liked to use lots of pictures; I've always been a visual thinker. More generally, that's an area which may well come up more as time goes by (Lynch and Woolgar 1988; Griesemer and Wimsatt 1989; Ruse and Taylor 1991; Wimsatt n.d.b.). As particular issues come up, maybe we'll respond to these.

---

13. I am only referring to Anglo-American philosophy of science here. In France, important philosophical work on chemistry was initiated by Bachelard (e.g., 1973; cf. Serres 1989).

Callebaut: How about the philosopher's responsibility to society, if I am allowed to use that old-fashioned phrase?

Ruse: I note that at this 1989 HPSSB meeting, we're still not very practical: there's nothing on acid rain, there's nothing on nuclear winter, there's nothing on the seal problem or a lot of practical ecological problems. You know, *philosophers tend to be pretty anally retentive people.* They like to do things which are safe, which are scholastic. So you get a whole industry working on the levels of selection problem. Elliott Sober publishes a very good book, and that gives people five years work, where they read a certain amount of fixed literature, and just *get at it* and work away on these things heads down—don't think about the external world! Evolutionary epistemology is now scholastic, and certainly levels of selection. On the other hand, we've got nothing on the *human genome* problem at this conference. We should have had a couple of biologists who are involved in it—one for, one against it, and a couple of philosophers talking about it. I don't get any papers submitted to me on these things.[14] I wish I could get people to write on these things. I think the philosophers should be much more concerned about these issues as well as *moral* issues.

Callebaut: But biologists, broadly speaking, may have a different assessment of priorities (e.g., Stent 1986, 1987) than philosophers. For instance, many of them would point to the strategic importance of molecular biology.

Lloyd: Part of the reason that philosophers have pounced on these issues in evolutionary theory is that there are conceptual problems, theoretical problems—the sort of thing that philosophers might make a contribution to. Now the thing is that there are such conceptual and theoretical questions in ecology, in developmental biology, in molecular biology, and in the interface between all of these fields and evolutionary biology. Students are already picking up on this: Mike Dietrich (1992), Paolo Palladino (1990, 1991) . . .

Callebaut: One doesn't see much of this yet.

Lloyd: I don't agree with that. If somebody is in there looking at ecology, they are doing something different. Sure, you got a lot of people working on eternal stuff. But there are also young people there; they are energetic, they are good, they are going to go out there and work on this stuff. Do I think philosophers should go out to the field? Absolutely. Again, this has to do with the technical expertise thing that I was talking about before. One of the reasons philosophers of biology have gotten so involved in, for instance, units of selection—and one of the reasons, I think, for some of the low-level quality papers that get written in the subject—is that they think they've got all the data they need. They think they've got all the biology they need in order to do the philosophical work. This is true for a number of papers that I've read on the issue of units of selection, which seem to assume that all you need to know is already set up, for instance, in Sober's (1984a) book.

14. Ruse is the editor of the journal *Biology and Philosophy*. The Human Genome Project was discussed at the 1991 meetings of the (by then) ISHPSSB in Evanston, Illinois. See, e.g., Tauber and Sarkar (1992).

Callebaut: Why do you laugh, Lisa?

Lloyd: I've come to realize that Elliott is not responsible for people taking his word as what is going on in the evolutionary biology area. I mean, he went in and did the footwork himself. I think his footwork is inadequate, so I do more footwork, and a different kind of footwork on a different kind of level of detail because of the principal question I'm interested in. The fact that philosophers who are writing about it think they are getting all the biology from each other means that it becomes an internal kind of philosophical debate, which is exactly the opposite of what is desirable. Michael Ruse is probably right, that it's safer. But it's also laziness.

Callebaut: Maybe it's even worse. Maybe some philosophers are really interested only in "the general conclusions."

Lloyd: People have different interests. Different projects should be done. I think all different levels of abstraction, different science, should be done. Look at van Fraassen's book on *Laws and Symmetry* (1989). I'm a big fan of his book, but half of it is metaphysics, OK? He's not talking about any particular science. I don't want to be perceived as being a fascist about the type of work that is interesting philosophically. I'm not!

Wimsatt: Too much of our research is in the foundational theories of evolutionary theory and genetics, or in taxonomy, which is maybe the oldest foundational theory that we have. What about development? What about physiology, physiological genetics, histology, behavioral ecology, quantitative genetics, functional morphology, etc.? Even if we're not going to study all of the special sciences (and why not?) we should pick a more representative sample.

Callebaut: Which unexplored, yet interesting areas do you see besides evolution, Dick?

Burian: Take muscle physiology, for example: it has fascinating results, and a fascinating metascientific account of what mechanistic explanation looks like that doesn't look quite the way physicists think it does.

Callebaut: Please elaborate.

Burian: Look at Hugh Huxley's work on the sliding filament model. Art Caplan has a little but first-class piece on that (1978a). It's a case of a not-derived-from-first-principles mechanism, which nonetheless, once one understands how it is that the fibrils lock one next to the other as muscle filaments slide past each other, shows how you can carry from the micro- to the macrolevel for some classes of explanation in a beautiful way, in a way that looks nicely reductionistic but doesn't fit any standard reductionist models. There is something deep and interesting to explain in that, and no others have looked at it. There are lots of things of that sort to be exploited. Physiology is basically an unknown discipline to philosophers and not very well known to historians. Molecular biology is relatively unpopulated, though people are discovering it finally. Genetics is somewhat less unpopulated. How many philosophers do ecology? And yet, talk about a discipline where levels-of-organization questions are raised! Paleontology shows the difficulties of extrapola-

"I'm as much in the biology department as I am in philosophy."

Richard M. Burian (b. Hanover, New Hampshire, 1941) has been professor of philosophy at Virginia Polytechnic Institute and State University in Blacksburg since 1983. He started out as a physics and math major at Reed College in Portland, Oregon (B.A., 1963), but soon found himself asking questions that were philosophers' questions about the foundations of knowledge. For graduate studies he went on to the University of Pittsburgh, which had the best department in the philosophy of science at the time. Burian wanted to "spend some serious time looking to one or two biological disciplines as a way of setting a comparative background to physics for conceptual change questions." It would take him eight years before he was to defend his Ph.D. thesis in Philosophy at Pitt in 1971.

Kent Mulligan

From 1967 to 1976 Burian taught at Brandeis, from 1977 to 1983 at Drexel. He also visited at Florida A&M University, Pitt, UC Davis, and—in 1976–77—at the Museum of Comparative Zoology at Harvard, where he officially worked with Gould and Mayr. There he really started philosophy of biology: "I almost went crazy in the first year. I took eight courses, from the first course in biology (which I had not taken in college) through a great variety of work in developmental biology, zoology, evolutionary biology; so that I would have some starting point understanding what it was about!" After that he would not be stopped. For almost a decade, he has been investing much of his research effort in a collaborative long-term project, "A Conceptual History of the Gene," the end of which is not in sight as it keeps on generating side projects. "It's also an *anti*-conceptual history in a sense—there is no linear story that one can tell of the internal development of a single discipline."

Burian has published extensively in philosophical, historical, and scientific journals and volumes and is the editor, with Robert Brandon, of *Genes, Organisms, and Populations* (1984). But his record as an organizer is equally impressive. He was the Assistant Director of Marjorie Grene's Summer Institute on Philosophy of Biology at Cornell in 1982. Burian recalls: "I think it was a sociological turning point, though perhaps not in terms of intellectual influence, for the formation of things in philosophy of biology in the U.S. There were some twenty people there of whom some ten are active now. The things that primarily Marjorie did were socially very important; they became the glue for the formation of working groups of a variety of sorts." Adds Wimsatt: "The group who were there that summer formally were working on learning to teach philosophy of biology in a way that appreciated the science. In so doing, there was also an appreciation of the history of nineteenth- and twentieth-century biology in the current work." By the sheer coincidence that Burian was a local arrangements person for the 1982 four society meeting of the HSS, the PSA, 4S, and SHOT, he called for an interest group lunch which charged Jane Maienschein, Bill Bechtel, and him to set up a meeting the following summer—the first of a series of biennial meetings that generated the International Society for the History, Philosophy and Social Studies of Biology (ISHPSSB). Although he tends to play down his own contribution, Burian has been the main driving force behind the subsequent summer conferences at Denison University in 1983, at St. Mary's College in Indiana in 1985, when he was hosting at Virginia Tech in 1987, and at the University of Western Ontario in Canada in 1989. With Stuart Kauffman, he also organized the Mountain Lake Research Conference on Evolution and Development (1984) that led to an important publication in the *Quarterly Review of Biology* (Maynard Smith et al. 1985). And he was active on APA (Eastern Division) and 4S program committees, chaired an APA liaison committee to the NSF, etc. etc. etc.

Well informed about the history and in almost daily touch with the current practice in the fields he is working on, the realist Burian has mixed feelings about the relativistic morals of the present constructivist vogue: "In cases like Andy Pickering's *Constructing Quarks* (1984a)—but he is not alone!—one finds a wonderful historical story told in interesting, indeed fascinating and controversial, detail. Let the controversy flourish: the story is wonderful! *But*—the series of morals that he draws are ill connected to the story! To a considerable extent, what I want to do is to resist the ill-connected morals while profiting from the marvelous history (Burian 1990a)."

I first met Dick at the Evolutionary Epistemology (ERISS II) Conference which anthropologist Rik Pinxten and I convened at Ghent University in 1984, and was soon to appreciate Jane Maienschein's dictum that "there aren't many others out there with Burian-levels of energy." My endeavors to interview Dick—who was instrumental in introducing me to several people in the American philosophy of biology community in 1985 (Callebaut 1986)—were definitely not born under a lucky star. After several aborted attempts to have this busiest of organizers sit down and talk, I succeeded in hiding him from view for an hour's conversation in London, Ontario, in June 1989. Alas, back home I discovered that the tape was almost inaudible. Even the heroic efforts of my friend Dré Broucke, an engineer and occasional disc jockey, proved of little or no avail. We had to do it all over. I was to meet Dick, who was flying into Paris on May 8, 1990, near Place de l'Étoile (he has no use for jet lags). Time went by smoothly—nothing beats Paris in the spring time—and I even got a nice tan on the terrace of the Brasserie de l'Empereur; but—no Burian! In the evening I was finally able to reach him at his friend Jean Gayon's place; it turned out he had been waiting patiently for me the whole afternoon at another terrace. Paris has so many! We finally made it.

tion and reductionistic arguments in ways that very few disciplines do. Yet there are relatively few philosophers that appreciate the depth of the issues it raises in that respect.

Callebaut: Immunology?

Burian: Ken Schaffner (1992) is about the only one I know of—there may be more—to have explored that . . .

Hull: . . . there is also Darden and Cain's (1989) discussion of immune systems . . .

Burian: . . . but immunology is, again, of great interest for the study of levels of organization . . .

Callebaut: . . . and of cognition . . .

Burian: . . . That's right, because it's self/non-self recognition. It necessarily involves at least molecules, cells, tissues and organs, and probably *system organization*—something the role of which is hard to understand when it is set into the middle of all of those concrete entities and processes at various levels on the above list. There is a huge number of things do be done.

Callebaut: Is there something you can do about the biases of philosophers as a journal editor, Michael?

Ruse: Again, I think that if certain things do come up, we'll respond to those. Where the thing will be in ten years time I don't know. Often movements and ideas have a life of their own; then they get formalized, even though perhaps they go on. But, you know, then they are like Christianity, perhaps: frankly, *its head is cut off although the body is still running around.* At the moment, philosophy of biology is an exciting area. I'm lucky. I'm sure it will be much more *established* in ten or twenty years. Whether it will be as much fun is another matter. If I were a businessman, what I would want to do is start my own firm and have fun doing it rather than go and work for General Motors. You've got to have General Motors; you can probably make a lot more money at it. But there are some people like myself who are much better as entrepreneurs! Socially, I hate laissez faire, but intellectually I love it!

10.6.2. *Progress after All?*

Callebaut: NIH director Bernadine Healy has interesting things to say about General Motors. You have written a big book on the concept of progress in evolutionary biology (Ruse 1991a). According to G. C. Williams (1966:34), "there is nothing in the basic structure of the theory of natural selection that would suggest the idea of any kind of cumulative progress." That seems to be the current majority viewpoint.

Ruse: The book takes on the whole question of values in science, and particularly the importance of nonepistemic values in evolutionary biology.[15] I certainly believe

---

15. At that level, Ruse thinks of himself as "very empathetic, if you like, to the strong sociologists of science."

that the empirical "facts" are important and that there can be genuine advance in science—although, to be frank, I tend to be somewhat of a pragmatist, a nonrealist of a kind. I think advance means getting one's theory more in tune with epistemic values like consilience[16] than progress towards knowledge of a metaphysical reality (1986a). But in addition, noncognitive or non-whatever-you-want-to-call-it values are very important in what drives science. Evolutionary theory in particular has in its history been one of grappling with the notion of progress. It is no exaggeration to say that the science is a mere epiphenomenon of human hopes of social progress. I don't mean that every evolutionist working on every problem has been thinking about it, but it *permeates* the theory. To a certain extent it's gone underground a bit, because in certain circles, progress isn't very fashionable. But—it's there! I think my book is going to make people, including scientists, really mad and want to respond. I think my book is going to make scientists rethink their thinking about evolutionary theory a lot, and they are going to see the extent to which values *do* permeate the theory that they've got.

Callebaut: Julian Huxley (1974:378) wrote that "after the disillusionment of the early twentieth century it has become fashionable to deny the existence of progress and to brand the idea of it as a human illusion, as it was fashionable in the optimism of the twentieth century to proclaim not only its existence but its inevitability" . . .

Ruse: . . . But it's still there to a great extent. We've got a sort of *anthropic principle* in evolutionary biology: we're part of the process ourselves, and so we cannot be entirely disinterested observers. Not only are we part of the process; we necessarily have to be the endpoint of the process. And we have to have abilities to ask questions like, "Is there progress in evolutionary biology?" All of those things "distort" or "color" the way that we think about evolution. And I think this gets into the science and makes people think in progressive or quasi-progressive sorts of ways.

10.6.3. *Evolutionary Epistemology: Start as Early as Possible*

Callebaut: Don, which future do you predict for evolutionary epistemology?

Campbell: Of course, being reflexive, I say, "This stem will continue to explore beyond what we can now anticipate!"

Callebaut: Quite a few evolutionary epistemologists are playing the same tune over and over again; and only a couple of them, I am afraid, are self-aware of this.

Campbell: It is probably true that those who have been in the field too long are repeating themselves too much and that it will get fresher views from people who are unaware of the literature (Cziko and Campbell 1990). There is a new popular book on the science of scientific discovery by someone who was originally a microbiologist/physiologist, Robert Scott Root-Bernstein (1989). You see in it the

---

16. The noncontradictory conjunction of empirical hypotheses on a combined data base.

enthusiasm and fanaticism of somebody who hits upon selection theory and really understands it. He has one half-hearted citation to me, but it is so wrong that it is clear that he hasn't read me and that it was added later. Gary Cziko is in education; he originally got this enthusiasm for selection theory from biochemical examples from friends (Cziko 1989a; cf. 1989b), and then hit upon my work. I would like to put in a plug for Ruth Millikan's (1984) book, which seems to me to do what so much needs to be done: *taking the evolutionary perspective and addressing a wide variety of problems in analytic philosophy that we haven't yet addressed.* Millikan is totally unaware of the literature on evolutionary epistemology; yet she does an excellent job with her emphasis on linguistics.

Callebaut: In your most recent work, you have attempted to reconcile evolutionary epistemology with a rather traditional sort of epistemology (Campbell and Paller 1989; Paller and Campbell 1990).

Campbell: Modern epistemologies from G. E. Moore to Gilbert Harman's "inference to the best explanation"[17] beg the question, assume inductive achievements; they start tackling the problem of knowledge by assuming tremendous bodies of knowledge, none of which is foundationally established. Evolutionary epistemologies are just one particularly focused way to get on with this epistemological agenda; they insist on starting as early as possible rather than with current commonsense knowledge.

Callebaut: So you think evolutionary epistemology can be improved . . .

Campbell: . . . by trying to translate all the epistemological issues into it. That would change evolutionary epistemology as well as probably minimize our differences from mainstream epistemology. But—if we could anticipate all of the future uses of evolutionary epistemology, we would be violating its very perspective!

10.6.4. *What about Technology?*

Callebaut: If my count is right, Karin, you have used the word "technology" only once during this whole interview. Do you think of technology and science as essentially different?

Knorr Cetina: No, I wouldn't want to make that distinction. Some of the general constructivist perspectives have been applied to technology (Bijker, Hughes, and Pinch 1987; Bijker and Law 1992). In my case, it's simply that—on the face of it—I haven't studied technologies. On the other hand, take particle physics, which is an eminently technological science. Most of what we see there is technology: it's technology being produced, or being handled, it's technology from which things are extracted—it's technology all over the place. There seems to be nothing which is not technology, because even when you look at so-called "physics analysis," computer programming and simulation is a heavy part of it. Even in sciences where you don't find the so-called "hardware technology," like in molecular ge-

17. Harman (1965); additional references in van Fraassen (1980:217n.9).

netics, where there's very little of it—centrifuges and autoradiograph apparatuses are the major instruments they use—there is a different kind of technology at work: animals and living processes are used as technical instruments. And as I said, social processes are used as technical instruments. So my notion of "technology" is not linked to hardware, but is linked to lots of other things—to the function something performs, or the effects something has in a process. Therefore I don't see the distinction as a hard one or even as a useful one.

10.6.5. *Centripetal/Centrifugal Preoccupations*

Callebaut: The relations between philosophy, history, and social studies of science reflect each of these disciplines' "centripetal preoccupations with its own interests, its own identity" (Manier 1980:1). Latour, for instance, has advised sociologists of science to "do their job, and enrich their own field" instead of answering philosophical questions. It seems to me that the ambition and ardor of sociologists of science in the battle for the largest piece of the science studies cake contrasts somewhat unfortunately with the more polite and tolerant attitude of at least some philosophers and historians. I take it that this reflects the circumstance that sociology is the youngest among those fields. You are a sociologist of science, yet you actually do talk to philosophers. How come you are different from many others of your ilk?

Knorr Cetina: It is the first time in my professional career that I see that happen: *the emergence of an incipient interdisciplinarity;* that philosophers, sociologists, and to a certain degree, historians of science are not only talking past each other, but that they talk to each other and start doing each other's research with the tools from their own disciplines, like when Ron Giere goes into a laboratory at his place and he is actually studying physicists at work as a philosopher, or when other philosophers, like Tom Nickles, are studying the context of discovery while keeping in mind sociological studies of knowledge production. I find that remarkable. And when I study a laboratory, I am actually looking at the philosophical literature and I try to make sense of philosophers' conclusions in terms of scientific practice as I see it at work. So there is a certain convergence, the emergence of an interdisciplinarity which rarely happens anywhere, although so many people call for it and try to implement it by political planning or things like that. Here it evolves and emerges out of a joint interest and some convergence of research on the same topic in three disciplines. That is an exciting thing about the whole development!

Callebaut: Bruno, since we talked in Paris nine months ago, your judgment about philosophers of science has become somewhat milder. Or is geographical location (we are in San Diego now) rather than time the relevant dimension here?

Latour: I now think there is room for a philosophy of science, contrary to what I said then (7.2.7). What made me change my mind was my interaction with a philosopher in the program in history and social studies of science here in San Diego. One role for philosophers, in softer sciences like evolutionary biology, is interaction with the scientists. One good sign of the usefulness of philosophy of science is the creation of a new hybrid role between theoretical biologists and philosophers of biology.

Callebaut: That's how several of the people in this book actually define their personal variety of naturalism.

Latour: This might not solve many of the philosophical questions, but at least it is an interesting way of doing science studies . . .

Callebaut: . . . Why? . . .

Latour: . . . because *intervening in the debate* is an important part of science studies. So this sort of philosophy of science is respectable. Same thing for the sort of work that Lisa Lloyd does, which is to try to *gut* the internal structure of the arguments to see how far they will carry. *If it doesn't neglect the connection with the rest of the network,* I take it as a very useful task, which surely is not tackled very seriously by most sociologists. So here is a form of collaboration which makes me less inclined to say that philosophers of science should disappear!

10.6.6. *The Deconstructible Subject*

Callebaut: As we have seen in previous chapters, you have made problematical the notion of "object," and you make rather wild suggestions in that respect (cf. Dagognet 1989). But what about the subject—isn't that concept equally problematic?

Latour: I'm less interested in that.

Callebaut: Fair enough. But is it obvious that one can talk about "human actors" without further ado, the way you tend to do?

Latour: No, I agree with you. That is clearly a *péché de jeunesse.* The project is difficult. The whole attribution of meaning to human actors is, of course, problematic. We have less difficulty with that in France than the English-speaking sociologists do, because we have been fed on Foucault and all these other people. *We take the deconstruction of the subject for granted.* I mean, we French, when we are five years old, we know that Man is dead, and God, etc. We don't have to learn that. "Actors with a strategy"—to us that's almost a game. For us it's tongue-in-cheek—we know that can be deconstructed immediately.

Callebaut: Classical sociology takes the actor to be more solid.

Latour: Especially in England, the human actor is supposed not to be deconstructible. We didn't pay enough attention—that's a mistake we made—to *that* "translation problem." The whole deconstruction-of-man business is not accepted as generally abroad as it is in France. In France one doesn't need to do that; someone who were to do it would be ridiculous. Right now there is a reaction against this; everybody is talking about *les droits de l'homme.* It's a completely reactionary scene in France: now people want to go back, abandon Foucault and all that. But for my generation this was done. Now what Foucault never did was to deconstruct the objects, and that's why our priority was there. But I agree with you that in the long run we ought to pay more attention to the subject as well. I just finished a book, *Aramis, ou l'amour des techniques* (1992b), which does that, because it is about subway systems and the study of technology and in that sense it's also a novel: the construction of an object/subject, which is a subway, a thing.

10.6.7. *Delegating Actions to Nonhumans*

Callebaut: I look forward to reading it (I just *love* everything on rails!)—but seriously?

Latour: What we did in the social studies of science, all things considered, is to reposition the notion of the actor. I would call "actor" the shifter, the redistributor, *the delegator of actions either to humans or to nonhumans.* In technology studies you can't start from a list of what humans are able to do as contrasted to what "mere things" will never be able to do, because the job of the engineer is to cross this boundary constantly and to reallocate skills and competences among "actants."

Callebaut: That sounds reasonable, but still freakish—doesn't it, Dick?

Burian: I do not take easily to treating the entities and processes that science "discovers" as agents, because there are so many overtones that are not met by those entities—intentionality and so on . . .

Callebaut: . . . Here, for once—*une fois n'est pas coutume*—David Hull could back Bruno; according to David, intentionality shouldn't keep us awake (cf. 7.2.4) . . .

Burian: Now there is more than I had recognized in these things and I need to study before I reach a clear judgment about them. But there are also different claims, and interestingly different ones: the claim that *the instruments and instrumentalities that scientists use play a more agent-like role,* and the claim that *experiments ought to be a proper object of study in science studies.* There is no question that both these latter claims are correct. Thus, there is something to be drawn from the analogy with agents, but it is less than Latour thinks it is; I am not sure how much of an insight it will yield.

Callebaut: Back to *Aramis.*

Latour: Now I will have more to say on the symmetric construction of humans and nonhumans. But it is true that we were never careful with the strategy business and human actors, because we never realized that it was that hard to deconstruct. That's not an excuse; it's not a philosophical argument; it's just a mistake. But you've got to understand that in France, the object is more a priority for us, because there the social sciences are completely "de-objectized," mainly because of Bourdieu.

Knorr Cetina: Latour wasn't always anti-Bourdieu. He had an article in which he really applied Bourdieu to science studies, trying to extract the "symbolic capitalism" from science (Latour and Woolgar 1982). Latour may consider himself to be anti-Bourdieu, but I don't think he really is! Clearly, his actor/network theory no longer uses Bourdieu's notions, but I don't see a great divide between the two.

Callebaut: The importance of the object has become so overriding in his recent work . . .

Knorr Cetina: . . . Yes; in that respect, Latour follows Baudrillard rather than Bourdieu. On the other hand, Bourdieu has never definitively excluded the object; he has made no point on that. He has studied objects a lot in his cultural investigations, for example. He has not included objects into his theories, but neither have many other people.

Latour: It's surprising that Karin says this, because the very beginning of what for Bourdieu is the scientificity of the science of sociology is *not to consider the content*. If you want to study sociologically, say, painting or pedagogy, the first thing to do is not to consider the content at all. Anti-Bourdieu? Personally, of course, I am not. But I don't think his approach is very useful. It's true that I wrote an article on this, and that it's not a very good one . . .

Knorr Cetina: It is correct that Bourdieu never considered the content of science in his sociology of science (e.g., Bourdieu 1975). But Bourdieu as a *cultural* sociologist has analyzed objects in interesting ways, and he brought the human body back into the focus of interest of a sociology that defined humans primarily as "subjectivities" (intentions, consciousness, mentalities). In that sense I would not consider Bourdieu as a traditionalist to be written off, even though I agree with Bruno that Bourdieu's sociology of science is orthogonal to, and beside the point of, the most interesting developments in recent sociology of science.

10.6.8. *Alien Science?*

Callebaut: Coming to the end of our journey through science studies, we are in for an even more freakish subject (object?): Extraterrestrial Intelligent Life. E. O. Wilson (1979:52–53)—a serious man—has written that "we should rescue the contemplation of other civilizations from science fiction. Real science tries to characterize not just the real world but all possible worlds. It identifies them within the much vaster space of all conceivable worlds studied by philosophers and mathematicians." I would like to extrapolate this to the metalevel: For philosophy of science to become truly naturalized, that is, a real science itself, how general would it have to be (cf. also Rorty 1988)? For those among us who want to remain some kind of empiricists, the question, "Could aliens have science?" (cf. Regis 1985) may not be irrelevant to that. Dudley and Helen, I am of course not expecting real *answers* from you on this score, but still you could try.

Shapere: How do we know that somebody living on a different star, on a planet somewhere else, wouldn't have developed a science, through processes of self-correction etc., which would be entirely different from our own? The view that I develop leaves that question totally open. *We don't know*. This is a part of the question of—we can't tell whether there is the one reality, because we don't understand enough, yet, about what this stuff we call reality is; that term doesn't even seem to fit it. It might be, and it might not be that there is some other civilization with such a different system. But we can't know that until we find out. In the absence of finding out that is one of those universal statements like "I may be dreaming," "I may be a brain in a vat," "There may be an entirely different possibility"—and that is no reason for doubting the views that we have developed. It is a mere possibility; though it is a possibility—but that cuts very little ice with regard to what we are going to accept and build on.

Callebaut: With the quote from Wilson and with Rosenberg's (1980) idea of "preemption of social science" by sociobiology in the back of my mind, let me rephrase my previous question as follows: would it take our encountering aliens to transcend the present contingency of the subject matter of science studies? (Are there things more approaching natural kinds *out there*?)

Longino: It's interesting, and I haven't thought about it in terms of contingency. But my own metaphysical view—which is sort of half-baked and not worked out—is that *all is contingency;* that for all we know, Nature is completely disorderly, and there just happens to be a match between our own needs for a certain amount of order in order to act at all and what occurs. I think the idea that what science is about is characterizing all possible worlds and that the laws are sort of necessary truths characterizing all possible worlds (*tries to suppress laughter*) is kind of a hopelessly arrogant ideal! I mean, it's not even an ideal, it's just hopelessly arrogant and completely blind to our own cognitive limitations as thinking, reasoning animals. *If that is what science is about, we can't do science.* I think what science is about is trying to characterize what order we can in order to accomplish certain social tasks. And that we have characterized order in terms of laws may itself be a historically contingent fact. This is something that for example Evelyn Fox Keller (1985) has talked about. She points out the theological origin of the notion of law. When laws of nature were first introduced as a subject of study, it was as the laws prescribed by God. It's post-Newton that we have detached the laws of nature from the laws of nature prescribed by God to the universe. But we still kept the concept of law and kept the notion of necessity inherent in that concept as our characterization of nature. But this may be just a historical accident. The encounter between earthlings and extraterrestrial scientists could raise critical discourse in science to new levels!

Callebaut: Michael, who does not mind being called "Wilson's lapdog" (Ruse 1990), just loves aliens.

Ruse: I've always been a bit sorry that today's philosophers seem to be so completely uninterested in extraterrestrials. I suppose, given the popularity of movies like *Star Wars,* they think any such interest would be a bit vulgar and beneath their dignity. But it wasn't always so. Aristotle, Kant, and my favorite philosopher of science, Whewell—to name but three—wrote on the "plurality of worlds" problem, as it used to be known. Some years ago, I suggested to Peter Machamer, who was then program chair, that we should have a PSA session on the topic, but my idea sank like a lead balloon. Yet there are all sorts of juicy problems, starting with the possibility of intellectual life elsewhere in the universe. My experience is that physicists believe in it and biologists don't.

Callebaut: A nice illustration of this contrast may be found in the *Omni* interviews with Cyril Ponnamperuma (a chemist, that is, almost a physicist), Francis Crick (who was originally trained as a physicist), and Ernst Mayr (Ponnamperuma 1984; Crick 1984a; Mayr 1984, 1985). Ponnamperuma uses a statistical argument (only a very small proportion of stars may have conditions around them that are suitable for life, but since there are 10^{18} stars, etc.) and invokes laboratory simulations of chemical evolution in the universe to claim that "the chances for life seem very great" and that "there must be intelligent life somewhere." Crick is playing with the idea of the extraterrestrial origin of life on Earth ("sperm from deep space"). Mayr, however, thinks the chances of the existence of extraterrestrial life are nil and refuses to bother with outer-space theories about the origin of life on Earth: "It's always some physicist who comes up with these totally nonsensical theories about biology."

Ruse: Partly this is a matter of self-interest. If such life is reasonable, then there is a good reason for heavy support of space probes. Partly this is a matter of one's understanding of the evolutionary process. If one thinks that evolution is essentially nonprogressive, then one will doubt the possibility of extraterrestrial intelligence.

Callebaut: I'm convinced that you also must have quite other reasons to be interested in this whole aliens business.

Ruse: Even if you don't believe in extraterrestrial intelligence—and I'm dubious—then if you're a naturalist—and I am—it's a great exercise to speculate on what kind of science and morality such beings might have. As Sir Ronald Fisher used to say, if you think about what it would be like to have three sexes, then you have better understanding of our existing two sexes. As an evolutionary epistemologist who's pointed towards a kind of pragmatism, a belief in a coherence theory of truth and a nonmetaphysical realism, I doubt very much that such beings would have anything recognizable as human science. Obviously, if they were blind and deaf and communicated through pheromones, they wouldn't have optics or acoustics. If they were water-snakes, they wouldn't care about Galileo's laws as much, but would care about hydrodynamics. But would they think that Newton's laws are true or that $5 + 7 = 12$? A Kantian would say yes. As a Darwinian evolutionist, I really don't see why their evolution should have taken them to *our* endpoint. They wouldn't believe that $5 + 7 = 12$. That's our mathematics, and wrong! But they might reason in an entirely different way.

Callebaut: What about alien morality?

Ruse: Alien morality is even more interesting, at least it is to me. I once wrote a paper called "Is rape wrong on Andromeda?" (Ruse 1985). I upset feminists and Christians by concluding "Not necessarily." Obviously, I'm not really recommending rape. I was pointing out how human morality is contingent on human evolved nature. If human females came into heat, our sexual morality would be quite different. More generally, I suspect that if extraterrestrials were social—and this might be an important prerequisite in evolving intelligence—then they would probably require some sort of reciprocation. But this wouldn't necessarily demand sensitivity to the Love Commandment or the Categorical Imperative. They might interact through a shared feeling of hostility, but as in the Cold War recognize that self-interest demands cooperation. Incidentally, all of these ideas are explored brilliantly by J. B. S. Haldane in his essay "Possible worlds" (1927, in Haldane 1930). It's funny how fashions change. Ideologically I suspect Haldane is quite OK for most of today's critics of human sociobiology, and yet in respects *he out-Wilson's Wilson.* That just confirms my suspicion that much of the sociobiology controversy was more personal than substantive. I know I'm not a fascist, right-wing ideologue, and because Ed Wilson and Dick Lewontin don't get on, I don't see why I should be labelled one. Although, I've had to cross chanting left-wing picket lines on occasion![18]

10.6.9. *Nothing Is Done Yet*

Callebaut: Would you venture to say something in general about the future of science studies, Karin?

Knorr Cetina: I think it is more a problem for philosophers than it is for sociologists at this point, because I believe sociologists or social scientists are on the right track! They just have to continue on that track. Whereas in the case of philosophers it's a much more open question what they should do, given that the normative and rationalist approach seems not very valuable and generates all kinds of problems, and given that the philosophical approach in general has been attacked by people like Rorty and others.

Callebaut: Something that may change, as far as social studies of science are concerned, I think, is the emphasis on debunking Science (cf. Latour on Canguilhem in 2.6.7).

Knorr Cetina: OK, but the debunking aspect was always a reception thing. It was never *intended* as debunking. When you go into the laboratory, you have an interest in the mechanisms through which knowledge is constituted. You don't have an

---

18. Looking back to the first decade of human sociobiology, the anthropologist Mildred Dickemann (1985:42) wrote: "In 10 or 20 years, the first decade of human sociobiology may well appear to have been just another, if slightly more agitated, part of the rebirth of the human sciences, as sciences in more than name." Ruse firmly believes that human sociobiology is moving forward today: "The interesting thing is that although Ed Wilson has done a lot to move it forward, I don't think his approach is where the action is. I've never been, for instance, particularly enthused by Lumsden and Wilson's work. I like the idea of epigenetic rules, but of course a lot of people have had that idea of innate principles. Their models—I leave it to the people who are more technical than I to criticize them. But somehow, it seems to me there was something not particularly clean-cut about the whole thing. I was always worried about it even on esthetic grounds. I've always felt that the way human sociobiology is going to move forwards is if the anthropologists work on it. And now people are. In fact in my recent book of essays, *The Darwinian Paradigm* (1989a), I've got one where I talk about some of the work which is being done by people like Monique Borgerhoff-Mulder (1988)—a student of Bill Lyons—and others. There is a new society just starting on this: the Human Behavior and Evolution Society."

interest in debunking. The kind of things that would really be detrimental to the scientists we have never published. For instance (*hesitating*), the handling of radioactivity is sometimes almost criminal. It certainly has forced us to stand behind their back and be protected by their body, because we found the risk associated with the handling of radioactivity unbearable. We've never published these things, because that would be hurting scientists, and it's not interesting from an epistemological point of view.

Callebaut: I was also thinking of the debunking of traditional philosophical views. At some point, that will have to come to an end; you can't go on repeating yourself indefinitely.

Knorr Cetina: Sure. We used to have a reference to philosophers' point of view, certainly in the introduction if not all over the place. Now it's no longer necessary to refer to them.

Latour: I'm worried that because of the weakness of philosophers and because of the sort of wishy-washy way in which philosophers of science handle the field studies, people in the social studies of science could believe that their work is finished. That worries me, because nothing is done yet. I'm very worried when I see them acting—as you mentioned in one of your questions—as if the paradigm and the background of science studies were already established. It's not because many philosophers of science are so weak that the work is done. The work just begins! In particular when you start to move the position of the objects—which is correlated to questions like, "What is the collective?" "What about description?" "What about method?"—the task is *enormous*! The tradition to do that is not there. Or it was there, but before Kant. So a huge amount of work in anthropology, history of philosophy . . . awaits us. It requests the best minds; *it should attract the best Ph.D. people now, the sort of people who, before, would have gone into the natural sciences.* I think they will come if we make science studies exciting enough. What will happen to the field? It might get stuck on the way, because of this hang-up on society, on the social. Collins's "social studies of science" track is a dead end if we do not redefine the social.

Knorr Cetina: What I am concerned about is that science studies itself produces accounts which seem to reify certain concepts. One problem I have with Bruno is that he reifies the concept of a "network" and attempts to explain everything in terms of linkages. That is either a tautology—you *define* everything in terms of linkages—in which case it is not very informative; or if you don't, you are not giving enough credit to the ontological flexibility of social reality, which might use networks for some purposes but not for others.

Callebaut: I guest that as an "irreductionist," Bruno should agree with you (cf. 5.4.2). My impression is that philosophy of science is reaching a dead end—no *really* big things are happening right now—and that most of the work in history of science remains quite traditional. And now you sociologists come and tell me that there's a danger in your shop too . . .

Latour: . . . That's because the postmodernism era ends, which is sad—not to me (1988d, 1990a, 1991, 1992a); people are disappointed. But if you think of the flip side of this, everything becomes interesting again! Philosophy, history, sociology, economics—all these disciplines are on the move again. No, I really think it's the end of the bad part of the social sciences. But one thing is clear to me: Not all of the disciplines will make it, and if philosophers of science do not seriously go back to history and to philosophy, they will be dead (*laughs*).

Callebaut: Ron and Ryan, would you venture to say something in general about the future of science studies?

Giere: My own view is that the cognitive sciences are going to provide a kind of inspiration point and a resource that could do now for current studies of science what logic did for logical empiricism: it provides a kind of resource, a tool to use to study science. But, you know, there are people who say this is a flash in the pan.

Tweney: I hate to predict, but I'll tell you what I *wish* happens: I wish the cognitive science of science will emerge with the kind of identity that the sociology, philosophy, and history of science have. I think that's necessary for the continued health of science studies in general, which has for too long ignored this side of it. In some circles it's getting very fashionable to talk about "cognitive approaches" in the absence of very much *concrete* knowledge of cognitive science, and that's unfortunate. It needs to be much more specific. There are many, many research programs that we need to begin working on. We need to spend less time on metatheory and more time on actually doing the work! We have talked about the metatheory for decades. We should get on with it!

Callebaut: If you allow me this comparison: sociology—general sociology—has always spent much of its time discussing its philosophy.

Tweney: Sure. Actually, they'll never give that up entirely. Sometimes I get the feeling that when we do actual work, when we produce concrete information, we're doing it merely as a *demonstration*. And the thing we really want to do is prove at a metalevel that this approach works. In some respects that applies to my work on Faraday. But that cannot be all we do! We have lots of approaches that work; let's use them! We can still argue about which is the best one, and where to apply this one and where to apply that one. Those discussions will never cease. But in the long run, the work that we do will remain long after people stop caring about the meta-arguments. They will still be looking at Rudwick's (1985) book. The problem, to use David Gooding's term, is that we degenerate into talk about talk about talk, that we spend so much time talking about each other's talk that we never bother talking about what's real.

Callebaut: I agree wholeheartedly.

POSTSCRIPT

WHEN I PONDER the outcome of my peregrination through science studies as it materialized in this book, several points crystallize which command some final comments.

1. *"Where are the arguments?"* Looking back on my role as go between among the naturalists, I cannot help comparing myself to Ronald Giere, who used his lab study to "overwhelm" the nonrealist with "data" he considers grist to the realist's mill. Most scientists, humanists, and science-policy makers, many philosophers working in different areas or traditions of philosophy, and a host of other people with an interest in matters discussed in this book (including quite a few empirical investigators of science who should know better) are simply not aware of the naturalistic stance in philosophy of science, its meaning and implications. The primary aim of this book has been to confront them with evidence—lots of evidence—concerning the actual *existence* of naturalistic philosophy of science. But if naturalism exists, it must be *possible,* I should think. We have sidestepped a transcendental question (Bhaskar 1979), then, in favor of an empirical "answer." We—most of the contributors, including the editor of the series in which this book is published, and I—want this book to function first and foremost as an eye opener. Now I am fully aware that this is not the only game in town. More specifically, I anticipate that the long empirical argument which I consider this book to be will barely impress epistemologists who, with Peter Geach, are convinced that "when we hear of some new attempt to explain reasoning or language or choice naturalistically, we ought to react as if we were told someone had squared the circle or proved $\sqrt{2}$ to be rational: only the mildest curiosity is in order—how well has the fallacy been concealed?" (quoted in Flew 1987:401). When, in times of crisis, it is no longer possible to run one's business as usual, one can still pretend—for a while. Replacing "categorical" by "instrumental" rationality is giving short shrift to a central part of more than two thousand years of Western intellectual culture; it is not likely to succeed soon, if ever (see, e.g., Flew 1987, Putnam 1990, or Engel 1992 for illustrations of this "inertia"). Likewise, one should not expect the deeply rooted resistance to the naturalization of man and society (see, e.g., Sanmartín, Simón, and García-Merita 1986) to disappear overnight.

2. *"Know thine enemy!"* When I started working on the project that resulted in this book, it was still a matter of debate among the "observers of the field" (who always observe from *somewhere*) whether "naturalization" was the right cue for grasping the significance of the transformations in post-Kuhnian philosophy of science. Today there is a quasi consensus that the converse of naturalistic approaches—foundational and justificatory types of philosophy—are floundering, whether one revels in this (e.g., Rorty 1979; Putnam 1990; Munévar 1991) or is irritated by it (e.g., Kekes 1983; Flew 1987; Schmaus, Segerstråle, and Jesseph 1992). "Softer" philosophical disciplines like aesthetics, ethics, or social philosophy, and deconstructionist approaches ("Nietzsche and Heidegger," as the one-liner goes) have come to the fore again. Rather than relegating epistemology, philosophy of science, and semantics to the background, they are amplifying and radicalizing the pragmatist and relativist tendencies that have attacked the very heart of these "harder" philosophical disciplines since "Kuhn." Hermeneuticists, holists, and neopragmatists (by no means *all* the conspirators were Continentals!) have joined forces in a "great coalition" in which French poststructuralists also have a say (cf. Forum für Philosophie Bad Homburg 1987:7–8). But what does all this tell us, if anything, about naturalism's health? The issue of the relationship between naturalism and these various other antifoundational tendencies has only been scratched at the surface.[1] Great coalitions are notoriously unstable, and it is tempting and actually quite easy to spot the fissures that will cause this monster coalition to implode soon. (Its unity is superficial, and disagreements have been sitting there from the beginning, waiting to be ventilated.) More to the point in our context is how the "Kant is dead" message of the deconstructionists,[2] voiced most clearly by Latour in this volume, screens off the very real differences between the older analytic tradition in philosophy (apriorist, hence "neo-Kantian" in a sense; cf. Kitcher) and the new, "neo-Humean" naturalism (cf. Ruse) which grew out of it. The French philosopher of mind, Pascal Engel, for one, while stressing the great value of what "the analytic dream" and "the naturalist awakening" have in common—a certain attitude toward and style of philosophizing reminiscent of Bechtel's "breath of fresh air of concrete, articulated arguments"[3]—also points to open problems of naturalistic philosophy in contradistinction to the analytic tradition. This takes us to my next point.

1. The suggestion that ethics can be "scientific" (Campbell at 10.1.3), to take just one example, must be horrifying to all those waving "Let us weaken science!" banners.

2. A telling illustration of this "post-critical" attitude is provided by Richard Rorty, who, in a newspaper interview (but is not the mass media where the postmodernism debate belongs?) conjures intellectuals on the left to stop being critical, since "nobody these days proposes any better than the market economy" (Uzan 1992; my translation). Unemployed from all over the world: relish this!

3. Viz., "être clair et précis dans ses arguments, écrire de manière à susciter des objections et des réponses, réviser ses théories quand on . . . montre qu'elles sont inadéquates ou partielles, et faire de la philosophie professionnellement, en se soumettant aux règles d'objectivité et d'intersubjectivité d'une communauté de chercheurs partageant les mêmes idéaux intellectuels et les mêmes critères" (Engel 1992:104). Engel reports that French philosophers of the current [his] generation who discovered these "trivial" prerequisites of philosophical work, regarded them as "new and extraordinary virtues."

3. *Conceptual analysis versus theory building.* It was suggested in chapters 2 and 3 that the demise of the analytic/synthetic distinction and the concomitant return of psychologism and sociologism made the analytic philosophers' cherished method of conceptual analysis problematic *to the extent that it was aprioristic.* But has not the naturalization of philosophy made conceptual analysis as such obsolete when the naturalist's conceptual reconstruction differs from the theoretical construction encountered in science—i.e., presumably, in the overwhelming majority of cases? The various statements concerning the relationship between "philosophy" and "theory" in this book tend to take this identity for granted, but pronunciamento is not proof. Engel (1992:109) suggests one way out. He tries to drive a wedge between the ontological thesis of naturalism ("there are only natural facts or entities") and its explanatory thesis ("there are no other explanations than those encountered in the sciences"). The former thesis he endorses; the latter he rejects, as he refuses to extend the reductive-explanatory ambition ("explaining away"), which he takes to characterize science in general, to the naturalistic reconstruction of science itself.[4] But Engel's solution hinges on his identification of naturalism with physicalist and materialist eliminativism and of naturalistic with causal explanation, which, as was argued before (chapter 1 and section 4.2, respectively), is unwarranted in general.

4. *"R-e-s-p-e-c-t"* (Otis Redding/Aretha Franklin). Like many other (participant) observers, I am struck by the level of animosity, and, at times, downright hostility within the science studies debate. In my optimistic moods I like to view this as a healthy state of affairs, not too different from the liveliness that characterizes any genuine (noncomplacent) intellectual discussion. At other times the strength of the centrifugal forces operating in the field worries me.[5] Not unlike the Flemish Movement in Belgium, whose leaders continue to fight all sorts of "discrimination of the Flemish" which no longer exist, a sort of *Eigendynamik* of the SSK branch of the sociology of science still has it fretting about being "beaten" by philosophers' or historians' sticks, long after it has lost its minority position.[6] Considering the current disarray in the history and philosophy of science, for which SSK is largely responsible,[7] this defensive attitude seems to me to be verging on paranoia.

---

4. One reason why he thinks that the project of conceptual reconstruction of the naturalist differs in kind from ordinary theory construction in science is that notions like "intentionality" or "justification," which naturalism wants to render in "scientifically acceptable" terms, "still need to be *interpreted*" (Engel 1992:111).

5. One rough-and-ready way to estimate the degree of intellectual and social cohesion (and, conversely, noncommunication) in the field is to plot interaction matrices displaying who refers to whom. Even for the group of people represented in this book, which one would expect to be somewhat more coherent than arbitrary different samples (cf. 6.1), one gets subclusters and a large amount of Simonian "near-decomposability."

6. See, e.g., David Bloor's review of Peter Galison (1987), in which he expresses his fear that "in the present climate of opinion, the charges that the author makes against sociology are all too likely to 'stand up in court'" (Bloor 1991:189).

7. This was made very clear, for instance, in the session on "What Has the History of Science to Say to the Philosophy of Science" at the 1992 PSA meeting in Chicago.

5. *A conspicuous absentee: The dismal science of science.* Finally, this book re-produces a bias that has puzzled me ever since I was the co-organizer of a conference on theory of knowledge and science policy (Callebaut et al. 1979): the absence from "science studies," as the field took shape in the last two decades, of a full-fledged economics of science. Apart from the odd interloper one meets at any scientific con-ference and the occasional review of a book produced in the other field, science studies and the economics of R&D (or, more broadly, the economics of innovation) make no contact, despite their obvious mutual relevance. As soon as philosophers grant the impossibility of free-floating ideas and the necessity of "embodiment," the knowledge they study becomes a commodity, that is, an object with economic properties (see, e.g., Fuller 1991a). Conversely, more and more economists who model science and technology (and even some who are interested in production, distribution, and con-sumption processes more generally) feel the need to look inside the black box econom-ics has traditionally taken "knowledge" to be. One candidate explanation for the lack of institutional communication between science studies and the economics of innova-tion is the highly normative character of the latter (cf. Fuller 1988; Rosenberg 1992a). Rumor has it that economics needs naturalizing too. It takes less than Cassandrian powers to predict that this next big battle will be at least as fierce as the one docu-mented in this book.

REFERENCES

Abir-Am, P. 1982a. The discourse of physical power and biological knowledge in the 1930s: A reappraisal of the Rockefeller Foundation's "policy" in molecular biology. *Soc. Stud. Sci.* 12:341–82.

———. 1982b. Essay review: How scientists view their heroes: Some remarks on the mechanism of myth construction. *J. Hist. Biol.* 15:281–315.

Achinstein, P. 1983. *The Nature of Explanation.* Oxford: Oxford University Press.

———. 1985. The pragmatic nature of explanation. In *PSA 1984* 2:275–92.

Achinstein, P., and S. Barker, eds. 1969. *The Legacy of Logical Positivism.* Baltimore: Johns Hopkins Press.

Ackermann, R. J. 1970. *The Philosophy of Science.* New York: Pegasus.

———. 1985. *Data, Instruments, and Theory: A Dialectical Approach to Understanding Science.* Princeton: Princeton University Press.

———. 1986. Consensus and dissensus in science. In *PSA 1986* 2:99–105.

Adams, E. M. 1960. *Ethical Naturalism and the Modern World-View.* Chapel Hill: University of North Carolina Press. Repr. Westport: Greenwood Press, 1973.

Adams, E. W. 1959. The foundations of rigid body mechanics and the derivation of its laws from those of particle mechanics. In L. Henkin, P. Suppes, and A. Tarski, eds., *The Axiomatic Method,* 250–65. Amsterdam: North-Holland.

Agazzi, E., ed. 1988. *Probability in the Sciences.* Dordrecht: Kluwer.

Alchian, A. 1950. Uncertainty, evolution, and economic theory. *J. Pol. Econ.* 57:211–22. Repr. in Alchian 1977.

———. 1977. *Economic Forces at Work.* Indianapolis: Westview Press.

Alexander, J. C., B. Giesen, R. Münch, and N. J. Smelser, eds. 1987. *The Micro-Macro Link.* Berkeley and Los Angeles: University of California Press.

Alexander, R. D. 1979. *Darwinism and Human Affairs.* Seattle: University of Washington Press.

———. 1987. *The Biology of Moral Systems.* New York: Aldine de Gruyter.

Allen, G. A. 1991. Review of Hull 1988a, 1989. *Isis* 82:698–704.

Altman, S., et al. 1987. Enzymatic cleavage of RNA by RNA. *Trends Biochem. Sci.* 11:515.

Anderson, J. 1991. Is human cognition adaptive? *BBS* 14:471–517.

Antonovics, J., N. C. Ellstrand, and R. N. Brandon. 1988. Genetic variation and environmental variation: Expectations and experiments. In L. D. Gottlieb and S. K. Jain, eds., *Plant Evolutionary Biology,* 275–303. London: Chapman and Hall.

Aoki, C., and P. Siekevitz. 1988. Plasticity in brain development. *Sci. Am.* (Dec.), 34–42.

Apostel, L. 1953. Logique et preuve. *Methodos* 5: 279–321.

———. 1989. *A Life History.* With I. Van Dooren. One of three unnumbered volumes of *The Philosophy of Leo Apostel,* ed. F. Vandamme and R. Pinxten. Ghent: Communication and Cognition.

Apostel, L., and W. Callebaut. 1978. Classification et typologie en démographie. In *Typologies et classifications en démographie,* 7–63. Liège: Ordina.

Apostel, L., and J. Vanlandschoot. 1988. *Interdisciplinariteit, wereldbeeldenkonstructie en diepe verspreiding als tegenzetten in een kultuurkrisis.* Brussels: Vrije Universiteit Brussel–Centrum Leo Apostel.

Arrow, K. 1974. *The Limits of Organization.* New York: Norton.

Arthur, W. 1984. *Mechanisms of Morphological Evolution.* New York: Wiley.

Ashby, W. R. 1952. *Design for a Brain.* New York: Wiley.

Ashmore, M. 1989. *The Reflexive Thesis: Wrighting Sociology of Scientific Knowledge.* Chicago: University of Chicago Press.

Aspray, W., and P. Kitcher, eds. 1988. *History and Philosophy of Modern Mathematics.* Minnesota Studies in the Philosophy of Science, vol. 11. Minneapolis: University of Minnesota Press.

Asquith, P. D., and H. E. Kyburg, Jr., eds. 1979. *Current Research in Philosophy of Science.* East Lansing: Philosophy of Science Association.

Atlan, H. 1986. *À tort et à raison: Intercritique de la science et du mythe.* Paris: Seuil.

———. *Tout, non, peut-être: Éducation et vérité.* Paris: Seuil.

Axelrod, R. 1984. *The Evolution of Cooperation.* New York: Basic Books.

Axelrod, R., and W. D. Hamilton. 1981. The evolution of cooperation. *Science* 211:1390–96.

Ayala, F. J., and T. Dobzhansky, eds. 1974. *Studies in the Philosophy of Biology.* London: Macmillan.

Ayer, A. J. 1936. *Language, Truth and Logic.* London: Gollancz. Repr. Harmondsworth: Penguin, 1971.

Bachelard, G. 1973. *Le pluralisme cohérent de la chimie moderne.* Paris: Vrin.

Baigrie, B. S. 1988. Why evolutionary epistemology is an endangereed species. *Soc. Epist.* 2:357–69.

Bajema, C. J. 1991. *Garrett James Hardin: Ecologist, educator, ethicist, and environmentalist. Population and Environment* 12:193–212.

Barash, D. 1977. *Sociobiology and Human Behavior.* New York: Elsevier.

———. 1979. *The Whisperings Within.* London: Penguin.

Barloy, J. J. 1980. *Lamarck contre Darwin.* Paris: Études vivantes.

Barnes, B. 1974. *Scientific Knowledge and Sociological Theory.* London: Routledge and Kegan Paul.

———. 1976. Natural rationality: A neglected concept in the social sciences. *Phil. Soc. Sci.* 6:115–26.

———. 1977. *Interests and the Growth of Knowledge.* London: Routledge and Kegan Paul.

———. 1981. On the conventional character of knowledge and cognition. *Phil. Soc. Sci.* 11:303–33.

———. 1982. *T. S. Kuhn and Social Science.* London: Macmillan.

———. 1985a. *About Science.* Oxford: Blackwell.

———. 1985b. Ethnomethodology as science. *Soc. Stud. Sci.* 15:751–62.

Barnes, B., and D. Bloor. 1982. Relativism, rationalism, and the sociology of knowledge. In Hollis and Lukes, 21–47.

Barnes, B., and D. Edge, eds. 1982. *Science in Context: Readings in the Sociology of Science.* Milton Keynes: Open University Press.

Barrow, J. D. 1991. *Theories of Everything: The Quest for Ultimate Explanation.* Oxford: Clarendon.

Bartlett, F. C. 1932. *Remembering: A Study in Experimental and Social Psychology.* Cambridge: Cambridge University Press. Repr. 1967.

Basalla, G. 1988. *The Evolution of Technology.* Cambridge: Cambridge University Press.

Bazerman, C. 1988. *Shaping Written Knowledge: The Genre and Activity of the Experimental Article in Science.* Madison: University of Wisconsin Press.

Beatty, J. 1987. Natural selection and the null hypothesis. In Dupré, 53–75.

Beauchamp, T. L., and A. Rosenberg. 1981. *Hume and the Problem of Causation.* Oxford: Oxford University Press.

Bechtel, W. 1982a. Two common errors in explaining biological and psychological phenomena. *Phil. Sci.* 49:549–74.

———. 1982b. Taking vitalism and dualism seriously: Toward a more adequate materialism. *Nature and System* 4:23–43.

———. 1983a. A bridge between cognitive science and neuroscience: The functional architecture of mind. *Phil. Stud.* 44:319–30.

———. 1983b. Forms of organization and the incompatibility of science. In N. Rescher, ed., *The Limits of Lawfulness,* 79–92. Lanham: University Press of America.

———. 1984a. The evolution of our understanding of the cell: A study in the dynamics of scientific progress. *Stud. Hist. Phil. Sci.* 15:309–56.

———. 1984b. Reconceptualizations and interfield connections: The discovery of the link between vitamins and coenzymes. *Phil. Sci.* 51:265–92.

———. 1984c. Autonomous psychology: What it should and should not entail. In *PSA 1984* 1:43–55.

———. 1985. Realism, instrumentalism, and the intentional stance. *Cogn. Sci.* 9: 473–97.

———, ed. 1986a. *Integrating Scientific Disciplines.* Dordrecht: Nijhoff.

———. 1986b. Biochemistry: A cross-disciplinary endeavor that discovered a distinctive domain. In Bechtel 1986a, 77–100.

———. 1986c. Building interlevel pathways: The discovery of the Embden-Meyerhof pathway and the phosphate cycle. In P. Weingartner and G. Dorn, eds., *Foundations of Biology,* 65–97. Vienna: Holder-Pichler-Tempsky.

————. 1986d. Teleological functional analyses and the hierarchical organization of nature. In N. Rescher, ed., *Current Issues in Teleology*, 26–48. Lanham: University Press of America.

————, ed. 1987. Issue on "Psycholinguistics as a Case of Cross-Disciplinary Research." *Synthese* 72:293–388.

————. 1988a. *Philosophy of Science: An Overview for Cognitive Science*. Hillsdale: Erlbaum.

————. 1988b. *Philosophy of Mind: An Overview for Cognitive Science*. Hillsdale: Erlbaum.

————. 1988c. Fermentation theory: Empirical difficulties and guiding assumptions. In Donovan et al., 163–80.

————. 1988d. New insights into the nature of science: What does Hull's evolutionary epistemology teach us? *Biol. Phil.* 3:157–64.

————. 1989a. Functional analyses and their justification. *Biol. Phil.* 4:159–62.

————. 1989b. An evolutionary perspective on the re-emergence of cell biology. In Hahlweg and Hooker, 433–57.

————. 1990. Toward making evolutionary epistemology into a truly naturalized epistemology. In Rescher, 63–77.

Bechtel, W., and A. A. Abrahamsen. 1990. Beyond the exclusively propositional era. *Synthese* 82:223–53. Repr. in Fetzer 1991.

————. 1991. *Connectionism and the Mind: A Philosophical Introduction to Parallel Processing in Networks*. Oxford: Blackwell.

Bechtel, W., and R. Richardson. 1983. Consciousness and complexity: Evolutionary perspectives on the mind-body problem. *Australas. J. Phil.* 61:378–95.

————. 1991. *A Model of Theory Development: Localization as a Strategy in Scientific Research*. Princeton: Princeton University Press.

————. 1992. Emergent phenomena and complex systems. In A. Beckermann, H. Flohr, and J. Kim, eds., *Emergence or Reduction? Essays on the Prospects of Nonreductive Physicalism*, 257–88. Berlin and New York: Walter de Gruyter.

Bechtel, W., and E. Stiffler. 1978. Quine and the epistemological nihilists. In *PSA 1978* 1:93–108.

Becker, G. 1976. *The Economic Approach to Human Behavior*. Chicago: University of Chicago Press.

Becker, G., and K. M. Murphy. 1988. A theory of rational addiction. *J. Pol. Econ.* 96:675–700.

Beckner, M. 1959. *The Biological Way of Thought*. New York: Columbia University Press.

Beer, A., and P. Beer, eds. 1975. *Kepler: Four Hundred Years*. New York: Pergamon.

Benacerraf, P., and H. Putnam. 1983. *Philosophy of Mathematics*. 2d ed. Cambridge: Cambridge University Press.

Berger, P. L., and T. Luckmann. 1966. *The Social Construction of Reality*. New York: Doubleday.

Berkson, W., and J. Wettersten. 1984. *Learning from Error: Karl Popper's Psychology of Learning*. La Salle: Open Court.

Bhaskar, R. 1979. *The Possibility of Naturalism: A Philosophical Critique of the Contemporary Human Sciences*. Brighton: Harvester Press.

————. 1986. *Scientific Realism and Human Emancipation*. London: Verso.

————. 1989. *Reclaiming Reality: A Critical Introduction to Contemporary Philosophy*. London: Verso.

Bijker, W. E., T. P. Hughes, and T. Pinch, eds. 1987. *The Social Construction of Technological Systems: New Directions in the Sociology and History of Technology*. Cambridge, Mass.: MIT Press.

Bijker, W. E., and J. Law, eds. 1992. *Constructing Networks and Systems*. Cambridge, Mass.: MIT Press.

Black, M. 1962. *Models and Metaphors*. Ithaca: Cornell University Press.

Blaug, M. 1980. *The Methodology of Economics: Or How Economists Explain*. Cambridge: Cambridge University Press.

Blom, T., W. Callebaut, and T. Nijhuis. 1989. Modalities and counterfactuals in history and the social sciences: Some preliminary reflections. In Special Issue on Counterfactuals and Modalities in History and the Social Sciences. *Philosophica* 44:3–14.

Bloom, A. 1987. *The Closing of the American Mind*. New York: Simon and Schuster.

Bloor, D. 1976. *Knowledge and Social Imagery*. London: Routledge and Kegan Paul.

————. 1981. The strengths of the Strong Programme. *Phil. Soc. Sci.* 11:199–213.

————. 1983. *Wittgenstein: A Social Theory of Knowledge*. New York: Columbia University Press.

————. 1989. Professor Campbell on models of language-learning and the sociology of science: A reply. In Fuller et al., 159–66.

————. 1991. Review of Galison 1987. *Soc. Stud. Sci.* 21:186–89.

Blute, M. 1979. Sociocultural evolutionism: An untried theory. *Behav. Sci.* 24:46–59.

Boden, M., ed. 1990. *The Philosophy of Artificial Intelligence*. Oxford: Oxford University Press.

Böhme, G., W. van den Daele, and W. Krohn. 1972. Alternativen in der Wissenschaft. *Z. Soziol.* 1:302–16.

Bohm, D. 1969a. Some remarks on the notion of order. In Waddington, vol. 2, 18–40.

———. 1969b. Further remarks on order. In Waddington, vol. 2, 41–60.

Bonevac, D. 1991. Semantics and supervenience. *Synthese* 87:331–61.

Borgerhoff-Mulder, M. 1988. Kipsigis bridewealth payments. In L. L. Betzig, M. Borgerhoff-Mulder, and P. W. Turke, eds., *Human Reproductive Behavior: A Darwinian perspective*, 65–82. Cambridge: Cambridge University Press.

Boudon, R. 1977. *Effets pervers et ordre social*. Paris: Presses Universitaires de France.

Bougnoux, D., J.-L. Le Moigne, and S. Proulx, dir. 1990. *Arguments pour une méthode (autour d'Edgar Morin)*. Paris: Seuil.

Boulding, K. E. 1981. *Evolutionary Economics*. London: Sage.

Bourdieu, P. 1968. *Le métier de sociologue*. With J.-C. Passeron and J.-C. Chamboredon. The Hague: Mouton. Paris: Bordas.

———. 1975. The specificity of the scientific field and the social conditions of the progress of reason. *Soc. Sci. Inform.* 14:19–47.

———. 1984. *Distinction: A Social Critique of the Judgement of Taste*. Trans. R. Nice. Cambridge, Mass.: Harvard University Press. French orig. pub. 1979.

Bowker, G., and B. Latour. 1987. A booming discipline short of discipline. *Soc. Stud. Sci.* 17:715–48.

Boyd, Richard N. 1981. Scientific realism and naturalistic epistemology. In *PSA 1980* 2:613–62.

———. 1984. The current status of scientific realism. In Leplin, 41–82.

Boyd, Robert, and P. J. Richerson. 1976. A simple dual inheritance model of the conflict between social and biological evolution. *Zygon* 11:254–62.

———. 1985. *Culture and the Evolutionary Process*. Chicago: University of Chicago Press.

———. 1987. Simple models of complex phenomena: The case of cultural evolution. In Dupré, 27–52.

Bradie, M. 1986. Assessing evolutionary epistemology. *Biol. Phil.* 1:401–50.

———. 1989. Evolutionary epistemology as naturalized epistemology. In Hahlweg and Hooker, 393–412.

Braithwaite, R. B. 1953. *Scientific Explanation*. Cambridge: Cambridge University Press.

Brandon, R. N. 1978a. Evolution. *Phil. Sci.* 45:96–109.

———. 1978b. Adaptation and evolutionary theory. *Stud. Hist. Phil. Sci.* 9:181–206. Repr. in Sober 1984b.

———. 1982. The levels of selection. In *PSA 1982* 1:315–23.

———. 1984. On the concept of the environment in evolutionary biology. *J. Phil.* 81:613–15.

———. 1985a. Phenotypic plasticity, cultural transmission, and human sociobiology. In Fetzer, 57–73.

———. 1985b. Grene on mechanism and reductionism: More than just a side issue. In *PSA 1984* 2:345–53.

———. 1985c. Adaptation explanations: Are adaptations for the good of replicators or interactors? In Depew and Weber, 81–96.

———. 1988. The levels of selection: A hierarchy of interactors. In Plotkin 1988a, 51–71.

———. 1990. *Adaptation and Environment*. Princeton: Princeton University Press.

Brandon, R. N., and R. M. Burian, eds. 1984. *Genes, Organisms and Populations: Controversies over the Units of Selection*. Cambridge, Mass.: MIT Press, Bradford Books.

Brandon, R. N., and N. Hornstein. 1986. From icons to symbols: Some speculations on the origins of language. *Biol. Phil.* 1:169–89.

Brandt, R. B. 1967. Parallels between epistemology and ethics. In Edwards, 3:6–8.

Brannigan, A. 1979. The reification of Mendel. *Soc. Stud. Sci.* 9:423–64.

Brewer, M. B., and B. E. Collins, eds. 1981. *Scientific Inquiry and the Social Sciences: A Volume in Honor of Donald T. Campbell*. San Francisco: Jossey-Bass.

Broad, W., and N. Wade. 1982. *Betrayers of the Truth*. New York: Simon and Schuster.

Brooks, D. R., and E. O. Wiley. 1988. *Evolution as Entropy: Toward a Unified Theory of Biology*. 2d ed. Chicago: University of Chicago Press.

Brown, J. R., ed. 1984. *Scientific Rationality: The Sociological Turn*. Dordrecht: Reidel.

Brunswik, E., and J. Kamiya. 1953. Ecological cue validity of "proximity" and other gestalt factors. *Am. J. Psychol.* 66:20–33.

Brush, S. G. 1974. Should the history of science be rated X? *Science* 183:1164–72.

Bunge, M. 1983. *Treatise on Basic Philosophy*. Vol. 5, *Epistemology and Methodology I: Exploring the World*. Dordrecht: Reidel.

Burian, R. M. 1978. A methodological critique of sociobiology. In Caplan 1978b, 376–95.

———. 1981–82. Human sociobiology and genetic determinism. *Phil. Forum* 13:43–66.

———. 1983. Adaptation. In M. Grene, ed., *Dimensions of Darwinism: Themes and Counterthemes in Twentieth Century Evolutionary Theory*, 287–314. Cambridge: Cambridge University Press.

———. 1984. Scientific realism and incommensurability: Some criticisms of Kuhn and Feyerabend. In R. S. Cohen and M. W. Wartofsky, eds., *Methodology, Metaphysics and the History of Science. In Memory of Benjamin Nelson,* 1–31. Dordrecht: Reidel.

———. 1985a. On conceptual change in biology: The case of the gene. In Depew and Weber, 21–42.

———. 1985b. Conceptual change, cross-theoretical explanation and the unity of science. *Synthese* 33:1–28.

———. 1986a. On integrating the study of evolution and development: Comments on Kauffman and Wimsatt. In Bechtel 1986a, 209–28.

———. 1986b. Why the Panda provides no comfort to the creationist. *Philosophica* 37:11–26.

———. 1987a. Realist methodology in contemporary genetics. In Nersessian, 195–209.

———. 1987b. How not to talk about conceptual change in science. In Pitt and Pera, 3–33.

———. 1987c. More than a marriage of convenience: On the inextricability of history and philosophy of science. *Phil. Sci.* 44:1–42.

———. 1990a. Review of Pickering 1984a. *Synthese* 82:163–74.

———. 1990b. La contribution française aux instruments de recherche dans le domaine de la génétique moléculaire. In J.-L. Fischer and W. Schneider, eds., *Histoire de la génétique.* Paris: Vrin.

———. 1992. How the choice of experimental organisms matters: Biological practices and discipline boundaries. *Synthese* 92:151–66.

———. N.d. Ontological progress in science. Unpublished paper.

Burian, R. M., and J. Gayon. 1990. Genetics after World War II: The laboratories at Gif. *Cah. Hist. CNRS* 7:25–48.

———. 1991. Un évolutionniste Bernardien à l'Institut Pasteur? Morphologie des Ciliés et évolution physiologique dans l'oeuvre d'André Lwoff. In M. Morange, éd., *L'Institut Pasteur: Contributions à son histoire,* 165–86. Paris: La Découverte.

Burian, R. M., J. Gayon, and D. Zallen. 1988. The singular fate of genetics in the history of French biology, 1900–1940. *J. Hist. Biol.* 21:357–402.

———. 1991. Boris Ephrussi and the synthesis of genetics and embryology. In S. Gilbert, ed., *A Conceptual History of Embryology,* 207–27. New York: Plenum Press.

Burian, R. M., and R. Richardson. 1991. Form and order in evolutionary biology: Stuart Kauffman's transformation of theoretical biology. In *PSA 1990* 2:267–87.

Bushkovitch, A. V. 1970. Philosophy of science as a model for all philosophy. *Phil. Sci.* 37:307–11.

Buss, L. W. 1987. *The Evolution of Individuality.* Princeton: Princeton University Press.

Butterfield, H. 1931. *The Whig Interpretation of History.* New York: Scribners.

———. 1940. *The Englishman and His History.* Cambridge: Cambridge University Press.

Butts, R. E. 1984. Philosophy of Science: 1934–1984. *Phil. Sci.* 51:1–2.

Callebaut, W. 1982. Reduction reassessed. In S. Rose, ed., *Against Biological Determinism,* 151–76. London: Allison and Busby.

———, ed. 1986. Current Issues in the Philosophy of Biology. *Philosophica* 37.

Callebaut, W., M. De Mey, R. Pinxten, and F. Vandamme, eds. 1979. *Theory of Knowledge and Science Policy.* 2 vols. Ghent: Communication and Cognition.

Callebaut, W., and R. Pinxten, eds. 1987. *Evolutionary Epistemology: A Multiparadigm Program.* Dordrecht: Reidel.

Callebaut, W., and J. P. Van Bendegem. 1982. The distributive approach to problem-solving in science: Prospects for General Systems Methodology. In R. Trappl, ed., *Cybernetics and Systems Research,* 51–56. Amsterdam: North-Holland.

Callon, M. 1986. Some elements of a sociology of translation: Domestication of the scallops and the fishermen of St. Brieuc Bay. In J. Law, ed., *Power, Action, and Belief: A New Sociology of Knowledge?,* 196–233. London: Routledge and Kegan Paul.

———, ed. 1989. *La science et ses réseaux: Genèse et circulation des faits scientifiques.* Paris: La Découverte/Conseil de l'Europe/Unesco.

———. 1992. Society in the making: The study of technology as a tool for sociological analysis. In Bijker and Law, 83–106.

Callon, M., and B. Latour. 1992. Don't throw the baby out with the Bath school! A reply to Collins and Yearley. In Pickering, ed., 343–68.

Campbell, D. T. 1956. Perception as substitute trial and error. *Psych. Rev.* 63:330–42.

———. 1959. Methodological suggestions from a comparative psychology of knowledge processes. *Inquiry* 2:152–82.

———. 1960. Blind variation and selective retention in creative thought as in other knowledge processes. *Psych. Rev.* 67:380–400.

———. 1965. Variation and selective retention in socio-cultural evolution. In H. R. Barringer, G. L. Blacksten, and R. W. Mack, eds., *Social Change in Developing Areas: A Reinterpretation of Evolutionary Theory,* 19–49. Cambridge, Mass.: Schenkman Press.

———. 1969. Ethnocentrism of disciplines and the

fish-scale model of omniscience. In M. Sherif and C. W. Sherif, eds., *Interdisciplinary Relationships in the Social Sciences,* 328–48. Chicago: Aldine.

———. 1972. On the genetics of altruism and the counter-hedonic components in human culture. *J. Soc. Issues* 28:21–37.

———. 1973. Ostensive instances and entitativity in language learning. In W. Gray and N. D. Rizzo, eds., *Unity Through Diversity,* pt. 2, 1043–57. New York: Gordon and Breach.

———. 1974a. Evolutionary epistemology. In P. A. Schilpp, ed., *The Philosophy of Karl Popper,* vol. 1, 412–63. LaSalle: Open Court.

———. 1974b. "Downward causation" in hierarchically organized biological systems. In Ayala and Dobzhansky, 179–86.

———. 1975. On the conflicts between biological and social evolution and between psychology and moral tradition. *Am. Psychol.* 30:1103–26.

———. 1979a. Comments on the sociobiology of ethics and moralizing. *Behav. Sci.* 24:37–45.

———. 1979b. A tribal model of the social system vehicle carrying scientific knowledge. *Knowledge* 2:181–201.

———. 1981a. Perspective on a scholarly career. In Brewer and Collins, 454–501. Repr. in Campbell 1988a.

———. 1981b. ERISS ["Epistemologically Relevant Internalist Sociology of Science"] conference. *4S Newsletter* 6 (3): 24–25.

———. 1983. The two distinct routes beyond kin selection to ultra-sociality: Implications for the humanities and social sciences. In D. L. Bridgeman, ed., *The Nature of Prosocial Development: Interdisciplinary Theories and Strategies,* 11–41. New York: Academic Press.

———. 1984. Can we be scientific in applied social science? In R. F. Connor, D. G. Attman, and C. Jackson, eds., *Evaluation Studies,* vol. 9, 26–48. Beverly Hills: Sage. Repr. in Campbell 1988a.

———. 1986a. Science's social system of validity-enhancing collective belief change and the problems of the social sciences. In Fiske and Shweder, 108–35.

———. 1986b. Science policy from a naturalistic sociological epistemology. In *PSA 1984* 2:14–26.

———. 1987a. Neurological embodiments of belief and the gap in the fit of phenomena to noumena. In A. Shimony and D. Nails, eds., *Naturalistic Epistemology: A Symposium of Two Decades,* 165–92. Dordrecht: Reidel.

———. 1987b. Selection theory and the sociology of scientific validity. In Callebaut and Pinxten, 139–58.

———. 1988a. *Methodology and Epistemology for Social Science: Selected Papers.* Ed. E. S. Overman. Chicago: University of Chicago Press.

———. 1988b. A general "selection theory" as implemented in biological evolution and in social belief-transmission-with-modification in science. *Biol. Phil.* 3:171–77.

———. 1988c. Popper and selection theory. *Soc. Epist.* 2:371–77.

———. 1989. Models of language learning and their implications for social-constructionist analyses of scientific belief. In Fuller et al., 153–58.

———. 1990a. Epistemological roles for selection theory. In Rescher, 1–19.

———. 1990b. The Meehlian corroboration-verisimilitude theory of science. *Psych. Inquiry.* 1:142–47.

———. 1990c. Levels of organization, downward causation, and the selection-theory approach to evolutionary epistemology. In E. Tobach and G. Greenberg, eds., *Scientific Methodology in the Study of Mind: Evolutionary Epistemology,* 1–15. Hillsdale: Erlbaum.

Campbell, D. T., and B. T. Paller. 1989. Extending evolutionary epistemology to "justifying" scientific beliefs (A sociological rapprochement with a fallibilist perceptual foundationalism?). In Hahlweg and Hooker, 231–57.

Canguilhem, G. 1988. *Ideology and Rationality in the History of the Life Sciences.* Trans. A. Goldhammer. Cambridge, Mass.: MIT Press.

Caplan, A. 1978a. Babies, bathwater, and derivational reduction. In *PSA 1978* 2:357–70.

———, ed. 1978b. *The Sociobiology Debate.* New York: Harper and Row.

———. 1981. Pick your poison: Historicism, essentialism, and emergentism in the definition of species. *BBS* 4:285–86.

———. 1984. Sociobiology as a strategy in science. *Monist* 67:143–60.

Caporael, L. R., R. M. Dawes, J. M. Orbell, and A. J. C. van de Kragt. 1989. Selfishness examined: Cooperation in the absence of egoistic incentives. *BBS* 12:683–739.

Carnap, R. 1929. *Der logische Aufbau der Welt.* Berlin: Im Weltkreis.

———. 1949. Logical foundations of the unity of science. In H. Feigl and W. Sellars, eds., *Readings in Philosophical Analysis,* 408–23. New York: Appleton-Century-Crofts. Orig. pub. 1938.

———. 1950. Empiricism, semantics and ontology. *Rev. Int. Phil.* 4:20–40. Repr. in *Meaning and Necessity,* 2d ed. Chicago: Phoenix Books, 1956.

———. 1963. Replies and systematic expositions. In P. A. Schilpp, ed., *The Philosophy of Rudolf Carnap*. The Library of Living Philosophers, vol. 11, 859–1013. La Salle: Open Court.

Carroll, L. 1898. *Sylvie and Bruno, Concluded*. London: Macmillan.

Cartwright, N. 1983. *How the Laws of Physics Lie*. Oxford: Oxford University Press.

———. 1986. Two kinds of teleological explanation. In Donagan, Perovich, and Wedin, 201–10.

———. 1989. *Nature's Capacities and Their Measurement*. Oxford: Clarendon.

Cartwright, N., and M. Jones, eds. 1992. *Varieties of Idealization*. Amsterdam: Rodopi.

Cavalli-Sforza, L. L., and M. W. Feldman. 1981. *Cultural Transmission and Evolution*. Princeton: Princeton University Press.

Cavallo, R. 1979. *The Role of Systems Methodology in Social Science Research*. The Hague: Nijhoff.

Cech, T. R. 1987. The chemistry of self-splicing RNA and RNA enzymes. *Science* 236:1532.

Chalmers, A. 1990. *Science and Its Fabrication*. Minneapolis: University of Minnesota Press.

Chandrasekaran, B. 1981. Natural and social system metaphors in distributed problem solving: Introduction to the issue. *IEEE Trans. SMC* 11:1–5.

Changeux, J.-P. 1985. *Neuronal Man: The Biology of Mind*. Oxford: Oxford University Press. Orig. pub. 1983.

Changeux, J.-P., and A. Connes. 1989. *Matière à pensée*. Paris: Odile Jacob.

Charnov, E. 1983. *The Theory of Sex Allocation*. Princeton: Princeton University Press.

Cherniak, C. 1986. *Minimal Rationality*. Cambridge, Mass.: MIT Press.

Chernyak, L., and A. I. Tauber. 1991. The dialectical self: Immunology's contribution. In Tauber, 109–56.

Chihara, C. S. 1990. *Constructibility and Mathematical Existence*. Oxford: Clarendon.

Chisholm, R. 1957. *Perceiving*. Ithaca, N.Y.: Cornell University Press.

———. 1982. *The Foundations of Knowing*. Minneapolis: University of Minnesota Press.

Chodorow, N. 1978. *The Reproduction of Mothering*. Berkeley and Los Angeles: University of California Press.

Chomsky, N. 1957. *Syntactic Structures*. The Hague: Mouton.

———. 1965. *Aspects of the Theory of Syntax*. Cambridge, Mass.: MIT Press.

———. 1966. *Cartesian Linguistics*. New York: Harper and Row.

———. 1975. *Reflections on Language*. New York: Random House. Orig. pub. 1957.

Church, R. 1984. Popper's 'World 3' and the problem of the printed line. *Australas. J. Phil.* 62:378–91.

Churchland, P. M. 1979. *Scientific Realism and the Plasticity of Mind*. Cambridge: Cambridge University Press.

———. 1984. *Matter and Consciousness: A Contemporary Introduction to the Philosophy of Mind*. Cambridge, Mass.: MIT Press.

———. 1989. *A Neurocomputational Perspective: The Nature of Mind and the Structure of Science*. Cambridge, Mass.: MIT Press.

Churchland, P. M., and P. S. Churchland. 1990. Could a machine think? *Sci. Am.* 262 (1): 32–37.

Churchland, P. M., and C. A. Hooker, eds. 1985. *Images of Science*. Chicago: University of Chicago Press.

Churchland, P. S. 1982. Mind-brain reduction: New light from the philosophy of science. *Neuroscience* 7:1041–47.

———. 1983. Is the visual system as smart as it looks? In *PSA 1982* 2:541–52.

———. 1986a. *Neurophilosophy: Toward a Unified Science of the Mind-Brain*. Cambridge, Mass.: MIT Press.

———. 1986b. Replies to comments [on 1986a]. *Inquiry* 29:241–72.

———. 1987. Epistemology in the age of neuroscience. *J. Phil.* 84:544–53.

Churchland, P. S., and P. M. Churchland. 1983. Stalking the wild epistemic engine. *Noûs* 17:5–22.

Churchland, P. S., and T. J. Sejnowski. 1988. Perspectives on cognitive neuroscience. *Science* 242:741–45.

———. 1992. *The Computational Brain*. Cambridge, Mass.: MIT Press, Bradford Books.

Churchman, C. W. 1971. *The Design of Inquiring Systems*. New York: Basic Books.

———. 1984. Early years of the Philosophy of Science Association. *Phil. Sci.* 51:20–22.

Cicourel, A. 1964. *Method and Measurement in Sociology*. New York: Free Press.

———. 1975. Discourse and text: Cognitive and linguistic processes in studies of social structure. *Versus: Quaderni di studi semiotici* 5 (Sept.-Dec.): 33–84.

Clark, A. J. 1989. *Microcognition: Philosophy, Cognitive Science, and Parallel Distributed Processing*. Cambridge, Mass.: MIT Press, Bradford Books.

Clarke, A., and J. Fujimura, eds. 1992. *The Right Tools for the Job in Twentieth Century Life Sciences: Materials, Techniques, Instruments,*

Models and Work Organization. Princeton: Princeton University Press.

Clement, J. 1989. Learning via model construction and criticism: Protocol evidence on sources of creativity in science. In J. A. Glover, R. R. Ronning, and C. F. Reynolds, eds., *Handbook of Creativity*, 341–82. New York: Plenum Press.

Clendinnen, L. J. 1989. Realism and the underdetermination of theory. *Synthese* 81:63–90.

Cohen, G. A. 1978. *Karl Marx's Theory of History: A Defence*. Oxford: Oxford University Press.

Cohen, J. 1976. Irreproducible results and the breeding of pigs (or nondegenerate limit random variables in biology). *BioScience* 26: 391–94.

Cohen, L. J. 1973. Is the progress of science evolutionary? *Brit. J. Phil. Sci.* 24:41–61.

———. 1981. Can human irrationality be experimentally demonstrated? *BBS* 4:317–70.

———. 1987. A note on the evolutionary theory of software development. *Brit. J. Phil. Sci.* 38:381–87.

Cohen, R. S., P. K. Feyerabend, and M. W. Wartofsky, eds. 1976. *Essays in Memory of Imre Lakatos*. Dordrecht: Reidel.

Collier, J. 1988. Supervenience and reduction in biological hierarchies. *Can. J. Phil.* 14: 209–34.

Collins, H. M. 1981a. Stages in the Empirical Programme of Relativism. *Soc. Stud. Sci.* 11: 3–10.

———. 1981b. What is TRASP? The radical programme as a methodological imperative. *Soc. Stud. Sci.* 11:215–24.

———. 1983. An empirical relativist programme in the sociology of scientific knowledge. In Knorr-Cetina and Mulkay, 85–113.

———. 1985. *Changing Order: Replication and Induction in Scientific Practice*. London: Sage. 2d ed., with a new afterword, Chicago: University of Chicago Press, 1992.

———. 1991. *Artificial Experts: Social Knowledge and Intelligent Machines*. Cambridge, Mass.: MIT Press.

———. 1992. Hubert L. Dreyfus, forms of life, and a simple test for machine intelligence. *Soc. Stud. Sci.* 22:726–39.

Collins, H. M., and T. Pinch. 1982. *Frames of Meaning: The Social Construction of Extraordinary Science*. London: Routledge and Kegan Paul.

Collins, H. M., and S. Yearley. 1992. Epistemological Chicken. In Pickering, ed., 301–26.

Colwell, R. K. 1981. Group selection is implicated in the evolution of female-based sex ratios. *Nature* 290:401–4.

Committee on the Conduct of Science, National Academy of Sciences [USA]. 1989. On being a scientist. *Proceedings of the National Academy of Sciences of the USA* 89:9053–74.

Commoner, B. 1961. In defense of biology. *Science* 133:1745–48.

Conrad, M. 1983. *Adaptability: The Significance of Variability from Molecule to Ecosystem*. New York: Plenum.

Cook, N. D. 1980. *Stability and Flexibility: An Analysis of Natural Systems*. Oxford and New York: Pergamon Press.

Cook, T. D., and D. T. Campbell. 1979. *Quasi-Experimentation: Design and Analysis for Field Settings*. Chicago: Rand McNally.

Corning, P. 1983. *The Synergism Hypothesis: A Theory of Progressive Evolution*. London: Blond and Briggs.

Corsi, P. 1988. *The Age of Lamarck: Evolutionary Theories in France, 1790–1830*. Revised and updated. Trans. J. Mandelbaum. Berkeley: University of California Press. Orig. pub. 1983.

Cosmides, L. 1985. Deduction or Darwinian Algorithms? An Explanation of the "Elusive" Content Effect on the Wason Selection Task. Ph.D. diss., Harvard University.

———. 1989. The logic of social exchange: Has natural selection shaped how humans reason? Studies with the Wason selection task. *Cognition* 31:187–276.

Cosmides, L., and J. Tooby. 1987. From evolution to behavior: Evolutionary psychology as the missing link. In Dupré, 277–306.

———. 1989. A computational theory of social exchange. *Ethology and Sociobiology* 10:51–97.

———. 1992a. Are humans good intuitive statisticians after all? Rethinking some conclusions from the literature on judgment under uncertainty. Report no. 3/92 of the Research Group on Biological Foundations of Human Culture at the Center for Interdisciplinary Research (1991–92), University of Bielefeld, Germany.

———. 1992b. From evolution to adaptations to behavior: Toward an integrated evolutionary psychology. Report no. 4/92 of the Research Group on Biological Foundations of Human Culture at the Center for Interdisciplinary Research (1991–92), University of Bielefeld, Germany.

Cottingham, J. 1983. Neo-naturalism and its pitfalls. *Philosophy* 58:455–70.

Cozzens, S. E., and T. F. Gieryn, eds. 1990. *Theories of Science in Society.* Bloomington: Indiana University Press.

Crane, D. 1972. *Invisible Colleges: Diffusion of Knowledge in Scientific Communities.* Chicago: University of Chicago Press.

Crick, F. 1984a. Sperm from deep space. In Weintraub, 20–33.

———. 1984b. Function of the thalamic reticular complex: The searchlight hypothesis. *Nat. Acad. Sci. USA. Proc. Biol. Sci.* 81:4586–90.

Culp, S., and P. Kitcher. 1989. Theory structure and theory change in contemporary molecular biology. *Brit. J. Phil. Sci.* 40:459–83.

Cushing, J. T., C. F. Delaney, and G. M. Gutting, eds. 1984. *Science and Reality.* Notre Dame: University of Notre Dame Press.

Cziko, G. A. 1989a. The clonal-selection production of antibodies as a stochastic, evolutionary model for cognitive development and learning. Ms., Department of Educational Psychology, University of Illinois, Champaign.

———. 1989b. Unpredictability and indeterminism in human behavior: Arguments and implications for educational research. *Educational Researcher* 18:17–25.

Cziko, G. A., and D. T. Campbell. 1990. Comprehensive evolutionary epistemology bibliography. *J. Soc. Biol. Struct.* 13:41–82.

Dagognet, F. 1989. *Éloge de l' objet: Pour une philosophie de la marchandise.* Paris: Vrin.

Danto, A. 1967. Naturalism. In Edwards, 5:488–50.

Darden, L. 1977. The heritage from Logical Positivism: A reassessment. In *PSA 1976* 2:242–58.

———. 1978. Discoveries and the emergence of new fields in science. In *PSA 1978* 1:149–60.

———. 1990. Diagnosing and fixing faults in theories. In Shrager and Langley, 319–46.

———. 1991. *Strategies for Theory Change: The Case of the Gene.* Oxford: Oxford University Press.

Darden, L., and J. A. Cain. 1989. Selection type theories. *Phil. Sci.* 56:106–29.

Darden, L., and N. Maull. 1977. Interfield theories. *Phil. Sci.* 44:43–64.

Darwin, C. 1839. *The Voyage of the Beagle.* Repr. New York: Anchor Books, 1962.

———. 1859. *On the Origin of Species by Means of Natural Selection, or, the Preservation of Favoured Races in the Struggle for Life.* London: Murray. Facsimile ed. with introduction by E. Mayr, Cambridge, Mass.: Harvard University Press, 1964.

———. 1871. *The Descent of Man, and Selection in Relation to Sex.* 2 vols. London: Murray.

Daston, L. 1992. Objectivity and the escape from perspective. *Soc. Stud. Sci.* 22:597–618.

Davidson, D. 1970. Mental events. In L. Foster and J. W. Swanson, eds., *Experience and Theory,* 79–101. Amherst: University of Massachussets Press.

Dawkins, R. 1976. *The Selfish Gene.* Oxford: Oxford University Press. New ed. 1989.

———. 1978. Replicator selection and the extended phenotype. *Z. Tierpsychol.* 47: 61–76. Repr. in Sober 1984b.

———. 1979. Twelve misunderstandings of kin selection. *Z. Tierpsychol.* 51:184–200.

———. 1982a. *The Extended Phenotype.* San Francisco: Freeman.

———. 1982b. The myth of genetic determinism. *New Scientist* 93 (7 January): 27–30.

———. 1986. *The Blind Watchmaker.* New York: W. W. Norton.

———. 1989. The evolution of evolvability. In C. Langton, ed., *Artificial Life: Santa Fe Institute Studies in the Sciences of Complexity.* Reading, Mass.: Addison-Wesley.

Dear, P. 1991. *The Literary Structure of Scientific Argument: Historical Studies.* Philadelphia: University of Pennsylvania Press.

de Beauvoir, S. 1953. *The Second Sex.* Trans. H. M. Parshley. New York: Knopf.

De Mey, M. 1982. *The Cognitive Paradigm.* Dordrecht: Reidel. 2d ed., with a new intro., Chicago: University of Chicago Press, 1992.

———. 1987. Answers in response to questions by Steve Fuller. *Soc. Epist.* 1:83–95.

———. 1989. Cognitive paradigms and the psychology of science. In Gholson et al., 275–95.

———. 1992. Scientific discovery: Cold fusion of ideas? *Int. Stud. Phil. Sci.* 6:23–27.

Dennett, D. C. 1969. *Content and Consciousness.* London: Routledge and Kegan Paul.

———. 1984. Computer models and the mind–A view from the East Pole. *Times Literary Supplement* (December 14), 1453–54.

———. 1987. *The Intentional Stance.* Cambridge, Mass.: MIT Press.

Depew, D., and B. Weber, eds. 1985. *Evolution at a Crossroads: The New Biology and the New Philosophy of Science.* Cambridge, Mass.: MIT Press.

Derlega, V., and J. Grzelak, eds. 1982. *Cooperation and Helping Behavior.* New York: Academic Press.

Desmond, A. 1989. *The Politics of Evolution: Morphology, Medicine, and Reform in Radical London.* Chicago: University of Chicago Press.

Desmond, A., and J. Moore. 1991. *Darwin.* London: Michael Joseph. Repr. Harmondsworth: Penguin, 1992.

DeVore, I., G. Goethals, and R. Trivers. 1973. *Exploring Human Nature*. Unit 1. Cambridge, Mass.: Educational Development Corporation.

Dewey, J., S. Hook, and E. Nagel. 1945. Are naturalists materialists? *J. Phil.* 42:515–30.

Dewsbury, D. A., ed. 1985. *Studying Animal Behavior: Autobiographies of the Founders*. Chicago: University of Chicago Press.

Dickemann, M. 1985. Human sociobiology: The first decade. *New Scientist* (10 October): 38–42.

Dietrich, M. 1992. Macromutation. In Keller and Lloyd, 194–201.

Dijksterhuis, E. J. 1961. *The Mechanization of the World Picture*. Trans. C. Dikshoorn. Oxford: Clarendon. Dutch orig. pub. 1950.

Dobzhansky, T. 1951. *Genetics and the Evolution of Species*. 3d ed. New York: Columbia University Press. 1st ed. 1937.

Dobzhansky, T., F. J. Ayala, G. L. Stebbins, and J. W. Valentine. 1977. *Evolution*. San Francisco: Freeman.

Döbert, R. 1981. The role of stage models within a theory of social evolution, illustrated by the European witch craze. In Jensen and Harré, 71–119.

Doell, R. G., and H. E. Longino. 1988. Sex hormones and human behavior: A critique of the linear model. *J. Homosexuality* 15 (3/4): 55–78.

Donagan, A., A. N. Perovich, Jr., and M. V. Wedin, eds. 1986. *Human Nature and Natural Knowledge*. Dordrecht: Reidel.

Donaldson, S. 1987. The orientation of Yang–Mills moduli spaces and 4–manifold topology. *J. Diff. Geom.* 126:397–427.

Donovan, A., L. Laudan, and R. Laudan, eds. 1988. *Scrutinizing Science: Empirical Studies of Scientific Change*. Dordrecht: Reidel.

Doolittle, W. F. 1988. Hierarchical approaches to genome evolution. *Can. J. Phil.* 14:101–33.

Doolittle, W. F., and C. Sapienza. 1980. Selfish genes, the phenotype paradigm and genome evolution. *Nature* 284:601–3.

Dosi, G., C. Freeman, R. Nelson, G. Silverberg, and L. Soete, eds. 1988. *Technical Change and Economic Theory*. London: Pinter.

Dresden, M. 1974. Reflection on "fundamentality and complexity." In C. P. Enz and J. Mehra, eds., *Physical Reality and Mathematical Description*, 133–66. Dordrecht: Reidel.

Dretske, F. 1981. *Knowledge and the Flow of Information*. Cambridge, Mass.: MIT Press, Bradford Books.

———. 1983. Precis of *Knowledge and the Flow of Information*. *BBS* 6:55–90. Repr. in Kornblith 1985a.

———. 1988. *Explaining Behavior: Reasons in a World of Causes*. Cambridge, Mass.: MIT Press, Bradford Books.

Dreyer, J. L. E. 1953. *A History of Astronomy from Thales to Kepler*. New York: Dover. Orig. pub. 1906.

Dreyfus, H. L. 1979. *What Computers Can't Do: The Limits of Artificial Intelligence*. 2d ed. New York: Harper and Row. 1st ed. 1972. Reedited as *What Computers Still Can't Do: A Critique of Artificial Reason*. Cambridge, Mass.: MIT Press, Bradford Books, 1992.

———. 1990. *Being-in-the-World: A Commentary on Heidegger's Being and Time, Division I*. Cambridge, Mass.: MIT Press, Bradford Books.

———. 1992. Response to Collins, *Artificial Experts*. *Soc. Stud. Sci.* 22:717–26.

Dreyfus, H., and S. Dreyfus. 1985. *Mind over Machine: The Power of Human Intuition and Expertise in the Era of the Computer*. New York: Macmillan.

Dreyfus, H., and P. Rabinow. 1982. *Michel Foucault: Beyond Structuralism and Hermeneutics*. With an afterword by M. Foucault. Hemel Hempstead, Herts.: Harvester Press, Wheatsheaf Books.

Duhem, P. 1953. *The Aim and Structure of Physical Theory*. Trans. P. P. Wiener. Princeton: Princeton University Press. Repr. New York: Atheneum, 1962. French orig. pub. 1906.

Dunn, R. 1987. *The Possibility of Weakness of Will*. Indianapolis: Hackett.

Dupré, J., ed. 1987. *The Latest on the Best: Essays on Evolution and Optimality*. Cambridge, Mass.: MIT Press, Bradford Books.

Durham, W. H. 1976. The adaptive significance of cultural behavior. *Human Ecology* 4:89–121.

———. 1979. Toward a coevolutionary theory of human biology and culture. In N. A. Chagnon and W. Irons, eds., *Evolutionary Biology and Human Social Behavior*, 39–59. North Scituate, Mass.: Duxbury Press.

Dvorak, J. C. 1992. Failing grades. *MacUser* 8:372.

Dyke, C. 1988. *The Evolutionary Dynamics of Complex Systems: A Study in Biosocial Complexity*. Oxford: Oxford University Press.

Earman, J. 1977. Till the end of time. In Earman et al., 109–33.

———, ed. 1983. *Testing Scientific Theories*. Minnesota Studies in the Philosophy of Science, vol. 10. Minneapolis: University of Minnesota Press.

Earman, J., C. Glymour, and J. Stachel, eds. 1977.

Foundations of Space-Time Theories. Minnesota Studies in the Philosophy of Science, vol. 8. Minneapolis: University of Minnesota Press.

Economist, The. 1986. Philosophy comes down from the clouds. (April 26): 101–5.

Edel, A. 1946. Is naturalism arbitrary? *J. Phil.* 43:141–52.

———. 1970. Science and the structure of ethics. In Neurath, Carnap, and Morris, 2:273–377.

Edelman, G. M. 1974. The problem of molecular recognition by a selective system. In Ayala and Dobzhansky, 45–56.

———. 1987. *Neural Darwinism*. New York: Basic Books.

———. 1988a. *Topobiology: An Introduction to Molecular Embryology*. New York: Basic Books.

———. 1988b. *The Remembered Present: A Biological Theory of Consciousness*. New York: Basic Books.

———. 1989. Topobiology. *Sci. Am.* (May), 44–52.

———. 1992. *Bright Air, Brillian Fire: On the Matter of Mind*. New York: Basic Books.

Edge, D., and M. J. Mulkay. 1976. *Astronomy Transformed*. New York: Wiley-Interscience.

Edwards, P., ed. 1967. *The Encyclopedia of Philosophy*. 8 vols. New York: Macmillan.

Efron, N. J., and M. Fisch. 1991. Science naturalized, science denatured: An evaluation of Ronald Giere's cognitivist approach to explaining science. *Hist. Phil. Life Sci.* 13:187–221.

Eldredge, N. 1985. *Unfinished Synthesis: Biological Hierarchies and Modern Evolutionary Thought*. Oxford: Oxford University Press.

———. 1986. Information, economics, and evolution. *Ann. R. Ecol.* 17:351–69.

———, ed. 1987. *The Natural History Reader in Evolution*. New York: Columbia University Press.

———. 1992. 'The two evolutionary theories' and modern evolutionary theory. *Synthese* 92:135–49.

Eldredge, N., and J. Cracraft. 1980. *Phylogenetic Patterns and the Evolutionary Process: Method and Theory in Comparative Biology*. New York: Columbia University Press.

Eldredge, N., and S. J. Gould. 1972. Punctuated equilibria: An alternative to phyletic gradualism. In T. J. M. Schopf, ed., *Models in Paleobiology*, 82–115. San Francisco: Freeman, Cooper.

Elster, J. 1975. *Leibniz et la formation de l'esprit capitaliste*. Paris: Aubier-Montaigne.

———. 1976. A note on hysteresis in the social sciences. *Synthese* 33:371–93.

———. 1978. *Logic and Society*. New York: Wiley.

———. 1979. *Ulysses and the Sirens: Studies in Rationality and Irrationality*. Cambridge: Cambridge University Press.

———. 1983a. *Explaining Technical Change*. Cambridge: Cambridge University Press.

———. 1983b. *Sour Grapes*. Cambridge: Cambridge University Press.

———. 1985. *Making Sense of Marx*. Cambridge: Cambridge University Press.

———. 1986a. *An Introduction to Karl Marx*. Cambridge: Cambridge University Press.

———, ed. 1986b. *The Multiple Self*. Cambridge: Cambridge University Press.

———, ed. 1986c. *Rational Choice*. Oxford: Blackwell.

———. 1989a. *The Cement of Society: A Study of Social Order*. Cambridge: Cambridge University Press.

———. 1989b. *Nuts and Bolts for the Social Sciences*. Cambridge: Cambridge University Press.

———. 1989c. *Solomonic Judgements: Studies in the Limitations of Rationality*. Cambridge: Cambridge University Press.

———. 1990. *Psychologie politique*. Paris: Éditions de Minuit.

Elster, J., and J. Roemer, eds. 1991. *Interpersonal Comparisons of Well-Being*. Cambridge: Cambridge University Press.

Enç, B. 1983. In defense of the identity theory. *J. Phil.* 80:279–98.

Engel, P. 1992. Le rêve analytique et le réveil naturaliste. *Le Débat* 72 (Nov.–Dec.): 104–14.

Engels, E.–M. 1989. *Erkenntnis als Anpassung? Eine Studie zur Evolutionären Erkenntnistheorie*. Frankfurt am Main: Suhrkamp.

Esposito, J. L. 1975. Remarks toward a general theory of organization. *Int. J. Gen. Syst.* 2:133–43.

Evans-Pritchard, E. E. 1937. *Witchcraft: Oracles and Magic Among the Azande*. Oxford: Clarendon.

Faia, M. 1986. *Dynamic Functionalism*. Cambridge: Cambridge University Press.

Fantini, B. 1987. L'entrée de la biologie moléculaire dans la définition de l'espèce. In *Histoire du concept d'espèce dans les sciences de la vie*, 285–301. Paris: Fondation Singer-Polignac.

———. 1988. Utilisation par la génétique moléculaire du vocabulaire de la théorie de l'information. In M. Groult, ed., *Transfert de vocabulaire dans les sciences*, 159–70. Paris: Éditions du Centre National de la Recherche Scientifique.

———. 1990. Jacques Monod et les origines de la

biologie moléculaire. *La Recherche* 21:180–87.

Feldman, M. W., and L. L. Cavalli-Sforza. 1976. Cultural and biological evolutionary processes, selection for a trait under complex transmission. *Theor. Pop. Biol.* 9:238–59.

———. 1979. Aspects of variance and covariance analysis with cultural inheritance. *Theor. Pop. Biol.* 15:276–307.

Festinger, L. 1964. *Conflict, Decision and Dissonance.* Stanford: Stanford University Press.

Fetzer, J. H., ed. 1985. *Sociobiology and Epistemology.* Dordrecht: Reidel.

———. 1990. Evolution, rationality, and testability. *Synthese* 82:423–39. Repr. in Fetzer 1991.

———, ed. 1991. *Epistemology and Cognition.* Dordrecht: Kluwer.

Feyerabend, P. K. 1962. Explanation, reduction, and empiricism. In H. Feigl and G. Maxwell, eds., *Scientific Explanation: Space and Time,* 28–97. Minnesota Studies in the Philosophy of Science, vol. 3. Minnepolis: University of Minnesota Press.

———. 1963. How to be a good empiricist: A plea for tolerance in matters epistemological. In B. Baumrin, ed., *The Delaware Seminar,* vol. 2, 3–39. New York: Interscience.

———. 1975. *Against Method: Outline of an Anarchist Theory of Knowledge.* London: New Left Books.

———. 1981. Rückblick. In H. P. Duerr, Hrsg., *Versuchungen: Aufsätze zur Philosophie Paul Feyerabends,* vol. 2, 320–72. Frankfurt am Main: Suhrkamp.

———. 1987. *Farewell to Reason.* London: Verso.

Feynman, R. P. 1985. *The Strange Theory of Light and Matter.* Princeton: Princeton University Press.

Field, H. 1973. Theory change and the indeterminacy of reference. *J. Phil.* 70:462–81.

Fine, A. 1984. The natural ontological attitude. In Leplin, 83–107.

———. 1986. *The Shaky Game: Einstein, Realism, and the Quantum Theory.* Chicago: University of Chicago Press.

Fisher, D. 1990. Boundary work and science: The relation between power and knowledge. In Cozzens and Gieryn, 98–119.

Fisher, R. A. 1930. *The Genetical Theory of Natural Selection.* Oxford: Clarendon.

Fisk, M. 1973. *Nature and Necessity: An Essay in Physical Ontology.* Bloomington: Indiana University Press.

Fiske, D. W., and R. A. Shweder, eds. 1986. *Metatheory in Social Science: Pluralisms and Subjectivities.* Chicago: University of Chicago Press.

Flew, A. G. N. 1967. *Evolutionary Ethics.* London: Macmillan.

———. 1987. Must naturalism discredit naturalism? In Radnitzky and Bartley, 401–21.

Fodor, J. 1974. Special sciences (or: the disunity of science as a working hypothesis). *Synthese* 28:97–115. Repr. in Fodor 1975.

———. 1975. *The Language of Thought.* New York: Crowell. Repr., Cambridge, Mass.: Harvard University Press, 1979.

———. 1983a. *The Modularity of Mind.* Cambridge, Mass.: MIT Press, Bradford Books.

———. 1983b. In *PSA 1982* 2:644–53.

Fogelman Soulié, F. dir. 1991. *Les théories de la complexité. Autour de l'oeuvre d'Henri Atlan.* Paris: Seuil.

Fontana, W. N.d. Algorithmic chemistry: A new approach to functional self-organization. Ms.

Forrester, J. W. 1975. *Collected Papers.* Cambridge, Mass.: Wright-Allen Press.

Forum für Philosophie Bad Homburg. 1987. *Philosophie und Begründung.* Frankfurt am Main: Suhrkamp.

Foster, P. L. 1992. Directed mutation: Between unicorns and goats. *J. Bacteriol.* 174:1711–16.

Foucault, M. 1972. *The Archaeology of Knowledge.* London: Tavistock. Orig. pub. 1969.

———. 1973. *The Order of Things: An Archaeology of the Human Sciences.* New York: Vintage. Orig. pub. 1966.

Fraenkel, A. A., Y. Bar-Hillel, and A. Levy. 1973. *Foundations of Set Theory.* Amsterdam: North-Holland.

Frankel, C. et al. 1986. *La science face au racisme.* Brussels: Complexe. Orig. pub. 1981.

Franklin, A. 1986. *The Neglect of Experiment.* Cambridge: Cambridge University Press.

Freudenthal, G. 1984. The role of shared knowledge in science: The failure of the constructivist programme in the sociology of science. *Soc. Stud. Sci.* 14:285–95.

Friedman, J. 1986. *Game Theory with Applications to Economics.* Oxford: Oxford University Press.

Friedman, Michael. 1974. Explanation and scientific understanding. *J. Phil.* 71:5–19.

———. 1983. *Foundations of Space-Time Theories: Relativistic Physics and Philosophy of Science.* Princeton: Princeton University Press.

Friedman, Milton. 1953. *Essays in Positive Economics.* Chicago: University of Chicago Press.

Fujimura, J. 1988. The molecular biological bandwagon in cancer research: Where social worlds meet. *Soc. Probl.* 35:261–83.

Fuller, S. 1988. *Social Epistemology.* Bloomington: Indiana University Press.

———. 1989. *Philosophy of Science and Its Discontents*. Boulder: Westview Press.

———. 1991a. Studying the propietary grounds of knowledge. *J. Soc. Behav. Personal.* 6:105–28.

———. 1991b. Is history and philosophy of science withering on the vine? *Phil. Soc. Sci.* 21:149–78.

———. 1992. Social epistemology and the research agenda of science studies. In Pickering, 390–428.

———. 1993. *Philosophy, Rhetoric, and the End of Knowledge: The Coming of Science and Technology Studies*. Madison: University of Wisconsin Press.

Fuller, S., M. De Mey, T. Shinn, and S. Woolgar, eds. 1989. *The Cognitive Turn: Sociological and Psychological Perspectives on Science*. Sociology of the Sciences Yearbook, vol. 13. Dordrecht: Kluwer.

Funtowizc, S., and J. Ravetz. 1990. *Uncertainty and Quality in Science for Policy*. Dordrecht: Kluwer.

Futuyama, D. J. 1982. *Science on Trial: The Case for Evolution*. New York: Pantheon Books.

Gärdenfors, P. 1990. An epistemic analysis of explanations and causal beliefs. *Topoi* 9: 109–24.

Galison, P. 1987. *How Experiments End*. Chicago: University of Chicago Press.

Gardner, M., ed. 1970. *The Annotated Alice*. Rev. ed. Harmondsworth: Penguin Books. 1st ed. 1960.

Garfinkel, H. 1967. *Studies in Ethnomethodology*. Englewood Cliffs: Prentice Hall.

Geertz, C. 1973. *The Interpretation of Cultures*. New York: Basic Books.

Georgescu-Roegen, N. 1971. *The Entropy Law and the Economic Process*. Cambridge, Mass.: Harvard University Press.

Gerson, E. M. 1983. Scientific work and social worlds. *Knowledge* 4:357–77.

———. 1990. A classification of heuristics. San Francisco: Tremont Research Institute technical report.

Ghiselin, M. T. 1974a. *The Economy of Nature and the Evolution of Sex*. Berkeley and Los Angeles: University of California Press.

———. 1974b. A radical solution to the species problem. *Systematic Zoology* 23:536–44.

———. 1981. Categories, life, and thinking. *BBS* 4:269–83.

Gholson, B., W. Shadish, R. A. Neimeyer, and A. C. Houts, eds. 1989. *Psychology of Science and Metascience*. Cambridge: Cambridge University Press.

Gibson, J. J. 1950. *The Perception of the Visual World*. Boston: Houghton Mifflin.

———. 1966. *The Senses Considered as a Perceptual System*. Boston: Houghton Mifflin.

———. 1979. *The Ecological Approach to Perception*. Boston: Houghton Mifflin.

Gibson, R. F. Jr. 1986. Translation, physics, and facts of the matter. In Hahn and Schilpp, 139–54.

———. 1988. *Enlightened Empiricism: An Examination of W. V. Quine's Theory of Knowledge*. Tampa: University of South Florida Press.

Giedymin, J. 1973. Antipositivism in contemporary philosophy of social science and humanities. *Brit. J. Phil. Sci.* 26:275–301.

Giere, R. N. 1973. History and philosophy of science: Intimate relationship or marriage of convenience? *Brit. J. Phil. Sci.* 24:282–97.

———. 1983. Testing theoretical hypotheses. In Earman, 269–98.

———. 1984a. *Understanding Scientific Reasoning*. 2d ed. New York: Holt, Rinehart and Winston. 1st ed. 1979.

———. 1984b. Toward a unified theory of science. In Cushing et al., 5–31.

———. 1985a. Philosophy of science naturalized. *Phil. Sci.* 52:331–56.

———. 1985b. Background knowledge in science: A naturalistic critique. In *PSA 1984* 2:664–71.

———. 1985c. Constructive realism. In Churchland and Hooker, 75–98.

———. 1986. Controversies involving science and technology: A theoretical perspective. In A. L. Caplan and H. T. Engelhardt, Jr., eds., *Scientific Controversies: Case Studies in the Resolution and Closure of Disputes in Science and Technology*, 125–50. Cambridge: Cambridge University Press.

———. 1988. *Explaining Science: A Cognitive Approach*. Chicago: University of Chicago Press.

———. 1989a. The units of analysis of science studies. In Fuller et al., 3–11.

———. 1989b. Scientific rationality as instrumental rationality. *Stud. Hist. Phil. Sci.* 20: 377–84.

———. 1989c. What does explanatory coherence explain? *BBS* 12:475–76.

———. 1990. Evolutionary models of science. In Rescher, 21–32.

———. 1991a. The implications of the cognitive sciences for the philosophy of science. In *PSA 1990* 2:419–30.

———. 1991b. Interpreting the philosophy of science. Review of Fuller 1989. *Stud. Hist. Phil. Sci.* 22:515–23.

1991c. Syntax, semantics and human interests. *Soc. Stud. Sci.* 21:150–153.

———, ed. 1992a. *Cognitive Models of Science*.

Minnesota Studies in the Philosophy of Science, vol. 15. Minneapolis: University of Minnesota Press.

———. 1992b. The cognitive construction of scientific knowledge. [Response to Pickering 1991.] *Soc. Stud. Sci.* 22:95–107.

Giere, R. N., and R. S. Westfall. 1973. *Foundations of Scientific Method: The Nineteenth Century.* Bloomington: Indiana University Press.

Gigerenzer, G., and D. Murray. 1986. *Cognition as Intuitive Statistics.* Hillsdale: Erlbaum.

Gilbert, G. N., and M. Mulkay. 1984. *Opening Pandora's Box: A Sociological Analysis of Scientists' Discourse.* Cambridge: Cambridge University Press.

Gillispie, C. C. 1959. *The Edge of Objectivity.* Princeton: Princeton University Press.

Glassmann, R. B., and W. C. Wimsatt. 1984. Evolutionary advantages and limitations of early plasticity. In R. Almli and S. Finger, eds., *Early Brain Damage,* vol. 1, 35–58. New York: Academic Press.

Glymour, C. 1977. Indistinguishable space-times and the fundamental group. In Earman et al., 50–60.

———. 1980. *Theory and Evidence.* Princeton: Princeton University Press.

———. 1982. Conceptual scheming, or confessions of a metaphysical realist. *Synthese* 51:169–80.

Gochet, P. 1986. *Ascent to Truth: A Critical Examination of Quine's Philosophy.* München: Philosophia.

Godfrey, L. R., and J. R. Cole., eds. 1983. *Scientists Confront Creationism.* New York: Norton.

Goffman, E. 1974. *Frame Analysis: An Essay on the Organization of Experience.* New York: Harper and Row.

Goldman, A. I. 1986. *Epistemology and Cognition.* Cambridge, Mass.: Harvard University Press.

———. 1987. Foundations of social epistemics. *Synthese* 73:109–44.

———. 1991. Social epistemics and social psychology. *Soc. Epist.* 5:121–25.

———. 1992. *Liaisons: Philosophy Meets the Cognitive and Social Sciences.* Cambridge, Mass.: MIT Press, Bradford Books.

Goldschmidt, R. 1940. *The Material Basis of Evolution.* New Haven: Yale University Press.

Gooding, D. 1985. "In nature's school": Faraday as an experimentalist. In Gooding and James, 105–35.

———. 1990. *The Making of Meaning.* Dordrecht: Nijhoff.

———. 1992. The Procedural turn; or, Why do thought experiments work? In Giere 1992a.

Gooding, D., and F. A. J. L. James, eds. 1985. *Faraday Rediscovered: Essays on the Life and Work of Michael Faraday, 1791–1867.* New York: Stockton.

Goodman, N. 1978. *Ways of Worldmaking.* Indianapolis: Hackett.

Goodwin, B. C., N. Holder, and C. C. Wylie. 1983. *Development and Evolution.* Cambridge: Cambridge University Press.

Gorman, M. E. 1991. Towards a psychology of science. Review of Gholson et al. 1989. *Soc. Stud. Sci.* 21:369–74.

Gorman, M. E., and B. Carlson. 1989. Can experiments be used to study science? *Soc. Epist.* 3:89–106.

Goudge, T. A. 1961. *The Ascent of Life.* Toronto: University of Toronto Press.

Gould, S. J. 1977. *Ontogeny and Phylogeny.* Cambridge, Mass.: Harvard University Press.

———. 1978. Sociobiology: The art of storytelling. *New Scientist* (16 Nov.), 530–33.

———. 1980. *Ever Since Darwin: Reflections in Natural History.* Harmondsworth: Penguin. Orig. pub. 1978.

———. 1981. *The Mismeasure of Man.* New York: Norton.

———. 1983. *The Panda's Thumb: More Reflections in Natural History.* Harmondworth: Penguin.

———. 1984. *Hen's Teeth and Horse's Toes: Further Reflections in Natural History.* Harmondsworth: Penguin.

———. 1985. *The Flamingo's Smile: Reflections in Natural History.* Harmondsworth: Penguin.

———. 1987. Freudian slip. *Natural History* 96, no. 2 (February): 14–21.

———. 1988. *An Urchin in the Storm.* Harmondsworth: Penguin.

———. 1990. *Wonderful Life: The Burgess Shale and the Nature of History.* New York: Norton.

———. 1991. *Bully for Brontosaurus: Reflections in Natural History.* New York: Norton.

Gould, S. J., and N. Eldredge. 1977. Punctuated equilibria: The tempo and mode of evolution reconsidered. *Paleobiology* 3:115–51.

Gould, S. J., and R. C. Lewontin. 1979. The spandrels of San Marco and the Panglossian paradigm: A critique of the adaptational programme. *Proceedings of the Royal Society of London* B 205:581–98. Repr. in Sober 1984b.

Gould, S. J., and E. Vrba. 1982. Exaptation: A missing term in the science of form. *Paleobiology* 8:4–15.

Grafen, A., ed. 1989. *Evolution and its Influence.* Oxford: Clarendon.

Granger, G.-G. 1955. *La Raison*. Paris: Presses Universitaires de France.

———. 1980. Sur l'unité de la science. *Fundamenta Scientiae* 1:199–214.

———. 1992. *La vérification*. Paris: Odile Jacob.

Greene, J. C. 1981. *Science, Ideology, and World View: Essays in the History of Evolutionary Ideas*. Berkeley and Los Angeles: University of California Press.

Gregory, M. S., A. Silvers, and D. Sutch, eds. 1979. *Sociobiology and Human Nature*. San Francisco: Jossey-Bass.

Grene, M. 1959. Two evolutionary theories. *Brit. J. Phil. Sci.* 9:110–27, 185–93.

———. 1969. Bohm's metaphysics and biology. In Waddington, vol. 2, 61–69.

———. 1987. Hierarchies in biology. *Am. Scient.* 75:504–9.

———. 1988. Hierarchies and behavior. In G. Greenberg and E. Tobach, eds., *Evolution of Social Behavior and Integrative Levels,* 3–17. Proc. Third T. N. Schneirla Conference. Hillsdale, N.J.: Erlbaum.

———. 1990. Evolution, "typology" and "population thinking." *Am. Phil. Q.* 27:237–44.

Griesemer, J. R. 1984. Presentations and the status of theories. In *PSA 1984* 1:102–14.

———. 1988. Genes, memes and demes. *Biol. Phil.* 3:179–84.

———. 1991. Material models in biology. In *PSA 1990* 2:79–93.

———. 1992. The informational gene and the substantial body: On the generalization of evolutionary theory by abstraction. In Cartwright and Jones,———.

Griesemer, J. R., and M. J. Wade. 1988. Laboratory models, causal explanation, and group selection. *Biol. Phil.* 3:67–96.

Griesemer, J. R., and W. C. Wimsatt. 1989. Picturing Weismannism: A case study of conceptual evolution. In Ruse 1989d, 75–137.

Grimshaw, J. 1986. *Feminist Philosophers: Women's Perspectives on Philosophical Traditions*. Brighton: Harvester Press, Wheatsheaf Books.

Gross, A. G. 1990. *The Rhetoric of Science*. Cambridge, Mass.: Harvard University Press.

Grove, J. W. 1980. Popper 'demystified': The curious ideas of Bloor (and some others) about World 3. *Phil. Soc. Sci.* 1980:173–80.

Gruber, H. E. 1981. *Darwin on Man: A Psychological Study of Scientific Creativity*. 2d ed. Chicago: University of Chicago Press.

Grünbaum, A. 1984. *The Foundations of Psychoanalysis: A Philosophical Critique*. Berkeley and Los Angeles: University of California Press.

———. 1986. Précis of Grünbaum 1984. *BBS* 9:217–84.

———. 1989. Why thematic kinship between events does not attest their causal linkage. *Epistemologia* 13:187–208.

Guichard, L., and R. Looijen. N.d. Some philosophical arguments against sociological eliminativism: Reconciling sociology and philosophy of science. Ms.

Gutting, G. 1984. Paradigms and hermeneutics: A dialogue on Kuhn, Rorty, and the social sciences. *Am. Phil. Q.* 21:1–15.

Guyot, K. 1987. Specious individuals. *Philosophica* 37:101–26.

Hacking, I. 1975. *Why Does Language Matter to Philosophy?* Cambridge: Cambridge University Press.

———, ed. 1981a. *Scientific Revolutions*. Oxford: Oxford University Press.

———. 1981b. Introduction. In Hacking 1981a, 1–5.

———. 1983. *Representing and Intervening: Introductory Topics in the Philosophy of Natural Science*. Cambridge: Cambridge University Press.

———. 1988. The participant irrealist at large in the laboratory. *Brit. J. Phil. Sci.* 39:277–94.

———. 1989. Extragalactic reality: The case of gravitational lensing. *Phil. Sci.* 56:555–81.

Hahlweg, K., and C. A. Hooker, eds. 1989. *Issues in Evolutionary Epistemology*. Albany: State University of New York Press.

Hahn, L. E., and P. A. Schilpp, eds. 1986. *The Philosophy of W. V. Quine*. The Library of Living Philosophers, vol. 18. La Salle: Open Court.

Haldane, J. B. S. 1930. *Possible Worlds, and Other Essays*. London: Chatto and Windus.

Hamilton, W. D. 1963. The evolution of altruistic behavior. *Am. Natural.* 97:354–56.

———. 1964. The genetical evolution of social behavior. *J. Theor. Biol.* 7:1–16, 17–51. Repr. in Caplan 1978b.

Hanson, N. R. 1958. *Patterns of Discovery*. Cambridge: Cambridge University Press.

———. 1963. *The Concept of the Positron: A Philosophical Analysis*. Cambridge: Cambridge University Press.

———. 1969. *Perception and Discovery: An Introduction to Scientific Inquiry*. San Francisco: Freeman, Cooper.

Haraway, D. J. 1985. Primatology is politics by other means. In *PSA 1984* 2:489–524.

Hardin, G. 1968. The tragedy of the commons. *Science* 162:1243–48.

Hardin, G., and J. Baden, eds. 1977. *Managing the Commons*. San Francisco: Freeman.

Harding, S. 1986. *The Science Question in Feminism*. Ithaca: Cornell University Press.

Harding, S., and M. Hintikka, eds. 1983. *Discovering Reality: Feminist Perspectives on Epistemology, Metaphysics, Methodology, and the Philosophy of Science*. Dordrecht: Reidel.

Harding, S., and J. F. O'Barr, eds. 1987. *Sex and Scientific Inquiry*. Chicago: University of Chicago Press.

Harman, G. 1965. The inference to the best explanation. *Phil. Rev.* 74:88–95.

———. 1983. Beliefs and concepts. [Comments on Loar 1983a.] In *PSA 1982* 2:654–61.

Harré, R. 1986. *Varieties of Realism: A Rationale for the Natural Sciences*. Oxford: Blackwell.

Harsanyi, J. 1976. *Essays on Ethics, Social Behavior, and Scientific Explanation*. Dordrecht: Reidel.

Hattiangadi, J. N. 1989. Basic Quine for social scientists. *Phil. Soc. Sci.* 19:461–81.

Haugeland, J., ed. 1981. *Mind Design: Philosophy, Psychology, Artificial Intelligence*. Montgomery, Vermont: Bradford Books.

———. 1982. Weak supervenience. *Amer. Phil. Quart.* 19:93–103.

Hausman, D. 1981. *Capital, Profits and Prices: An Essay on the Philosophy of Economics*. New York: Columbia University Press.

———, ed. 1984. *The Philosophy of Economics: An Anthology*. Cambridge: Cambridge University Press.

Hawking, S. 1985. Is the end in sight for theoretical physics? Inaugural lecture as Lucasian Professor in Mathematics, Cambridge University. Repr. in J. Boslough, *Stephen Hawking's Universe*, 119–39. New York: Avon Books, 1989.

———. 1988. *A Brief History of Time: From the Big Bang to Black Holes*. New York: Bantam Books.

Hayek, F. A. 1967. The results of human action but not of human design. In *Studies in Philosophy, Politics, and Economics*, 96–105. London: Routledge and Kegan Paul.

———. 1972. The primacy of the abstract. In A. Koestler and J. R. Smythies, eds., *Beyond Reductionism: New Perspectives in the Life Sciences*, 309–33. London: Hutchinson.

Healy, B. 1992. Is this your father's NIH? and other strategic questions. *Science* 257:312–13, 414–15.

Hebb, D. O. 1949. *The Organization of Behavior*. New York: Wiley.

Heidegger, M. 1962. *Being and Time*. Trans. J. Macquarrie and E. S. Robinson. New York: Harper. German orig. pub. 1927.

Heil, J. 1983. *Perception and Cognition*. Berkeley and Los Angeles: University of California Press.

Hempel, C. G. 1965. *Aspects of Scientific Explanation and Other Essays in the Philosophy of Science*. New York: Free Press.

———. 1966. *Philosophy of Natural Science*. Englewood Cliffs: Prentice-Hall.

Henderson, D. K. 1990. On the sociology of science and the continuing importance of epistemologically couched accounts. *Soc. Stud. Sci.* 20:113–48.

Henderson, L. J. 1958. *The Fitness of the Environment*. Boston: Beacon Press. Orig. pub. 1913.

Hernnstein, R. 1988. A behavioural alternative to utility maximization. In S. Maital, ed., *Applied Behavioural Economics*, vol.1, 3–60. London: Wheatsheaf.

Hesse, M. 1966. *Models and Analogies in Science*. Notre Dame: University of Notre Dame Press.

———. 1980. *Revolutions and Reconstructions in the Philosophy of Science*. Brighton: Harvester Press.

Heyes, C. M. 1987. Cognisance of consciousness in the study of animal knowledge. In Callebaut and Pinxten, 105–36.

———. 1989. Uneasy chapters in the relationship between psychology and epistemology. In Gholson et al., 115–37.

———. 1991. Who's the horse? A response to Corlett. *Soc. Epist.* 5:127–34.

Hill, C. S. 1984. In defense of type materialism. *Synthese* 59:295–320.

Hirsch, G. 1985. Les mathématiques et les "révolutions scientifiques." *Bull. Soc. Math. Belg.* 37:19–62.

Hodge, M. J. S. 1989. Darwin's theory and Darwin's argument. In Ruse 1989d, 163–82.

Hofstadter, D. R. 1981. Mathematical chaos and strange attractors. *Sci. Am.* (November), Mathematical Games section. Repr. with postscript in *Metamagical Themas: Questing for the Essence of Mind and Pattern*. Harmondsworth: Penguin Books, 1986.

Hofstadter, D. R., and D. C. Dennett. comp., arr. 1981. *The Mind's I: Fantasies and Reflections on Self and Soul*. New York: Basic Books.

Holland, J. H. 1992a. *Adaptation in Natural and Artificial Systems*. Cambridge, Mass.: MIT Press.

———. 1992b. Genetic algorithms. *Sci. Am.* 267 (July): 44–50.

Holland, J. H., K. J. Holyoak, R. E. Nisbett, and P. R. Thagard. 1986. *Induction: Processes of Influence, Learning and Discovery*. Cambridge, Mass.: MIT Press.

Hollis, M., and S. Lukes, eds. 1982. *Rationality and Relativism*. Cambridge, Mass.: MIT Press.

Hollis, M., and E. J. Nell. 1975. *Rational Economic Man: A Philosophical Critique of Neo-Classical Economics*. Cambridge: Cambridge University Press.

Holmes, F. L. 1974. *Claude Bernard and Animal Chemistry: The Emergence of a Scientist*. Cambridge, Mass.: Harvard University Press.

Holton, G. 1973. *Thematic Origins of Scientific Thought: Kepler to Einstein*. Cambridge, Mass.: Harvard University Press.

———. 1986. *The Advancement of Science, and its Burdens*. Cambridge: Cambridge University Press.

Holzner, B., K. Knorr, and H. Strasser, eds. 1982. *The Political Realization of Social Science Knowledge and Research: Toward New Scenarios*. Cambridge, Mass. and Würzburg: Physica.

Hones, M. J. 1991. Scientific realism and experimental practice in high-energy physics. *Synthese* 86: 29–76.

Hooker, C. A. 1987. *A Realistic Theory of Science*. Albany: State University of New York Press.

Hooker, C. A., J. J. Leach, and E. F. McClellen, eds. 1978. *Foundations and Applications of Decision Theory*. 2 vols. Dordrecht: Reidel.

Horan, B. 1989. Functional explanations in sociobiology. *Biol. Phil.* 4:131–58.

Howe, H., and J. Lyne. 1992. Gene talk in sociobiology. *Soc. Epist.* 6:109–63.

Howson, C., and P. Urbach. 1989. *Scientific Reasoning: The Bayesian Approach*. LaSalle, Ill.: Open Court.

Hull, D. L. 1965. The effects of essentialism on taxonomy: Two thousand years of stasis. *Brit. J. Phil. Sci.* 15:314–26, 16:1–18.

———. 1967. Certainty and circularity in evolutionary taxonomy. *Evolution* 21:174–89.

———. 1968. The operational imperative: Sense and nonsense in operationism. *Systematic Zoology* 17:438–57.

———. 1969. What philosophy of biology is not. *J. Hist. Biol.* 2:241–68; *Synthese* 20:157–84.

———. 1973. *Darwin and his Critics: The Reception of Darwin's Theory of Evolution by the Scientific Community*. Cambridge, Mass.: Harvard University Press. Repr. Chicago: University of Chicago Press, 1983.

———. 1974. *Philosophy of Biological Science*. Englewood Cliffs: Prentice-Hall.

———. 1976a. Are species really individuals? *Systematic Zoology* 25:174–91.

———. 1976b. Informal aspects of theory reduction. In *PSA 1974*, 653–70.

———. 1978. A matter of individuality. *Phil. Sci.* 45:335–60. Repr. in Sober 1984b.

———. 1979. Philosophy of biology. In Asquith and Kyburg, 421–35.

———. 1981. Units of evolution: A metaphysical essay. In Jensen and Harré, 23–44. Brighton: Harvester Press. Repr. in Brandon and Burian 1984.

———. 1982. The naked meme. In Plotkin, 273–327.

———. 1983. Exemplars and scientific change. In *PSA 1982* 2:479–503.

———. 1984. Platonic fish scales. Abstracts, Conference on Integrating Scientific Disciplines, 11–12. Atlanta: Georgia State University.

———. 1985a. Darwinism as a historical entity. In Kohn, 773–812.

———. 1985b. Openness and secrecy in science: The origins and limitations. *Sci. Technol. Human Values* 10:4–13.

———. 1987. Genealogical actors in ecological roles. *Biol. Phil.* 2:168–84.

———. 1988a. *Science as a Process: An Evolutionary Account of the Social and Conceptual Development of Science*. Chicago: University of Chicago Press.

———. 1988b. A mechanism and its metaphysics: An evolutionary account of the social and conceptual development of science. *Biol. Phil.* 3:123–55. Author's response. *Biol. Phil.* 3:241–63.

———. 1988c. Progress in ideas of progress. In Nitecki, 27–48.

———. 1988d. Interactors versus vehicles. In Plotkin 1988a, 19–50.

———. 1989. *The Metaphysics of Evolution*. Stony Brook: State University of New York Press.

———. 1992. An evolutionary account of science: A response to Rosenberg's Critical Notice. *Biol. Phil.* 7:229–36.

Hull, D. L., P. Tessner, and A. Diamond. 1978. Planck's principle. *Science* 202:717–23. Repr. in Hull 1989.

Husserl, E. 1962. *Die Krisis der europäischen Wissenschaften und die transzendentale Phänomenologie. Eine Einleitung in die phänomenologische Philosophie*. Ed. W. Biemel. 2d ed. The Hague: Nijhoff.

Hutchison, T. W. 1960. *The Significance and Basic Postulates of Economic Theory*. New York: A. M. Kelley. Orig. pub. 1938.

Huxley, J. 1974. *Evolution: The Modern Synthesis*. 3d ed. Ed. J. R. Baker. London: Allen and Unwin. 1st ed. 1942.

Jackson, F. 1982. Epiphenomenal qualia. *Phil. Quart.* 32:127–36.

Jacob, F. 1977. Evolution and tinkering. *Science* 196:1161–66.

Jacquard, A. 1978. *Éloge de la différence: la génétique et les hommes.* Paris: Seuil.

————. 1989. *Idées vécues.* With H. Amblard. Paris: Flammarion.

————. 1991. *L'héritage de la liberté: De l'animalité à l'humanitude.* Paris: Seuil.

Janzen, D. H. 1980. What is coevolution? *Evolution* 34:611–12.

Jeffrey, R. C. 1983. Bayesianism with a human face. In Earman, 133–56.

Jensen, U. L., and R. Harré, eds. 1981. *The Philosophy of Evolution.* Brighton: Harvester Press.

Johnson-Laird, P. N. 1983. *Mental Models: Towards a Cognitive Science of Language, Inference, and Consciousness.* Cambridge: Cambridge University Press.

Johnston, T. 1981. Selective costs and benefits in the evolution of learning. *Advances in the Study of Behavior* 12:65–106.

Kahneman, D., B. Slovic, and A. Tversky, eds. 1982. *Judgement under Uncertainty.* Cambridge: Cambridge University Press.

Kane, J. 1984. *Beyond Empiricism: Michael Polanyi Reconsidered.* Bern: Peter Lang.

Kary, C. 1982. Can Darwinian inheritance be extended from biology to epistemology? In *PSA 1982* 1:356–69.

Kauffman, S. 1969. Metabolic stability and epigenesis in randomly constructed genetic networks. *J. Theor. Biol.* 22:437–67.

————. 1972. Articulation of parts explanations in biology and the rational search for them. In *PSA 1970,* 257–72.

————. 1974. The large scale structure and dynamics of gene control circuits: An ensemble approach. *J. Theor. Biol.* 44:167–82.

————. 1986. A framework to think about evolving genetic regulatory systems. In Bechtel 1986a, 165–84.

————. 1992. *The Origins of Order: Self Organization and Selection in Evolution.* Oxford: Oxford University Press.

Keene, G. B. 1990. The irrelevance of classical logic. *Phil. Quart.* 41:76–82.

Kekes, J. 1983. Philosophy, historicism, and foundationalism. *Philosophia* 13:213–33.

Keller, E. F. 1983. *A Feeling for the Organism: The Life and Work of Barbara McClintock.* San Francisco: Freeman.

————. 1985. *Reflections on Gender and Science.* New Haven: Yale University Press.

————. Between language and science: The question of directed mutation in molecular genetics. *Perspect. Biol. Med.* 35:292–306.

Keller, E. F., and E. S. Lloyd, eds. 1992. *Keywords in Evolutionary Biology.* Cambridge, Mass.: Harvard University Press.

Kim, J. 1978. Supervenience and nomological incommensurables. *Am. Phil. Q.* 15:149–56.

————. 1982. Psychophysical supervenience. *Phil. Stud.* 41:41–70.

Kimura, M. 1983. *The Neutral Theory of Molecular Evolution.* Cambridge: Cambridge University Press.

Kincaid, H. 1988. Supervenience and explanation. *Synthese* 77:251–81.

Kitcher, P. 1976. Explanation, conjunction and unification. *J. Phil.* 73:207–12.

————. 1978. Theories, theorists and theoretical change. *Phil. Rev.* 87:518–47.

————. 1981. Explanatory unification. *Phil. Sci.* 48:507–31.

————. 1982. *Abusing Science: The Case Against Creationism.* Cambridge, Mass.: MIT Press.

————. 1983. *The Nature of Mathematical Knowledge.* Oxford: Oxford University Press.

————. 1985a. *Vaulting Ambition: Sociobiology and the Quest for Human Nature.* Cambridge, Mass.: MIT Press.

————. 1985b. Two approaches to explanation. *J. Phil.* 82:632–39.

————. 1987a. Why not the best? In Dupré, 77–102.

————. 1987b. Précis of *Vaulting Ambition. BBS* 10:61–100.

————. 1987c. The transformation of human sociobiology. In *PSA 1986* 2:63–74.

————. 1988a. Mathematical progress. *Rev. Int. Phil.* 4:518–40.

————. 1988b. The animal within: Biology and the social sciences. *L.S.E. Quart.* 2:339–59.

————. 1988c. Imitating selection. In M.-W. Ho and S. W. Fox, eds., *Evolutionary Process and Metaphors,* 295–318. New York: Wiley.

————. 1988d. Mathematical naturalism. In Aspray and Kitcher, eds., 293–325.

————. 1989. Explanatory unification and the causal structure of the world. In Kitcher and Salmon, 410–505.

————. 1990a. Developmental decomposition and the future of human behavioral ecology. *Phil. Sci.* 57:96–117.

————. 1990b. The division of cognitive labor. *J. Phil.* 87:5–22.

————. 1992a. The naturalists return. *Phil. Rev.* 101:53–114.

————. 1992b. Authority, deference, and the role of individual reason. In McMullin 1992.

————. 1993. *The Advancement of Science.* Oxford: Oxford University Press.

Kitcher, P., and W. Salmon. 1987. Van Fraassen on explanation. *J. Phil.* 84:315–30.

————, eds. 1989. *Scientific Explanation.* Minnesota Studies in the Philosophy of Science, vol. 13. Minneapolis: University of Minnesota Press.

Klamer, A. 1984. *Conversations with Economists*. Totowa: Rowman and Allanheld.

Klayman, J., and Y.-W. Ha. 1987. Confirmation, disconfirmation, and information in hypothesis testing. *Psych. Rev.* 94:211–28.

Kleiman, D. 1977. Monogamy in mammals. *Quart. Rev. Biol.* 52:39–69.

Kline, M. 1972. *Mathematical Thought from Ancient to Modern Times*. Oxford: Oxford University Press.

Knorr, K. D. 1979. Tinkering toward success: Prelude to a theory of scientific practice. *Theory and Society* 8:347–76.

———. 1980. The scientist as an analogical reasoner: A critique of the metaphor theory of innovation. In Knorr, Krohn, and Whitley, 25–52.

Knorr, K. D., M. Haller, and H. G. Zilian. 1979. *Erkenntnis- und Verwertungsbedingungen sozialwissenschaftlicher Forschung*. Wien: Jugend und Volk.

Knorr-Cetina, K. D. 1981a. *The Manufacture of Knowledge: An Essay on the Constructivist and Contextual Nature of Science*. Oxford: Pergamon.

———. 1981b. Social and scientific method, or what do we make of the distinction between the natural and the social sciences? *Phil. Soc. Sci.* 11:335–59.

———. 1982. Scientific communities or transepistemic arenas of research. *Soc. Stud. Sci.* 12:101–30.

———. 1983. New developments in science studies: The ethnographic challenge. *Can. J. Sociol.* 8:153–77.

———. 1984. The fabrication of facts: Toward a microsociology of scientific knowledge. In N. Stehr and V. Meja, eds., *The Sociology of Knowledge*, 223–44. New Brunswick: Transaction Books.

———. 1985. Germ warfare. [Review of Latour 1984.] *Soc. Stud. Sci.* 15:577–85.

Knorr Cetina, K. D. 1987. Evolutionary epistemology and sociology of science. In Callebaut and Pinxten, 179–201.

———. 1989. *Mikrosoziale Theorien*. Weinheim: Juventa.

———. 1993a. *Epistemic Cultures: How Scientists Make Sense*. Forthcoming.

———. 1993b. *Mikrosoziologie*. Forthcoming.

Knorr-Cetina, K. D., and A. V. Cicourel, eds. 1982. *Advances in Social Theory and Methodology: Toward an Integration of Micro- and Macrosociologies*. London: Routledge and Kegan Paul.

Knorr-Cetina, K. D., R. Krohn, and R. Whitley, eds., 1980. *The Social Process of Scientific Investigation*. Dordrecht: Reidel.

Knorr-Cetina, K. D., and M. Mulkay, eds. 1983. *Science Observed: Perspectives on the Social Study of Science*. London: Sage.

Kochen, M. 1979. Fragmentation and integration in the growth of science and community development. In Callebaut et al., 522–35.

Kohlberg, L. 1973. *Collected Papers on Moral Development and Moral Education*. Cambridge, Mass.: Harvard University Press.

Kohn, D., ed. 1985. *The Darwinian Heritage*. Princeton: Princeton University Press.

Koppelberg, D. 1987. *Die Aufhebung der analytischen Philosophie: Quine als Synthese von Carnap und Neurath*. Frankfurt am Main: Suhrkamp.

Kornblith, H., ed. 1985a. *Naturalizing Epistemology*. Cambridge, Mass.: MIT Press, Bradford Books.

———. 1985b. Introduction: What is naturalistic epistemology? In Kornblith 1985a, 1–13.

Kornfeld, W. A., and C. E. Hewitt. 1981. The scientific community metaphor. *IEEE Trans. SMC* 11:24–33.

Kourany, J. 1979. The nonhistorical basis of Kuhn's theory of science. *Nature and System* 1:46–59.

Koyré, A. 1954. Influence of philosophical trends on the formulation of scientific theories. In P. Frank, ed., *The Validation of Scientific Theories*, 177–87. Boston: Beacon.

———. 1958. *From the Closed World to the Infinite Universe*. Evanston: Harper and Row.

———. 1968. *Metaphysics and Measurement: Essays in Scientific Revolution*. Cambridge, Mass.: Harvard University Press.

———. 1973. *The Astronomical Revolution*. Ithaca: Cornell University Press. French orig. pub. 1961.

Krajewski, W. 1977. *Correspondence Principle and Growth of Science*. Dordrecht: Reidel.

Krantz, D. L., and L. Wiggins. 1973. Personal and impersonal channels of recruitment in the growth of theory. *Human Development* 16:133–56.

Krohn, W., and G. Küppers. 1989. Selforganization: A new approach to evolutionary epistemology. In Hahlweg and Hooker, 151–70.

Krohn, W., G. Küppers, and H. Nowotny, eds. 1990. *Selforganization: Portrait of a Scientific Revolution*. Dordrecht: Kluwer.

Krüger, L., L. Daston, and M. Heidelberger, eds. 1987. *The Probabilistic Revolution*. Vol. 1, *Ideas in History*. Cambridge, Mass.: MIT Press, Bradford Books.

Krüger, L., G. Gigerenzer, and M. S. Morgan, eds.

1987. *The Probabilistic Revolution.* Vol. 2, *Ideas in the Sciences.* Cambridge, Mass.: MIT Press, Bradford Books.

Kuhn, T. S. 1970. *The Structure of Scientific Revolutions.* 2d ed. Chicago: University of Chicago Press. 1st ed. 1962.

———. 1977. *The Essential Tension.* Chicago: University of Chicago Press.

———. 1979. History of science. In Asquith and Kyburg, 121–28.

———. 1980. The halt and the blind. *Brit. J. Phil. Sci.* 31:181–92.

———. 1983. Commensurability, comparability, communicability. In *PSA 1982* 2:669–88.

———. 1987. What are scientific revolutions? In Krüger, Daston, and Heidelberger, 7–22.

———. 1991. The road since structure. In *PSA 1990,* 2:3–13.

———. N.d. Scientific development and lexical change. The Thalheimer Lectures, The Johns Hopkins University, 12 to 19 November 1984. Typescript.

Kulkarni, D., and H. A. Simon. 1988. The processes of scientific discovery: The strategy of experimentation. *Cognitive Science* 12:139–76.

———. 1990. Experimentation in machine discovery. In Shrager and Langley, 255–73.

Kurtz, P. 1990. *Philosophic Essays in Pragmatic Naturalism.* Buffalo: Prometheus.

LaFollette, M. C., ed. 1983. *Creationism, Science, and the Law: The Arkansas Case.* Cambridge, Mass.: MIT Press.

Lakatos, I. 1970. Falsification and the methodology of scientific research programmes. In Lakatos and Musgrave, 91–196. Repr. in Lakatos 1978a.

———. 1971. History of science and its rational reconstruction. In *PSA 1970*, 91–136.

———. 1976. *Proofs and Refutations.* Ed. J. Worrall and E. Zahar. Cambridge: Cambridge University Press.

———. 1978a. *The Methodology of Scientific Research Programmes.* Vol. 1 of *Philosophical Papers.* Cambridge: Cambridge University Press.

———. 1978b. *Mathematics, Science and Epistemology.* Vol. 2 of *Philosophical Papers.* Cambridge: Cambridge University Press.

Lakatos, I., and A. Musgrave, eds. 1970. *Criticism and the Growth of Knowledge.* Cambridge: Cambridge University Press.

Lamarck, J. B. 1984. *Zoological Philosophy: An Introduction with Regard to the Natural History of Animals.* Trans. H. Elliott. With introductory essays by D. L. Hull and R. W. Burckhardt, Jr. Chicago: University of Chicago Press.

Lamb, D., and S. M. Easton. 1984. *Multiple Discovery.* [Amersham]: Avebury.

La Mettrie, J. O. de. 1960. *L'homme machine.* A. Vartanian, ed. Princeton: Princeton University Press. Published anonymously, Leyden: Luzac, 1747.

Langley, P., H. A. Simon, G. L. Bradshaw, and J. M. Zytkow. 1987. *Scientific Discovery: Computational Explorations of the Creative Process.* Cambridge, Mass.: MIT Press.

Latour, B. 1980. The three little dinosaurs, or a sociologist's nightmare. *Fund. Sci.* 1:79–85.

——— 1983. Give me a laboratory and I will move the world. In Knorr-Cetina and Mulkay, 141–74.

———. 1984. *Les Microbes: guerre et paix, suivi de Irréductions.* Paris: A.-M. Métaillé.

———. 1987. *Science in Action: How to Follow Scientists and Engineers through Society.* Milton Keynes: Open University Press.

———. 1988a. *The Pasteurization of France.* Cambridge, Mass.: Harvard University Press. Revised and amplified trans. by A. Sheridan and J. Law of Latour 1984.

———. 1988b. The Enlightenment without the Critique: An introduction to Michel Serres's Philosophy. In J. Griffith, ed., *Contemporary French Philosophy,* 83–97. Cambridge: Cambridge University Press.

———. 1988c. The politics of explanation: An alternative. In Woolgar 1988a, 155–77.

———. 1988d. Clothing the naked truth. In H. Lawson and L. Appignanesi, eds., *Dismantling Truth: Reality in the Post-Modern World,* 101–26. London: Weidenfeld and Nicolson.

———. 1988e. A relativist account of Einstein's relativity. *Soc. Stud. Sci.* 18:3–44.

———. 1988f. Mixing humans and nonhumans together: The sociology of a door-closer. *Social Problems* 35:298–310.

———. 1989. Pasteur et Pouchet: hétérogenèse de l'histoire des sciences. In Serres, 423–45.

———. 1990a. Postmodern? No simply amodern! Steps towards an anthropology of science. *Stud. Hist. Phil. Sci.* 21:145–71.

———. 1990b. The force and the reason of experiment. In LeGrand, 49–80. Dordrecht: Kluwer.

———. 1991. The impact of science studies on political philosophy. *Sci. Technol. Hum. Val.* 16:3–19.

———. 1992a. *Nous n'avons jamais été modernes.* Paris: La Découverte. English trans. *We Have Never Been Modern.* Cambridge, Mass.: Harvard University Press, 1993.

———. 1992b. *Aramis, ou l'amour des techniques.* Paris: La Découverte.

———. 1992c. One more turn after the social turn: Easing science studies into the non-modern world. In McMullin, ed.

Latour, B., and S. Strum. 1986. Human social origins: Please tell us another origin story! *J. Biol. Soc. Struct.* 9:169–87.

Latour, B., and S. Woolgar. 1979. *Laboratory Life: The Social Construction of Scientific Facts.* Los Angeles: Sage. Rev. ed., *Laboratory Life: The Construction of Scientific Facts.* Princeton: Princeton University Press, 1986.

———. 1982. The cycle of credibility. In Barnes and Edge, 35–43.

Latsis, S. J., ed. 1976. *Method and Appraisal in Economics.* Cambridge: Cambridge University Press.

Laudan, L. 1972. Commentary: Science at the bar—causes for concern. *Sci. Technol. Hum. Val.* 7 (41): 16–19. Repr. in LaFollette 1983 and Ruse 1988c.

———. 1977. *Progress and its Problems: Towards a Theory of Scientific Growth.* Berkeley and Los Angeles: University of California Press.

———. 1979. Historical methodologies: An overview and manifesto. In Asquith and Kyburg, 40–54.

———. 1981a. A confutation of convergent realism. *Phil. Sci.* 48:19–49. Repr. in Leplin 1984.

———. 1981b. The pseudo-science of science? *Phil. Soc. Sci.* 11:173–98.

———. 1982a. More on Bloor. *Phil. Soc. Sci.* 12:71–74.

———. 1982b. A note on Collins' blend of relativism and empiricism. *Soc. Stud. Sci.* 12:131–32.

———. 1984. *Science and Values: The Aims of Science and Their Role in Scientific Debate.* Berkeley and Los Angeles: University of·California Press.

———. 1987a. Progress or rationality? The prospects for normative naturalism. *Am. Phil. Q.* 24:19–31.

———. 1987b. Methodology's prospects. In *PSA 86* 2:347–54.

———. 1987c. Relativism, naturalism and reticulation. *Synthese* 71:221–34.

———. 1988. Are all theories equally good? A dialogue. In Nola, 117–39.

———. 1990a. *Science and Relativism: Some Key Controversies in the Philosophy of Science.* Chicago: University of Chicago Press.

———. 1990b. Aim-less epistemology? *Stud. Hist. Phil. Sci.* 21:315–22.

Laudan, R., L. Laudan, and A. Donovan. 1988. Testing theories of scientific change. In Donovan et al., 3–44.

Le Grand, H., ed. 1990. *Experimental Inquiries: Historical, Philosophical, and Social Studies of Experimentation in Science.* Dordrecht: Kluwer.

Lehrer, K. 1974. *Knowledge.* Oxford: Clarendon Press.

Lehrer, K., and C. Wagner. 1981. *Rational Con-*

sensus in Science and Society. Dordrecht: Reidel.

Leibenstein, H. 1976. *Beyond Economic Man: A New Foundation for Microeconomics.* Cambridge, Mass.: Harvard University Press.

Lenat, D. 1978. The ubiquity of discovery. *Artificial Intelligence* 9:257–85.

Leontief, W. 1971. Theoretical assumptions and nonobserved facts. *Am. Econ. Rev.* 61:1–7.

Leplin, J., ed. 1984. *Scientific Realism.* Berkeley and Los Angeles: University of California Press.

Levi, I. 1967. *Gambling With Truth.* New York: Knopf.

———. 1986. *Hard Choices.* Cambridge: Cambridge University Press.

Levins, R. 1966. The strategy of model-building in population biology. *Am. Scient.* 54:421–31. Repr. in Sober 1984b.

———. 1968. *Evolution in Changing Environments.* Princeton: Princeton University Press.

———. 1975. Evolution in communities near equilibrium. In M. L. Cody and J. Diamond, eds., *Ecology and Evolution of Communities,* 16–48. Cambridge, Mass.: Harvard University Press.

Levins, R., and R. C. Lewontin. 1985. *The Dialectical Biologist.* Cambridge, Mass.: Harvard University Press.

Levins, R., and R. MacArthur. 1966. The maintenance of genetic polymorphism in a spatially heterogeneous environment: Variations on a theme by Harold Levene. *Am. Natural.* 100:585–89.

Lévi-Strauss, C., and D. Eribon. 1990. *De près et de loin. Suivi d'un entretien "Deux ans après."* Paris: Odile Jacob.

Levy, M. J. 1982. Review of Knorr-Cetina 1981a. *Knowledge* 4:147–54.

Lévy-Garboua, L., ed. 1979. *Sociological Economics.* London: Sage.

Lewin, B. 1987. *Genes.* 3d ed. New York: Wiley.

Lewontin, R. C. 1961. Evolution and the theory of games. *J. Theor. Biol.* 1:382–403. Repr. in M. Grene and E. Mendelsohn, eds., *Topics in the Philosophy of Biology.* Dordrecht: Reidel, 1976.

———. 1963. Models, mathematics, and metaphors. *Synthese* 15:222–44.

———. 1965. Selection in and of populations. In J. A. Moore, ed., *Ideas in Modern Biology,* 299–311. Garden City: National History Press.

———. 1967. The principle of historicity in evolution. In P. S. Moorhead and M. M. Kaplan, eds., *Mathematical Challenges to the Neo-Darwinian Interpretation of Evolution,* 81–88. Philadelphia: Wistar Institute Press.

————. 1968. The concept of evolution. In E. Shils, ed., *International Encyclopedia of the Social Sciences,* vol. 5, 202–10. New York: Macmillan, Free Press.

————. 1969. The bases of conflict in biological explanation. *J. Hist. Biol.* 2:35–45.

————. 1970. The units of selection. *Ann. R. Ecol.* 1:1–23.

————. 1974a. *The Genetic Basis of Evolutionary Change.* New York: Columbia University Press.

————. 1974b. The analysis of variance and the analysis of causes. *Am. J. Human Genetics* 26:400–11.

————. 1974c. Darwin and Mendel: The materialist revolution. In J. Neyman, ed., *The Heritage of Copernicus: Theories "More Pleasing to the Mind,"* 166–83. Cambridge, Mass.: MIT Press.

————. 1976. Sociobiology: A caricature of Darwinism. In *PSA 1976* 2:22–31.

————. 1977. Biological determinism as a social weapon. In *Biology as a Social Weapon,* 6–18. Minneapolis: Burgess.

————. 1978. Adaptation. *Sci. Am.* 239 (September): 212–30.

———— 1979. Work collectives: Utopian and otherwise. *Radic. Sci. J.* no. 8:133–37.

————. 1981. Sleight of hand. Review of Lumsden and Wilson 1981. *The Sciences* (July/August): 23–26.

————. 1982a. *Human Diversity.* Redding, Conn.: Scientific American, Freeman.

————. 1982b. Organism and environment. In Plotkin, 151–70.

————. 1983a. Darwin's revolution. *New York Review of Books* 30:21–27.

————. 1983b. Biological determinism. *Tanner Lectures on Human Values* 4:149–83. Salt Lake City: University of Utah Press.

————. 1985. Population genetics. *Ann. Rev. Gen.* 19:81–102.

————. 1987. The shape of optimality. In Dupré, 151–59.

————. 1989. The evolution of cognition. In D. Osherson, ed., *An Invitation to Cognitive Science,* vol. 3, 229–46. Cambridge, Mass.: MIT Press.

————. 1990. Fallen angels. [Review of Gould 1990.] *New York Review of Books* 37 (June 14): 3–7.

————. 1991. Review of Lloyd [1988.] *Biol. Phil.* 6:461–66.

————. 1992. Gene talk on target. *Soc. Epist.* 6:179–82.

Lewontin, R. C., and M. J. D. White. 1960. Interaction between inversion polymorphisms of two chromosome pairs in the grasshopper *Moraba scurra. Evolution* 14:116–29.

Lieberman, P. 1984. *The Biology and Evolution of Language.* Cambridge, Mass.: Harvard University Press.

Lighthill, J. 1986. The recently recognized failure of predictability in Newtonian dynamics. *Proceedings of the Royal Society of London* A 407:35–50.

Little, D. 1986. *The Scientific Marx.* Minneapolis: University of Minnesota Press.

Lloyd, E. A. 1983. The nature of Darwin's support for the theory of natural selection. *Phil. Sci.* 50:112–29.

————. 1984. A semantic approach to the structure of population genetics. *Phil. Sci.* 51:242–64.

————. 1986a. Thinking about models in evolutionary theory. *Philosophica* 37:87–100.

————. 1986b. Evaluation of evidence in group selection debates. In *PSA 1986* 1:483–93.

————. 1987a. Confirmation of evolutionary and ecological models. *Biol. Phil.* 2:277–93.

————. 1987b. Response to Sloep and van der Steen. *Biol. Phil.* 2:23–26.

————. 1988. *The Structure and Confirmation of Evolutionary Theory.* Westport, Conn.: Greenwood Press.

————. 1989. A structural approach to defining units of selection. *Phil. Sci.* 56:395–418.

————. N.d.a. *All About Eve: Bias in Evolutionary Explanations of Women's Sexuality.* Cambridge, Mass.: Harvard University Press.

————. N.d.b. Different questions. Ms.

Lloyd, E. A., and S. J. Gould. N.d. Species selection on variability. *Proc. Nat. Acad. Sci. USA.* In press.

Loar, B. 1983a. Must beliefs be sentences? In *PSA 1982* 2:627–43.

————. 1983b. Reply to Fodor and Harman. In *PSA 1982* 2:662–66.

Locke, J. 1959. *An Essay Concerning Human Understanding.* A. C. Fraser, ed. 2 vols. New York: Dover. Orig. pub. 1690.

Lockery, S. R., G. Wittenberg, W. B. Kristan, Jr., and G. W. Cottrell. 1989. Function of identified interneurons in the leech elucidated using neural networks trained by back-propagation. *Nature* 340:468–71.

Longino, H. E. 1986. Science overrun: The threat to freedom by external control. In M. Goggin et al., ed., *Governing Science and Technology in a Democracy,* 57–73. Knoxville: University of Tennessee Press.

————. 1989. Feminist critiques of rationality: Critiques of science or philosophy of science? *Women's Stud. Int. Forum* 12:261–69.

———. 1990. *Science as Social Knowledge: Values and Objectivity in Scientific Inquiry.* Princeton: Princeton University Press.

Longino, H. E., and R. Doell. 1983. Body, bias and behavior: A comparative analysis of reasoning in two areas of biology. *Signs* 9:206–27. Repr. in Harding and O'Barr 1987.

Lorenz, K. 1969. Innate bases of learning. In K. Pribram, ed., *On the Biology of Learning,* 13–93. New York: Harcourt, Brace and World.

———. 1982. Kant's doctrine of the a priori in the light of contemporary biology. In Plotkin, 121–43. German orig. pub. 1941.

Losee, J. 1977. Limitations of an evolutionist philosophy of science. *Stud. Hist. Sci.* 8:349–52.

———. 1980. *A Historical Introduction to the Philosophy of Science.* Oxford: Oxford University Press.

Luce, R. D., and H. Raiffa. 1957. *Games and Decisions.* New York: Wiley.

Lugg, A. 1983. Explaining scientific beliefs: The rationalist's strategy re-examined. *Phil. Soc. Sci.* 13:265–78.

Luhmann, N. 1990a. *Die Wissenschaft der Gesellschaft.* Frankfurt am Main: Suhrkamp.

———. 1990b. The cognitive program of constructivism and a reality that remains unknown. In Krohn et al., 64–85.

Lumsden, C. J., and E. O. Wilson. 1981. *Genes, Mind, and Culture: The Coevolutionary Process.* Cambridge, Mass.: Harvard University Press.

———. 1982. Précis of *Genes, Mind, and Culture. BBS* 5:1–37.

———. 1983. *Promethean Fire: Reflections on the Origin of Mind.* Cambridge, Mass.: Harvard University Press.

Lycan, W. 1985. Epistemic value. *Synthese* 64:137–64.

———. 1987. *Consciousness.* Cambridge, Mass.: MIT Press, Bradford Books.

———, ed. 1990. *Mind and Cognition: A Reader.* Oxford: Blackwell.

Lynch, M., and S. Woolgar, eds. 1990. *Representation in Scientific Practice.* Cambridge, Mass.: MIT Press. Orig. pub. 1988.

Lynch, W. T., and E. R. Fuhrman. 1991. Recovering and expanding the normative: Marx and the new sociology of scientific knowledge. *Sci. Technol. Hum. Val.* 16:233–48.

MacArthur, R. H., and E. O. Wilson. 1967. *Theory of Island Biogeography.* Princeton: Princeton University Press.

McClelland, J. L., D. E. Rumelhart, and the PDP Research Group. 1986. *Parallel Distributed Processing: Explorations in the Microstructure of Cognition.* Vol. 2, *Psychological and Biological Models.* Cambridge, Mass.: MIT Press, Bradford Books.

McClelland, J. L., and D. E. Rumelhart. 1988. *Explorations in Parallel Distributed Processing: A Handbook of Models, Programs, and Exercises.* Cambridge, Mass.: MIT Press, Bradford Books.

McCloskey, D. 1983. The rhetoric of economics. *J. Econ. Lit.* 21:481–517.

———. 1986. *The Rhetoric of Economics.* Madison: University of Wisconsin Press.

Machlup. F. 1964. Professor Samuelson on theory and realism. *Am. Econ. Rev.* 54:733–36.

McKenzie, R. B., and G. Tullock. 1978. *The New World of Economics: Explorations into the Human Experience.* 2d ed. Homewood: Irwin. 1st ed. 1975.

Mackie, J. 1978. The law of the jungle. *Philosophy* 53:553–73.

McMullin, E. 1983. Values in science. In *PSA 1982* 2:3–28.

———. 1984. Stability and change in science. *New Ideas Psychol.* 2:9–19.

———, ed. 1988. *Construction and Constraint: The Shaping of Scientific Rationality.* Notre Dame: University of Notre Dame Press.

———, ed. 1992. *The Social Dimension of Science.* Notre Dame: University of Notre Dame Press.

Maffie, J. 1991. What is social about social epistemics? *Soc. Epist.* 5:101–10.

Magee, B. 1973. *Popper.* London: Fontana.

Maienschein, J. 1990. From the President. *ISHPSSB Newsletter* no. 2 (Spring): 1.

Malthus, T. R. 1989. *An Essay on the Principles of Population.* Ed. P. James. 2 vols. Cambridge: Cambridge University Press. Orig. pub. 1798.

Manheim, J. H. 1964. *The Genesis of Point Set Topology.* New York: Macmillan.

Manier, E. 1978. History, philosophy and sociology of science: A family romance. *Stud. Hist. Phil. Sci.* 11:1–24.

Mannison, D., M. A. McRobbie, and R. Routley, eds. 1980. *Environmental Philosophy.* Canberra: Australian National University.

Manser, A. R. 1965. The concept of evolution. *Philosophy* 40:18–34.

Margolis, H. 1982. *Selfishness, Altruism, and Rationality.* Cambridge: Cambridge University Press.

———. 1987. *Patterns, Thinking, and Cognition: A Theory of Judgment.* Chicago: University of Chicago Press.

Margolis, J. 1986. *Pragmatism without Foundations: Reconciling Realism and Relativism.* New York: Blackwell.

Margolis, J., M. Krausz, and R. M. Burian, eds.

1986. *Rationality, Relativism and the Human Sciences.* Dordrecht: Nijhoff.

Matthew, P. 1831. *On Naval Timber and Arboriculture.* London: Longman. Repr. in H. L. McKinney, *Lamarck to Darwin: Contributions to Evolutionary Biology 1809–1859,* Lawrence, Kans.: Colorado Press, 1971.

Maturana, H. R., and F. Varela. 1980. *Autopoiesis and Cognition: The Realization of the Living.* Dordrecht: Reidel.

Maull, N. 1977. Unifying science without reduction. *Stud. Hist. Phil. Sci.* 8:143–62. Repr. in Sober 1984b.

Maynard Smith, J. 1975. *The Theory of Evolution.* 3d ed. Harmondsworth: Penguin. 1st ed. 1958.

———. 1976. Group selection. *Quart. Rev. Biol.* 51:277–83. Repr. with a preface in Brandon and Burian 1984, 238–49.

——— 1978. *The Evolution of Sex.* Cambridge: Cambridge University Press.

———. 1982. *Evolution and the Theory of Games.* Cambridge: Cambridge University Press.

———. 1984. The evolution of animal intelligence. In C. Hookway, ed., *Minds, Machines and Evolution,* 63–71. Cambridge: Cambridge University Press.

———. 1986. *The Problems of Biology.* Oxford: Oxford University Press.

———. 1987a. How to model evolution. In Dupré, 119–31.

———. 1987b. Reply to Sober. In Dupré, 147–49.

———. 1988. Mechanisms of advance. Review of Hull 1988a. *Science* 242:1182–83.

Maynard Smith, J., R. Burian, S. Kauffman, P. Alberch, J. Campbell, B. Goodwin, R. Lande, D. Raup, and L. Wolpert. 1985. Developmental constraints and evolution: A perspective from the Mountain Lake Conference on Development and Evolution. *Quart. Rev. Biol.* 60: 265–87.

Maynard Smith, J., and N. Warren. 1982. Review of Lumsden and Wilson 1981. *Evolution* 36:620–27.

Mayo, D., and N. Gilinsky. 1987. Models of group selection. *Phil. Sci.* 54:515–38.

Mayr, E. 1942. *Systematics and the Origin of Species.* New York: Columbia University Press.

———. 1959. Typological versus population thinking. In "Darwin and the evolutionary theory in biology," in *Evolution and Anthropology: A Centennial Appraisal,* 409–12. Washington, D.C.: The Anthropological Society of Washington. Repr. in Mayr 1976 and in Sober 1984b.

———. 1961. Cause and effect in biology. *Science* 134:1501–6. Repr. in Mayr 1976.

———. 1963. *Animal Species and Evolution.*

Cambridge, Mass.: Harvard University Press, Belknap Press.

———. 1969. Footnotes in the philosophy of biology. *Phil. Sci.* 36:197–202.

———. 1976. *Evolution and the Diversity of Life: Selected Essays.* Cambridge, Mass.: Harvard University Press, Belknap Press.

———. 1982. *The Growth of Biological Thought: Diversity, Evolution, and Inheritance.* Cambridge, Mass.: Harvard University Press, Belknap Press.

———. 1983. Comments on David Hull's paper on exemplars and type specimens. In *PSA 1982* 2:504–11.

———. 1984. Darwinian flights. In Weintraub, 36–55.

———. 1985. The problem of extraterrestrial intelligent life. In Regis, 23–30. Repr. in May 1988, 67–74.

———. 1988. *Toward a New Philosophy of Biology: Observations of an Evolutionist.* Cambridge, Mass.: Harvard University Press, Belknap Press.

———. 1990. When is historiography whiggish? *J. Hist. Ideas* 51:301–9.

Merton, R. K. 1959. *Social Theory and Social Structure.* Glencoe: Free Press.

———. 1973. *The Sociology of Science.* Ed. N. Storer. Chicago: University of Chicago Press. Orig. pub. 1942.

Michaels, C. F., and C. Carello. 1981. *Direct Perception.* Englewood Cliffs: Prentice-Hall.

Miller, A. 1991. Have incommensurability and causal theory of reference anything to do with actual science?—Incommensurability, no; causal theory, yes. *Int. Stud. Phil. Sci.* 5:97–108.

Miller, H. B. 1992. Not the only game in town: Zoöepistemology and ontological pluralism. *Synthese* 93:25–37.

Miller, J. G. 1978. *Living Systems.* New York: McGraw-Hill.

Millikan, R. 1984. *Language, Thought, and Other Biological Categories.* Cambridge, Mass.: MIT Press, Bradford Books.

Miner, V., and H. E. Longino, eds. 1987. *Competition: A Feminist Taboo?* New York: Feminist Press at the City University of New York.

Minsky, M., and S. Papert. 1969. *Perceptrons: An Introduction to Computational Geometry.* Cambridge, Mass.: MIT Press.

Mirowski, P. 1988a. *Against Mechanism: Protecting Economics From Science.* Totawa: Rowman & Littlefield.

———, ed. 1988b. Special issue of *Social Concept* on biological analogies in economics.

———. 1989. *More Heat Than Light: Economics*

as Social Physics; Physics as Nature's Economics. Cambridge: Cambridge University Press.

Mitroff, I. 1974. *The Subjective Side of Science: A Philosophical Inquiry into the Psychology of the Apollo Moon Scientists*. Amsterdam: Elsevier.

———. 1976. Integrating the philosophy and the social psychology of science, or a plague of two housed divided. In Cohen et al., 529–48.

Monastyrsky, M. 1987. *Riemann, Topology, and Physics*. Boston: Birkhäuser.

Montagu, A., ed. 1980. *Sociobiology Examined*. Oxford: Oxford University Press.

Moore, G. E. 1903. *Principia Ethica*. Cambridge: Cambridge University Press. Repr. 1952.

Morin, E. 1970. *Journal de Californie*. Paris: Seuil.

———. 1977–1991. *La méthode*. 4 vols. Paris: Seuil.

———. 1985. On the definition of complexity. In S. Aida et al., *The Science and Praxis of Complexity*, 62–68. Tokyo: United Nations University.

Moyer, D. F. 1979. Revolution in science: The 1919 eclipse test of general relativity. In P. Bursunoglu, A. Perlmutter, and L. F. Scott, eds., *On the Path of Albert Einstein*, 55–101. New York: Plenum Press.

Mulkay, M. 1979. *Science and the Sociology of Knowledge*. London: Allen and Unwin.

———. 1985. *The Word and the World: Explorations in the Form of Sociological Analysis*. London: Allen and Unwin.

———. 1988. Don Quixote's double: A self-exemplifying text. In Woolgar 1988a, 81–100.

———. 1991. *Sociology of Science: A Sociological Pilgrimage*. Bloomington: Indiana University Press.

Mulkay, M., and G. Gilbert. 1981. Putting philosophy to work: Karl Popper's influence on scientific practice. *Phil. Soc. Sci.* 11:389–407.

———. 1982a. What is the ultimate question? Some remarks in defence of the analysis of scientific discourse. *Soc. Stud. Sci.* 12:309–19.

———. 1982b. Accounting for error: How scientists construct their social world when they account for correct and incorrect belief. *Sociology* 16:165–83.

Munévar, G., ed. 1991. *Beyond Reason: Essays in the Philosophy of Paul Feyerabend*. Dordrecht: Kluwer.

Munz, P. 1985. *Our Knowledge of the Growth of Knowledge: Popper or Wittgenstein?* London: Routledge and Kegan Paul.

Murphy, A. E. 1945. Review of J. Dewey et al., *Naturalism and the Human Spirit*. *J. Phil.* 42:1400–17.

Murphy, G. G. S. 1969. On counterfactual propositions. *Hist. Theory* Beiheft 9:14–38.

Murphy, J. G. 1982. *Evolution, Morality and the Meaning of Life*. Totowa: Rowman and Littlefield.

Musil, R. 1953–1960. *The Man Without Qualities*. 3 vols. Trans. E. Wilkins and E. Kaiser. London: Seeker and Warburg. German orig. 1930–1943.

———. 1982. *On Mach's Theories*. Trans. K. Mulligan. Munich: Philosophia. German orig. pub. 1908.

Mynatt, C. R., M. E. Doherty, and R. D. Tweney. 1977. Confirmation bias in a simulated research environment. *Q. J. Exp. Psych.* 29:85–95. Repr. in P. N. Johnson-Laird and P. C. Wason, eds., *Thinking*. Cambridge: Cambridge University Press, 1977.

———. 1978. Consequences of confirmation and disconfirmation in a simulated research environment. *Q. J. Exp. Psych.* 30:395–406.

Nagel, E. 1961. *The Structure of Science*. New York: Harcourt, Brace, and World.

———. 1963. Assumptions in economic theory. *Am. Econ. Rev.: Papers and Proc.* 53:211–19.

———. 1977. Teleology revisited. *J. Phil.* 74:261–301. Repr. in *Teleology Revisited and Other Essays in the Philosophy and History of Science*. New York: Columbia University Press, 1979.

Nagel, T. 1974. What it is like to be a bat? *Phil. Rev.* 83:435–51. Repr. in *Mortal Questions*. Cambridge: Cambridge University Press, 1979, and in Hofstadter and Dennett 1981.

———. 1986. *The View From Nowhere*. Oxford: Oxford University Press.

Neisser, U. 1967. *Cognitive Psychology*. New York: Appleton

———. 1975. *Cognition and Reality*. San Francisco: Freeman.

Nelson, R. J. 1984. Naturalizing intentions. *Synthese* 61:174–203.

———. 1987. *The Logic of Mind*. Dordrecht: Kluwer.

Nelson, R. R. 1987. *Understanding Technical Change as an Evolutionary Process*. Amsterdam: North-Holland.

Nelson, R. R., and S. G. Winter. 1982. *An Evolutionary Theory of Economic Change*. Cambridge, Mass.: Harvard University Press, Belknap Press.

Nersessian, N. J. 1984. *Faraday to Einstein: Constructing Meaning in Scientific Theories*. Dordrecht: Nijhoff.

———, ed. 1987. *The Process of Science*. Dordrecht: Nijhoff.

———. 1988. Reasoning from imagery and analogy in scientific concept formation. In *PSA 1988*, 1:41–47.

Neurath, O., R. Carnap, and C. Morris, eds. 1955–1970. *Foundations of the Unity of Science. Toward an International Encyclopedia of Unified Science.* Vol. 1, nos. 1–10 (1955). Vol. 2, nos. 1–9 (1970). Chicago: University of Chicago Press.

Newell, A., and H. A. Simon. 1972. *Human Problem Solving.* Englewood Cliffs: Prentice-Hall.

Newton, I. 1952. *Opticks.* Repr. of 4th ed. (1730). New York: Dover.

Newton-Smith, W. H. 1981. *The Rationality of Science.* London: Routledge and Kegan Paul.

Nickles, T. 1973. Two concepts of intertheoretic reduction. *J. Phil.* 70:181–201.

———. 1976. Theory generalization, problem reduction, and the unity of science. In *PSA 1974*, 31–74.

———. 1977. Heuristics and justification in scientific research: Comments on Shapere. In Suppe, 571–89.

———. 1978. Scientific problems and constraints. In *PSA 1978*, 1:134–48.

———. 1980a. *Scientific Discovery, Logic, and Rationality.* Dordrecht: Reidel.

———. 1980b. *Scientific Discovery: Case Studies.* Dordrecht: Reidel.

———. 1980c. Scientific problems: Three empiricist models. In *PSA 1980* 1:3–19.

———. 1981. What is a problem that we may solve it? *Synthese* 47:85–118.

———. 1983. Justification as discoverability. In P. Weingartner et al., eds., *Proceedings, 7th Int. Congress on Logic, Methodology and Philosophy of Science*, vol. 6, 157–60. Salzburg: Hutteger.

———. 1984a. Positive science and discoverability. In *PSA 1984* 1:13–27.

———. 1984b. Justification as discoverability II. *Philosophia Naturalis* 21:563–76.

———. 1985. Beyond divorce: Current status of the discovery debate. *Phil. Sci.* 52:177–206.

———. 1986. Remarks on the use of history as evidence. *Synthese* 69:253–66.

———. 1987a. Twixt method and madness. In Nersessian, 41–67.

———. 1987b. Lakatosian heuristics and epistemic support. *Brit. J. Phil. Sci.* 38:181–205.

———. 1987c. From natural philosophy to metaphilosophy of science. In R. Kargon and P. Achinstein, eds., *Kelvin's Baltimore Lectures and Modern Theoretical Physics*, 507–41. Cambridge, Mass.: MIT Press.

———. 1988. Questioning and problems in philosophy of science: Problem-solving versus directly truth-seeking epistemologies. In M.

Meyer, ed., *Questions and Questioning.* Berlin: De Gruyter.

———. 1989a. Integrating the science studies disciplines. In Fuller et al., 225–56.

———. 1989b. Justification and experiment. In D. Gooding, T. Pinch, and S. Schaffer, eds., *The Uses of Experiment: Studies of Experiment in the Natural Sciences*, 299–333. Cambridge: Cambridge University Press.

———. 1989c. Truth or consequences? Generative versus consequential justification in science. In *PSA 1988* 2:393–405.

———. 1990a. Discovery logics. *Philosophica* 45:7–32.

———. 1990b. Good science is bad history. In E. McMullin, ed., *Social Aspects of Science.* Notre Dame: University of Notre Dame Press.

Nicolis, G., and I. Prigogine. 1977. *Self-organization in Nonequilibrium Systems: From Dissipative Structures to Order Through Fluctuations.* New York: Wiley.

Nitecki, M. H., ed. 1984. *Extinctions.* Chicago: University of Chicago Press.

———, ed. 1988. *Evolutionary Progress.* Chicago: University of Chicago Press.

Nola, R., ed. 1988. *Relativism and Realism in Science.* Dordrecht: Kluwer.

Nozick, R. 1974. *Anarchy, State and Utopia.* New York: Basic Books.

———. 1981. *Philosophical Explanations.* Cambridge, Mass.: Harvard University Press, Belknap Press.

Oakley, D. A., and H. C. Plotkin, eds. 1979. *Brain, Behaviour and Evolution.* London: Methuen.

Odling-Smee, F. J., and H. C. Plotkin. 1984. Evolution: Its levels and its units. *BBS* 7:318–20.

O'Driscoll, Jr., G. P. 1977. *Economics as a Coordination Problem: The Contributions of Friedrich A. Hayek.* Kansas City: Sheed, Andrews and McMeel.

Odum, E. P. 1971. *Fundamentals of Ecology.* 3d ed. Philadelphia: Saunders.

Odum, H. T. 1971. *Environment, Power, and Society.* New York: Wiley-Interscience.

———. 1982. *Systems Ecology: An Introduction.* New York: Wiley.

Oeser, E. 1987. Zusammenfassender Kommentar. In Riedl and Wuketits, 274–76.

Olby, R. C. 1974. *The Path to the Double Helix.* New York: Macmillan.

———. 1979. Mendel no Mendelian? *Hist. Sci.* 17:53–72.

Oldroyd, D. 1990. David Hull's evolutionary model of science. *Biol. Phil.* 4:473–87.

Oliver, W. D. 1949. Can naturalism be materialistic? *J. Phil.* 46:608–15.

Oppenheim, P., and H. Putnam. 1958. Unity of science as a working hypothesis. In H. Feigl, M. Scriven, and G. Maxwell, eds., *Concepts, Theories, and the Mind-Body Problem*, 3–36. Minnesota Studies in the Philosophy of Science, vol. 2. Minneapolis: University of Minnesota Press.

Oruka, H. O., ed. 1990a. *Sage Philosophy: Indigeneous Thinkers and Modern Debate on African Philosophy*. Leyden: E. J. Brill.

———. 1990b. Introduction. In Oruka 1990a, 1–11.

Oster, G. F., and E. O. Wilson. 1978. *Caste and Ecology in the Social Insects*. Princeton: Princeton University Press.

Ostrom, V. 1974. *The Intellectual Crisis in American Public Administration*. Rev. ed. Alabama: University of Alabama Press.

Paley, W. 1835–39. *Paley's Natural Theology Illustrated*. Introductory and concluding volumes by Henry Lord Brougham. Ed. by Henry Lord Brougham and Sir Charles Bell. 5 vols. London: Knight.

Palladino, P. 1990. Ecological theory and pest control practice: A study of the institutional and conceptual dimensions of a scientific debate. *Soc. Stud. Sci.* 20:255–81.

———. 1991. Defining ecology: Ecological theories, mathematical modeling, and applied biology in the 1960s and 1970s. *J. Hist. Biol.* 24:223–43.

Paller, B. T., and D. T. Campbell. 1990. Maxwell and van Fraassen on observability, reality, and justification. In M. L. Maxwell and C. W. Savage, eds., *Science, Mind and Psychology: Essays on Grover Maxwell's World View*, 121–54. Lanham, Maryland: University Press of America.

Papandreou, A. 1958. *Economics as a Science*. Chicago: Lippincott.

Papineau, D. 1988. Does the sociology of science discredit science? In Nola, 37–57.

Parfit, D. 1984. *Reasons and Persons*. Oxford: Clarendon.

Pattee, H., ed. 1973. *Hierarchy Theory: The Challenge of Complexity*. New York: Braziller.

Pera, M., and W. Shea, eds. 1991. *Persuading Science: The Art of Scientific Rhetoric*. Canton, Mass.: Science History Publications.

Perovich Jr., A. N. 1986. Genius, scientific method, and the stability of synthetic a priori principles. In Donagan et al., 327–39.

Peters, R. H. 1976. Tautology in evolution and ecology. *Am. Naturalist* 110:1–12.

Peterson, D. 1990. *Wittgenstein's Early Philosophy: Three Sides of the Mirror*. New York: Harvester, Wheatsheaf.

Phelps, E. S., ed. 1975. *Altruism, Morality, and Economic Theory*. New York: Russell Sage Foundation.

Piaget, J. 1918. *Recherche*. Lausanne: La Concorde. Summary in English in Piaget 1977, 42–50.

———. 1971. *Biology and Knowledge*. Trans. B. Walsh. Chicago: University of Chicago Press. French orig. pub. 1967.

———. 1977. *The Essential Piaget*. Ed. H. E. Gruber and J. Vonèche. New York: Basic Books.

———. 1978. *Behavior and Evolution*. Trans. D. Nicholson-Smith. New York: Random House. French orig. pub. 1973.

———. 1980. *Adaptation and Intelligence: Organic Selection and Phenocopy*. Trans. S. Eames. Chicago: University of Chicago Press. French orig. pub. 1974.

Piaget, J., and R. Garcia. 1983. *Psychogénèse et histoire des sciences*. Paris: Flammarion.

Piattelli-Palmarini, M. 1986. The rise of selection theories: A case study and some lessons from immunology. In W. Demopoulos and A. Marras, eds., *Language Learning and Concept Acquisition: Foundational Issues*, 117–30. Norwood, N.J.: Ablex.

Pickering, A. 1984a. *Constructing Quarks: A Sociological History of Particle Physics*. Chicago: University of Chicago Press.

———. 1984b. Against putting the phenomena first. *Stud. Hist. Phil. Sci.* 15:85–117.

———. 1990. Making plans: On the goals of scientific practice. In Le Grand 215–39.

———. 1991. Philosophy naturalized a bit. Review of Giere 1988. *Soc. Stud. Sci.* 21:575–84.

———, ed. 1992. *Science as Practice and Culture*. Chicago: University of Chicago Press.

Pinch, T. 1986. *Confronting Nature: The Sociology of Solar Neutrino Detection*. Dordrecht: Reidel.

Pinker, S., and P. Bloom. 1990. Natural language and natural selection. *BBS* 13:707–84.

Pitt, J., ed. 1986. The role of history in and for philosophy. *Synthese* 67 (1).

Pitt, J. C., and M. Pera, eds. 1987. *Rational Change in Science*. Dordrecht: Reidel.

Plotkin, H. C. 1979. Learning in a Carabid beetle (*Pterostichus melanarius*). *Animal Behaviour* 27:567–75.

———. 1981. Changes in the behaviour of a Carabid beetle (*Pterostichus melanarius*) following exposure to food and water. *Animal Behaviour* 29:1245–51.

———, ed. 1982. *Learning, Development and Culture: Essays in Evolutionary Epistemology*. New York: Wiley.

———. 1987a. Evolutionary epistemology as science. *Biol. Phil.* 2:295–313.

———. 1987b. Evolutionary epistemology and the synthesis of biological and social science. In Callebaut and Pinxten, 75–96.

———, ed. 1988a. *The Role of Behavior in Evolution.* Cambridge, Mass.: MIT Press.

———. 1988b. An evolutionary epistemological approach to the evolution of intelligence. In H. J. Jerison and I. Jerison, eds., *Intelligence and Evolutionary Biology,* 73–91. Berlin: Springer.

———. 1988c. Intelligence and evolutionary epistemology. *Human Evolution* 3:437–48.

———. 1991. The testing of evolutionary epistemology. Review of Edelman 1987. *Biol. Phil.* 6:481–97.

Plotkin, H. C., and F. J. Odling-Smee. 1979. Learning, change, and evolution: An enquiry into the teleonomy of learning. *Advances in the Study of Behavior* 10:1–41.

———. 1981. A multiple-level model of evolution and its implications for sociobiology. *BBS* 4:225–68.

———. 1982. Learning in the context of a hierarchy of knowledge gaining processes. In Plotkin 1982, 117–34.

Polanyi, M. 1962. *Personal Knowledge: Towards a Critical Philosophy.* Corr. ed. Chicago: University of Chicago Press. Orig. pub. 1958.

Ponnamperuma, C. 1984. Seeds of life. In Weintraub, 2–19.

Popper, K. R. 1935. *Logik der Forschung.* Wien: Julius Springer.

———. 1945. *The Open Society and its Enemies.* 2 vols. London: Routledge.

———. 1957. *The Poverty of Historicism.* 2 vols. London: Routledge and Kegan Paul; Boston: Beacon.

———. 1959. *The Logic of Scientific Discovery.* London: Hutchison; New York: Basic Books.

———. 1963. *Conjectures and Refutations: The Growth of Scientific Knowledge.* London: Routledge.

———. 1970. Normal science and its dangers. In Lakatos and Musgrave, 51–58.

———. 1972. *Objective Knowledge: An Evolutionary Approach.* Oxford: Clarendon.

———. 1974. Scientific reduction and the incompleteness of all science. In Ayala and Dobzhansky, 259–84.

———. 1975. The rationality of scientific revolutions. In R. Harré, ed., *Problems of Scientific Revolution,* 72–101. Oxford: Clarendon. Repr. in Hacking 1981a.

———. 1976. *Unended Quest: An Intellectual Bibliography,* Glasgow: Fontana/Collins. Orig. pub. 1974.

———. 1987. Die erkenntnistheoretische Position der Evolutionären Erkenntnistheorie. In Riedl and Wuketits, 29–37.

Popper, K. R., and J. Eccles. 1977. *The Self and Its Brain: An Argument for Interactionism.* Berlin: Springer.

Potter, J. 1988. What is reflexive about discourse analysis? The case of reading readings. In Woolgar 1988a, 37–52.

Prelli, L. J. 1989. *A Rhetoric of Science: Inventing Scientific Discourse.* Columbia: University of South Carolina Press.

Priest, G. R., R. Routley, and J. Norman, eds. 1989. *Paraconsistent Logic: Essays on the Inconsistent.* Munich: Philosophia.

Prigogine, I., and I. Stengers. 1984. *Order Out of Chaos: Man's New Dialogue with Nature.* London: Heinemann.

Primas, H. 1981. *Chemistry, Quantum Mechanics and Reductionism: Perspectives in Theoretical Chemistry.* Berlin: Springer.

Provine, W. 1986. *Sewall Wright and Evolutionary Biology.* Chicago: University of Chicago Press.

———. 1988. Progress in evolution and meaning of life. In Nitecki, 49–74.

Puccia, C., and R. Levins. 1985. *Qualitative Modeling of Complex Systems: An Introduction to Loop Analysis and Time Averaging.* Cambridge, Mass.: Harvard University Press.

Pulliam, H. R., and C. Dunford. 1980. *Programmed to Learn.* New York: Columbia University Press.

Putnam, H. 1962. What theories are not. In E. Nagel, P. Suppes, and A. Tarski, eds., *Logic, Methodology and Philosophy of Science,* 240–51. Stanford: Stanford University Press. Repr. in Putnam 1975a.

———. 1973. Explanation and reference. In G. Pearce and P. Maynard, eds., *Conceptual Change,* 199–221. Dordrecht: Reidel. Repr. in Putnam 1975a.

———. 1975a. *Mathematics, Matter and Method. Philosophical Papers,* vol. 1. Cambridge: Cambridge University Press.

———. 1975b. *Mind, Language and Reality. Philosophical Papers,* vol. 2. Cambridge: Cambridge University Press.

———. 1975c. Introduction: Philosophy of language and the rest of philosophy. In Putnam 1975b, vii–xvii.

———. 1975d. Language and reality. In Putnam 1975b, 272–90.

———. 1978. *Meaning and the Moral Sciences.* London: Routledge and Kegan Paul.

———. 1979. Philosophy of mathematics. In Asquith and Kyburg, 386–98.

———. 1981. *Reason, Truth and History.* Cambridge: Cambridge University Press.

———. 1990. *Realism with a Human Face.* Ed. J. Conant. Cambridge, Mass.: Harvard University Press.

Quine, W. V. 1951. Two dogmas of empiricism. *Phil. Rev.* 60:20–43. Repr. in Quine 1953 and in Kornblith 1985a.

———. 1953. *From a Logical Point of View.* Cambridge, Mass.: Harvard University Press.

———. 1960. *Word and Object.* Cambridge, Mass.: MIT Press.

———. 1969a. *Ontological Relativity and Other Essays.* New York: Columbia University Press.

———. 1969b. Epistemology naturalized. In Quine 1969a, 69–90.

———. 1973. *The Roots of Reference.* La Salle: Open Court.

———. 1975. The nature of natural knowledge. In S. Guttenplan, ed., *Mind and Language,* 67–81. Oxford: Clarendon.

———. 1986. Reply to Roger F. Gibson, Jr. In Hahn and Schilpp, 155–57.

Quinn, P. 1984. The philosopher of science as expert witness. In Cushing et al., 32–53.

———. 1987. Comments on Laudan [1987b]. In *PSA 1986,* 2:355–58.

Rachlin, H., R. Battalio, J. Kagel, and L. Green. 1981. Maximization theory in behavioral psychology. *BBS* 4:371–417.

Radnitzky, G., and W. W. Bartley, III., eds. 1987. *Evolutionary Epistemology, Theory of Rationality, and the Sociology of Knowledge.* La Salle: Open Court.

Radnitzky, G., and P. Bernholz, eds. 1987. *Economic Imperialism: The Economic Approach Applied Outside the Traditional Areas of Economics.* New York: Paragon House.

Raes, K., and J. Vanlandschoot, eds. 1984. *Afbraak en opbouw: Dialogen met Leo Apostel.* Brussels: VUB Press.

Ramsey, J. L. 1990. Metastable States: The Justification of Approximative Procedures in Chemical Kinetics, 1923–1947. Ph.D. diss., University of Chicago.

Ramsey, W., and S. Stich. 1990. Connectionism and three levels of nativism. *Synthese* 82:177–205. Repr. in Fetzer 1991.

Rasmussen, N. 1987. A new model of developmental constraints as applied to the *Drosophila* system. *J. Theor. Biol.* 127:217–93.

———. 1991. The decline of recapitulationism in early twentieth-century biology: Disciplinary conflict and consensus on the battleground of theory. *J. Hist. Biol.* 24:51–98.

Raup, D. M.. and J. J. Sepkoski. 1984. Periodicity of extinctions in the geological past. *Proc. Nat. Acad. Sci. USA* 81:801–5.

Ravetz, J. R. 1977. Criticisms of science. In I. Spiegel-Rösing and D. de Solla Price, eds., *Science, Technology and Society,* 71–89. London: Sage.

———. 1980. Ideologische Überzeugungen in der Wissenschafstheorie. In H. P. Duerr, Hrsg., *Versuchingen: Aufsätze zur Philosophie Paul Feyerabends,* 13–34. Frankfurt am Main: Suhrkamp.

Rawls, J. 1971. *A Theory of Justice.* Cambridge, Mass.: Harvard University Press.

Read, S. 1988. *Relevant Logic.* Oxford: Blackwell.

Reed, E. 1988. *James J. Gibson and the Psychology of Perception.* New Haven: Yale University Press.

Regis, E., ed. 1985. *Extraterrestrials: Science and Alien Intelligence.* Cambridge: Cambridge University Press.

Reichenbach, H. 1951. *The Rise of Scientific Philosophy.* Berkeley and Los Angeles: University of California Press.

Rescher, N. 1977. *Methodological Pragmatism: A Systems-Theoretic Approach to the Theory of Knowledge.* Oxford: Blackwell.

———, ed. 1990. *Evolution, Cognition, and Realism: Studies in Evolutionary Epistemology.* Lanham: University Press of America.

———. 1991. *A Useful Inheritance: Evolutionary Aspects of the Theory of Knowledge.* Savage. Md.: Rowan and Littlefield.

Restivo, S. 1983. *The Social Relations of Physics, Mysticism, and Mathematics.* Dordrecht: Reidel.

Richards, R. J. 1973. Substantitive and methodological teleology in Aristotle and some logical empiricists. *Thomist* 37:702–33.

———. 1976. James Gibson's passive theory of perception: A rejection of the doctrine of specific nerve energies. *Phil. Phenomenol. Res.* 37:218–33.

———. 1977. The natural selection model of conceptual evolution. *Phil. Sci.* 44:494–501.

———. 1983. Why Darwin delayed, or interesting problems and models in the history of science. *J. Hist. Behav. Sci.* 19:45–53.

———. 1986. A defense of evolutionary ethics. With responses by C. J. Cela-Conde, A. Gewirth, W. Hughes, L. Thomas, and R. Trigg; and a rejoinder. *Biol. Phil.* 1:265–354.

———. 1987. *Darwin and the Emergence of Evolutionary Theories of Mind and Behavior.* Chicago: University of Chicago Press.

———. 1988. The moral foundations of the idea of evolutionary progress: Darwin, Spencer, and the neo-Darwinians. In Nitecki, 149–68.

————. 1989. Dutch objections to evolutionary ethics. *Biol. Phil.* 4:331–43.

————. 1992. *The Meaning of Evolution: The Morphological Construction and Ideological Reconstruction of Darwin's Theory.* Chicago: University of Chicago Press.

Richardson, R. C. 1980. Reductionistic research programs in psychology. In *PSA 1980* 1:171–83.

————. 1982. Grades of organization and the units of selection controversy. In *PSA 1982* 1:324–40.

Riedl, R. 1978. *Order in Living Organisms: A Systems Analysis of Evolution.* Trans. R. P. S. Jefferies. New York: Wiley. German orig. pub. 1975.

————. 1984. *Biology of Knowledge: The Evolutionary Basis of Reason.* New York: Wiley. German orig. pub. 1980.

Riedl, R., and F. Wuketits, eds. 1987. *Die Evolutionäre Erkenntnistheorie: Bedingungen, Lösungen, Kontroversen.* Berlin: Parey.

Robbins, L. 1935. *An Essay on the Nature and Signifance of Economic Science.* 2d ed. London: Macmillan.

Root-Bernstein, R. S. 1989. *Discovering.* Cambridge, Mass.: Harvard University Press.

Rorty, R., ed. 1967. *The Linguistic Turn: Recent Essays in Philosophical Method.* Chicago: University of Chicago Press.

————. 1979. *Philosophy and the Mirror of Nature.* Oxford: Blackwell.

————. 1988. Is natural science a natural kind? In McMullin, 49–74.

Ros, A. 1991. Der Status der Wirklichkeit in der Genese kognitiver Strukturen bei Jean Piaget. Paper read at the Freiburger Arbeitstage für Soziologie, Freiburg, Germany, October 1991.

Rose, H., and S. Rose. 1980. Against an oversocialized conception of science. In Callebaut et al., 173–87.

Rose, S., ed. 1982a. *Against Biological Determinism.* London: Allison and Busby.

————. 1982b. *Towards a Liberatory Biology.* London: Allison and Busby.

Rose, S., L. Kamin, and R. C. Lewontin. 1984. *Not in our Genes: Biology, Ideology and Human Nature.* Harmondsworth: Penguin.

Rosenberg, A. 1972. Friedman's "methodology" for economics: A critical examination. *Phil. Soc. Sci.* 2:15–30.

————. 1976. *Microeconomic Laws: A Philosophical Analysis.* Pittsburgh: University of Pittsburgh Press.

————. 1978a. Hollis and Nell: Rationalist economic men. *Phil. Soc. Sci.* 8:87–98.

————. 1978b. The supervenience of biological concepts. *Phil. Sci.* 45:368–86. Repr. in Sober 1984b.

————. 1980. *Sociobiology and the Preemption of Social Science.* Baltimore: Johns Hopkins University Press.

————. 1983a. If economics isn't science, what is it? *Phil. Forum* 14:296–315.

————. 1983b. Fitness. *J. Phil.* 80:457–73.

————. 1985. *The Structure of Biological Science.* Cambridge: Cambridge University Press.

————. 1986. Ignorance and disinformation in the philosophy of biology: A reply to Stent. *Biol. Phil.* 1:461–71.

————. 1987. Is there really "juggling," "artifice," and "trickery" in *Genes, Mind, and Culture*? *BBS* 10:80–83.

————. 1988a. *The Philosophy of Social Science.* Boulder: Westview Press; Oxford: Oxford University Press.

————. 1988b. Economics is too important to be left to the rhetoricians. *Econ. Phil.* 4:129–49.

————. 1988c. Rhetoric is not important enough for economists to bother about. *Econ. Phil.* 4:173–75.

————. 1988d. Is the theory of natural selection a statistical theory? *Can. J. Phil.* 14:187–207.

————. 1988e. Grievous faults in *Vaulting Ambition*? [Review of Kitcher 1985a.] *Ethics* 98:43–68.

————. 1989. From reductionism to instrumentalism? In Ruse 1989d, 245–62.

————. 1991. The biological justification of ethics. *Soc. Phil. Pol.* 8:86–101.

————. 1992a. *Economics—Mathematical Politics or Science of Diminishing Returns?* Chicago: University of Chicago Press.

————. 1992b. Selection and science. Critical notice of [Hull 1988a.] *Biol. Phil.* 7:217–28.

————. 1992c. Neo-classical economics and evolutionary theory: Strange bedfellows? In *PSA 1992*, 1:174–83.

Rosenthal, M., and P. Yudin. 1967. *A Dictionary of Philosophy.* Moscow: Progress.

Rostow, W. 1960. *The Stages of Economic Growth: A Non-Communist Manifesto.* Cambridge: Cambridge University Press.

Roughgarden, J. 1979. *Theory of Population Genetics and Evolutionary Ecology: An Introduction.* New York: Macmillan.

Rudwick, M. J. S. 1985. *The Great Devonian Controversy: The Making of Scientific Knowledge among Gentlemanly Specialists.* Chicago: University of Chicago Press.

Rumelhart, D. E., J. L. McClelland, and the PDP Research Group. 1986. *Parallel Distributed*

Processing: Explorations in the Microstructure of Cognition. Vol. 1, *Foundations.* Cambridge, Mass.: MIT Press, Bradford Books.

Ruse, M. 1973. *The Philosophy of Biology.* London: Hutchinson.

———. 1977. Karl Popper's philosophy of biology. *Phil. Sci.* 44:638–61.

———. 1979a. *The Darwinian Revolution: Science Red in Tooth and Claw.* Chicago: University of Chicago Press.

———. 1979b. *Sociobiology: Sense or Nonsense?* Dordrecht: Reidel. 2d ed., with new Afterword, 1984.

———. 1981. *Is Science Sexist? And Other Problems in the Biomedical Sciences.* Dordrecht: Reidel.

———. 1982. *Darwinism Defended. A Guide to the Evolution Controversies.* Reading, Mass.: Addison-Wesley.

———. 1984. The morality of the gene. *Monist* 67:167–99.

———. 1985. Is rape wrong on Andromeda? An introduction to extraterrestrial evolution, science, and morality. In Regis, 43–78.

———. 1986a. *Taking Darwin Seriously: A Naturalistic Approach to Philosophy.* Oxford: Blackwell.

———. 1986b. Evolutionary ethics: A phoenix arisen. *Zygon* 21:95–112.

———. 1986c. Sociobiology moves along. *Phil. Soc. Sci.* 16:141–49.

———. 1988a. *Homosexuality: A Philosophical Inquiry.* Oxford: Blackwell.

———. 1988b. *Philosophy of Biology Today.* Albany: State University of New York Press.

———, ed. 1988c. *But is it Science? The Philosophical Question in the Creation/Evolution Controversy.* Buffalo: Prometheus Books.

———. 1989a. *The Darwinian Paradigm: Essays on its History, Philosophy, and Religious Implications.* London: Routledge.

———. 1989b. The view from somewhere: A critical defense of evolutionary epistemology. In Hahlweg and Hooker, 185–228.

———. 1989c. Booknotes. *Biol. Phil.* 4:503–8.

———, ed. 1989d. *What the Philosophy of Biology Is: Essays Dedicated to David Hull.* Dordrecht: Kluwer.

———, ed. 1989e. *Readings in the Philosophy of Biology.* New York: Macmillan.

———. 1990. Booknotes. "E. O. Wilson's lapdog Michael Ruse on sociobiology." *Biol. Phil.* 5:259–63.

———. 1991a. *Molecules to Men. The Concept of Progress in Biology.* Cambridge, Mass.: Harvard University Press.

———. 1991b. Evolutionary ethics and the search for predecessors: Kant, Hume, and all the way back to Aristotle? *Soc. Phil. Pol.* 8:59–85.

Ruse, M., and P. J. Taylor, eds. 1991. Special issue on pictorial representation in biology. *Biol. Phil.* 6 (2).

Ruse, M., and E. O. Wilson. 1986. Ethics as applied science. *Philosophy* 61:173–92.

Russell, B. 1946. *A History of Western Philosophy and Its Connection With Political and Social Circumstances from the Earliest Times to the Present Day.* New York: Simon and Schuster.

Russell, E. S. 1982. *Form and Function: A Contribution to the History of Animal Morphology.* With a new Introduction by George V. Lauder. Chicago: University of Chicago Press. Orig. pub. 1916.

Ryan, A. 1991. When it's rational to be irrational. [Review of Elster 1989a, b, c.] *New York Review of Books* 38 (October 10): 19–22.

Ryle, G. 1949. *The Concept of Mind.* London: Hutchinson.

Sackman, H. 1967. *Computers, Systems Science, and Evolving Society.* New York: Wiley.

Sahlins, M. 1976. *The Use and Abuse of Biology.* Ann Arbor: University of Michigan Press.

Salmon, W. 1971. *Statistical Explanation and Statistical Relevance.* Pittsburgh: University of Pittsburgh Press.

———. 1984. *Scientific Explanation and the Causal Structure of the World.* Princeton: Princeton University Press.

———. 1989. *Four Decades of Scientific Explanation.* Minneapolis: University of Minnesota Press. Repr. of Salmon's review essay in Kitcher and Salmon 1989.

Salomon, J.-J. 1973. *Science and Politics.* London: Macmillan.

Salthe, S. N. 1985. *Evolving Hierarchical Systems: Their Structure and Representation.* New York: Columbia University Press.

———. 1989. Comments on some possible meanings of the phrase "evolutionary paradigm." In M. K. Hecht, ed., *Evolutionary Biology at the Crossroads,* 174–76. Flushing, N.Y.: Queens College Press.

Samuelson, P. A. 1963. Problems of methodology—discussion. *Am. Econ. Rev.: Papers and Proc.* 53:232–36.

———. 1964. Theory and realism: A reply. *Am. Econ. Rev.* 54:736–40.

———. 1965. Professor Samuelson on theory and realism. *Am. Econ. Rev.* 55:1162–72.

———. 1975. Maximum principles in analytical economics. *Synthese* 31:323–44.

Sanmartín, J. 1987. *Los nuevos redentores: Reflexiones sobre la ingeniería genética, la sociobio-*

logía y el mundo feliz que nos prometen. Barcelona: Anthropos.

Sanmartín, J., V. Simón, and L. García-Merita, comps. 1986. *La Sociedad Naturalizada: Genética y Conducta.* Valencia: Tirant Lo Blanch.

Sarkar, S. 1988. Natural selection, hypercycles and the origin of life. In *PSA 1988* 1:197–206.

———. 1989. Reductionism and Molecular Biology: A Reappraisal. Ph.D. diss., University of Chicago.

———. 1991a. Lamarck *contre* Darwin, reduction *versus* statistics: Conceptual issues in the controversy over directed mutagenesis in bacteria. In Tauber, 235–71.

———. 1991b. Reduction and functional explanation in molecular biology. *Uroboros* 1:67–94.

———, ed. 1992a. *The Founders of Evolutionary Genetics: A Centenary Appraisal.* Dordrecht: Kluwer.

———, ed. 1992b. Carnap: A Centenary Reappraisal. *Synthese* 93 (1–2).

———. 1992c. Models of reduction and categories of reductionism. *Synthese* 91:167–94.

———. 1992d. Science, philosophy, and politics in the work of J. B. S. Haldane, 1992–1937. *Biol. Phil.* 7:385–409.

———. N.d. *The Philosophy and History of Molecular Biology: New Perspectives.* Dordrecht: Kluwer. Forthcoming.

Sarton, G. A. 1927–47. *Introduction to the History of Science.* 3 vols. Baltimore: Williams and Wilkins.

———. 1952. *A Guide to the History of Science.* New York: Ronald Press.

Schaffner, K. F. 1967. Approaches to reduction. *Phil. Sci.* 34:137–47.

———. 1976. Reductionism in biology: Prospects and problems. In *PSA 1974,* 613–32.

———. 1992. *Discovery and Explanation in the Biomedical Sciences.* Forthcoming.

Schank, J., and W. C. Wimsatt. 1986. Generative entrenchment and evolution. In *PSA 1986* 2:33–60.

Scheffler, I. 1967. *Science and Subjectivity.* New York: Bobbs-Merrill.

Schelling, T. C. 1978. *Micromotives and Macrobehavior.* New York: Norton.

Schlanger, J. E. 1971. *Les métaphores de l'organisme.* Paris: Vrin.

Schmaus, W., U. Segerstråle, and D. Jesseph. 1992. The 'Hard Program' in the sociology of scientific knowledge: A manifesto. *Soc. Epistemol.* 6:243–65.

Schmitt, F. 1991. Social epistemology and social cognitive psychology. *Soc. Epistemol.* 5:111–20.

Schoemaker, P. J. H. 1991. The quest for optimal-

ity: A positive heuristic of science? *BBS* 14:205–45.

Schull, J. 1990. Are species intelligent? *BBS* 13:63–108.

Scriven, M. 1959. Explanation and prediction in evolutionary theory. *Science* 130:477–82.

Searle, J. 1980. Minds, brains, and programs. *BBS* 3:417–57.

———. 1983. *Intentionality: An Essay on the Philosophy of Mind.* Cambridge: Cambridge University Press.

———. 1985. *Minds, Brains and Science.* Cambridge, Mass.: Harvard University Press.

———. 1989. Reply to Elkanan Motzkin. *New York Review of Books* 36 (February 16): 45.

———. 1991. Consciousness, explanatory inversion, and cognitive science. *BBS* 13:585–642.

———. 1992. *The Rediscovery of the Mind.* Cambridge, Mass.: MIT Press, Bradford Books.

Seavey, C., P. Katz, and S. Rosenberg Zalk. 1975. Baby X: The effect of gender labels on adult response to infants. *Sex Roles* 1:103–9.

Segerstråle, U. 1985. Colleagues in conflict: An "in vivo" analysis of the sociobiology controversy. *Biol. Phil.* 1:53–88.

———. 1992. Reductionism, "bad science," and politics: A critique of anti-reductionist reasoning. *Politics and the Life Sciences* 11:199–214.

Sejnowski, T. J., C. Koch, and P. S. Churchland. 1988. Computational neuroscience. *Science* 241:1299–1306.

Sellars, R. W. 1944. Does naturalism need ontology? *J. Phil.* 41:686–94.

Serres, M. 1987. *Statues.* Paris: Bourin.

———, dir. 1989. *Éléments d'histoire des sciences.* Paris: Bordas.

———. 1992. *Éclaircissements: Entretiens avec Bruno Latour.* Paris: François Bourin.

Shapere, D. 1965. *Philosophical Problems of Natural Science.* New York: Macmillan.

———. 1969. Towards a post-positivistic interpretation of science. In Achinstein and Barker, 115–60. Repr. in Shapere 1984.

———. 1974. *Galileo: A Philosophical Study.* Chicago: University of Chicago Press.

———. 1982. The concept of observation in science and philosophy. *Phil. Sci.* 49:485–525.

———. 1984. *Reason and the Search for Knowledge.* Dordrecht: Reidel.

———. 1985. Objectivity, rationality, and scientific change. In *PSA 1984* 2:627–62.

———. 1987a. Method in the philosophy of science and epistemology: How to inquire about inquiry and knowledge. In Nersessian, 1–39.

———. 1987b. External and internal factors in the development of science. *Sci. Technol. Stud.* 4:1–23.

———. 1989a. Modern physics and the philosophy of science. In *PSA 1988* 2:201–10.

———. 1989b. Evolution and continuity in scientific change. *Phil. Sci.* 56:419–37.

———. 1990. The origin and nature of metaphysics. *Phil. Topics* 18:163–74.

———. 1991a. The universe of modern science and its philosophical exploration. In E. Agazzi and A. Cordero, eds., *Philosophy and the Origin and Evolution of the Universe,* 87–202. Dordrecht: Kluwer.

———. 1991b. On deciding what to believe and how to talk about nature. In Pera and Shea, 89–103.

———. 1991c. Leplin on essentialism. *Phil. Sci.* 58:655–77.

———. 1992. Astronomy and antirealism. Forthcoming in *Phil. Sci.*

———. N.d.a. *The Mechanical Philosophy of Nature.* In progress.

———. N.d.b. On the introduction of new ideas in science. In J. Leplin, ed., *Innovation in Science.* Forthcoming.

Shapin, S. 1982. History of science and its sociological reconstructions. *Hist. Sci.* 20:157–211.

Shapin, S., and S. Schaffer. 1985. *Leviathan and the Air Pump: Hobbes, Boyle and the Experimental Life.* Princeton: Princeton University Press.

Shimony, A. 1971. Perception from an evolutionary point of view. *J. Phil.* 68:571–83.

———. 1989a. The non-existence of a principle of natural selection. *Biol. Phil.* 4:255–73.

———. 1989b. Reply to Sober. *Biol. Phil.* 4:281–86.

Shimony, A., and D. Nails, eds. 1987. *Naturalistic Epistemology: A Symposium of Two Decades.* Dordrecht: Reidel.

Shrader, D. 1980. The evolutionary development of science. *Rev. Metaphys.* 34:273–96.

Shrader-Frechette, K. 1989. Idealized laws, antirealism, and applied science: A case study in hydrogeology. *Synthese* 81:329–52.

Shrager, J., and P. Langley, eds. 1990. *Computational Models of Scientific Discovery and Theory Formation.* San Mateo, Calif.: Morgan Kaufmann.

Siegel, H. 1989. Philosophy of science naturalized? Some problems with Giere's naturalism. *Stud. Hist. Phil. Sci.* 20:365–75.

———. 1990. Laudan's normative naturalism. *Stud. Hist. Phil. Sci.* 21:295–313.

Simon, H. A. 1963. Problems of methodology—discussion. *Am. Econ. Rev.: Papers and Proc.* 53:229–31.

———. 1967. Reply to Professor Binkley's comments. In N. Rescher, ed., *The Logic of Decision and Action,* 32–33. Pittsburgh: University of Pittsburgh Press.

———. 1973a. The organization of complex systems. In Pattee, 3–27. Repr. in Simon 1977a.

———. 1973b. The structure of ill-structured problems. *Artificial Intelligence* 4:181–201. Repr. in Simon 1977a.

———. 1977a. *Models of Discovery and Other Topics in the Methods of Science.* Dordrecht: Reidel.

———. 1977b. How complex are complex systems? In *PSA 1976* 2:507–22.

———. 1979a. *Models of Thought.* New Haven: Yale University Press.

———. 1979b. Rational decision making in business organizations. *Am. Econ. Rev.* 69:493–513.

———. 1981. *The Sciences of the Artificial.* 2d ed. Cambridge, Mass.: MIT Press. 1st ed. 1969.

———. 1982. *Models of Bounded Rationality.* 2 vols. Cambridge, Mass.: MIT Press.

———. 1983. *Reason in Human Affairs.* Stanford: Stanford University Press.

———. 1991a. *Models of My Life.* New York: Basic Books.

———. 1991b. Comments on the Symposium on "Computer discovery and the sociology of scientific knowledge." *Soc. Stud. Sci.* 21:143–56.

———. 1992a. *Economics, Bounded Rationality, and the Cognitive Revolution.* Ed. M. Egidi and R. Marris. Aldershot: Elgar.

———. 1992b. Scientific discovery as problem solving. *Int. Stud. Phil. Sci.* 6:1–14.

Sinha, C. 1988. *Language and Representation: A Socio-Naturalistic Approach to Human Development.* New York and London: Harvester, Wheatsheaf.

Sklar, L. 1982. Saving the noumena. *Phil. Topics* 13:89–110.

———. 1984. *Space, Time, and Space-Time.* Berkeley and Los Angeles: University of California Press.

Skolimowski, H. 1986. Quine, Ajdukiewicz, and the predicament of 20th-century philosophy. In Hahn and Schilpp, 463–91.

Slezak, P. 1989a. Scientific discovery by computer as empirical refutation of the Strong Programme. *Soc. Stud. Sci.* 19:563–600.

———. 1989b. Computers, contents and causes: Replies to my respondents. *Soc. Stud. Sci.* 19:671–95.

———. 1991. Review of Collins 1990. *Soc. Stud. Sci.* 21:175–201.

Slote, M. 1989. *Beyond Optimizing: A Study of Rational Choice.* Cambridge, Mass.: Harvard University Press.

Smokler, H. 1990. Are theories of rationality empirically testable? *Synthese* 82:297–306.

Smolensky, P. 1988. On the proper treatment of connectionism. *BBS* 11:1–23.

Sneed, J. 1971. *The Logical Structure of Mathematical Physics.* Dordrecht: Reidel.

Sober, E. 1975. *Simplicity.* Oxford: Clarendon Press.

———. 1978. Psychologism. *J. Theory Soc. Behavior* 8:165–91.

———. 1981a. The evolution of rationality. *Synthese* 46:95–120.

———. 1981b. Revisability, a priori truth, and evolution. *Australas. J. Phil.* 59:68–85.

———. 1981c. Holism, individualism, and the units of selection. In *PSA 1980* 2:93–121. Repr. in Sober 1984b.

———. 1984a. *The Nature of Selection: Evolutionary Theory in Philosophical Focus.* Cambridge, Mass.: MIT Press, Bradford Books.

———, ed. 1984b. *Conceptual Issues in Evolutionary Biology: An Anthology.* Cambridge, Mass.: MIT Press, Bradford Books.

———. 1985a. Darwin on natural selection: A philosophical perspective. In Kohn, 867–99.

———. 1985b. Methodological behaviorism, evolution, and game theory. In Fetzer, 181–200.

———. 1987. Comments on Maynard Smith's "How to model evolution." In Dupré, 133–45.

———. 1988a. *Reconstructing the Past: Parsimony, Evolution, and Inference.* Cambridge, Mass.: MIT Press, Bradford Books.

———. 1988b. What is evolutionary altruism? In B. Linsky and M. Matthen, eds., *New Essays on Philosophy and Biology,* 75–99. *Can. J. Phil.* supplementary vol. 14.

———. 1989a. What is psychological egoism? *Behaviorism* 17:89–102.

———. 1989b. Is the theory of natural selection unprincipled? A reply to Shimony. *Biol. Phil.* 4:275–79.

———. 1990. *Core Questions in Philosophy.* New York: Macmillan.

———. 1991a. *Philosophy of Biology.* Boulder: Westview Press.

———. 1991b. Organisms, individuals, and units of selection. In Tauber, 275–96.

———. 1992. The evolution of altruism: Correlation, cost, and benefit. *Biol. Phil.* 7:177–87.

Sober, E., and R. C. Lewontin. 1982. Artifact, cause, and genic selection. *Phil. Sci.* 49:147–76. Repr. in Sober 1984b.

Sohn-Rethel, A. 1970. *Geistige und körperliche Arbeit.* Frankfurt am Main: Suhrkamp.

———. 1975. Science as alienated consciousness. *Radic. Sci. J.* nos. 2/3: 72–101.

Sommerhoff, G. 1950. *Analytical Biology.* Oxford: Oxford University Press.

Sorell, T. 1991. *Scientism: Philosophy and the Infatuation with Science.* London and New York: Routledge.

Sperber, D. 1990. Anthropology and psychology: Towards an epidemiology of representation. *Man* (n.s.) 20:73–89.

Stanley, S. 1981. *The New Evolutionary Timetable.* New York: Basic Books.

Stanzione, M. 1990. *Epistemologie Naturalizzate.* Rome: Bagatto Libri.

Star, L. 1989. *Regions of the Mind: Brain Research and the Quest for Scientific Certainty.* Stanford: Stanford University Press.

———. N.d. Layered space, formal representations and long-distance control: The politics of information. Forthcoming in *Fund. Sci.*

Star, L., and E. Gerson. 1986. The management and dynamics of anomalies in scientific work. *Sociol. Quart.* 28:147–69.

Star, L., and J. R. Griesemer. 1989. Institutional ecology, 'translations,' and boundary objects: Amateurs and professionals in Berkeley's Museum of Vertebrate Zoology, 1907–1939. *Soc. Stud. Sci.* 19: 387–420.

Stebbins, G. L. 1950. *Variation and Evolution in Plants.* New York: Columbia University Press.

Stebbins, G. L., and F. J. Ayala. 1981. Is a new evolutionary synthesis necessary? *Science* 213:967–71.

Steele, E. J. 1981. *Somatic Selection and Adaptive Evolution: On the Inheritance of Acquired Characters.* 2d ed. Chicago: University of Chicago Press. Orig. pub. 1979.

Stegmüller, W. 1969–1973. *Probleme und Resultate der Wissenschaftstheorie und Analytischen Philosophie.* 4 vols. Berlin: Springer.

———. 1976. *The Structure and Dynamics of Theories.* Berlin: Springer. Orig. pub. 1973.

———. 1979. *The Structuralist View of Theories: A Possible Analogue of the Bourbaki Programme in Physical Sciences.* Berlin: Springer.

Stein, H. 1983. On the present state of the philosophy of quantum mathematics. In *PSA 1982* 2:563–81.

Stengers, I., and J. Schlanger. 1991. *Les concepts scientifiques: Invention et pouvoir.* Paris: Gallimard. Orig. pub. 1988.

Stent, G., ed. 1980. *Morality as a Biological Phenomenon: Presuppositions of Sociobiological Research.* Berkeley and Los Angeles: University of California Press.

———. 1986. Glass bead game. Review of Rosenberg 1985. *Biol. Phil.* 1:227–47.

———. 1987. A reply to Alexander Rosenberg. *Biol. Phil.* 2:375.

Sterelny, K., and P. Kitcher. 1988. The return of the gene. *J. Phil.* 85:339–61.

Sterman, J. D. 1988. Deterministic chaos in models of human behavior: Methodological issues and experimental results. *Systems Dynamics Review* 4:148–78.

Stich, S. 1983. *From Folk Psychology to Cognitive Science: The Case Against Belief.* Cambridge, Mass.: MIT Press, Bradford Books.

———. 1985. Could man be an irrational animal? Some notes on the epistemology of rationality. *Synthese* 64:115–35. Repr. in Kornblith 1985a.

Strawson, P. F. 1985. *Skepticism and Naturalism: Some Varieties.* The Woodbridge Lectures 1983. New York: Columbia University Press.

Strum, S., and B. Latour. 1987. The meaning of social: From baboons to humans. *Soc. Sci. Inform.* 26:783–802.

Suppe, F., ed. 1977. *The Structure of Scientific Theories.* 2d ed. Urbana: University of Illinois Press. 1st ed. 1974.

———. 1979. Theory structure. In Asquith and Kyburg, 317–38.

———. 1988. *The Semantic Conception of Theories and Scientific Realism.* Urbana: University of Illinois Press.

Suppes, P. 1967. What is a scientific theory? In S. Morgenbesser, ed., *Philosophy of Science Today,* 55–67. New York: Meridian.

———. 1968. The desirability of formalization in science. *J. Phil.* 65:651–64.

———. 1979. The role of formal methods in the philosophy of science. In Asquith and Kyburg, 16–27.

Swanson, D. 1990. Medical literature as a potential source of new knowledge. *Bull. Med. Libr. Assoc.* 78:29–37.

———. N.d. Combining complementary literatures: A discovery process. Ms.

Swinburne, R. 1984. Analytic/synthetic. *Am. Phil. Q.* 21:31–42.

Symons, D. 1979. *The Evolution of Human Sexuality.* Oxford: Oxford University Press.

Tanner, N., and A. Zihlman. 1976. Women in evolution, Part I: Innovation and selection in human origins. *Signs* 1:585–608.

Tauber, A. I., ed. 1991. *Organism and the Origins of Self.* Dordrecht: Kluwer.

Tauber, A. I., and S. Sarkar. 1992. The Human Genome Project: Has blind reductionism gone too far? *Perspec. Biol. Med.* 35:220–35.

Taylor, P. 1985. Construction and Turnover of Multi-species Communities: A Critique of Approaches to Ecological Complexity. Ph.D. diss., Harvard University.

Tennant, N. 1988. Theories, concepts and rationality in an evolutionary account of science. *Biol. Phil.* 3:224–31.

Thagard, P. 1988. *Computational Philosophy of Science.* Cambridge, Mass.: MIT Press, Bradford Books.

———. 1989. Explanatory coherence. *BBS* 12:435–67.

Thom, R. 1990. *Apologie du logos.* Paris: Hachette.

Thomas, D. 1979. *Naturalism and Social Science.* Minneapolis: University of Minnesota Press.

Thompson, D'Arcy. 1917. *On Growth and Form.* Cambridge: Cambridge University Press. Abridged ed. 1961.

Thompson, P. 1989a. *The Structure of Biological Theories.* Albany: State University of New York Press.

———. 1989b. Philosophy of biology under attack: Stent vs. Rosenberg. *Biol. Phil.* 4:345–51.

Thuillier, P. 1981. *Les biologistes vont-ils prendre le pouvoir? La sociobiologie en question.* Brussels: Complexe.

Tibbets, P. 1986. The sociology of scientific knowledge: The constructivist thesis and relativism. *Soc. Stud. Sci.* 16:39–57.

Tierney, A. J. 1986. The evolution of learned and innate behavior. *Animal Learning and Behavior* 14:339–48.

Tinbergen, N. 1968. On war and peace in animals and man. *Science* 160:1411–18. Repr. in Caplan 1978b.

Tolman, E. C., and E. Brunswik. 1935. The organism and the causal texture of the environment. *Psych. Rev.* 42:43–77.

Tooby, J., and L. Cosmides. 1988. The evolution of war and its cognitive foundations. *Proceedings of the Institute for Evolutionary Studies* 88:1–15.

———. 1990. On the universality of human nature and the uniqueness of the individual: The role of genetics and adaptation. *J. of Personality* 58:17–68.

Toulmin, S. E. 1961. *Foresight and Understanding: An Enquiry into the Aims of Science.* New York: Harper and Row.

———. 1967. The evolutionary development of natural science. *Am. Scient.* 55:456–71.

———. 1969. From logical analysis to conceptual history. In Achinstein and Barker, 25–53.

———. 1972. *Human Understanding.* Princeton: Princeton University Press.

Trigg, R. 1978. The sociology of knowledge. *Phil. Soc. Sci.* 8:289–98.

Trivers, R. L. 1971. The evolution of reciprocal altruism. *Quart. Rev. Biol.* 40:35–57. Repr. in Caplan 1978b.

Tuomela, R. 1989. Collective action, superveni- ence, and constitution. *Synthese* 80:243–66.

Tversky, A., and D. Kahneman. 1974. Judgment under uncertainty: Heuristics and biases. *Science* 185:1124–31.

Tweney, R. D. 1985. Faraday's discovery of induc- tion: A cognitive approach. In Gooding and James, 189–209.

———. 1987. Procedural representation in scien- tific thinking. In *PSA 1986* 2:336–44.

———. 1989. A framework for the cognitive psy- chology of science. In Gholson et al., 342–69.

———. 1992. Serial and parallel processing in scientific discovery. In Giere 1992a.

Tweney, R. D., and M. E. Doherty. 1983. Ration- ality and the psychology of inference. *Synthese* 57:139–61.

Tweney, R. D., M. E. Doherty, and C. R. Mynatt, eds. 1981. *On Scientific Thinking.* New York: Columbia University Press.

Uebel, T. E. 1991. Neurath's programme for natu- ralistic epistemology. *Stud. Hist. Phil. Sci.* 22:623–46.

Urbanek, A. 1988. Morpho-physiological pro- gress. In Nitecki, 195–216.

Uzan, M. 1992. Un entretien avec Richard Rorty. *Le Monde,* March 3, p. 2.

Vandenbrande, R. 1979. La critique littéraire: Un espace intermédiaire. In Callebaut et al., 183– 89.

Van Fraassen, B. C. 1970. On the extension of Beth's semantics of physical theories. *Phi. Sci.* 37:325–39.

———. 1980. *The Scientific Image.* Oxford: Clar- endon.

———. 1985. Empiricism in the philosophy of sci- ence. In Churchland and Hooker, 245–308.

———. 1989. *Laws and Symmetry.* Oxford: Clar- endon.

Van Parijs, P. 1981. *Evolutionary Explanation in the Social Sciences: An Emerging Paradigm.* Totowa: Rowman and Littlefield.

Van Valen, L. 1973. A new evolutionary law. *Evo- lutionary Theory* 1:1–30.

Vaughan, W., and R. Herrnstein. 1987. Stability, melioration, and natural selection. In L. Green and J. Kagel, eds., *Advances in Behavioral Economics,* vol. 1. Norwood, N.J.: Ablex.

Varela, F. 1979. *Principles of Biological Auton- omy.* New York: North-Holland.

Veyne, P. 1984. *Writing History.* Middletown, Conn.: Wesleyan University Press.

Vidal, F. 1987. Jean Piaget and the liberal Protes- tant tradition. In M. G. Ash and W. Woodward, eds., *Psychology in 20th-century Thought and Society,* 271–94. Cambridge: Cambridge Uni- versity Press.

Vienna Circle. 1973. *Wissenschaftliche Weltauffas- sung: Der Wiener Kreis.* English translation of the 1929 German orig. Dordrecht: Reidel.

Vollmer, G. 1975. *Was können wir wissen?* Bd. 1, *Die Natur der Erkenntnis.* Stuttgart: Hirzel.

von Bertalanffy, L. 1968. *General Systems Theory: Foundations, Development, Applications.* Har- mondsworth: Penguin.

von Ditfurth, H. 1987. Evolution and Transzen- denz. In Riedl and Wuketits, 258–73.

von Mises, L. 1949. *Human Action: A Treatise on Economics.* New Haven: Yale University Press.

Vrba, E. 1984. What is species selection? *Syst. Zool.* 33:318–28.

Vrba, E., and N. Eldredge. 1984. Individuals, hierarchies and processes: Towards a more complete evolutionary theory. *Paleobiology* 10:146–71.

Vrba, E., and S. J. Gould. 1986. The hierarchical expansion of sorting and selection: Sorting and selection cannot be equated. *Paleobiology* 12:217–28.

Waddington, C., ed. 1968–1972. *Towards a Theo- retical Biology.* 4 vols. Edinburgh: Edinburgh University Press.

———. 1975. *The Evolution of an Evolutionist.* Edinburgh: Edinburgh University Press.

Wade, M. 1976. Group selection among labora- tory populations of *Tribolium. Proceedings of the National Academy of Sciences (USA)* 73: 4604–7.

———. 1977. An experimental study of group se- lection. *Evolution* 31:134–53.

———. 1978. A critical review of the models of group selection. *Quart. Rev. Biol.* 53:101–14.

Wade, N. 1978. Sociobiology: Troubled birth for new discipline. In Caplan, 325–32. Orig. in *Science* 191 (1978): 1151–55.

Walton, David. 1991. The units of selection and the bases of selection. *Phil. Sci.* 58:417–35.

Walton, Douglas N. 1985. Are circular arguments necessarily viscious? *Amer. Phil. Quart.* 22:263–74.

Wartofsky, M. W. 1979. *Models: Representation and the Scientific Understanding.* Dordrecht: Reidel.

Wason, P. C. 1966. Reasoning. In B. M. Foss, ed., *New Horizons in Psychology,* 135–51. Har- mondsworth: Penguin.

———. 1983. Realism and rationality in the selec- tion task. In J. S. B. Evans, ed., *Thinking and Reasoning: Psychological Approaches,* 44–75. London: Routledge and Kegan Paul.

Wason, P. C., and P. N. Johnson-Laird. 1972. *The Psychology of Reasoning: Structure and Con- tent.* Cambridge, Mass.: Harvard University Press.

Watson, J. D. 1968. *The Double Helix: A Personal Account of the Discovery of the Structure of DNA*. New York: Atheneum.

Weber, B. H., D. J. Depew, and J. D. Smith, eds. 1988. *Entropy, Information, and Evolution: New Perspectives on Physical and Biological Evolution*. Cambridge, Mass.: MIT Press, Bradford Books.

Weber, R. 1986. *Dialogues with Scientists and Sages: The Search for Unity*. London and New York: Routledge and Kegan Paul.

Weintraub, E. R. 1985. *General Equilibrium Analysis: Studies in Appraisal*. Cambridge: Cambridge University Press.

Weintraub, P., ed. 1984. *The Omni Interviews*. New York: Ticknor and Fields.

Weizenbaum, J. 1976. *Computer Power and Human Reason: From Judgment to Calculation*. San Francisco: Freeman.

Werskey, G. 1978. *The Visible College*. London: Allen Lane–Penguin Books. Repr. London: Free Association Books, 1988.

Whewell, W. 1834. Review of Mary Somerville, *On the Connexion of the Physical Sciences* (1834). *Quart. Rev.* 51:54–68.

Wilkins, B. T. 1989. Review of Berkson and Wettersten 1984. *Synthese* 78:357–58.

Williams, G. C. 1966. *Adaptation and Natural Selection: A Critique of Some Current Evolutionary Thought*. Princeton: Princeton University Press.

———. 1975. *Sex and Evolution*. Princeton: Princeton University Press.

Williams, M. 1970. Deducing the consequences of evolution: A mathematical model. *J. Theor. Biol.* 29:343–85.

Williams, P. 1990. Evolved ethics re-examined: The theory of Robert J. Richards. *Biol. Phil.* 4:451–57.

Williams, R., and J. Law. 1980. Beyond the bounds of credibility. *Fund. Sci.* 1:295–315.

Wilson, D. S. 1975. A general theory of group selection. *Proc. Nat. Acad. Sci. (USA)* 72:143–46.

———. 1977. Structured demes and the evolution of group advantageous traits. *Am. Natural.* 111:157–85.

———. 1980. *The Natural Selection of Populations and Communities*. Menlo Park: Benjamin Cummings.

———. 1983. Individual and group selection: A historical and conceptual review. *Ann. R. Ecol.* 14:159–88.

———. 1987. Altruism in Mendelian populations derived from sibling groups: The haystack model revisited. *Evolution* 41:1059–70.

Wilson, D. S., and R. K. Colwell. 1981. Evolution of sex ratio in structured demes. *Evolution* 35:882–97.

Wilson, E. B. 1896. *The Cell in Development and Inheritance*. London: Macmillan. 2d ed. 1900.

Wilson, E. O. 1975. *Sociobiology: The New Synthesis*. Cambridge, Mass.: Harvard University Press.

———. 1978. *On Human Nature*. Cambridge, Mass.: Harvard University Press.

———. 1979. Comparative social theory. *Tanner Lectures on Human Values* 1:49–73. Salt Lake City: University of Utah Press.

Wimsatt, W. C. 1970. Review of Williams 1966. *Phil. Sci.* 37:620–29.

———. 1972. Teleology and the logical structure of function statements. *Stud. Hist. Phil. Sci.* 3:1–80.

———. 1974. Complexity and organization. In *PSA 1972*, 67–82.

———. 1976a. Reductive explanation: A functional account. In *PSA 1974*, 671–710. Repr. in Sober 1984b.

———. 1976b. Reductionism, levels of organization, and the mind-body problem. In G. G. Globus, G. Maxwell, and I. Savodnik, eds., *Consciousness and the Brain: A Scientific and Philosophical Inquiry*, 205–67. London: Plenum Press.

———. 1979. Reduction and reductionism. In Asquith and Kyburg, 352–77.

———. 1980a. Reductionistic research strategies and their biases in the units of selection controversy. In Nickles 1980b, 213–59. Repr. in Sober 1984b.

———. 1980b. Randomness and perceived-randomness in evolutionary biology. *Synthese* 43:287–329.

———. 1981a. Units of selection and the structure of the multi-level genome. In *PSA 1980*, 2:122–83.

———. 1981b. Robustness, reliability, and overdetermination. In Brewer and Collins, 124–63.

———. 1986a. Developmental constraints, generative entrenchment, and the innate-acquired distinction. In Bechtel 1986a, 185–208.

———. 1986b. Forms of aggregativity. In Donagan, Perovich, and Wedin, 259–91.

———. 1986c. Heuristics and the study of behavior. In Fiske and Shweder, 293–314.

———. 1987. False models as means to truer theories. In M. Nitecki and A. Hoffmann, eds., *Neutral Models in Biology*, 23–55. Oxford: Oxford University Press.

———. N.d.a. Generative entrenchment, scientific change, and the analytic-synthetic distinction. Drafts around since 1983, and forthcoming, probably in *Biology and Philosophy,* as a discussion paper.

———. N.d.b. Taming the dimensions: Visualisation in science. Ms.

Wimsatt, W. C., and J. C. Schank. N.d. Why Use Models in Biology? Ms.

Winch, P. 1958. *The Idea of a Social Science and its Relation to Philosophy.* London: Routledge and Kegan Paul; Atlantic Highlands: Humanities Press.

Winckelgren, I. 1992. How the brain "sees" borders where there are none. *Science* 256:1520–21.

Winograd, T., and F. Flores. 1986. *Understanding Computers and Cognition: A New Foundation for Design.* Norwood, N.J.: Ablex.

Winter, S. 1964. Economic 'natural selection' and the theory of the firm. *Yale Econ. Rev.* 4:255–72.

Wittgenstein, L. 1953. *Philosophical Investigations.* Trans. G. E. M. Anscombe. Oxford: Blackwell.

———. 1961. *Tractatus logico-philosophicus.* Trans. D. F. Pears and B. MacGuinness. London: Routledge and Kegan Paul. German orig. pub. 1921.

Wolman, B. B. 1981. *Contemporary Theories and Systems in Psychology.* 2d ed. New York and London: Plenum Press.

Woodger, J. H. 1937. *Axiomatic Method in Biology.* Cambridge: Cambridge University Press.

———. 1952. *Biology and Language: An Introduction to the Methodology of the Biological Sciences, Including Medicine.* Cambridge: Cambridge University Press.

Woolgar, S. 1981. Interests and explanation in the social study of science. *Soc. Stud. Sci.* 11:365–94.

———, ed. 1988a. *Knowledge and Reflexivity: New Frontiers in the Sociology of Science.* London: Sage.

———. 1988b. Reflexivity is the ethnographer of the text. In Woolgar 1988a, 14–34.

Worrall, J. 1976. Imre Lakatos (1922–1974): Philosopher of mathematics and philosopher of science. In Cohen et al., 1–8.

Wright, L. 1976. *Teleological Explanations: An Etiological Analysis of Goals and Functions.* Berkeley and Los Angeles: University of California Press.

Wright, R. A., ed. 1984. *African Philosophy: An Introduction.* 3d ed. Lanham: University Press of America.

Wright, S. 1984. The roles of mutation, inbreeding, crossbreeding, and selection in evolution. In Brandon and Burian, 29–39. Orig. pub. 1932.

———. 1986. *Evolution: Selected Papers.* Chicago: University of Chicago Press.

Wuketits, F., ed. 1984. *Concepts and Approaches in Evolutionary Epistemology.* Dordrecht: Reidel.

———. 1987. Hat die Biologie Kant mißverstanden? In W. Lüttersfels, Hrsg., *Transzendentale oder evolutionäre Erkenntnistheorie?,* 33–50. Darmstadt: Wissenschaftliche Buchgesellschaft.

———. 1990. *Evolutionary Epistemology and Its Implications for Humankind.* Albany: State University of New York Press.

Wussing, H. 1984. *The Genesis of the Abstract Group Concept.* Cambridge, Mass.: MIT Press.

Wynne-Edwards, V. C. 1962. *Animal Dispersion in Relation to Social Behavior.* Edinburgh: Oliver and Boyd.

Young, R. M. 1971. Evolutionary biology and ideology: Then and now. *Sci. Stud.* 1:177–206.

———. 1985. *Darwin's Metaphor.* Cambridge: Cambridge University Press.

Zallen, D. T. 1989. The Rockefeller Foundation and French research. *Cah. Hist. CNRS* 5:35–58.

Zihlman, A. L. 1978. Women and evolution, Part II: Subsistence and social organization among early hominids. *Signs* 4:4–20.

Ziman, J. 1960. Scientists: Gentlemen or players? *The Listener* no. 68:599–607.

———. 1985. CUDOS and PLACE. *EASST Newsletter* 4 (2): 5–6.

Zimmerman, M. E. 1990. *Heidegger's Confrontation with Modernity: Technology, Politics, and Art.* Bloomington: Indiana University Press.

INDEX

523